Skeletal Muscle Mechanics

FROM MECHANISMS TO FUNCTION

Skeletal Muscle Mechanics

FROM MECHANISMS TO FUNCTION

Edited by
W. HERZOG
University of Calgary, Calgary

JOHN WILEY & SONS, LTD.
Chichester · New York · Weinheim · Brisbane · Singapore · Toronto

Baffins Lane, Chichester,
West Sussex PO19 1UD, UK

National 01243 779777
International (+44) 1243 779777
e-mail (for orders and customer service enquiries): cs-books@wiley.co.uk
Visit our Home Page on: http://www.wiley.co.uk or http://www.wiley.com

Other Wiley Editorial Offices

John Wiley & Sons, Inc., 605 Third Avenue,
New York, NY 10158-0012, USA

WILEY-VCH Verlag GmbH, Pappelallee 3,
D-69469 Weinheim, Germany

Jacaranda Wiley, Ltd., 33 Park Road, Milton,
Queensland 4064, Australia

John Wiley & Sons (Asia) Pte, Ltd., 2 Clementi Loop #02-01,
Jin Xing Distripark, Singapore 129809

John Wiley & Sons (Canada), Ltd., 22 Worcester Road,
Rexdale, Ontario M9W 1L1, Canada

British Library Cataloguing in Publication Data

A catalogue record for this book is available from the British Library

ISBN 0-471-49238-8

Printed from camera ready copy supplied by the author
and bound in Great Britain by Antony Rowe Ltd, Chippenham.
This book is printed on acid-free paper responsibly manufactured from sustainable forestry,
in which at least two trees are planted for each one used for paper production.

Contents

Contributors

Rachid Ait-Haddou
Human Performance Laboratory, Faculty of Kinesiology, Room 205, Physical Education B Building, University of Calgary, 2500 University Drive N.W., Calgary, Alberta, Canada T2N 1N4

Todd L. Allinger
Institute for Sport Science and Medicine, The Orthopedic Specialty Hospital, 5848 South Fashion Boulevard (300 East), Murray, Utah, 84107 USA

Vasilios Baltzopoulos
TEFAA, University of Thessaly, Karyes, Trikala 42100, Greece

Reinhard Blickhan
Institute of Sport Science, Biomechanics Group, Friedrich-Schiller-University, Seidelstraße 20, D-07749, Jena, Germany

Robert M. Campy
Department of Physical Education, Health, and Recreation; Eastern Washington University, 526 - 5 Street, Mail Stop 4, Cheney, Washington, 99004 USA

Alan J. Coelho
Department of Physical Education, Health, and Recreation; Eastern Washington University, 526 - 5 Street, Mail Stop 4, Cheney, Washington, 99004 USA

Denise M. Connelly
School of Rehabilitation Therapy, Room 210, Louise D. Acton Building, Queen's University, Kingston, Ontario, Canada K7L 3N6

Christophe Cornu
Université de Technologie de Compiègne, UMR CNRS 6000, Biomécanique et Génie Biomédical, B.P. 20529, 60205 Compiègne cedex, France

Jack Tigh Dennerlein
Harvard School of Public Health, 665 Huntington Avenue, Boston, Massachusetts, 02115 USA

John E. Drace
Diagnostic Radiology Center, Veterans Affairs, Palo Alto Health Care System, Palo Alto, California, 94304 USA

Maarten R. Drost
Department of Movement Science, Faculty of Health Science, Maastricht University, P.O. Box 616, 6200 MD, Maastricht, The Netherlands

Marcelo Epstein
Department of Mechanical Engineering, Faculty of Engineering, Room 403, Mechanical Engineering Building, University of Calgary, 2500 University Drive N.W., Calgary, Alberta, Canada T2N 1N4

Tetsuo Fukunaga
Department of Life Science (Sports Sciences), University of Tokyo, Komaba 3-8-1, Meguro, Tokyo 153-8902, Japan

Francis Goubel
Université de Technologie de Compiègne, UMR CNRS 6000, Biomécanique et Génie Biomédical, B.P. 20529, 60205 Compiègne cedex, France

David Hawkins
Human Performance Laboratory, Department of Exercise Science, University of California (Davis), Davis, California, 95616 USA

Walter Herzog
Human Performance Laboratory, Faculty of Kinesiology, Room 205, Physical Education B Building, University of Calgary, 2500 University Drive N.W., Calgary, Alberta, Canada T2N 1N4

Matthijs K. C. Hesselink
Department of Movement Science, Faculty of Health Science, Maastricht University, P.O. Box 616, 6200 MD, Maastricht, The Netherlands

Takao Hijikata
Faculty of Medicine, Gunma University School of Medicine, Maebashi, Gunma, Japan

Ellen H. Hollander
c/o H.E.D.J. ter Keurs, Health Sciences Centre, Faculty of Medicine, University of Calgary, 3330 Hospital Drive N.W., Calgary, Alberta, Canada T2N 4N1

Peter A. Huijing
Interfaculty of Physical Education, Free University, 1007 MC Amsterdam, The Netherlands

Glenn D. Hunter
School of Physiotherapy and Occupational Therapy (SPOT), Faculty of Health and Social Care, University of the West of England, Glenside Campus, Blackberry Hill, Stapleton, Bristol, BS16 1DD England

Andrew F. Huxley
Manor Field, 1 Vicarage Drive, Grantchester, Cambridge, CB39NG England

Hiroaki Kanehisa
Department of Life Science (Sports Sciences), University of Tokyo, Komaba 3-8-1, Meguro, Tokyo 153-8902, Japan

Stephen L. Katz
22441 Lassen Street, Chatsworth, California, 91311 USA

Yasuo Kawakami
Department of Life Science (Sports Sciences), University of Tokyo, Komaba 3-8-1, Meguro, Tokyo 153-8902, Japan

John F. Kramer
Schools of Physical Therapy and Kinesiology, Faculty of Health Sciences, University of Western Ontario, Room 1400, Elborn College, 1201 Western Road, London, Ontario, Canada N6G 1H1

Keitaro Kubo
Department of Life Science (Sports Sciences), University of Tokyo, Komaba 3-8-1, Meguro, Tokyo 153-8902, Japan

Kenya Kumagai
Department of Sports Sciences, Tokyo Metropolitan University, 1-1 Minami-Ohsawa, Hachioji-shi, Tokyo, 192-0397, Japan

Daniel Lambertz
Université de Technologie de Compiègne, UMR CNRS 6000, Biomécanique et Génie Biomédical, B.P. 20529, 60205 Compiègne cedex, France

Amir Landesberg
Department of Biomedical Engineering, Technion - Israel Institute of Technology, Technion City, Haifa 32000, Israel

Robson Lemos
Department of Computer Science, Faculty of Science, Room 247, Math Sciences Building, University of Calgary, 2500 University Drive N.W., Calgary, Alberta, Canada T2N 1N4

Leonid Livshitz
Department of Biomedical Engineering, Technion - Israel Institute of Technology, Technion City, Haifa 32000, Israel

Mascha Maenhout
Department of Materials Technology, Faculty of Mechanical Engineering, Eindhoven University of Technology, P.O. Box 513, 5600 MB, Eindhoven, The Netherlands

Constantinos N. Maganaris
Department of Life Science (Sports Sciences), University of Tokyo, Komaba 3-8-1, Meguro, Tokyo 153-8902, Japan

M. Polly McGuigan
The Royal Veterinary College, Hawkshead Road, North Mymms, Hatfield, Hertfordshire, AL9 7TA England

Petra Meier
Institute of Sport Science, Biomechanics Group, Friedrich-Schiller-University, Seidelstraße 20, D-07749, Jena, Germany

Masahito Miura
c/o H.E.D.J. ter Keurs, Health Sciences Centre, Faculty of Medicine, University of Calgary, 3330 Hospital Drive N.W., Calgary, Alberta, Canada T2N 4N1

Cees W. J. Oomens
Department of Materials Technology, Faculty of Mechanical Engineering, Eindhoven University of Technology, P.O. Box 513, 5600 MB, Eindhoven, The Netherlands

George Pappas
Diagnostic Radiology Center and Rehabilitation R&D Center, Veterans Affairs, Palo Alto Health Care System, Palo Alto, California, 94304 USA

Chantal Pérot
Université de Technologie de Compiègne, UMR CNRS 6000, Biomécanique et Génie Biomédical, B.P. 20529, 60205 Compiègne cedex, France

Danny M. Pincivero
Human Performance and Fatigue Laboratory, Department of Physical Therapy, Eastern Washington University, 526 - 5 Street, Mail Stop 4, Cheney, Washington, 99004 USA

Michelle M. Porter
Faculty of Physical Education and Recreation Studies, University of Manitoba, 115 Frank Kennedy Centre, Winnipeg, Manitoba, Canada R3T 2N2

Dilson E. Rassier
Health Sciences Center (02), Laboratory of Physiology, University of Vale do Rio dos Sinos (UNISINOS), Avenida Unisinos 050, Caixa Postal 275, CEP: 93022-000, São Leopoldo, Rio Grande do Sul, Brazil

Hans H. C. M. Savelberg
Department of Human Movement Science, Faculty of Health Sciences, Maastricht University, P.O. Box 616, 6200 MD, Maastricht, The Netherlands

Robert E. Shadwick
MBRD-0204, Scripps Institute of Oceanography, La Jolla, California, 93029 USA

Frances T. Sheehan
School of Engineering, Mechanical Engineering Department, Room 102 Pangborn Hall, The Catholic University of America, 620 Michigan Avenue N.E., Washington, D.C., 20064 USA

Jonathan Spriggs
c/o SOAR Research, Long Island University, 122 Ashland Place #1A, Brooklyn, New York, 11201 USA

Bruno D. Stuyvers
c/o H.E.D.J. ter Keurs, Health Sciences Centre, Faculty of Medicine, University of Calgary, 3330 Hospital Drive N.W., Calgary, Alberta, Canada T2N 4N1

Esther Suter
Human Performance Laboratory, Faculty of Kinesiology, Room 205, Physical Education B Building, University of Calgary, 2500 University Drive N.W., Calgary, Alberta, Canada T2N 1N4

Henk E. D. J. ter Keurs
Cardiovascular Research Group, Health Sciences Centre, Faculty of Medicine, University of Calgary, 3330 Hospital Drive N.W., Calgary, Alberta, Canada T2N 4N1

Mark H. C. ter Keurs
c/o H.E.D.J. ter Keurs, Health Sciences Centre, Faculty of Medicine, University of Calgary, 3330 Hospital Drive N.W., Calgary, Alberta, Canada T2N 4N1

Ton J. van den Bogert
Department of Biomedical Engineering, Cleveland Clinic Foundation, 9500 Euclid Avenue (ND-20), Cleveland, Ohio, 44195 USA

Anthony A. Vandervoort
Schools of Physical Therapy and Kinesiology, Faculty of Health Sciences, University of Western Ontario, Room 1400, Elborn College, 1201 Western Road, London, Ontario, Canada N6G 1H1

Alan M. Wilson
The Royal Veterinary College, Hawkshead Lane, North Mymms, Hatfield, Hertfordshire, AL9 7TA England

Brian Wyvill
Department of Computer Science, Faculty of Science, Room 247, Math Sciences Building, University of Calgary, 2500 University Drive N.W., Calgary, Alberta, Canada T2N 1N4

George I. Zahalak
Departments of Biomedical Engineering and Mechanical Engineering, Washington University, Campus Box 1185, 1 Brookings Drive, St. Louis, Missouri, 63130 USA

Preface

In August of 1999, I organized a small conference on the mechanics of skeletal muscle contraction and *in vivo* function in a small town at the eastern edge of the Canadian Rocky Mountains: Canmore. The conference (called the *Canmore Symposium on Skeletal Muscle*) was an experiment aimed at bringing together scientists from vastly different areas of skeletal muscle research: biophysicists interested in the molecular and cellular events of contraction; engineers and mathematicians focusing on the modelling of skeletal muscle contraction and function; and biologists, biomechanists, zoologists, and others primarily interested in the *in vivo* function and control of muscle. The idea was to see if these researchers from different areas could interact and provide stimulating information across their research areas. It was obvious that such an interaction required tremendous generosity from the people involved in order to accept the diverging technical and philosophical approaches to muscle research used by the different groups. It was particularly satisfying for me to see how each of the sessions retained the character representative of the specific research area, but despite—or because of that—was able to unite the researchers and produce stimulating discussions across disciplinary boundaries.

However, when considering any system on vastly differing scales, as we did in Canmore and are doing in this book, one must ask the question: How useful is it to compare a system across levels, and is it possible to make feasible predictions from the known molecular behaviour of a system to the system behaviour itself, or vice versa? Or, to paraphrase one of my engineering friends: Can you use quantum mechanics to calculate the deflection of a beam when subjected to a constant force? Probably not, but simple beam mechanics provides an accurate solution quite readily.

Therefore, is it useful for a researcher interested in the molecular events of muscular contraction to know about the structure and function of a muscle in the living system? Or, viewed the other way around, how important is it for a functional biomechanist who is concerned with the force production of a muscle group during locomotion to know the detailed molecular events of active force production? Philosophically, it is probably very important, as knowledge in a related area might prove useful in the formulation of meaningful scientific questions and the interpretation of results in a global manner. However, practically, knowledge about the molecular events of muscle contraction are unlikely to help in identifying the function of a muscle during movement. Accepting the above arguments, one is driven to the conclusion that a book with the current content is philosophically exciting but probably not very practical, at least not in its entirety, and as viewed by any given individual. As was the conference, so is this book an experiment in crossing the boundaries from the molecular events of contraction to *in vivo* functions of muscles; an experiment that I do hope will turn out to be stimulating, constructive, and unifying.

The book is divided into four parts. The first part deals with the molecular mechanisms of muscle contraction. It contains a historical overview of cross-bridge mechanics based on the Wartenweiler Memorial Lecture given by Sir Andrew Huxley. The second part focuses on the mechanical and mathematical modelling of skeletal muscle, featuring Dr. George Zahalak's work on the two-state cross-bridge model as a possible link between molecular and macroscopic muscle mechanics, which was presented in a keynote lecture at the Canmore Symposium. The third part deals with the *in vivo* function of skeletal muscle, featuring a particularly large number of contributions on non-invasive approaches aimed at revealing structural changes of *in vivo* contracting human skeletal muscle: ultrasonography and cine phase-contrast magnetic resonance imaging. Finally, the fourth part deals with *in vivo* approaches in animal studies, demonstrating how much insight into muscle function may be gained by direct muscle-force measurements and the *in situ* testing of muscle and limb properties.

This book was written and edited in the last few months of the second millennium. As such, it represents a historical record of ongoing research in selected areas of skeletal muscle research. It is my sincere hope that this book may stimulate young scientists to do research in muscle mechanics, and may provide senior scientists with a glimpse of muscle research that they may not have encountered before.

Walter Herzog
September 1999

Acknowledgements

The Killam Foundation: This book was written and edited while I was on a research fellowship sponsored by The Killam Foundation. The Killam Foundation also provided financial support for the production of this book.

I would also like to thank my assistant, Holly Hanna; our technical writer, Claire Huene; and for figure editing, Jorge Palafox.

I would like to thank Professor Andrew Huxley, Timothy Leonard, and Jane Mactaggart. They had a profound influence on this book, and they contributed to it in significant but very different ways.

Walter Herzog

I Mechanisms of Muscle Contraction

1 Considerations on the Mechanisms of Muscular Contraction

W. HERZOG
Human Performance Laboratory, University of Calgary, Calgary, Alberta, Canada

The origins of suggesting that muscles are the organ of force and movement production may be traced to ancient Greece. Aristotle (384-322 B.C., *De Motu Animalium*) described the interrelation of breath, brain, and blood vessels in the production of movement. According to Aristotle, muscle contraction was associated with the entering of the *spiritus animalius* into the muscle. The discovery that muscle is the true organ of voluntary movement must be accredited to Galen (129-201 A.D., *De Tremore*). His detailed dissection of muscles, and his discovery that arteries contain blood, not air or spirits, are considered the first attempts to establish a science of muscles (myology). Vesalius (1543, *De Humano Corporis Fabrica*) discovered that the contractile power resides in the muscle substance. Swammerdam (1663, c.f. Needham, 1971) showed with elegant experiments that muscular volume was preserved during muscle contraction; Croone (1664, *De Ratione Motus Musculorum*) concluded from nerve section experiments that the brain must send a signal to the muscles to cause contraction, and Stensen (1664, *De Musculis et Glandulis observationem specimen*; 1667, *Elementorum Myologiae Specimen sen Musculi Descripto Geometrica*) provided detailed descriptions of muscular and tendinous structures. Van Leeuwenhoek performed light microscopic examinations and discovered the regular cross-striation patterns in skeletal muscles. His description of muscular structure dominated the following century (c.f. Needham, 1971).

In the past 100 years, ideas about how muscles contract changed several times. In chapter 2, Huxley talks about three scientific revolutions. In the early part of the 20th century, it was believed that contraction occurred by a folding of long protein chains that was triggered by a liberation of lactic acid upon activation. However, in the early 1930s, it was demonstrated that perfect contractions could be obtained without lactic acid. The idea that protein folding and, later, more specifically folding or shortening of the myosin filaments (A-bands) caused contraction persisted. However, A.F. Huxley and Niedergerke (1954) and H.E. Huxley and Hanson (1954) observed that thick myofilament length remained virtually constant for a variety of contractile conditions, possibly with the exception of when the av-

Skeletal Muscle Mechanics: From Mechanisms to Function. Edited by W. Herzog.

erage sarcomere length became very small, i.e., below about 1.6-1.8 μm. These observations gave rise to the sliding filament theory. This theory assumes that length changes in muscle occur through a relative sliding of the two sets of filaments. However, this theory does not indicate where the forces for the relative sliding come from.

A.F. Huxley (1957) formulated the 'cross-bridge theory' of muscle contraction. This theory postulates that there are independent force generators arranged in a systematic way on one set of filaments (the myosin) that interact with specific and systematically arranged attachment sites on the other (the actin) filament. Once attached, force production might occur through a variety of mechanisms: shortening of some element of the cross-bridge, rotation of the cross-bridge head or part of it, etc. The 1957 theory was proposed using structural, physiological, and biological arguments, and was formulated mathematically to contain mechanical and biochemical considerations.

Despite a variety of attempts to propose alternative theories of muscle contraction, and despite a large number of variations of the original two-state model proposed by Huxley (1957), the basic features of the original model are still accepted by the majority of scientists in this area of research.

However, as pointed out by A.F. Huxley on many occasions, and most recently in his Wartenweiler Memorial Lecture at the XVII Conference of the International Society of Biomechanics (see chapter 2), there are many unexplained phenomena and observations, and there are many unknowns in the detailed explanation of how muscular contraction is produced.

Muscle mechanics research and the mechanisms underlying muscular contraction have gone through the changes normally associated with a scientific field. A discovery is made, and the corresponding theoretical framework to explain the findings is produced. Then, much research is performed to essentially extend the field of knowledge in an incremental way. Finally, observations are made that are incompatible with the theoretical framework, and the old framework is replaced by a new one.

It appears to me that the big conceptual discoveries and observations, as well as the corresponding theoretical framework in muscle mechanics, were made in the 1950s to 1970s. Since then, the field has been in a state of incremental discovery and small adjustments. In the past six years, new techniques have become available that are now used systematically in many laboratories. These techniques allow for the evaluation of the independent force-generator theory at the molecular level. It will be of interest to see whether the cross-bridge paradigm will hold up to the discoveries that are made on this particular structural level.

I believe that the field of muscle mechanics might be headed toward a scientific revolution. With the new technical possibilities, we should rapidly gain insight into the molecular interactions of the S1 myosin head with actin, the relationship between the mechanics and biochemistry of contraction, and the precise mechanical properties of the molecules involved in active and passive force production. Muscle mechanics and scientific research dealing with the mechanisms of muscle contraction should be exciting, fast-paced, and possibly full of surprises in the very near future.

REFERENCES

Huxley, A.F. (1957). Muscle structure and theories of contraction. *Progress in Biophysics and Biophysical Chemistry* **7**, 255–318.

Huxley, A.F., and Niedergerke, R. (1954). Structural changes in muscle during contraction: Interference microscopy of living muscle fibres. *Nature* **173**, 971–973.

Huxley, H.E., and Hanson, J. (1954). Changes in cross-striations of muscle during contraction and stretch and their structural interpretation. *Nature* **173**, 973–976.

Needham, D.M. (1971). *Machina Carnis.* Cambridge University Press, Cambridge, England.

2 Cross-bridge Action: Present Views, Prospects, and Unknowns

A. F. HUXLEY
Trinity College, Cambridge, England

INTRODUCTION

The following chapter on ideas about the mechanics of action of cross-bridges in muscle is closely based on the Wartenweiler Lecture that I gave at the XVII Conference of the International Society of Biomechanics on August 8, 1999, in Calgary, Alberta, Canada.

It is partly historical, tracing the development of present ideas during this century, but it also emphasizes the many uncertainties that still exist. In it, I concentrate on experiments on intact muscle, especially single fibres, both because in the end we need to know how living muscle works and because almost all my work on muscle has been on isolated single fibres in a fully living state, and I am therefore best equipped to speak about such work. A result is that I say relatively little about chemical and structural observations.

THREE REVOLUTIONS

Research on muscle has progressed very much in the way proposed by T.S. Kuhn (1962) in his book *The Structure of Scientific Revolutions*. Most of the time, new discoveries, however important, are within the existing framework of ideas, but occasionally there are revolutions when a new observation or a new idea makes current ideas obsolete.

At the beginning of the century, a theory was developed—the lactic acid theory—according to which contraction was caused by folding of long protein chains running along each muscle fibre. In the resting state, these chains were supposed to be kept extended by repulsion between negative charges on the proteins. When a fibre was excited, a certain amount of lactic acid was liberated inside each fibre and hydrogen ions from the acid neutralized the negative charges and allowed the chains to fold up by thermal agitation. In the 1920s, this theory was almost universally held.

The first use of the word 'revolution' in connection with muscle was made by A.V. Hill (Figure 2.1), the grand old man of muscle and nerve physiology. I got to

Skeletal Muscle Mechanics: From Mechanisms to Function. Edited by W. Herzog.

know him through his son, with whom I overlapped as an undergraduate at Trinity College, Cambridge, before World War II, and he had a great influence on me, both personally and through his scientific writings. He died at the age of 90, a little over 20 years ago. In 1932, he published a review entitled 'The Revolution in Muscle Physiology'. The revolution in question was the demise of the lactic acid theory. This was brought about by the observation by the Danish physiologist Lundsgaard (1930) that a muscle poisoned with iodoacetic acid could perform a substantial number of normal contractions without producing any lactic acid. The part played by lactic acid in the theory was replaced by a reaction that was already known to occur during muscle contraction: the breakdown of phosphorylcreatine, PCr (Fiske and Subbarow, 1927; Eggleton and Eggleton, 1927). It was then assumed that the PCr reaction was somehow coupled to the folding of the protein filaments, although no mechanism for this link was known. With the discovery of adenosine triphosphate (ATP) and the recognition that the only reaction undergone by PCr was the rephosphorylation of ADP to ATP, the hydrolysis of PCr suffered the same fate as the production of lactic acid, and was relegated to the position of a back-up reaction. The direct cause of folding of the protein chains was then assumed to be the hydrolysis of ATP to ADP and inorganic phosphate.

Figure 2.1. Professor A.V. Hill, F.R.S., 1886-1977.

A second revolution was the recognition that the liberation of energy during contraction is not an all-or-none consequence of excitation, and that the rates of the chemical reactions underlying contraction are controlled by the mechanical conditions. The basis of this idea is what is known as the Fenn Effect (Fenn, 1923). Wallace Fenn, working in the laboratory of A.V. Hill, measured the heat and work produced in a twitch of frog muscle and found that the total energy liberated was greater if the muscle was allowed to shorten and do work than if it was held isometrically. Much more extensive results were obtained, with improved apparatus, by A.V. Hill (1938; Figure 2.2) in tetanic contractions, confirming Fenn's conclusion and making it quantitative. This established the important point that the total amount of energy liberated in a given time was not, as had been generally supposed, a fixed quantity of which a fraction would appear as work, but increased severalfold when the muscle was allowed to shorten at moderate speed. This implied that the rate of the chemical reactions that produced the energy was partly controlled by the mechanical situation. It is ironic that Hill (1924) drew attention to Fenn's result in his Nobel Lecture at the end of 1923, but admitted later (Hill, 1965, p. 84) that his thinking about the contraction process did not take it into account until after his own later experiments. I suspect that, even now, the influence of mechanical conditions on the chemical events in muscle is not given the prominence it deserves; certainly the detailed mechanisms are unknown.

On the first page of his 1923 paper, Fenn said: 'Having reached this rather novel point of view [that more energy is liberated when shortening is permitted], it was a surprise to find that in its essentials it was not new but was urged by A. Fick 30 years ago, without however a satisfactory experimental basis. For this reason, the history of the subject is of interest'. He goes on to point out that the first observation of this kind was made by Rudolf Heidenhain in 1864. These old observations were referred to again by A.V. Hill (1965, p. 9).

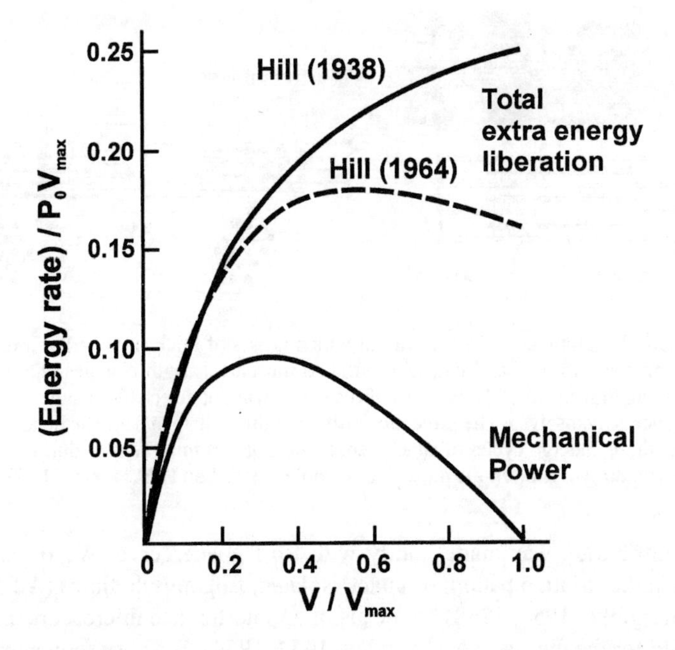

Figure 2.2. Rates of energy liberation in excess of steady rate during an isometric contraction, as determined by A.V. Hill.

The third revolution was the recognition that length changes in a muscle fibre take place not by folding of long protein chains, but by sliding of two sets of filaments past one another without appreciable length change in either. It is a double coincidence that this conclusion was reached independently at the same time (early 1953) by two groups using different methods, and that one group included Hugh Huxley and the other included me; we have the same surname but are not detectably related.

The first published suggestion of sliding filaments is in the 1953 paper by Hugh Huxley, on the basis of transmission electron micrographs of transverse sections of frog muscle. Figure 2.3 is based on a diagram in this paper. Jointly with Jean Hanson, who died relatively young in 1973, he followed this up by observations on separated myofibrils with the phase contrast microscope, showing that the high refractive index of the A-bands is due to the presence there of myosin (Hanson and H.E. Huxley, 1953), and that changes of the band pattern during active and passive length changes are as expected on the basis of sliding filaments (H.E. Huxley and Hanson, 1954). The location of myosin in the A-bands was also shown at the same time by Wilhelm Hasselbach (1953) in the laboratory of H.H. Weber.

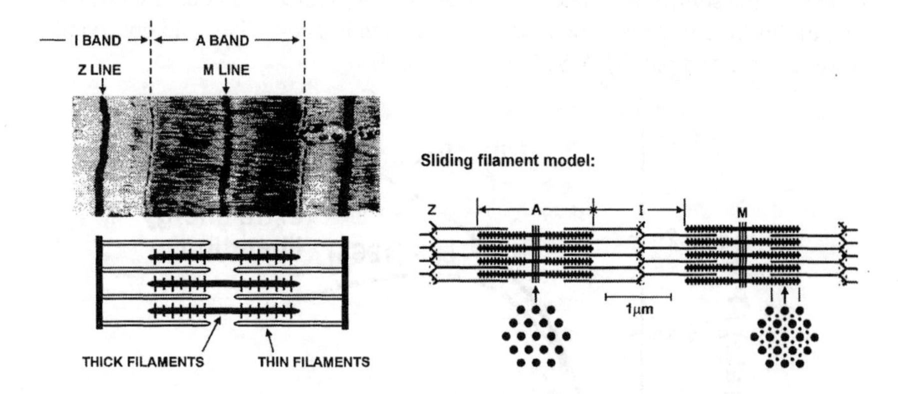

Figure 2.3. Diagram showing the way in which arrays of thick (myosin) and thin (actin) filaments give rise to the band pattern of striated muscles. Length changes take place by a relative sliding movement of the two sets of filaments, without detectable change in the lengths of the filaments themselves. The projections from the thick filaments are the heads of myosin molecules; these undergo cycles of attachment to the adjacent thin filament during which they exert force tending to shorten the muscle as a whole. (Based on H.E. Huxley, 1953.)

My contribution was made jointly with Rolf Niedergerke. We observed the changes in the striation pattern of intact, isolated, frog muscle fibres (A.F. Huxley and Niedergerke, 1954, 1958) by means of an interference microscope that I had developed for the purpose (A.F. Huxley, 1952, 1954). We very soon saw that the width of each A-band stayed constant as the fibre was lengthened or shortened, a surprising result since the textbooks of the time stated that length changes took place principally by changes in width of the A-bands. This constancy of width of the A-bands suggested that their high refractive index was due to rodlets running from one side of each band to the other, while the existence of a second set of filaments was suggested by the appearance of a narrow dense line at the centre of each

A-band when the fibre shortened actively to a sarcomere length of about 1.8 μm, which we attributed to crumpling or overlapping of the ends of a second set of filaments traversing the I-bands and extending into the adjacent A-bands. The well-known 'contraction bands' at the level of the Z-lines appeared when the fibre shortened further, and were evidently caused by overlapping or crumpling of the ends of the thick filaments when they collided.

It was only after we had made these observations that I followed up a suggestion by Rolf Niedergerke and found that much of what we had discovered had been common knowledge in the late 19th century, but had been lost after 1900 as interest switched from structures visible with the light microscope to chemical events (reviewed in A.F. Huxley 1977, 1980). As far as I know, however, no one at that time suggested sliding filaments.

WHAT MAKES THE FILAMENTS SLIDE?

The evidence for sliding filaments raised the further question: What makes them slide? A huge number of theories have been published. In a review a quarter of a century ago (A.F. Huxley, 1974), I classified these into eight categories, and if each of these is subdivided we would get something like a hundred theories. I shall not attempt to go into all of them, but shall concentrate on what has survived.

Two important aspects of nearly all theories now current are based on experiments that had been done long before the sliding theory was proposed. Robert Ramsey and Sibyl Street in the United States (one of the many husband and wife teams that have made important contributions to muscle research) were the first to dissect out single fibres from frog muscle in an undamaged condition, and they used these to record the tension generated during isometric tetani when the fibre was stretched to different initial lengths (Ramsey and Street, 1940). They found that the developed tension was a maximum when the fibre was close to its slack length, corresponding to a sarcomere length of about 2 μm. Active tension decreased on either side of this optimum length. In particular, their results show the striking feature that tension declined in a nearly linear fashion with increase of fibre length above the optimum, reaching zero at about double the slack length. There was no ready explanation for this observation on the then-current theory that tension was generated by folding of continuous protein chains. On the other hand, Niedergerke and I pointed out (A.F. Huxley and Niedergerke, 1954) that it would receive at least a qualitative explanation on the sliding filament theory if it is supposed that contributions to the relative force between a thick and an adjacent thin filament are provided by active sites ('independent force generators') uniformly distributed along their zone of overlap. In this case, it would follow that the number of these force generators and, therefore, the total active force would be proportional to the length of the overlap zone, which would in turn decline linearly as length is increased above the optimum.

However, the length at which active tension reached zero appeared to be too great to be explained in this way, since zero tension ought to be reached when the sarcomere length is equal to the sum of the lengths of the thick (about 1.5 μm) and thin (2 μm) filaments, i.e., about 3.5 μm, but in the experiments of Ramsey and Street (1940), zero tension was reached at a sarcomere length of about 4.0 μm. The explanation of this discrepancy was provided by Lee Peachey and me when we tried to repeat Ramsey and Street's measurements (A.F. Huxley and Peachey, 1959, 1961). We noticed that when an isolated fibre is stretched, the end regions do not elongate as much as the middle, so that there is still overlap near the ends of a fibre, although the average sarcomere length may be such that no overlap would be expected. Specifically, we found that the sarcomere length in the middle region of a fibre, above which shortening fails to develop, is closely equal to the sum of the lengths of the thick and the thin filaments. At these fibre lengths, the end regions of the fibre still shortened, as was to be expected since there was still overlap of thick and thin filaments there. These results showed that once overlap between the two types of filament is lost, there is no active contraction.

This non-uniformity of sarcomere length made it difficult to measure isometric tension at different lengths. What was needed was a device to hold constant the length of a uniform region in the middle of a fibre. Al Gordon, Fred Julian, and I (1966) put two markers on the fibre at the ends of a uniform region and used an optical-electronic device to measure continuously the distance between them, together with a servo-control mechanism pulling on one end of the fibre to hold this distance constant. When measuring tension with sarcomere length controlled in this way, we found that the tension rise on stimulation was closely proportional to overlap of the two types of filament. Figure 2.4 shows the resulting curve of active tension against sarcomere length. Tension reaches zero at about 3.6 μm, which turned out, within experimental error, to be the length with just no overlap, taking into account more accurate data on the lengths of the thin (2.05 μm) and the thick (1.6 μm) filaments.

CYCLIC ACTION OF CROSS-BRIDGES

If the idea of muscle tension being developed by 'independent force generators' distributed throughout each zone of overlap of thick and thin filaments is correct, it is natural to think that each of them acts repeatedly during a contraction, since no interaction between a site on the myosin filament and one on the actin filament can be expected to exist throughout the amount of relative movement that can occur in a single contraction (more than 0.5 μm). Such cyclic operation of the active sites had in fact been proposed on quite different grounds some years before sliding filaments were proposed.

The relationship between speed of shortening and total rate of energy liberation (heat plus work) reported by A.V. Hill (1938) was a rectangular hyperbola, as shown in Figure 2.2, although Hill (1964) corrected this later. Dorothy Needham, the lady half of another of those highly successful husband-and-wife teams, pointed out the analogy between the relation found by Hill (1938) and the Michaelis-Menten relation between the rate of an enzyme reaction and the concentration of

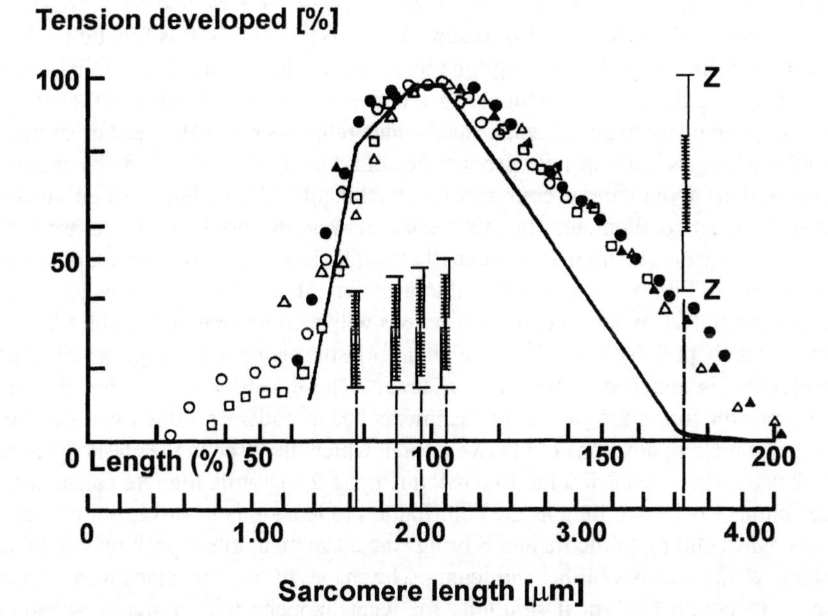

Figure 2.4. Tension developed by an isolated frog muscle fibre on stimulation at different lengths. Points: Ramsey and Street (1940). Line: Ford, A.F. Huxley, and Simmons (1966).

substrate (Needham, 1950, and personal communication). This suggested that the active sites in a muscle operated in a cyclic manner, in the same way that any enzyme causing a chemical reaction acts cyclically, i.e., the products are liberated, another substrate molecule is bound, and this is then reacted on. This was a very important suggestion, but the idea is difficult or impossible to reconcile with continuous filaments folding progressively, and it may be for this reason that it received little attention at the time. However, it fits well with the sliding filament idea.

The deviation from a hyperbolic relation between speed of shortening and rate of energy liberation shown by A.V. Hill (1964; see Figure 2.2) suggested another feature of cross-bridge action that is now widely accepted, namely that the state of a cross-bridge when it is first formed is a weak attachment of the myosin head to an actin site, which may be broken without completing the hydrolysis of an ATP molecule and, therefore, without a contribution to the measured energy liberation (A.F. Huxley, 1973). Another explanation for the drop in rate of energy liberation has also been proposed (Barclay, 1999).

A TWO-STATE THEORY

Soon after the sliding-filament theory was established, I developed a theory of contraction incorporating these three features, i.e., independent force generators

distributed throughout each overlap zone, cyclic action, and rate constants dependent on the mechanical situation (A.F. Huxley, 1957). It is kinetic in character, and the particular assumptions made about the chemical and mechanical events during the cycle were highly speculative and are not essential to the theory. The objective was to simulate the steady-state relations between speed of shortening, tension, and rate of heat production found by A.V. Hill (1938) by making assumptions about the rate constants for attachment and detachment of interacting sites on the two filaments. In this theory, it was proposed that there are little projections from the myosin filament that can be in either of two states, detached (as shown in Figure 2.5), or attached to a site on the thin filament so as to form a cross-bridge. When attached, the projections will exert tension on the thin filament and tend to pull it along, because of an elastic element through which each projection is attached to the thick filament. The essential idea is that the rate constant for formation of these attachments has a moderate value over a certain range of relative positions of the two sites in which the force in the elastic element makes a positive contribution to tension (Figure 2.6), while the rate constant for detachment is low so long as the contribution to tension is positive, but becomes large when sliding of the filaments brings the attachment into a position where the elastic element resists further shortening. The shapes of these relations were chosen so as to obtain analytical solutions for what happens with a steady speed of shortening. The theory gave a very satisfactory fit to the relations found by A.V. Hill (1938) between speed of shortening, tension, and rate of heat production.
So, for these aspects of steady-state behaviour of muscle, the theory worked well. On the other hand, it failed completely on events that happened on a much faster time scale (see section on transient responses, below), so the theory became merely a step towards more complex theories containing more than one state of attached bridges, which were able to explain rapid events by transitions between these states without actual detachment. However, the three broad features of the 1957 theory (an elastic element in the cross-bridge, and dependence of rates of attachment and detachment on relative positions of the sites) are probably correct, and have been incorporated into most of the theories that have been taken to a point where they can be tested against experimental observations. However, I was astonished and, I must admit, gratified to hear just yesterday, at the Canmore Symposium on Skeletal Muscle, that the two-state theory is still used extensively for biomechanical modelling of the overall behaviour of muscle in the living animal, where many geometrical complications of fibre arrangement are involved, but the events are slow enough that the theory gives an adequate description of the force-velocity and energy relations.

Among the most beautiful electron micrographs of the time were those of H.E. Huxley (1957). Figure 2.7 shows a more recent micrograph of the same type, i.e., a longitudinal section containing only a single layer of thick and thin filaments. In these sections, one can clearly see connections ('cross-bridges') between the thick and the thin filaments. Similar projections are also seen on the thick (myosin) filaments in the H-zone, where there are no actin filaments, while there are no such

1957 Cross-Bridge:

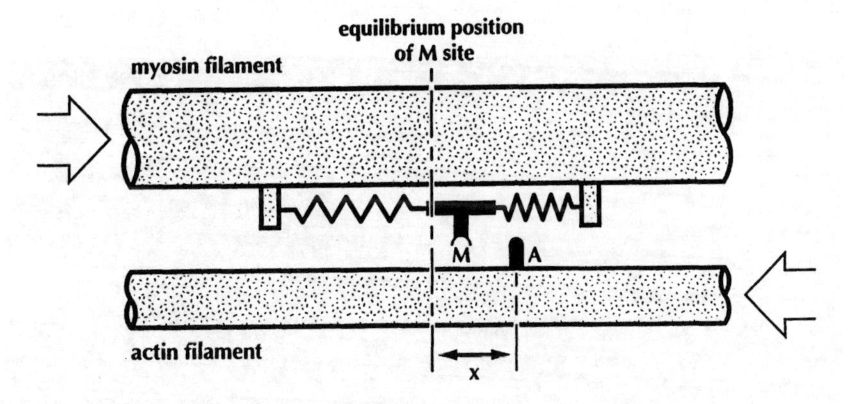

Figure 2.5. Schematic diagram of a cross-bridge as postulated in the theory of A.F. Huxley (1957).

Rate Functions:

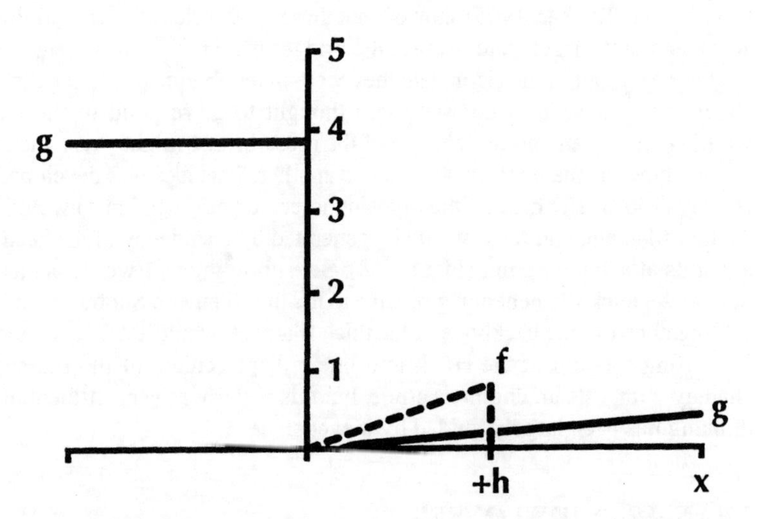

Figure 2.6. Rate coefficients for attachment (f) and detachment (g) of a myosin head to actin, in the theory of A.F. Huxley (1957). Abscissa x as defined in Figure 2.5. The unit of the ordinate scale is the value of $(f+g)$ at $x=h$.

projections on the actin filaments in the I-bands where there are no myosin filaments. This showed that the cross-bridges are indeed part of myosin molecules, and it was natural to suppose that these uniformly distributed connections between the filaments were the independent force generators. This has been almost universally accepted ever since.

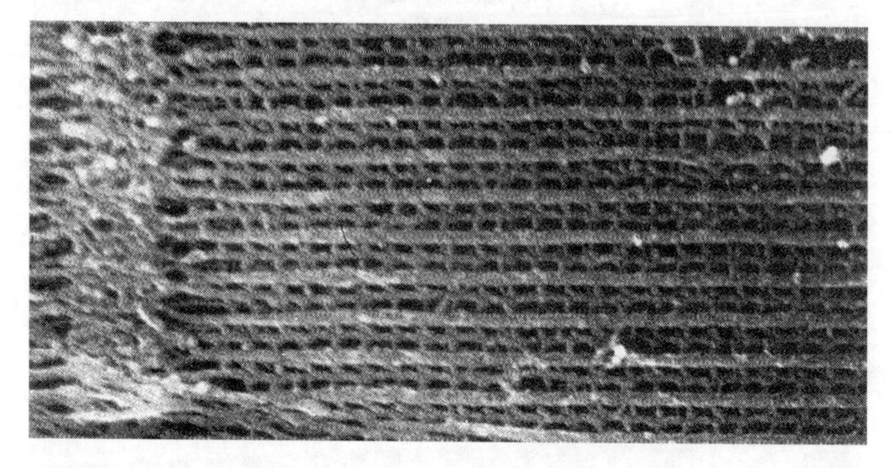

Figure 2.7. Electron micrograph of insect flight muscle, showing a layer of alternately thick and thin filaments. (Reproduced from Trombitás et al., 1986, with the permission of Academic Press.)

The next step came when another group of electron microscopists (Reedy, Holmes, and Tregear, 1965) showed that these cross-bridges are at different angles at rest and in the rigor state. At rest, the bridges were more or less perpendicular to the fibre axis, but in the rigor state they were in an oblique position (tilted through about 45°). Since the rigor state was thought to correspond to the end of the working stroke, an angular change of the cross-bridge might cause the sliding of the two types of filaments relative to one another. This idea was developed by H.E. Huxley (1969). The head of the myosin molecule (subfragment 1) would attach to the thin filament, and force would be generated by a tendency of the head to rotate about its attachment point (Figure 2.8). So, subfragment 1 would act as a crank, pulling the thick filament along relative to the thin filament. Subfragment 2, joining subfragment 1 to the backbone of the thick filament, would act as a connecting rod, converting rotation of the crank into linear displacement of the filaments. This analogy with a steam engine is crude, but this paper was very influential, and my thinking has been largely guided by it ever since.

TRANSIENT RESPONSES

The next step towards understanding the action of cross-bridges was to record the responses of muscle fibres to sudden changes of either load or length. The earliest

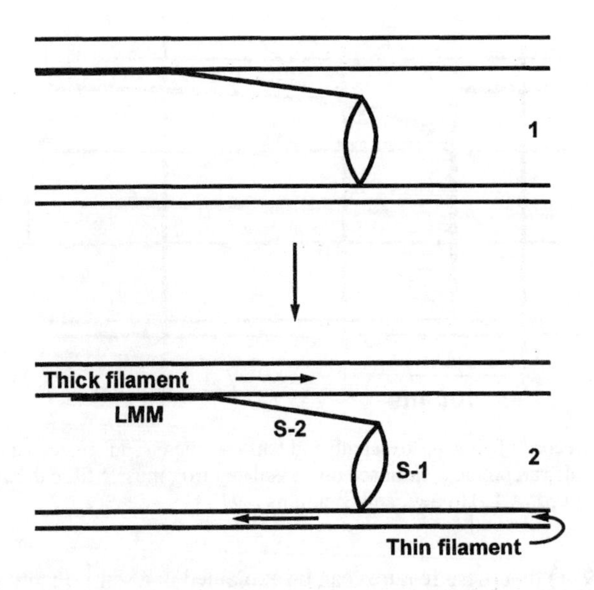

Figure 2.8. Diagram of mechanism of force generation by a cross-bridge, proposed by H.E. Huxley (1969).

experiments of this type were made by Richard Podolsky (1960; Civan and Podolsky, 1966). The load on a fibre or small bundle of fibres was suddenly reduced during a tetanic contraction, and the time course of the consequent shortening was recorded. The preparation shortened slightly simultaneously with the load change, and this was followed by a heavily damped oscillation superposed on the steady shortening that continued so long as the load was reduced. Some recordings of ours showed much more lightly damped oscillations when the load was kept very close to the isometric tension (Armstrong, A.F. Huxley, and Julian, 1966). Most of our experiments, however, have been of the converse type, in which the length of the fibre is suddenly either increased or decreased and the time course of the tension changes is recorded (e.g., Ford, A.F. Huxley, and Simmons, 1977). The main features of the response to shortening by about 6 nm per half-sarcomere are shown in Figure 2.9. Tension drops simultaneously with the shortening step and then recovers part of the way toward the original tension in the first millisecond or two. The later recovery of tension occurs in two phases, but there is no actual oscillation, which is an indication that length change, rather than tension change, is the factor directly affecting the cross-bridges. This was one of the reasons why we concentrated on length steps rather than load steps. The other reason was that the non-linear behaviour of the fibre makes it much more difficult to obtain good step changes of load than of length.

There are striking non-linearities in the early tension recovery, as regards both its extent, as shown in Figure 2.10, and its speed. We showed (A.F. Huxley and

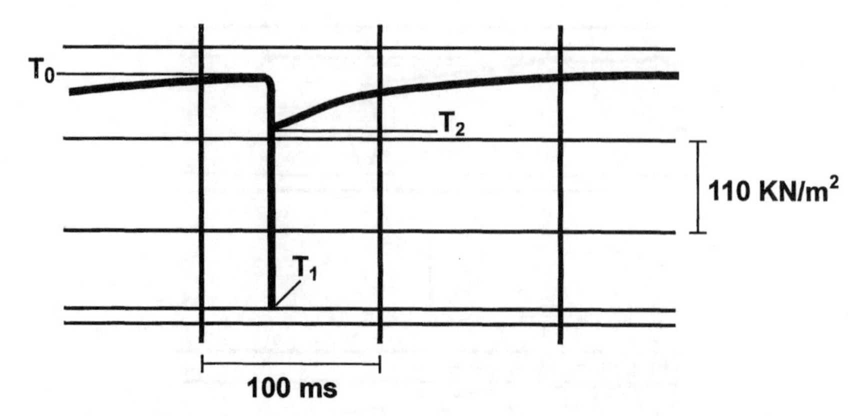

Figure 2.9. Record of changes of length and tension when a sudden decrease of length by 6 nm in each half-sarcomere is imposed on an isolated frog muscle fibre during contraction at 0°C. (From Ford, A.F. Huxley, and Simmons, 1977.)

Simmons, 1971) that these features can be explained in a semi-quantitative manner by assuming that each cross-bridge contains an elastic element (as in the theory of A.F. Huxley, 1957) and can undergo a reversible stepwise transition, which stretches the elastic element. The state corresponding to the higher tension is favoured by the difference of chemical potential between the two states, while increase in tension naturally favours the state with the lower tension. An equilibrium with rapid switching between these states is established and, when the fibre length is suddenly changed, this equilibrium is disturbed by the change in tension in the elastic element. The early tension recovery corresponds to the re-establishment of equilibrium, and the rate at which this occurs depends on the direction and size of the step, since work done against the elastic element forms part of the activation energy of the transition. More recently, we obtained good quantitative agreement with experiment (A.F. Huxley and Tideswell, 1996), taking into account recordings with much better time resolution than we had in 1971, and also allowing for the compliance that has now been shown to exist in the filaments themselves.

If we now define T_0 as the tension just before the length step, T_1 as the extreme tension reached during the step, and T_2 as the tension reached in the early recovery (Figure 2.9), we showed that T_2 is nearly as great as T_0 for releases up to about 5 nm per half-sarcomere (Figure 2.10). With larger releases, T_2 drops away, reaching zero (no early recovery) with a release of about 13 nm. T_1 varied almost linearly with the size of the step, reaching zero with a release of about 6 nm, but this value would be smaller with more rapid steps, and we estimated that it would be about 4 nm with an instantaneous step. The early recovery is about six times faster after a release of 6 nm per half-sarcomere than after a small release or stretch. These non-linearities are, of course, tiresome to those who prefer linear phenomena that can be conveniently treated by Fourier analysis. My advice, however, is: treasure your non-linearities, since they are likely to point to something interesting.

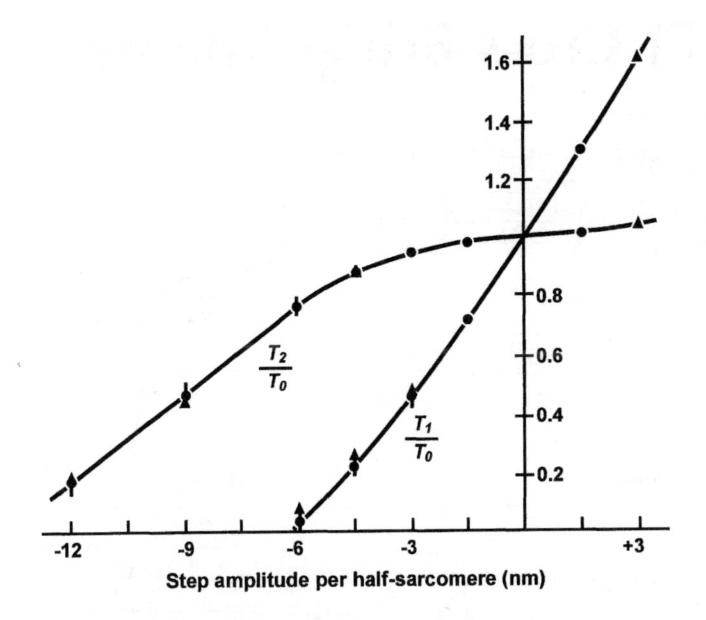

Figure 2.10. Values of T_1 (the extreme tension reached after the step; see Figure 2.9) and T_2 (the tension reached during the early tension recovery after the step) as functions of the size of step. (From Ford, A.F. Huxley, and Simmons, 1977.)

As shown in Figure 2.11, Bob Simmons and I (1971) pictured the stepwise changes in the cross-bridge as rotations of subfragment 1 about its attachment to the thin filament, as proposed by Hugh Huxley (1969), and subfragment 2 was drawn as the elastic element. However, the key features of our theory were simply that shortening happens in steps and that there is an elastic element somewhere in the cross-bridge. It is not important to the theory whether the steps are steps in the angle of the myosin head or, for instance, in the length of subfragment 2, and although we drew the elastic element in subfragment 2, the theory would be the same if the elasticity were in the bending of the head itself. Drawing the system as in Figure 2.11 has the advantage of separating the parts that undergo the stepwise and elastic changes, and this makes it easier to think about them and their interactions. The transient responses give no evidence about the structures that provide the stepwise and elastic characteristics.

CONFORMATIONAL CHANGES

The first indication that the steps might not be simple rotations of the whole myosin head came from the determination of the atomic structures of the myosin head and of the actin monomer by X-ray crystallography (Rayment et al., 1993a; and Kabsch

1971 Cross-Bridge Theory:

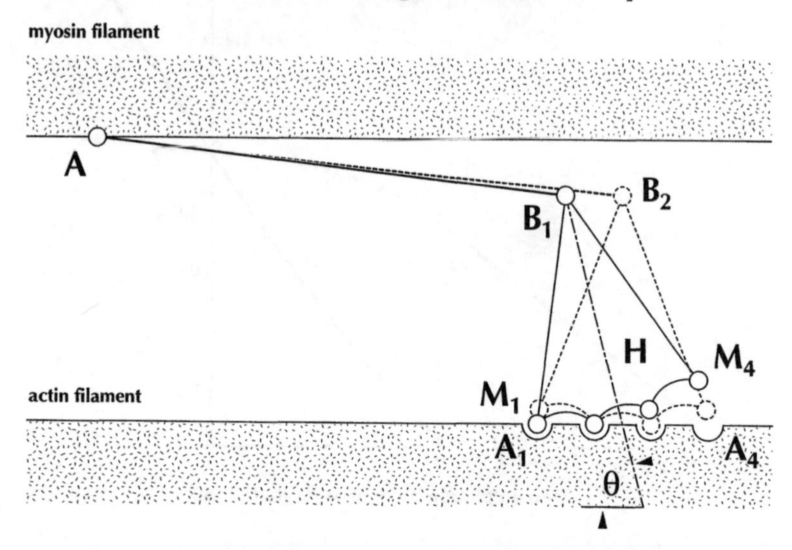

Figure 2.11. Diagram to illustrate the assumptions in the theory of A.F. Huxley and Simmons (1971). The bond between M_3 and A_3 is stronger than that between M_1 and A_1, favouring position B_2 over B_1. Likewise, the bond between M_4 and A_4 is stronger than that between M_2 and A_2. The link AB is elastic.

et al., 1990, respectively), and their fitting together by electron microscopy (Rayment et al., 1993b). These studies suggested that the tilting might involve only a part of the myosin head. Figure 2.12 shows a representation of the myosin head and its catalytic domain as it attaches to the actin filament. This attachment is thought to be rigid, and the stepwise change is thought to be due to a conformational change by which the 'lever arm' (the light-chain binding domain) rotates about a hinge within the myosin head recognized in the X-ray reconstruction. This idea is widely accepted at the present time, though I suspect that part of the working stroke may be due to rotation about the point of attachment (see below). It is widely thought that most of the cross-bridge compliance is in the bending of the head, but the evidence for this is not direct.

The reality of this change of angle at the hinge has been confirmed by several recent observations. Carolyn Cohen and her group (Dominguez et al., 1998) have used X-ray crystallography to determine the structure of a smooth muscle myosin before and after the working stroke, and showed a change of angle at the hinge of 70°, enough to account for the size of stroke estimated from mechanical transient experiments (10-12 nm, Ford et al., 1977). A change in the angle, though not its magnitude, has also been shown by luminescence energy transfer (Getz et al., 1998) and by fluorescence energy transfer (Suzuki et al., 1998), by attaching an

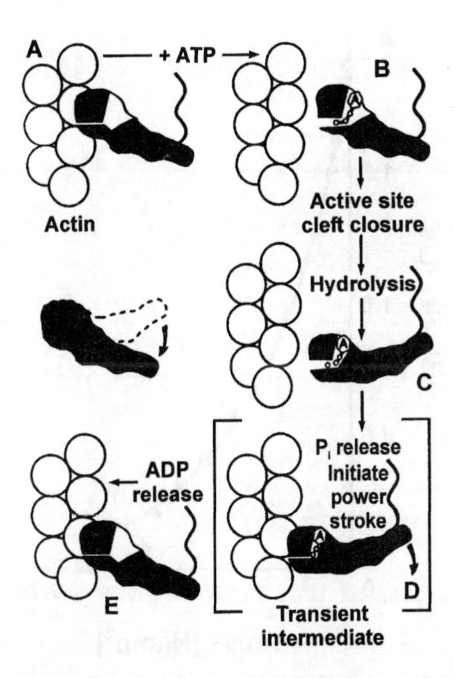

Figure 2.12. Diagram illustrating cycle based on X-ray crystallography. (Reprinted with permission from Rayment et al., 1993b. Copyright 1993 American Association for the Advancement of Science.)

emitting molecule on one side of the hinge and an absorber on the other. These methods have the advantage over X-ray crystallography that the myosin is in solution, without the possibility of being constrained by the crystal lattice to adopt a particular form. Yet another piece of evidence for a hinge was provided by Dobbie et al. (1998), based on changes in the 14.3 nm meridional X-ray reflection from single living muscle fibres at rest, during contraction, and in rigor.

I will just mention two phenomena that have not yet been explained on this theory of the mechanism of contraction. The first of these (Figure 2.13) was shown by Edman et al. (1976) in Sweden: they found that the force-velocity curve of single frog fibres did not exactly follow the hyperbolic relation described by A.V. Hill (1938), but showed much lower velocities under loads near to the isometric tension, a region difficult to investigate.

The second is the 'repriming phenomenon' described by Lombardi et al. (1992) in Florence. They applied two small shortening steps to a single frog fibre during an otherwise isometric contraction. When the interval between them was less than 2 ms or so (at 4°C), the force following the quick phase of recovery was close to what it was when the same total amount of shortening was applied in a single step, but when the interval was, say, 10 ms, the force of recovery was considerably greater (Figure 2.14).

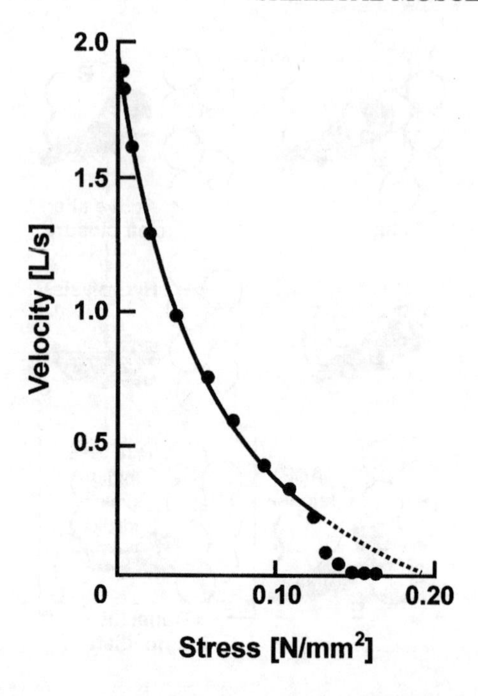

Figure 2.13. Force-velocity curve found by Edman (points) contrasted with the rectangular hyperbola described by Hill (1938, line). (Reproduced from Edman, 1988, with the permission of The Physiological Society.)

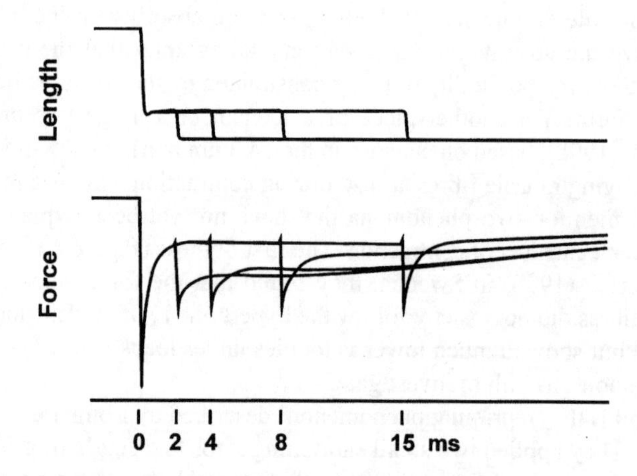

Figure 2.14. 'Repriming' phenomenon described by Lombardi et al. (1992). Upper traces show sarcomere length; lower traces show tension. A shortening step of 5 nm per half-sarcomere is followed, after a variable interval, by a second step of 2 nm per half-sarcomere. The tension reached in the early recovery phase increases as the interval is increased above

2 ms. Frog fibre, 4°C. (Reproduced from Lombardi et al., 1992, with the permission of Macmillan Publishers Ltd.)

As I said, there is no definite explanation for these two phenomena, but not for want of possible explanations. For Edman's force-velocity curve there are two explanations in print and I have a third in mind, and for 'repriming' there are three already in print.

SURPRISES

We have been told, this evening, about several unexpected observations on muscle. Ten years ago I would have said that it was totally impossible to record the tension generated by a single myosin molecule, but this is now a routine procedure in many laboratories. It was first achieved by Jim Spudich and his collaborators at Stanford (Finer et al., 1994). Figure 2.15 is a diagram of their technique. A single actin filament is attached at its ends to two polystyrene beads, held in light traps, that can be manipulated by deflecting the laser beams entering the condenser of the microscope. The filament is brought into contact with a silica bead coated with heavy meromyosin at a density insufficient to support continuous movement of an actin filament. Longitudinal displacement of the filament is recorded by projecting a greatly enlarged image of one of the polystyrene beads onto a differential photocell device. Sudden displacements were seen and were maintained for a period that, on average, was inversely proportional to the concentration of ATP (Figure 2.16). Estimates from different laboratories for the displacement produced by a single attachment vary between about 5 and 15 nm.

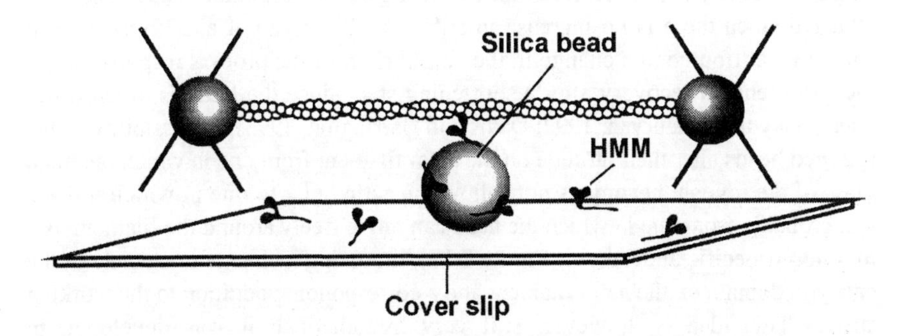

Figure 2.15. Diagram of arrangement for recording tension steps generated by single interactions between a myosin molecule and an actin filament. The filament is stuck at its ends to two polystyrene beads about 1 μm in diameter that are held in light traps, i.e., the foci of laser beams focussed by a microscope objective of high numerical aperture. (Reproduced from Finer et al., 1994, with the permission of Macmillan Publishers Ltd.)

Several new complications have turned up this year. Veigel et al. (1999) showed in single-molecule experiments on two types of very slow non-muscle myosin that displacement or force was generated in two steps. This has not yet been observed with muscle myosin, but it is possible that the observed single step really consists of two that are too close in time to be resolved. This question should be settled either by using slower myosins or by improvements in technique. There is no evidence yet whether both of the steps are driven by conformational changes within the myosin head, or whether one is of a different character, perhaps a tilt of the catalytic domain relative to the actin filament.

Another surprise came from Carolyn Cohen's group. Previously, only two attitudes of the lever arm relative to the catalytic domain had been recognized by X-ray crystallography, but Houdusse et al. (1999) showed a third, in which the angle at the hinge was even greater than in rigor, so that it corresponds to a greater sliding movement in the direction of shortening. It is quite uncertain what part this state plays in contraction: it may represent an additional step in the working stroke, but another possibility is that it is not due to an active conformational change but is brought about by sliding induced by other cross-bridges, and is the process that allows ATP to become bound and to cause dissociation of the attachment to actin.

It has long been known that the layer lines due to the long helix of actin (repeat distance 36 nm) are weak unless the muscle is in rigor and myosin heads are attached all along the actin helix. On going from rest to contraction, there is almost no accentuation of the layer lines (H.E. Huxley et al., 1982). This observation has been taken as evidence that only a small proportion of myosin heads are attached to actin during contraction. But recent work by two Russians, with collaborators in Britain, shows that the rise of tension that follows a sudden rise in temperature is accompanied by an accentuation of the first actin layer line (Bershitsky et al., 1997; Tsaturyan et al., 1999). This cannot be attributed to additional heads becoming attached since there is no increase in stiffness (Tsaturyan et al., 1999). Also it cannot be attributed to a change in the equilibrium of the process responsible for the early tension recovery after a shortening step, since the latter is much faster (Bershitsky and Tsaturyan, 1992; Davis and Harrington, 1993). A possibility is that attached heads alter their attitude on the actin filament from one in which the main mass of the myosin head does not follow the actin helix to one in which it does, i.e., probably from one in which the head can move freely around the filament axis to a stereospecific attachment (Figure 2.17), involving a change in attitude of the catalytic domain on the actin filament and a corresponding addition to the working stroke. This idea is, however, still very hypothetical; it was developed in conversation among Sergey Bershitsky, Andrey Tsaturyan, and me. If it is correct, it would be natural to suppose that the change of attachment involves formation of a hydrophobic bond, since it is favoured by a rise of temperature. In this case, it would resemble the step to the final state in the scheme of Diaz Baños et al. (1996, Figure 7). There is electron microscope evidence that, in asynchronous insect flight muscle, a myosin head may attach to actin in one or other of two attitudes (Schmitz

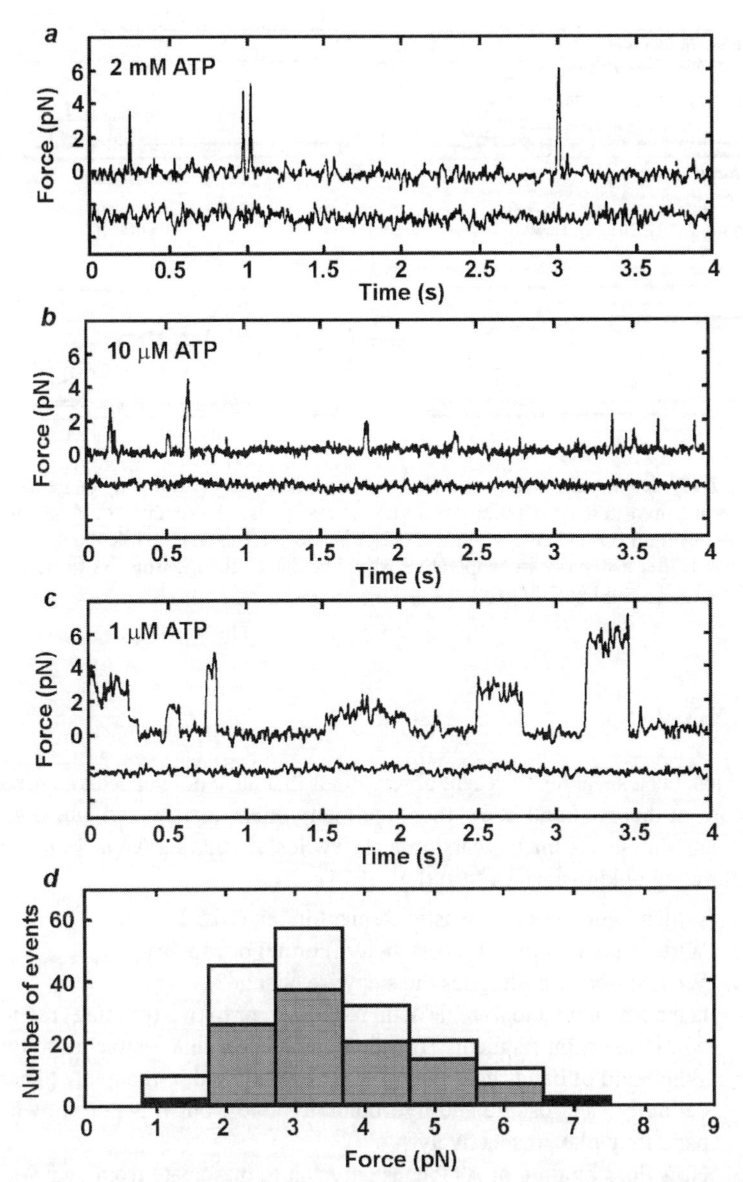

Figure 2.16. Force generated by single interactions between myosin and actin at various concentrations of ATP. (Reproduced from Finer et al., 1994, with the permission of Macmillan Publishers Ltd.)

et al., 1997; Taylor et al., 1999), which may correspond to the low- and high-temperature states postulated here.

I am sure that there are more surprises in store for us.

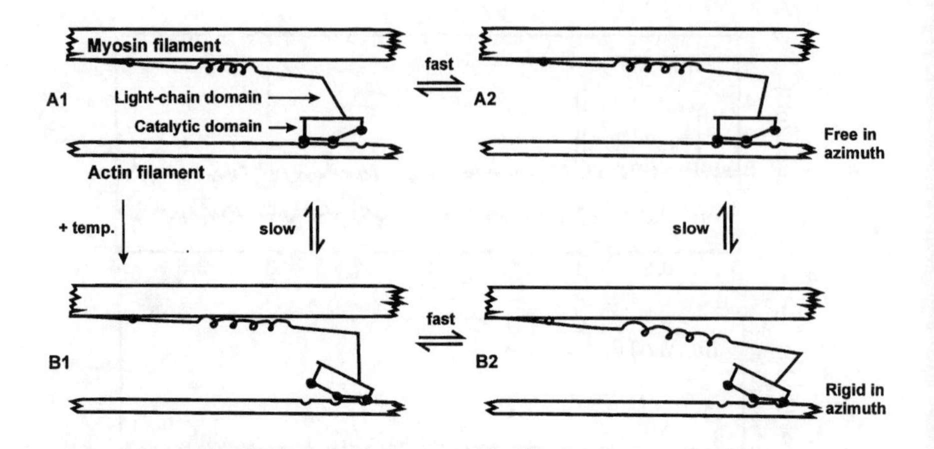

Figure 2.17. Schematic diagram showing two ways in which tension may be generated by changes in a myosin head. The early recovery after stepwise shortening would be due to re-establishing equilibrium between A1 and A2 or between B1 and B2, while the much slower tension rise after a step rise in temperature would be due to change from A1 to B1 or A2 to B2. (From A.F. Huxley, 2000.)

UNKNOWNS

I have just been speaking of recent observations that have not yet been incorporated into current theory. In addition, there are many other unanswered questions that have been with us for many years. In a review lecture that I gave in 1973, I listed the following unknowns (A.F. Huxley, 1974):

1. Which structure is the elastic element in the cross-bridge?
2. What is the nature of this elasticity [normal or rubbery]?
3. What structure undergoes the stepwise change?
4. Is the attachment to a single actin monomer, or to two (or more) monomers within the actin filament? [It is now fairly clear that it attaches to two.]
5. What kind of bonds hold the myosin head to the thin filament? [There are certainly electrostatic and hydrophobic bonds, but it is not known what parts they play, respectively.]
6. How does binding of ATP cause myosin to dissociate from actin?
7. What is the significance of the fact that each myosin molecule has two heads?

So, most of these problems listed in 1973 are still unsolved. In addition, the following problems have arisen since then:

1. What fraction of heads are active at any one time in an isometric contraction? [Estimates vary between 15 and 50%.]

2. Is the attachment probability uniform within each repeat of the actin helix? [There is evidence that it does vary in some muscles (Taylor et al., 1999).]
3. What is the origin of the rise of tension that occurs with a delay after a stretch?
4. What is the basis of the oscillatory contraction of asynchronous insect flight muscle?
5. What is the amount of force from a single head? [Estimates vary between 2 and 10 pN.]
6. What is the size of the step? [Estimates vary between 5 and 15 nm.]

At the end of my lecture in 1973, I said (and I repeat it now): 'Quite apart from the fact that all these things are unknown, the whole history of muscular contraction during the last half century shows that even when a set of ideas seems to be well established, there is a large chance that it will be overthrown by some unexpected discovery'.

PROSPECTS

But there are good prospects. There will certainly be major developments in techniques already in use, such as measurement of single-molecule events pioneered by Yanagida and his group: stepwise motion along actin (Kitamura et al., 1999), and binding and release of fluorescent ATP analogues (Funatsu et al., 1995; Ishijima et al., 1998). A big advance in measurement of changes of orientation of the myosin head is the use of probes whose orientation on the protein is known (Corrie et al., 1999); another advance is the use of indicators of steps in the hydrolysis of ATP, such as the release of P_i (He et al., 1997). Unfortunately, few of these methods can be used on intact muscle. In single-molecule experiments, the relative positions of actin and myosin are not aligned as they are in intact muscle, and methods requiring the penetration of large molecules into the filament lattice can only be used on skinned preparations, where it is likely that irregular local shortening occurs and conditions are abnormal in many respects.

Site-directed mutation, used with immense success in many other fields, has hardly been attempted on the myosin heavy chain. Another approach that I expected would be widely used as soon as the atomic structures of actin and myosin had been solved is the calculation of atomic interactions, both between actin and myosin and between different parts of the myosin molecule, but I am so far aware of only one attempt in this direction (Diaz Baños et al., 1996).

We must be on our guard against repeating a disaster that happened at the beginning of the 20th century. The rise of classical biochemistry—of inestimable importance for the understanding of muscle in the long run—had the immediate effect of concentrating interest on molecular events, and led to almost complete disregard of events on a larger scale, such as the movements of the thick and thin filaments that show up in the striation pattern. The mass of excellent information gathered by light microscopy in the 19th century was lost and had to be rediscovered in the 1950s by electron microscopy and phase and interference light microscopy. There is a real danger now of interest switching exclusively to single-molecule and intramolecular events. Let us make sure that things that were common knowledge 20 years ago are not forgotten and have to wait another 20 years before they are rediscovered.

REFERENCES

Armstrong, C.M., Huxley, A.F., and Julian, F.J. (1966). Oscillatory responses in frog skeletal muscle fibres. *Journal of Physiology* **186**, 26–27*P*.

Barclay, C.J. (1999). A weakly coupled version of the Huxley crossbridge model can simulate energetics of amphibian and mammalian muscle. *Journal of Muscle Research and Cell Motility* **20**, 163–176.

Bershitsky, S.Y., and Tsaturyan, A.K. (1992). Tension responses to Joule temperature jump in skinned rabbit muscle fibres. *Journal of Physiology* **447**, 425–448.

Bershitsky, S.Y., Tsaturyan, A.K., Bershitskaya, O.N., Mashanov, G.I., Brown, P., Burns, R., and Ferenczi, M.A. (1997). Muscle force is generated by myosin heads stereospecifically attached to actin. *Nature* **388**, 186–190.

Civan, M.M., and Podolsky, R.J. (1966). Contraction kinetics of striated muscle fibres following quick changes in load. *Journal of Physiology* **184**, 511–534.

Corrie, J.E.T., Brandmeier, B.D., Ferguson, R.E., Trentham, D.R., Kendrick-Jones, J., Hopkins, S.C., van der Heide, U.A., Goldman, Y.E., Sabido-David, C., Dale, R.E., Criddle, S., and Irving, M. (1999). Dynamic measurement of myosin light-chain-domain tilt and twist in muscle contraction. *Nature* **400**, 425–430.

Davis, J.S., and Harrington, W.F. (1993). A single order-disorder transition generates tension during the Huxley-Simmons Phase 2 in muscle. *Biophysical Journal* **65**, 1886–1898.

Diaz Baños, F.G., Bordas, J., Lowy, J., and Svensson, A. (1996). Small segmental rearrangements in the myosin head can explain force generation in muscle. *Biophysical Journal* **71**, 576–589.

Dobbie, I., Linari, M., Piazzesi, G., Reconditi, M., Koubassova, N., Ferenczi, M.A., Lombardi, V., and Irving, M. (1998). Elastic bending and active tilting of myosin heads during muscle contraction. *Nature* **396**, 383–387.

Dominguez, R., Freyzon, Y., Trybus, K.M., and Cohen, C. (1998). Crystal structure of a vertebrate smooth muscle myosin motor domain and its complex with the essential light chain: Visualization of the pre-power stroke state. *Cell* **94**, 559–571.

Edman, K.A.P. (1988). Double-hyperbolic force-velocity relation in frog muscle fibres. *Journal of Physiology* **404**, 301–321.

Edman, K.A.P., Mulieri, L.A., and Scubon-Mulieri, B. (1976). Non-hyperbolic force-velocity relationship in single muscle fibres. *Acta Physiologica Scandinavica* **98**, 143–156.

Eggleton, P., and Eggleton, G.P. (1927). The inorganic phosphate and a labile form of organic phosphate in the gastrocnemius of the frog. *Biochemical Journal* **21**, 190–195.

Fenn, W.O. (1923). A quantitative comparison between the energy liberated and the work performed by the isolated sartorius muscle of the frog. *Journal of Physiology* **58**, 175–203.

Finer, J.T., Simmons, R.M., and Spudich, J.A. (1994). Single myosin molecule mechanics: piconewton forces and nanometre steps. *Nature* **368**, 113–119.

Fiske, C.H., and Subbarow, Y. (1927). The nature of the 'inorganic phosphate' in voluntary muscle. *Science* **65**, 401–403.

Ford, L.E., Huxley, A.F., and Simmons, R.M. (1977). Tension responses to sudden length change in stimulated frog muscle fibres near slack length. *Journal of Physiology* **269**, 441–515.

Funatsu, T., Harada, Y., Tokunaga, M., Saito, K., and Yanagida, T. (1995). Imaging of single fluorescent molecules and individual ATP turnovers by single myosin molecules in aqueous solution. *Nature* **374**, 555–559.

Getz, E.B., Cooke, R., and Selvin, P.R. (1998). Luminescence resonance energy transfer measurements in myosin. *Biophysical Journal* **74**, 2451–2458.

Gordon, A.M., Huxley, A.F., and Julian, F.J. (1966). The variation in isometric tension with sarcomere length in vertebrate muscle fibres. *Journal of Physiology* **184**, 170–192.

Hanson, J., and Huxley, H.E. (1953). Structural basis of the cross-striations in muscle. *Nature* **172**, 530–532.

Hasselbach, W. (1953). Elektronenmikroskopische Untersuchungen an Muskelfibrillen bei totaler and partieller Extraktion des L-Myosins. *Zeitschrift für Naturforschung* **8b**, 449–454.

He, Z-H., Chillingworth, R.K., Brune, M., Corrie, J.E.T., Trentham, D.R., Webb, M.R., and Ferenczi, M.A. (1997). ATPase kinetics on activation of rabbit and frog permeabilized isometric muscle fibres: a real time phosphate assay. *Journal of Physiology* **501**, 125–148.

Hill, A.V. (1924). The mechanism of muscular contraction (Nobel lecture). In: *Les Prix Nobel en 1923*. Norstedt, Stockholm, Sweden.

Hill, A.V. (1932). The revolution in muscle physiology. *Physiological Reviews* **12**, 56–67.

Hill, A.V. (1938). The heat of shortening and the dynamic constants of muscle. *Proceedings of the Royal Society of London B* **126**, 136–195.

Hill, A.V. (1964). The effect of load on the heat of shortening of muscle. *Proceedings of the Royal Society of London B* **159**, 297–318.

Hill, A.V. (1965). *Trails and Trials in Physiology*. Arnold, London, England.

Houdusse, A., Kalabokis, V.N., Himmel, D., Szent-Györgyi, A.G., and Cohen, C. (1999). Atomic structure of scallop myosin subfragment S1 complexed with MgATP: A novel conformation of the myosin head. *Cell* **97**, 459–470.

Huxley, A.F. (1952). Applications of an interference microscope. *Journal of Physiology* **117**, 52–53*P*.

Huxley, A.F. (1954). A high-power interference microscope. *Journal of Physiology* **125**, 11–13*P*.

Huxley, A.F. (1957). Muscle structure and theories of contraction. *Progress in Biophysics and Biophysical Chemistry* **7**, 255–318.

Huxley, A.F. (1973). A note suggesting that the cross-bridge attachment during muscle contraction may take place in two stages. *Proceedings of the Royal Society of London B* **183**, 83–86.

Huxley, A.F. (1974). Muscular contraction (review lecture). *Journal of Physiology* **243**, 1–43.

Huxley, A.F. (1977). Looking back on muscle. In: *The Pursuit of Nature*, by A.L. Hodgkin et al., Cambridge University Press, Cambridge, England, pp. 23–64.

Huxley, A.F. (1980). *Reflections on Muscle*. University Press, Liverpool, England.

Huxley, A.F. (2000). Mechanics and models of the myosin motor. *Philosophical Transactions of the Royal Society of London B* **355**, 433–440.

Huxley, A.F., and Niedergerke, R. (1954). Interference microscopy of living muscle fibres. *Nature* **173**, 971–973.

Huxley, A.F., and Niedergerke, R. (1958). Measurement of the striations of isolated muscle fibres with the interference microscope. *Journal of Physiology* **144**, 403–425.

Huxley, A.F., and Peachey, L.D. (1959). The maximum length for contraction in striated muscle. *Journal of Physiology* **146**, 55–56*P*.

Huxley, A.F., and Peachey, L.D. (1961). The maximum length for contraction in vertebrate striated muscle. *Journal of Physiology* **156**, 150–165.

Huxley, A.F., and Simmons, R.M. (1971). Proposed mechanism of force generation in striated muscle. *Nature* **233**, 533–538.

Huxley, A.F., and Tideswell, S. (1996). Filament compliance and tension transients in muscle. *Journal of Muscle Research and Cell Motility* **17**, 507–511.

Huxley, H.E. (1953). Electron microscope studies of the organisation of the filaments in striated muscle. *Biochimica et Biophysica Acta* **12**, 387–394.

Huxley, H.E. (1957). The double array of filaments in cross-striated muscle. *Journal of Biophysical and Biochemical Cytology* **3**, 631–648.

Huxley, H.E. (1969). The mechanism of muscular contraction. *Science* **164**, 1356–1366.

Huxley, H.E., Faruqi, A.R., Kress, M., Bordas, J., and Koch, M.H.J. (1982). Time-resolved X-ray diffraction studies of the myosin layer-line reflections during muscle contraction. *Journal of Molecular Biology* **158**, 637–684.

Huxley, H.E., and Hanson, J. (1954). Changes in the cross-striations of muscle during contraction and stretch and their structural interpretation. *Nature* **173**, 973–976.

Ishijima, A., Kojima, H., Funatsu, T., Tokunaga, M., Higuchi, H., Tanaka, H., and Yanagida, T. (1998). Simultaneous observation of individual ATPase and mechanical events by a single myosin molecule during interaction with actin. *Cell* **92**, 161–171.

Kabsch, W., Mannherz, H.G., Suck, D., Pai, E.F., and Holmes, K.C. (1990). Atomic structure of the actin: DNase I complex. *Nature* **347**, 37–44.

Kitamura, K., Tokunaga, M., Iwane, A.H., and Yanagida, T. (1999). A single myosin head moves along an actin filament with regular steps of 5.3 nanometres. *Nature* **397**, 129–134.

Kuhn, T.S. (1962). *The Structure of Scientific Revolutions*. University of Chicago Press, Chicago, Illinois, USA.

Lombardi, V., Piazzesi, G., and Linari, M. (1992). Rapid regeneration of the actin-myosin power stroke in contracting muscle. *Nature* **355**, 638–641.

Lundsgaard, E. (1930). Untersuchungen über Muskelkontraktionen ohne Milch-säurebildung. *Biochemische Zeitschrift* **217**, 162–177.

Needham, D.M. (1950). Myosin and adenosinetriphosphate in relation to muscle contraction. *Biochimica et Biophysica Acta* **4**, 42–49. (Reprinted as *Metabolism and Function*, Elsevier, 1950.)

Podolsky, R.J. (1960). Kinetics of muscular contraction: the approach to the steady state. *Nature* **188**, 666–668.

Ramsey, R.W., and Street, S.F. (1940). The isometric length-tension diagram of isolated skeletal muscle fibers of the frog. *Journal of Cellular and Comparative Physiology* **15**, 11–34.

Rayment, I., Holden, H.M., Whittaker, M., Yohn, C.B., Lorenz, M., Holmes, K.C., and Milligan, R.A. (1993b). Structure of the actin-myosin complex and its implications for muscle contraction. *Science* **261**, 58–65.

Rayment, I., Rypniewski, W.R., Schmidt-Bäse, K., Smith, R., Tomchick, D.R., Benning, M.M., Winkelmann, D.A., Wesenberg, G., and Holden, H.M. (1993a). Three-dimensional structure of myosin subfragment-1: A molecular motor. *Science* **261**, 50–58.

Reedy, M.K., Holmes, K.C., and Tregear, R.T. (1965). Induced changes in orientation of the cross-bridges of glycerinated insect flight muscle. *Nature* **207**, 1276–1280.

Schmitz, H., Reedy, M.C., Reedy, M.K., Tregear, R.T., and Taylor, K.A. (1997). Tomographic three-dimensional reconstruction of insect flight muscle partially relaxed by AMPPNP and ethylene glycol. *Journal of Cell Biology* **139**, 695–707.

Suzuki, Y., Yasunaga, T., Ohkura, R., Wakabayashi, T., and Sutoh, K. (1998). Swing of the lever arm of a myosin motor at the isomerization and phosphate-release steps. *Nature* **396**, 380–383.

Taylor, K.A., Schmitz, H., Reedy, M.C., Goldman, Y.E., Franzini-Armstrong, C., Sasaki, H., Tregear, R.T., Poole, K., Lucaveche, C., Edwards, R.J., Chen, L.F., Winkler, H., and Reedy, M.K. (1999). Tomographic 3D reconstruction of quick-frozen, Ca^{2+}-activated contracting insect flight muscle. *Cell* **99**, 421–431.

Trombitás, K., Baatsen, P.H.W.W., and Pollack, G.H. (1986). Rigor bridge angle: Effects of applied stress and preparative procedure. *Journal of Ultrastructure and Molecular Structure Research* **97**, 39–49.

Tsaturyan, A.K., Bershitsky, S.Y., Burns, R., and Ferenczi, M.A. (1999). Structural changes in the actin-myosin cross-bridges associated with force generation induced by temperature jump in permeabilized frog muscle fibers. *Biophysical Journal* **77**, 354–372.

Veigel, C., Coluccio, L.M., Jontes, J.D., Sparrow, J.C., Milligan, R.A., and Molloy, J.E. (1999). The motor protein myosin-I produces its working stroke in two steps. *Nature* **398**, 530–533.

3 Cellular and Molecular Muscle Mechanics

W. HERZOG
Human Performance Laboratory, University of Calgary, Calgary, Alberta, Canada

INTRODUCTION

A single muscle fibre is a cell. Single fibre research was started systematically in the 1940s and 1950s. Ramsey and Street (1940) demonstrated the dependence of maximal isometric force on fibre length. A. Huxley and Niedergerke (1954) showed in a single-fibre preparation that A-band width remained nearly constant for a variety of contractile conditions. This latter study, together with the research by H. Huxley and Hanson (1954), gave rise to the sliding filament theory, that is, the idea that muscle contraction and shortening (lengthening) was accomplished by a relative sliding of the thick (myosin) and thin (actin) myofilaments.

Ever since the sliding filament theory, there has been a search for the molecular mechanisms of contraction, or, what is the molecular motor that drives the relative sliding of the two sets of filaments? One of the first explanations was given by A. Huxley (1957), who formulated, based on structural information and the force-velocity experiments by Hill (1938), a mechanism of contraction. This mechanism, which is now typically referred to as the cross-bridge theory or the independent force-generator theory, was formulated mathematically to give a quantitative description of the mechanical and energetic events underlying muscle contraction. In its most simple form, the cross-bridge theory states that there is a set of uniformly arranged side pieces (cross-bridges) on the thick filament that can attach to specific sites on the neighbouring thin filament. Once attached, these cross-bridges undergo some conformational change that produces a sliding of the thin filament, relative to the thick filament, towards the centre of the sarcomere. The cross-bridge then detaches and is ready to start its attachment and force cycle anew.

In the 1957 theory, A. Huxley assumed two states for the cross-bridge: attached (to the actin filament) and detached. The rate constants of attachment and detachment were chosen such that the force-velocity relationship found by Hill (1938) could be matched. By restricting the model to two states, and by assuming that one ATP was required per cross-bridge cycle, several known phenomena at that time could not be predicted. For example, the heat produced by stretching a muscle was too high, and so was the maximal eccentric force predicted by this par-

Skeletal Muscle Mechanics: From Mechanisms to Function. Edited by W. Herzog.

ticular model. By further assuming that all cross-bridges were identical, that cross-bridges were independent of each other, and that their attachment/detachment rates were history-independent, the steady-state forces predicted for given, constant contractile conditions were always the same, independent of the muscle's contractile history; a prediction that was known to be not correct at that time (Abbott and Aubert, 1952). However, even to date, it is not known whether these history-dependent phenomena are related to structural non-uniformities (and thus are independent of the molecular mechanisms of force production), or if they are associated with molecular mechanisms that should be represented in the theory governing muscular contraction and force production.

Another observation that could not be explained satisfactorily with the two-state cross-bridge model was the fast force transients that followed a quick length change, or the fast length transients that followed a quick force change. A. Huxley and Simmons (1971) proposed a cross-bridge model that could account for these fast transients observed experimentally and simultaneously retained the slow force-velocity properties of the two-state model. The 1971 model consisted of more than two distinct cross-bridge states: one detached and several attached states. Assuming that the attached states had different binding energies, and further assuming that the rate constants between the various attached states were very fast compared to the rate constants for cross-bridge attachment and detachment, satisfactory predictions could be made for the fast transients observed following a length or a force step.

However, the experiments by A. Huxley and Simmons (1971), and their corresponding biological and mathematical explanation using a multi-state cross-bridge model, made the understanding of the precise molecular events of contraction even more complex than it already was. Many questions remained:

- How do the cross-bridges attach to the actin filament, and how are these multiple attachment states coupled to the corresponding biochemical events (ATP hydrolysis) that produce the energy for the mechanical steps?
- How is force and the relative sliding of the two sets of filaments produced?
- What is the energetic cost of a 'cross-bridge cycle', how much force can a cross-bridge produce, and how far can the cross-bridge 'pull' the thin filament past the thick filament for a single actomyosin interaction?

Between the 1950s and the early 1990s, all these questions were approached with methods relying on isolated fibre preparations or myofibrillar preparations. The whole approach to answering the above questions changed, however, with the description of the molecular structure of the myosin subfragment 1 (S1, the cross-bridge head) and the actin attachment site by Rayment et al. (1993a, 1993b), and the research on isolated cross-bridge-actin interactions pioneered by Finer et al. (1994).

STRUCTURAL CONSIDERATIONS

Rayment et al. (1993a) described the three-dimensional structure of the head portion of myosin, usually referred to as subfragment 1. This part of the myosin molecule contains the actin and nucleotide binding sites, as well as the essential and regulatory light chains (Figure 3.1). The structure of subfragment S1 was determined by single X-ray diffraction.

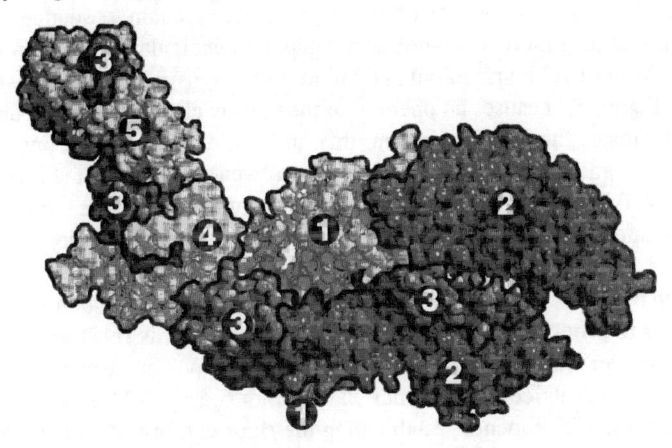

Figure 3.1. Space-filling representation of all atoms in the myosin subfragment 1 model. Segments labelled 1, 2, and 3 represent parts of the heavy chains, while those labelled 4 and 5 show the essential and regulatory light chains, respectively. The proposed actin binding site is located at the lower right-hand corner (2). The active nucleotide site is on the opposite side from the proposed actin binding surface; segment 1. (Reprinted with permission from Rayment et al., 1993a. Copyright 1993 American Association for the Advancement of Science.)

In order to obtain X-ray quality crystals of S1, it was necessary to subject the protein to a mild modification of the lysine residues by reductive methylation. The exact effect of this modification, although likely not severe, remains unknown. Myosin S1 was prepared by digestion with papain in the presence of $MgCl_2$. Using this non-specific approach, proteolytic breaks are produced in the regulatory and essential light chains. Also, partial phosphorylation of the regulatory light chains was produced in this approach. In order to remove the heterogeneities arising from the proteolysis and the phosphorylation of the light chains, Rayment et al. (1993a) used a purification protocol.

The structure of S1 was determined in a two-stage procedure: the positions of the metal binding sites were determined using X-ray data sets with a 4.5Å resolution; then these data were filled in with Synchrotron radiation results to produce a 2.8Å resolution.

Figure 3.1 shows a space-filling representation of all atoms in the myosin S1 model. Segments labelled 1, 2, and 3 represent parts of the heavy chains, while those labelled 4 and 5 show the essential and regulatory light chains, respectively. The proposed actin binding surface is located at the lower right-hand corner (2). The active nucleotide site is on the opposite side from the proposed actin binding surface, segment 1 in Figure 3.1.

In a further study, Rayment et al. (1993b) revealed the structure of the actin-myosin complex, and speculated on a molecular mechanism of contraction. Starting from the rigor conformation, Rayment et al. (1993b) suggested that the narrow cleft that splits the 50-kD segments of the myosin heavy-chain sequence into two domains is closed (Figure 3.2A, horizontal gap, perpendicular to the actin filament axis). Addition of ATP and initial ATP binding to myosin at the active site (segment 1 in Figure 3.1) causes an opening of the narrow cleft between the upper and lower domains of the 50-kD segment; this, in turn, disrupts the 'strong' binding between actin and myosin but still allows for a 'weak' attachment (Figure 3.2B). The final ATP binding to myosin causes a closure of the nucleotide binding pocket and a corresponding configurational change of the myosin molecule. ATP is now hydrolyzed (Figure 3.2C). Rebinding of myosin to actin can now occur, presumably in multiple steps. The gap between the upper and lower domain closes in this process to produce strong binding, and phosphate, P, is released. This event starts the power stroke (Figure 3.2D). During the power stroke, the myosin molecule reverses its conformational change induced by ATP binding, and the active site pocket is reopened, establishing the rigor conformation (Figure 3.2E). The cross-bridge cycle can then be restarted.

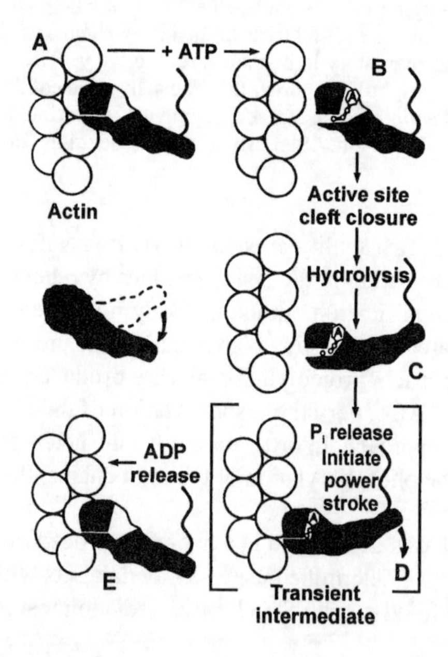

Figure 3.2. Proposed molecular mechanism of contraction. A) Rigor conformation. The narrow cleft that splits the 50-kD segments of the myosin heavy-chain sequence into two domains is closed (horizontal gap, perpendicular to the actin filament axis). B) Addition of ATP, and initial binding of ATP to the active site, causes an opening of the narrow cleft and disrupts the strong binding between actin and myosin, but still allows for weak binding. The actin and myosin dissociate. C) The final binding of ATP to myosin causes a closure of the

nucleotide binding pocket and a corresponding configurational change of the myosin molecule. ATP is now hydrolyzed. D) Myosin can now reattach to actin, presumably in multiple steps. The narrow cleft closes to produce strong binding. Phosphate, P, is released, and the power stroke starts. E) During the power stroke, the myosin molecule reverses its conformation change induced by ATP binding, and the active site pocket is reopened, establishing the rigor conformation. The cross-bridge cycle can now start all over again. (Reprinted with permission from Rayment et al., 1993b. Copyright 1993 American Association for the Advancement of Science.)

When analyzing the contractile events proposed by Rayment et al. (1993b), several limitations of the approach must be remembered. These include that the myosin S1 model was obtained after reductive methylation which has an as-yet unknown effect on the structure. Furthermore, the structures of myosin S1 and the actin binding site are incomplete; the myosin head likely contains flexible elements but, after crystallization, these elements are fixed in an unknown configuration (unknown relative to the working stroke). Similarly, the actomyosin configuration was assumed to closely resemble that in rigor.

Nevertheless, and with all the limitations mentioned above, the research by Rayment et al. (1993a, 1993b) has opened the doors for theoretical investigations aimed at modelling the possible interactions of myosin S1 with actin using the structural elements provided in these two studies.

SINGLE MYOSIN MECHANICS

The studies on the structure of myosin S1 and the actin attachment site (Rayment et al., 1993a, 1993b) suggested, from a structural point of view, the possible mechanism of contraction. However, the actual movement produced by such interactions, the corresponding forces, and the relationship of the mechanics with the biochemical steps cannot be studied using single X-ray diffraction. In 1994, Finer et al. showed the first results of single myosin S1 forces and steps using a double system of optical traps.

Finer et al. (1994) attached silica beads to a microscope cover slip. The cover slip was coated with skeletal muscle heavy meromyosin (HMM) at a low density to allow for single attachments of cross-bridges to actin. Polystyrene beads coated with N-ethylmaleimide (NEM)-treated HMM were attached to actin filaments. An actin filament with two beads attached near its end was then caught and suspended in two optical traps (Figure 3.3). The image of one of the beads was projected to a photodiode detector for position detection. The actin filament was then pulled taut (with a force of about 2 pN) and was lowered to the silica bead with the HMM. Now, a single (or few) HMM molecule(s) could interact and attach to the actin filament. When a cross-bridge head attached to the actin, a rapid transient movement of the silica bead along the axis of actin (typically in the same direction) could be observed. By keeping the stiffness of the optical traps high enough to decrease the noise caused by Brownian motion, but small enough that a myosin molecule could produce a full displacement (i.e., about 0.02 pN/nm per trap), consistent displacement traces were observed.

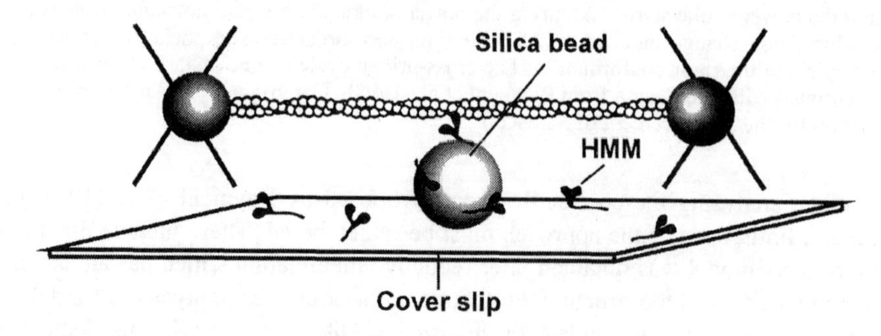

Figure 3.3. Schematic illustration of single HMM interaction with an actin filament. The silica bead on the cover slip is coated with skeletal muscle heavy meromyosin (HMM). Coated polystyrene beads are attached to the ends of actin. The actin filament with its two beads is caught and suspended in two optical traps. The suspended actin filament is then lowered to the silica bead, and single HMM interactions with the actin filament are now possible. (Reprinted by permission from *Nature*, Finer et al., 1994. Copyright 1994 Macmillan Magazines Ltd.)

The average size found for (presumably) single myosin cross-bridge steps was about 11±2.4 nm, independent of the ATP concentration (1 μM to 2 mM ATP) and independent of the total trap stiffness (0.014 to 0.08 pN/nm).

In order to measure the forces produced by single HMM molecules, the trap stiffness was increased to 6 pN/nm. Movements of the bead from the centre of the optical trap were proportional to the force applied on the actin filament. For the force measurements, such movements of the bead relative to the trap were prevented by a feedback position system that moved the optical trap (when force was applied to the bead) by the exact amount required to keep the bead stationary. Therefore, the displacement of the trap became a measure of the applied force. The magnitude of forces measured by (presumably) single HMM interactions with actin were 1-7 pN and they averaged 3.4±1.2 pN. The force magnitudes were not affected by ATP concentration (Figure 3.4).

One of the limitations of the study by Finer et al. (1994) was that it could not be determined with confidence whether or not a given mechanical event was produced by one or more HMM molecules or by one or both beads of the HMM molecule. Molloy et al. (1995) addressed this issue using two optical traps in essentially the same way as Finer et al. (1994), and measured the interactions of HMM molecules (two-headed cross-bridges) and myosin subfragment 1 (S1, single-headed cross-bridges) with actin. Molloy et al. (1995) found that the average working strokes of S1 and HMM were comparable (about 4 nm) and were much smaller than those found by Finer et al. (1994) for HMM (about 11 nm). Also, the average force values for both S1 and HMM interactions with actin were low (about 1.7 pN) compared to those of Finer et al. (1994).

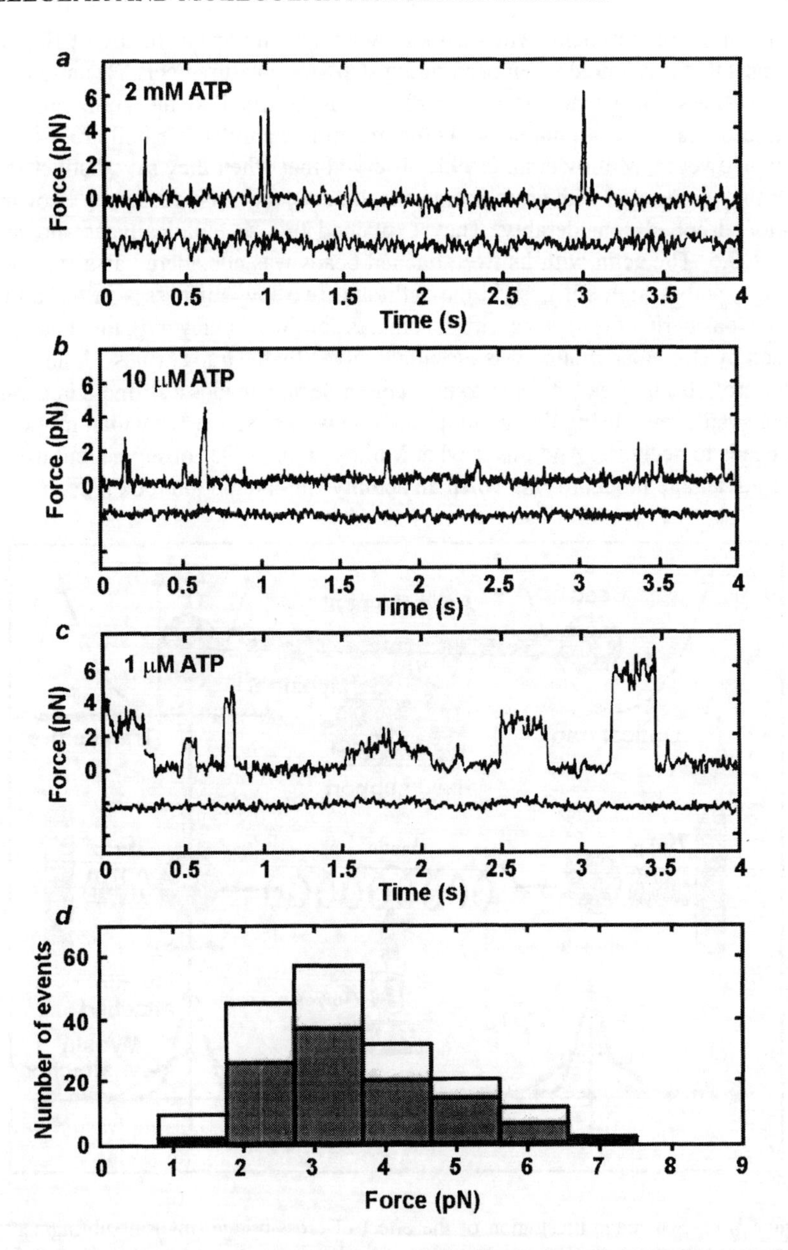

Figure 3.4. Forces produced by single HMM molecules interacting with actin filaments at different concentrations of ATP. The force magnitudes were not affected by ATP concentration, however, the interaction times increased with decreasing ATP concentrations. The cross-bridge forces ranged from 1-7 pN with an average of 3.4±1.2 pN. (Reprinted by permission from *Nature*, Finer et al., 1994. Copyright 1994 Macmillan Magazines Ltd.)

One of the big problems when working with optical traps is the low stiffness of the traps that is required when performing step-size measurements. Because of the low stiffness, Brownian motion of the beads attached to the actin molecule produces a large positional noise with a magnitude similar to small cross-bridge steps. However, Molloy et al. (1995) observed that when they saw distinct steps (that they associated with cross-bridge attachment), the noise caused by Brownian motion dropped considerably. They explained this schematically as shown in Figure 3.5. The actin with its two attached beads was considered as a system of springs, with a high stiffness in the actin and two low stiffnesses in the optical traps. Considering the actin as stiff in compression (which they did), the noise level caused by Brownian motion was essentially given by the trap stiffness. If an HMM or S1 cross-bridge head attaches to the actin molecule, it adds a stiffness in parallel to the system and, using that assumption, one would expect Brownian motion of the beads to decrease. And this is what Molloy et al. (1995) observed and used to their advantage in identifying when an actomyosin interaction took place.

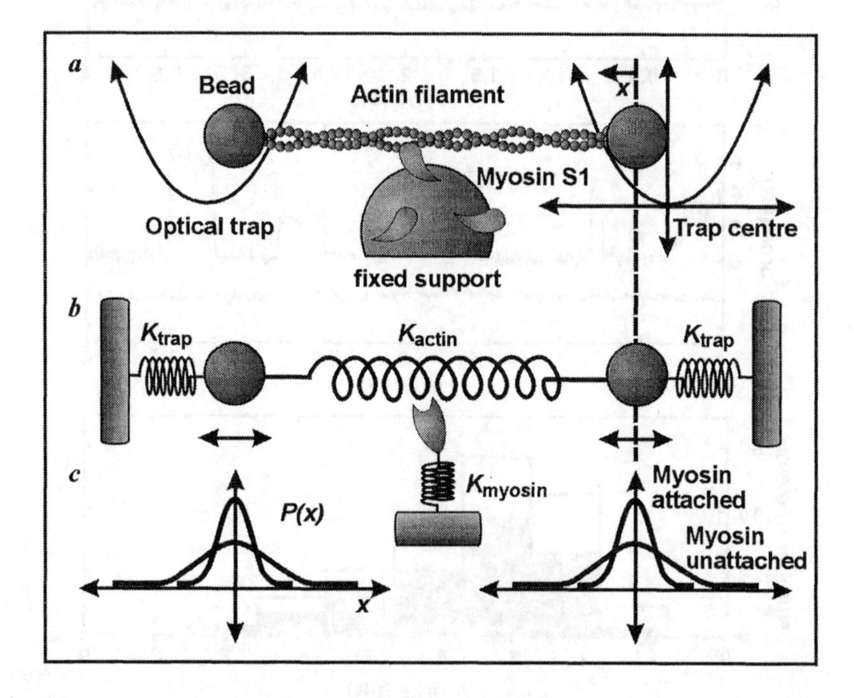

Figure 3.5. Schematic illustration of the effect of cross-bridge (myosin subfragment 1) attachment to actin on the Brownian motion of the beads fixed at the end of actin. a) Actin filament with beads attached near its ends. The beads are held by two optical traps. They move from the centre of the trap because of Brownian motion. The S1 molecules are attached to a fixed support. b) Brownian motion of the beads is reduced because of the stiffness in the two optical traps (S1 is not attached). When S1 attaches, it adds an additional stiffness to the actin system, further restricting the range of Brownian motion. c) Schematic illustration of the

expected location distribution of the actin beads when the myosin subfragment 1 is attached (narrow distribution) and when it is unattached (broad distribution). (Reprinted by permission from *Nature*, Block, 1995. Copyright 1995 Macmillan Magazines Ltd.)

A point that deserves attention regarding the results by Molloy et al. (1995) is that the myosin subfragment 1 (S1) has a shorter neck than the double-headed HMM construct. Assuming that the neck region of the cross-bridge is a lever that determines the step size of a single cross-bridge stroke, one would have expected a smaller average step size for the S1 compared to the HMM experiments. However, this was not the case. Yanagida (August 1999, personal communication) also found that step sizes remained about the same for molecules with normal neck lengths and molecules in which the neck region was reduced to 20% of the normal length. These results could be explained in a variety of ways, for example, if the cross-bridge stroke was not associated with an angular displacement of the neck region, but a shortening of some part of the cross-bridge molecule. Also, the reduced length of the neck region in some molecular preparations might be offset by a correspondingly increased angular displacement of that region during force production. Finally, there always exists the possibility for errors in the measurements. After all, the step-size measurements by Finer et al. (1994) and Molloy et al. (1995) differ by a factor of almost 3 (11 nm versus 4 nm). Similarly, the corresponding HMM forces also differ by a factor of 2 (3.4 pN versus 1.7 pN), suggesting that the final word has not been spoken.

One way in which the cross-bridge theory has been adapted to account for the smaller-than-expected heat and force production during eccentric contractions is the idea that cross-bridges might be mechanically torn from the actin attachment sites, once cross-bridge elongation and force become large. Nishizaka et al. (1995) used an optical trap method, as described above, to measure the force required to pull a single HMM molecule from its attachment on the actin filament. The average unbinding force obtained was 9.2±4.4 pN, which, they argued, was a few times larger than the force produced during contraction, but was an order of magnitude smaller than the intermolecular forces. Nishizaka et al. (1995) also found that HMM attachment times decreased with increasing stress on the actomyosin interaction site. They suggested that pulling on the actin filament tends to distort the actomyosin interface and causes a breaking of the actomyosin bonds sequentially from one end. One of the unresolved questions regarding the results by Nishizaka et al. (1995) is: what happens to these cross-bridges that are 'mechanically torn' from the actin attachment sites, rather than 'biochemically released'? It appears quite possible that these torn-off cross-bridges are in different biochemical and mechanical states than normally released cross-bridges and, possibly, their attachment characteristics are different from normal. Possibly, these differences in attachment characteristics could explain some of the history-dependent phenomena observed during muscular contraction. Specifically, it appears feasible that force enhancement following stretching might be explained by the changed attachment properties of torn-off cross-bridges (Herzog, 1998).

SINGLE MYOSIN ATP TURN-OVER EVENTS

The ability to manipulate single actin filaments with optical traps or microneedles, and to measure the position of the ends of actin with photodiode detectors, allowed for measurement of displacement and forces exerted by individual myosin molecules. An important aspect of muscular contraction, i.e., actomyosin interaction, is the coupling of the energy delivering biochemical processes (ATP hydrolysis) with the mechanical energy produced (work and heat). Funatsu et al. (1995) refined epifluorescence and total internal reflection microscopy to visualize single fluorescent dye molecules. Using this technique, they were able to detect individual ATP turn-over reactions in single fluorescently labelled myosin molecules in aqueous solution using a video-camera and light microscopy. Measurements of ATP turn-over reactions were made using HMM and single S1 molecules.

In order to detect individual Mg-ATP turn-over events by single S1 molecules, S1 molecules were labelled with Cy5 and fixed on a quartz slide surface (Figure 3.6a). The turn-over events were directly detected by observing the association, hydrolysis, and dissociation of an ATP analogue labelled with Cy3. The background (noise) fluorescence was small because the illumination was focussed near the quartz slide surface with the S1 molecule (Figure 3.6a). When the ATP analogue with the Cy3 label attached to the Cy5-S1 molecule, it could be detected as a fluorescent spot (Figure 3.6c). When the Cy3-ATP analogue was associated with the Cy5-S1 molecule (whose position was known, Figure 3.6b), Brownian motion was reduced (giving a focussed rather than blurred spot), the Cy3-ATP was at the position of the S1 molecule, and it remained stationary for a discrete period of time.

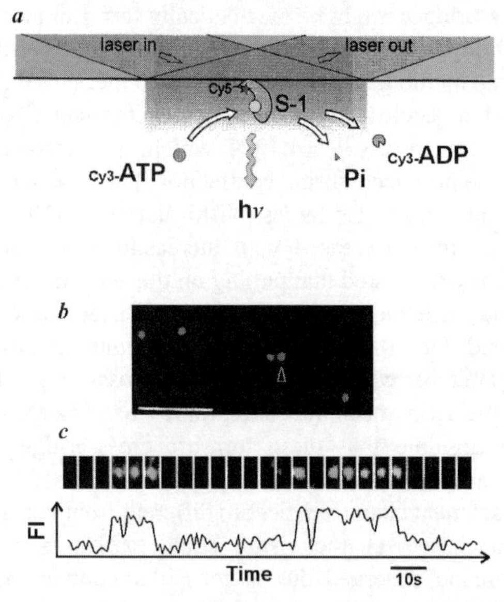

Figure 3.6. ATP turn-over reaction with a single myosin subfragment 1. a) S1 molecules were labelled with Cy5 and fixed on a quartz slide surface. The turn-over events could be

directly detected by observing the association, hydrolysis, and dissociation of an ATP analogue labelled with Cy3. The background noise was kept small by focussing the illumination near the quartz slide surface. b) The position of the Cy5-S1 molecule was known (arrow). c) When the ATP analogue attached to the Cy5-S1 molecule, it was detected as a fluorescent spot. During attachment, the Brownian motion of the Cy3-ATP was reduced, thus giving a focussed rather than blurred spot, and the Cy3-ATP analogue remained stationary at the known location of Cy5-S1 for a discrete period of time. (Reprinted by permission from *Nature*, Funatsu et al., 1995. Copyright 1995 Macmillan Magazines Ltd.)

Funatsu et al. (1995) did not report results from direct measurements of single cross-bridge forces in conjunction with the ATP turn-over events. Needless to say, future experiments that could reveal simultaneously the mechanics, biochemistry, and structural changes of force generation in single myosin molecules might answer many questions about the basic mechanisms of muscle contraction. Such experiments have not been done to date.

MYOFILAMENT COMPLIANCE

One of the cornerstones of the cross-bridge theory was the observation that myofilament lengths remained virtually unchanged during muscle contraction (Huxley and Niedergerke, 1954; Huxley and Hanson, 1954), therefore, it had been assumed that thick and thin filaments were essentially rigid (Huxley, 1957; Huxley and Simmons, 1971). This assumption has several important mechanical implications. For example, if the myofilaments are rigid, compliance in actomyosin interactions are directly and exclusively associated with the cross-bridges. If all the compliance is in the cross-bridges, stiffness values in fibre preparations should directly reflect the number of attached cross-bridges. Finally, rigid filaments imply that the sliding speed of actin relative to a thick myofilament directly reflects the speed of shortening of the sarcomere. In compliant myofilaments, this would not be correct, as there could be movement of an actin attachment site relative to the myosin cross-bridge without shortening or stretch of the sarcomere.

The fact that thin-filament compliance accounts for as much as 50% of the total sarcomere compliance was detected independently in different laboratories. Kojima et al. (1994) used single actin filaments attached to the tip of a flexible glass needle of known stiffness and performed 10 nm sinusoidal vibrations to determine actin compliance. They made measurements with pure actin, which they found to be more compliant than the values obtained for actin decorated with tropomyosin. Kojima et al. (1994) found that a 1.0 μm actin/tropomyosin molecule had a compliance of 0.015 μm per 1pN, and therefore would be stretched by about 0.23% during isometric contraction. This corresponds to an elongation of the actin molecule of about 2.3 nm during isometric contraction; an apparently negligible amount. However, realizing that a quick 4 nm shortening step is sufficient to reduce the isometric force in a half-sarcomere to zero (Ford and Simmons, 1977), it becomes obvious that 2.3 nm corresponds to more than 50% of the sarcomere compliance.

Huxley et al. (1994) and Wakabayashi et al. (1994) used X-ray diffraction to determine actin compliance. The positions of the 2.7 nm repeat patterns of the actin monomers, and the off-axial reflections arising from the helical structure (5.1 μm and 5.9 μm) of the filament were measured by illuminating muscle with intense and highly collimated Synchrotron radiation. The diffraction patterns were recorded on a storage phosphor imaging plate. Huxley et al. (1994) and Wakabayashi et al. (1994) found increases in the actin-based reflections of 0.2-0.3% on stimulation of the muscles, extensions that are in perfect agreement with the results obtained by Kojima et al. (1994). Similar results were found for the myosin compliance, suggesting that both sets of filaments contribute significantly to sarcomere compliance.

Despite the experimental evidence confirming actin and myosin compliance, few theoretical models have evaluated the implications of filament compliance on cross-bridge mechanics. Mijailovich et al. (1996) reformulated Huxley's two-state cross-bridge theory (1957) by accounting explicitly for filament compliance. They found that even slight filament extensibility caused non-uniform displacements of the two sets of filaments relative to one another in the overlap region, that cross-bridge strain must vary systematically within the filament overlap region, and that local shortening speeds (at constant sarcomere length) caused a force decrease compared to the situation with rigid filaments. This latter result may be attributed to the local shortening force-velocity relationship encountered in compliant myofilaments during force production. Using their model, Mijailovich et al. (1996) could explain that stiffness must lead force production in muscle contraction. They also determined that filament compliance immediately meant that cross-bridge stiffness had to be significantly higher than previously assumed (at least by a factor of 2), and that fibre stiffness was dissociated from the number of attached cross-bridges.

Forcinito et al. (1997) confirmed the theoretical results of Mijailovich et al. (1996) in a modelling approach that represented actomyosin interaction with a ladder-structure. This model was powerful insofar as its mathematical description was simple, it could be arbitrarily extended in length and three-dimensionality, and it allowed for studying the effects of cross-bridge attachment distribution on stiffness. The model retained, in the limit, the characteristics of the previously published continuum model by Ford et al. (1981), but demonstrated that Ford's model was not adequate in situations where only a few cross-bridge attachments were made.

Finally, Forcinito et al. (1997) demonstrated that it was impossible to determine the number of attached cross-bridges from stiffness measurements, as largely different stiffness values could be obtained for the same number of attached cross-bridges depending on the exact distribution of the attached cross-bridges.

However, it appears that filament compliance was an important and well-researched phenomenon for approximately three years and, by now, has almost been forgotten, despite the considerable significance filament compliance may have on actomyosin interaction and contraction. I personally feel that elasticity and compliance of the molecular structures associated with contraction and force production may, at present, be underestimated, and may hold the key to several questions about unexplained phenomena of the molecular interaction of myosin and actin.

THE MOLECULAR SPRING, TITIN

Titin, sometimes referred to as connectin, is a giant filamentous polypeptide, consisting primarily of about 300 immunoglobulin (Ig) and related fibronectin type III (FNIII) repeats, and a unique proline (P), glutamate (E), valine (V), and lysine (K)-rich (PEVK) domain. Titin spans each half-sarcomere and is anchored to the Z-line and the thick filament reaching all the way to the M-band (Figure 3.7). Titin is thought to play a basic role in maintaining sarcomere structural integrity and producing passive force when muscle sarcomeres are stretched. It also is believed to provide a molecular scaffold for thick filament formation.

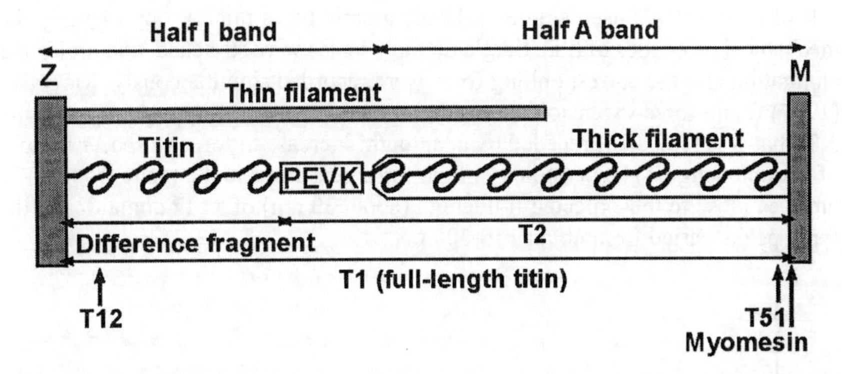

Figure 3.7. Schematic illustration of titin and its association with other structures within a half-sarcomere. Titin spans from the Z-line to the M-band. It consists of about 300 immunoglobulin and fibronectin type III repeats and a proline (P), glutamate (E), valine (V), and lysine (K)-rich (PEVK) domain. (Reproduced from Kellermayer et al., 1997, with the permission of the American Association for the Advancement of Science.)

It has been argued that titin stabilizes the thick filament in the centre of the sarcomere when, upon contraction, small asymmetries in pulling forces are produced on the two halves of the thick filament. Furthermore, titin is assumed to produce many of the passive forces that come about in most muscles, somewhere in the plateau region or the descending limb of the force-length relationship. Such passive forces might help stabilize what has been labelled the unstable (Hill, 1953), or softening behaviour of active muscle force on the descending limb of the force-length relationship. Finally, once a sarcomere is stretched beyond thick and thin filament overlap, cross-bridge attachments become impossible, and the forces required to re-establish myofilament overlap are thought to come (primarily) from the passive elastic forces of the (highly) stretched titin.

Despite the apparent importance of titin, the way it accomplishes its functional role has not been fully resolved. If titin really provides most of the passive force in a stretched muscle, it should have several distinct characteristics: first, for centring the thick filament upon contraction, it should provide a low-level stiffness. The stiffness needs to be low because the asymmetries in active force acting on the thick filaments would presumably not be large, and the muscle (sarcomere) should still be stretchable within its normal range of physiologic lengths, without encountering much passive resistance from titin. Second, at some point titin should become stiff at a very fast rate in order to prevent large stretches of muscle against external forces that might cause injury. Finally, titin must accomplish its passive force at different lengths in different muscles, as the passive forces measured in different muscles occur at distinctly different sarcomere lengths. For example, in cardiac muscle, passive force is known to be high at about optimal sarcomere length, whereas in many skeletal muscles, passive force is negligible at optimal sarcomere length.

Rief et al. (1997) used single-molecule atomic force microscopy to study the mechanical properties of titin. Single titin molecules were stretched, and molecular elongation and the corresponding force were recorded simultaneously. Rief et al. (1997) found force-extension curves that showed a sawtooth-type pattern (Figure 3.8) that was typically preceded by a 'smooth' increase in force (a spacer region) of variable length. The periodicity of the force peaks was in the range from 25-28 nm, i.e., close to the expected full length (about 30 nm) of an Ig domain, and the force peaks varied from about 150-300 pN.

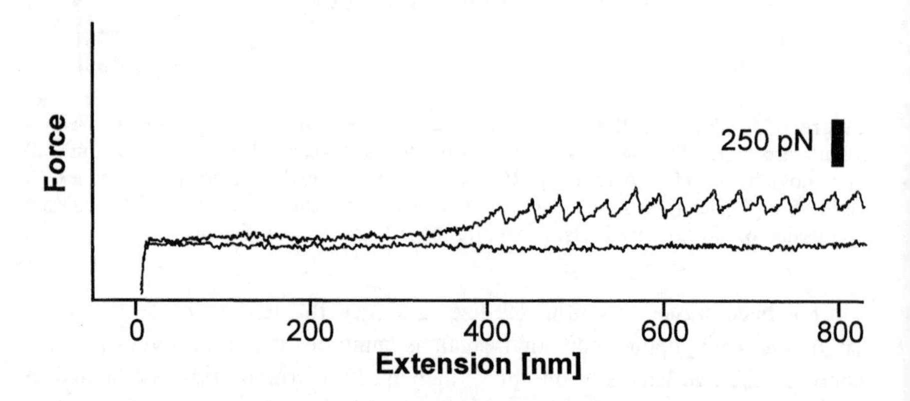

Figure 3.8. Force-extension curve of a single titin molecule obtained using atomic force microscopy. The force-extension curve shows a sawtooth-type pattern that is preceded by a smooth increase in force. The periodicity for the force peaks is about 25-28 nm, and the force peak magnitudes were in the range of 150-300 pN. The smooth increase in force preceding the sawtooth pattern was associated with an elongation of the PEVK region. The sawtooth pattern

was associated with an unfolding of the Ig domains. (Reproduced from Rief et al., 1997, with the permission of the American Association for the Advancement of Science.)

In order to test whether the peaks of the sawtooth pattern indeed reflect an unravelling of the Ig domains, two model recombinant titin fragments were constructed, consisting of either four (Ig4) or eight (Ig8) Ig segments in the I-band region (the flexible region) of titin. All traces obtained with these Ig segments showed a strict 25 nm periodicity. Furthermore, the force peaks were found to increase from the first to the last (Figure 3.9), but the stiffness was found to decrease from force peak to force peak (Figure 3.9). Finally, when these Ig segments were released to one-half of their fully stretched length (but not to their resting length) and then pulled again, the sawtooth force-extension curve was not apparent. The sawtooth pattern was only fully recovered after complete relaxation of the titin to its resting length.

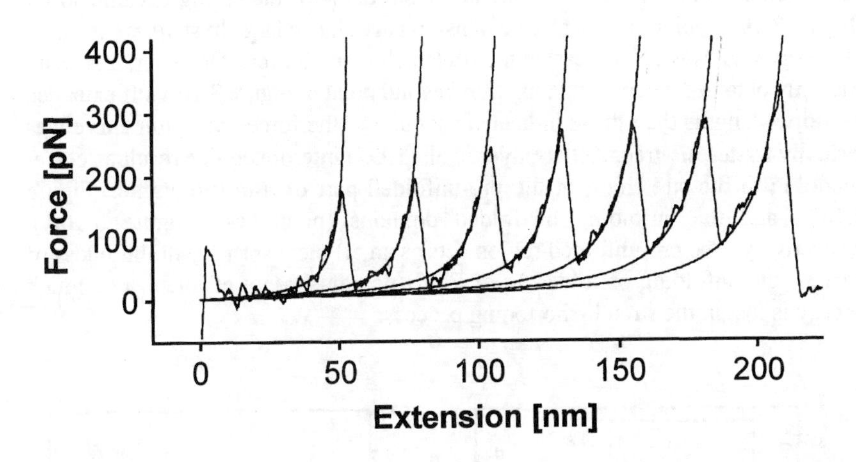

Figure 3.9. Force-extension curve of a recombinant titin fragment consisting of eight Ig segments. The curve shows a strict 25 nm periodicity. Force peaks increase from the first to the last sawtooth pattern, but the stiffness decreases. The force and stiffness on the ascending part of the sawtooth pattern likely reflect the molecular forces that must be overcome to produce unfolding of one Ig domain. The force decrease following the peak reflects the sudden 25 nm elongation of the molecule once the molecular unfolding forces have been overcome (Reproduced from Rief et al., 1997, with the permission of the American Association for the Advancement of Science.)

Based on their results, Rief et al. (1997) concluded that each sawtooth pattern corresponded to an unfolding of one Ig domain: the force on the ascending part reflecting the molecular forces that must be overcome to produce the unfolding, the force decrease following the peak reflecting the fact that the molecular spring was now 'abruptly' elongated by about 25 nm. The increasing peak forces with each subsequent sawtooth pattern were assumed to reflect increasingly stronger molecular bonds of the folded Ig domains, therefore the 'weakest' domain was unfolded first, followed by increasingly 'stronger' Ig domains. The relative 'smooth' increase in force preceding the sawtooth patterns was associated with a different molecular mechanism, possibly an elongation of the PEVK region, although this assumption was not rigorously tested. Finally, it took a full shortening of titin before a sawtooth pattern force-extension curve could be observed after the molecule had been stretched. Partial shortening did not produce partial sawtooth patterns, therefore it appears that a refolding of the Ig domains only occurs at or near the titin 'resting' length or, in terms of force, at very low titin forces.

Kellermayer et al. (1997) used an optical trap method to obtain force-extension properties of single titin molecules. The results by Kellermayer et al. (1997) were similar to those of Rief et al. (1997), however, they discussed two additional results that appear of functional importance within the context of this book. First, Kellermayer et al. (1997) observed that the stretch-shortening curve of titin contained a distinct change in slope in the stretch part: the spring became softer (Figure 3.10A, point c). The force-extension curve had a large hysteresis (i.e., a lot of energy was 'wasted' as heat) in the stretch-shortening cycle. However, this result was only obtained when elongating titin beyond point c (Figure 3.10A). If extended to shorter lengths than those indicated by point c, the force-extension curve was virtually hysteresis-free. Kellermayer et al. (1997) interpreted this result as corresponding to the stretching of the pre-unfolded part of titin (points a-c, Figure 3.10A) and the unfolding of folded domains (points c-d, Figure 3.10A), respectively. The pre-unfolded region acts as an efficient spring with little loss of energy, the unfolding of folded Ig or FNIII domains is not efficient, i.e., much energy is lost in the stretch-shortening process.

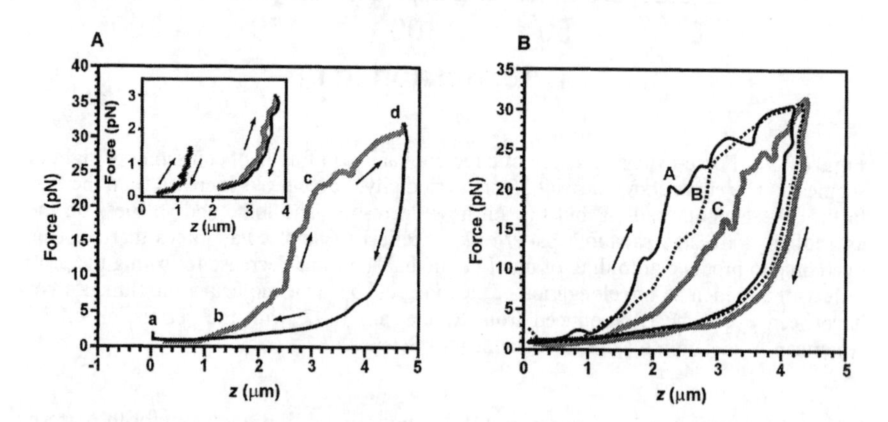

Figure 3.10. Stretch-shortening curves for single titin molecules obtained using optical trap methods. A) The stretch-shortening curve has a large hysteresis. The extension curve shows an increasing stiffness initially, but stiffness decreases when the extension reaches a limiting value (point c). If stretch-shortening tests are performed with extensions smaller than the

limiting value, titin is essentially elastic, showing little if any hysteresis (insets). B) Repeated stretching of titin decreases the stiffness of the molecule (stiffness decreases from curves A to B to C). (Reproduced from Kellermayer et al., 1997, with the permission of the American Association for the Advancement of Science.)

Second, Kellermayer et al. (1997) observed that repeated stretching of titin decreased the stiffness of the molecule (Figure 3.10B); this was associated with an increase in length of the pre-unfolded region (the efficient region) because previously folded domains will not easily refold when titin is subjected to repeated stretch-shortening cycles. Therefore, repeated stretching or large stretching of a muscle may increase titin's pre-unfolded region. As a consequence, muscle stiffness would decrease and passive loss of energy would be reduced. Both these functional implications should enhance the performance of muscular contraction, and suggest that titin may be a spring of 'instantaneous', adjustable stiffness, length, and efficiency—a property that would be of great functional importance.

Marszalek et al. (1999) engineered single proteins that had multiple copies of Ig domains of human cardiac titin. They then stretched these engineered molecules using atomic force microscopy. This approach allowed them to identify the precise mechanics of Ig domains without contamination of the mechanical results from 'other' domains of the naturally occurring titin. Marszalek et al. (1999) found an initial extension before the unfolding of each Ig domain. This initial extension was reversible, which means it recovered 'elastically' when released, and it extended titin's domain by about 15% of its slack length. Dynamic molecular simulations revealed that the initial extension may be associated with the rupture of two hydrogen bonds at a specific site. Disruption of these hydrogen bonds eliminated the initial, reversible extension, lending support to the proposed mechanism: an unfolding intermediate. This previously unrecognized component of titin elasticity may play an important functional role in intact muscle.

Obviously, the precise molecular mechanisms determining titin's unique force-extension properties are not completely known. However, it is generally recognized that titin plays an important functional role in passive and contracting skeletal muscle. It has been suggested, as shown above, that titin can adapt its resting length and stiffness based on the contractile history of the muscle. Also, it has been suggested that titin may act as a bi-directional spring, and may, based on the chemical environment, exhibit different force-extension properties. Needless to say, the investigation of titin's properties is of interest in terms of revealing the fundamentals of biological springs and the functional role of this molecule in skeletal and cardiac muscle mechanics.

HISTORY DEPENDENCE OF FORCE PRODUCTION

As has been demonstrated quite convincingly above, titin appears to have distinct history-dependent properties. Its stiffness and reversible operating length can differ from one stretch-shortening cycle to the next. One of the basic questions in muscle mechanics, and one that has eluded convincing explanation, is the phenomenon of history dependence of active force production. Abbott and Aubert (1952) were among the first to show, in whole muscle, that active forces were depressed and enhanced following shortening and stretching, respectively, compared to the corresponding forces obtained for purely isometric contractions. This phenomenon has typically been associated with structural non-uniformities in muscle, particularly sarcomere length non-uniformities. However, it appears odd that sarcomere length non-uniformities may be called upon to explain both a decrease and an increase in force from the isometric force, depending on the contractile history.

Granzier and Pollack (1989) performed a study in which they shortened frog single fibres under two conditions: first, no attention was given to sarcomere non-uniformities that might develop during fibre shortening; and second, sarcomere non-uniformities were prevented (as well as this can be done with today's techniques) using a so-called 'sarcomere clamp'. Granzier and Pollack (1989) found large force depressions following muscle shortening for both conditions, and concluded that force depression could and did occur even in the absence of (large) sarcomere non-uniformities. Interestingly, in the only other study in which this precise mechanism was investigated systematically, exactly the opposite result was found, that is, sarcomere length non-uniformities that developed during single-fibre shortening were directly proportional to the observed magnitude of force depression (Edman et al., 1993). The reason why these two studies are so important in the context of the present chapter is because the results by Edman et al. (1993) would suggest that history dependence of active force production could be explained (exclusively) with structural non-uniformities. The results by Granzier and Pollack (1989), in contrast, would suggest that history-dependent phenomena of active force production may be associated with molecular events of actomyosin interaction.

Aside from the study by Granzier and Pollack (1989), there is other evidence that history-dependent active force production might be associated with molecular events surrounding contraction. Sugi and Tsuchiya (1988) performed a study on single fibres in which they evaluated force depression and stiffness following fibre shortening. They found a direct proportionality between the amount of force depression and the decrease in stiffness, suggesting that the number of attached cross-bridges was directly proportional to the amount of observed force depression. We have observed that such force depressions are long-lasting and not transient (Herzog et al., 1998), therefore the questions arises: why should cross-bridge attachment be decreased following muscle shortening, and why is this phenomenon not transient, but persists? These questions cannot be answered at present.

Recently, we demonstrated that force depression following shortening and force enhancement following stretch were not commutative (Epstein and Herzog, 1998). When performing stretch-shortening and shortening-stretch cycles with entire cat soleus, it was observed that a stretch preceding shortening had a negligible (if any) effect on the force depression observed after shortening. However, shortening before the stretch influenced (or even offset) the force enhancement following stretch in a dose-dependent manner, that is, the larger the amount of shortening relative to the following stretch, the less force enhancement was observed. Once the amount of shortening equalled the amount of stretch, no force enhancement was observed; once the amount of shortening exceeded the amount of stretch, a net force depression effect persisted.

In concluding this chapter, I would like to emphasize something that is repeated in Professor Huxley's Wartenweiler Memorial Lecture, presented in Chapter 2 of this book. He observed, and I concur, that despite the advances in techniques and the possibility for measuring the mechanics of isolated actomyosin interactions and isolated molecular structures, there are a number of unanswered questions. Interestingly, many of these questions are the same as those posed 20 or more years ago. We do not know, at present, the detailed events underlying the molecular mechanisms of muscular contractions and force production. There exists, in my opinion, the strong possibility that new findings with the available single-molecule techniques may make us rethink the fundamental assumptions and theories of muscular contraction and force production.

REFERENCES

Abbott, B.C., and Aubert, X.M. (1952). The force exerted by active striated muscle during and after change of length. *Journal of Physiology* **117**, 77–86.

Block, S.M. (1995). One small step for myosin. *Nature* **378**, 132–133.

Edman, K.A.P., Caputo, C., and Lou, F. (1993). Depression of tetanic force induced by loaded shortening of frog muscle fibres. *Journal of Physiology* **466**, 535–552.

Epstein, M., and Herzog, W. (1998). *Theoretical Models of Skeletal Muscle: Biological and Mathematical Considerations*. John Wiley & Sons, Ltd., New York.

Finer, J.T., Simmons, R.M., and Spudich, J.A. (1994). Single myosin molecule mechanics: Piconewton forces and nanometre steps. *Nature* **368**, 113–119.

Forcinito, M., Epstein, M., and Herzog, W. (1997). Theoretical considerations on myofibril stiffness. *Biophysical Journal* **72**, 1278–1286.

Ford, L.E., Huxley, A.F., and Simmons, R.M. (1977). Tension responses to sudden length change in stimulated frog muscle fibres near slack length. *Journal of Physiology* **269**, 441–515.

Ford, L.E., Huxley, A.F., and Simmons, R.M. (1981). The relation between stiffness and filament overlap in stimulated frog muscle fibres. *Journal of Physiology* **311**, 219–249.

Funatsu, T., Harada, Y., Tokunaga, M., Saito, K., and Yanagida, T. (1995). Imaging of single fluorescent molecules and individual ATP turnovers by single myosin molecules in aqueous solution. *Nature* **374**, 555–559.

Granzier, H.L.M., and Pollack, G.H. (1989). Effect of active pre-shortening on isometric and isotonic performance of single frog muscle fibres. *Journal of Physiology* **415**, 299–327.

Herzog, W. (1998). History dependence of force production in skeletal muscle: A proposal for mechanisms. *Journal of Electromyography and Kinesiology* **8**, 111–117.

Herzog, W., Leonard, T.R., and Wu, J.Z. (1998). Force depression following skeletal muscle shortening is long lasting. *Journal of Biomechanics* **31**, 1163–1168.

Hill, A.V. (1938). The heat of shortening and the dynamic constants of muscle. *Proceedings of the Royal Society of London B* **126**, 136–195.

Hill, A.V. (1953). The mechanics of active muscle. *Proceedings of the Royal Society of London B* **141**, 104–117.

Huxley, A.F. (1957). Muscle structure and theories of contraction. *Progress in Biophysics and Biophysical Chemistry* **7**, 255–318.

Huxley, A.F., and Niedergerke, R. (1954). Structural changes in muscle during contraction: Interference microscopy of living muscle fibres. *Nature* **173**, 971–973.

Huxley, A.F., and Simmons, R.M. (1971). Proposed mechanism of force generation in striated muscle. *Nature* **233**, 533–538.

Huxley, H.E., and Hanson, J. (1954). Changes in cross-striations of muscle during contraction and stretch and their structural interpretation. *Nature* **173**, 973–976.

Huxley, H.E., Stewart, A., Sosa, H., and Irving, T. (1994). X-ray diffraction measurements of the extensibility of actin and myosin filaments in contracting muscles. *Biophysical Journal* **67**, 2411–2421.

Kellermayer, M.S.Z., Smith, S.B., Granzier, H.L., and Bustamante, C. (1997). Folding-unfolding transitions in single titin molecules characterized with laser tweezers. *Science* **276**, 1112–1116.

Kojima, H., Ishijima, A., and Yanagida, T. (1994). Direct measurement of stiffness of single actin filaments with and without tropomyosin by *in vitro* nanomanipulation. *Proceedings of the National Academy of Sciences USA* **91**, 12962–12966.

Marszalek, P.E., Lu, H., Li, H., Carrion-Vazquez, M., Oberhauser, A.F., Schulten, K., and Fernandez, J.M. (1999). Mechanical unfolding intermediates in titin molecules. *Nature* **402**, 100–103.

Mijailovich, S.M., Fredberg, J.J., and Butler, J.P. (1996). On the theory of muscle contraction: Filament extensibility and the development of isometric force and stiffness. *Biophysical Journal* **71**, 1475–1484.

Molloy, J.E., Burns, J.E., Kendrick-Jones, J., Tregear, R.T., and White, D.C.S. (1995). Movement and force produced by a single myosin head. *Nature* **378**, 209–212.

Nishizaka, T., Miyata, H., Yoshikawa, H., Ishiwata, S., and Kinosita, K.J. (1995). Unbinding force of a single motor molecule of muscle measured using optical tweezers. *Nature* **377**, 251–254.

Ramsey, R.W., and Street, S.F. (1940). The isometric length-tension diagram of isolated skeletal muscle fibers of the frog. *Journal of Cellular Composition* **15**, 11–34.

Rayment, I., Holden, H.M., Whittaker, M., Yohn, C.B., Lorenz, M., Holmes, K.C., and Milligan, R.A. (1993a). Structure of the actin-myosin complex and its implications for muscle contraction. *Science* **261**, 58–65.

Rayment, I., Rypniewski, W.R., Schmidt-Bäse, K., Smith, R., Tomchick, D.R., Benning, M.M., Winkelmann, D.A., Wesenberg, G., and Holden, H.M. (1993b). Three-dimensional structure of myosin subfragment-1: A molecular motor. *Science* **261**, 50–58.

Rief, M., Gautel, M., Oesterhelt, F., Fernandez, J.M., and Gaub, H.E. (1997). Reversible unfolding of individual titin immunoglobulin domains by AFM. *Science* **276**, 1109–1112.

Sugi, H., and Tsuchiya, T. (1988). Stiffness changes during enhancement and deficit of isometric force by slow length changes in frog skeletal muscle fibres. *Journal of Physiology* **407**, 215–229.

Wakabayashi, K., Sugimoto, Y., Tanaka, H., Ueno, Y., Takezawa, Y., and Amemiya, Y. (1994). X-ray diffraction evidence for the extensibility of actin and myosin filaments during muscle contraction. *Biophysical Journal* **67**, 2422–2435.

Yanagida, T. (1999). Simultaneous observation of individual ATPase and mechanical events by a single myosin molecule during interaction with actin. *Canmore Symposium on Skeletal Muscle*, August 6-7, 1999, Canmore, Alberta, Canada.

4 The Effect of Sarcomere Length on the Force-cytosolic [Ca^{2+}] Relationship in Intact Rat Cardiac Trabeculae

HENK E. D. J. TER KEURS, ELLEN H. HOLLANDER, AND MARK H. C. TER KEURS

Faculty of Medicine, University of Calgary, Calgary, Alberta, Canada

ABSTRACT

Starling's Law of the Heart results from length-dependent sensitivity of the myofilaments to Ca^{2+}. We have studied the effects of sarcomere length (SL) on maximal force and on the force-cytosolic Ca^{2+} ($[Ca^{2+}]_i$) relationship in intact rat cardiac trabeculae. Force was measured using a silicon strain gauge in 26 right ventricular trabeculae dissected from rat heart, SL was measured using light diffraction techniques, and $[Ca^{2+}]_i$ was measured using Fura-2 fluorescence. Tetani were elicited using stimuli at 12 Hz in HEPES buffered saline solution (pH 7.4, 26°C) with varied [Ca^{2+}] in the solution ($[Ca^{2+}]_o$) (0.2-16 mM), in the presence of cyclopiazonic acid (30 μM) to inhibit the sarcoplasmic reticulum Ca^{2+} uptake. The relationship between maximal tetanic force (F_{max}) and SL was close to identical to the ascending limb of the force-length relationship based on the cross-bridge theory of muscle contraction for skeletal muscle, provided that the dimensions of the filaments for the rat are used and provided that forces opposing shortening below slack length are incorporated in the relationship. F_{max} was 107±10 mN/mm^2 at SL=2.15 μm. Force-$[Ca^{2+}]_i$ relations during the tetanus plateau revealed an increase of EC_{50} from 0.65±0.04 μM at SL = 2.15 μm to 0.98±0.09 μM at SL=1.85 μm without significant change in the Hill coefficient (4 and 5, respectively).

The correspondence between the observed data and the theoretical relationship suggests that there is no difference between cardiac and skeletal muscle in the mechanisms that determine force at saturating [Ca^{2+}]. This study confirms in intact muscle that stretching of cardiac sarcomeres over the range of lengths where only double overlap of the actin filaments plays a role leads to increase of the apparent sensitivity of the contractile system to Ca^{2+} ions. Comparison of the data in this study with the data from skinned-fibre experiments shows that the

Skeletal Muscle Mechanics: From Mechanisms to Function. Edited by W. Herzog.

mechanisms underlying the length dependence of Ca^{2+} sensitivity are the same, despite the loss of soluble cytosolic proteins from the skinned fibre. These data provide a rational framework of mechanisms in intact cardiac muscle underlying Frank Starling's Law of the Heart.

INTRODUCTION

It is well known that during systole the ventricle can be characterized by a unique end-systolic pressure volume relationship (ESPVR), which depends little on the volume change during ejection (Suga et al., 1986; Sagawa et al., 1977; Sagawa et al., 1988; Burkhoff et al., 1987; Wannenburg et al., 1992; Burkhoff et al., 1991; Hunter, 1989; Maughan et al., 1985). Hence, the ESPVR dictates the interrelationship between stroke volume and end-diastolic volume, which gave rise to Starling's Law of the Heart (Le Peuch et al., 1979). Since the ESPVR is a geometric transform of the force-length relationship of the cardiac sarcomere, it follows that Starling's Law is based on the properties of the myofilaments. Many studies on intact, isolated cardiac muscle have provided support for this fundamental hypothesis (Jewell, 1977). Later, it was shown in skinned cardiac trabeculae, during steady activation at controlled levels of free calcium ($[Ca^{2+}]$), that sarcomere length (SL) dependence of force output observed in intact cardiac muscle (ter Keurs, 1980) results from length-dependent sensitivity of the contractile system to calcium (Hibberd and Jewell, 1982; Kentish et al., 1986). Kentish et al. (1986) initiated a search for the mechanism for this length dependence of Ca^{2+} sensitivity by showing that the force-SL relationships in intact and skinned cardiac muscle were similar. The latter appeared to be due to the effect of stretch in the skinned fibre to both increase maximal force and shift the sigmoidal force-$[Ca^{2+}]$ relation leftward. These effects suggest an increase of the sensitivity of the contractile apparatus for Ca^{2+} with stretch. The finding by Allen and collaborators (Allen and Kurihara, 1982) that stretch does not affect the peak of the intracellular Ca^{2+} ($[Ca^{2+}]_i$) transient reported by aequorin is consistent with this concept (Allen and Kurihara, 1982). This finding was later confirmed by studies using the fluorescent probe Fura-2 (Backx and ter Keurs, 1993). The Fura-2 studies showed that stretch increases the rate of decline of the $[Ca^{2+}]_i$ transient, which is also in agreement with the prediction that stretch increases Ca^{2+} binding to Troponin C. The observation that rapid (<10 ms) stretch of skinned fibres causes both a slow increase in force and a slow decrease of $[Ca^{2+}]_i$ surrounding the myofilaments suggested that not SL by itself, but force development, determines Ca^{2+} binding (Allen and Kentish, 1985). Conversely, it has been observed (Allen and Kurihara, 1982; Backx and ter Keurs, 1993; Housmans et al., 1983) that the force decrease during rapid shortening of contracting muscle is accompanied by an increase of $[Ca^{2+}]_i$, suggesting that shortening reduces force and thereby makes the contractile system less sensitive to Ca^{2+}. Ca^{2+}-sensitive regulation of force is controlled by

Troponin C on cardiac actin (cTnC). Hence, SL and/or force have to feed back onto the properties of cTnC. Mammalian cardiac muscle operates only on the ascending limb of the force-SL relationship at lengths (*in vivo*, usually 1.7-2.15 μm [Rodriguez et al., 1992]; *in vitro*, maximally 1.5-2.3 μm [ter Keurs et al., 1980; ter Keurs et al., 1988; ter Keurs et al., 1987]) at which the actin filaments (1.15 μm long) are always in double overlap. Hence, length-dependent sensitivity of the contractile system for Ca^{2+} may be due to SL-dependent Ca^{2+} affinity of Troponin C (TnC) (Gulati et al., 1990) with TnC as the length sensor (Babu et al., 1988), or variation of Ca^{2+}-affinity of Troponin C as a result of feedback of force generated by the cross-bridge. Force generated by the cross-bridges might be sensitive to SL because varied overlap between actin and myosin would result in variation of the number of cross-bridges and/or cross-bridge force exerted on actin. Alternatively, by influencing the myofilament spacing, SL could affect the probability of cross-bridge attachment as well as force development by the attached cross-bridge. The observation that, after expression of skeletal muscle TnC (sTnC) in mouse cardiac muscle, SL still regulates the Ca^{2+}-sensitivity of force development (McDonald et al., 1995) makes it unlikely that cTnC is the length sensor for Ca^{2+} sensitivity. Similarly, the fact that cTnC present in soleus muscle does not confer length-dependent Ca^{2+} sensitivity to the soleus muscle suggests that other properties of the contractile filaments may be involved (Wang and Fuchs, 1994).

Biochemical experiments have strengthened the notion that the attachment of cross-bridges increases Ca^{2+} binding in cardiac and skeletal muscle (Brandt et al., 1987; Zot and Potter, 1989; Metzger and Moss, 1991; Hannon et al., 1991). Fuchs and collaborators have shown that the amount of Ca^{2+} bound to the thin filament is sensitive to lattice spacing of the myofilaments (Wang and Fuchs, 1995) and, consequently, cross-bridge formation (Hofmann and Fuchs, 1987a) and force development (Hofmann and Fuchs, 1987b). They also showed that Ca^{2+} sensitivity of the cardiac contractile system to double overlap of the actin filaments is small (Fuchs and Wang, 1996). These studies created a foundation for models that suggest a role for force development by the cross-bridges in length dependence of Ca^{2+} activation of the mammalian cardiac sarcomere, and that faithfully predict mechanical properties of the cardiac ventricle (Landesberg and Sideman, 1994a, 1994b; Landesberg et al., 1994; Landesberg, 1996).

Whether alterations in myofilament Ca^{2+} activation with length indeed occur in intact muscle fibres has yet to be determined. Gao et al. (1994) have shown marked differences in the EC_{50} for [Ca^{2+}] of force development in skinned versus intact muscle fibres. Even after correction of the [Mg^{2+}] in the activating solutions, skinned fibres require substantlally higher [Ca^{2+}] to reach half-maximal force, indicating that skinning renders the fibres less sensitive to Ca^{2+} than intact muscles (Gao et al., 1994). This may be due to the loss of essential cytosolic components, as well as differences in the myofilament lattice spacing owing to the skinning process (Gao et al., 1994), or to the fact that the duration of activation in skinned fibres is orders of magnitude longer than in the normal cardiac twitch. If the EC_{50} of the force-[Ca^{2+}] relationship in skinned muscle is indeed substantially higher than in intact muscle, as suggested by Gao et al. (1994), then the response to stretch may also differ between intact muscles and skinned fibres. Finally, the relationship between force and sarcomere length at a saturating [Ca^{2+}] in intact fibres is unknown. Cardiac muscle operates on the ascending limb of the force-length relationship of the sarcomere. Hence, the relationship between force and length of the sarcomeres is of importance, because it might provide information about the role of the actin-myosin interaction under these constraints. The purpose of this study is to evaluate the effect of stretch of intact muscle on the sensitivity of the contractile system for Ca^{2+} ions and on maximal force development.

METHODS

Lewis Brown Norway F_1 rats were anaesthetized with ether and the hearts were quickly removed and perfused through the aorta with a modified Krebs-Henseleit solution (see below). Thin, unbranched trabeculae running from the atrio-ventricular ring to the right ventricular free wall were selected for the experiments.

Force was measured using a modified silicon semiconductor strain gauge (Model AE-801, Sensonor, Horten, Norway) with a carbon extension arm and a stainless-steel wire bent in a U shape to hold the ventricular end of the muscle. A hook attached to a motor arm pierced the valvular end of the muscle. The motor was used to control muscle length.

SL was measured using the optical diffraction technique as described previously (ter Keurs et al., 1980). The beam of a 15-mW He-Ne laser was directed through the central part (300 μm diameter) of the muscle. The series arrangement of the sarcomeres within the muscle act as a grating, creating orders of diffracted light, the relative spacing of which is related to the SL. The diffracted light of the first order of the diffraction pattern was scanned twice per millisecond, using a 512-element linear photodiode array (Reticon, Sunnyvale, CA). The median of the first order was detected and its position was used to calculate SL during each scan.

Fura-2 fluorescence was measured in order to assess the cytosolic [Ca^{2+}] according to the techniques developed by Backx and ter Keurs (1993).

RESULTS

The following sections provide information about the outcomes of our study.

TETANI IN THE PRESENCE OF CYCLO-PIAZONIC ACID (CPA)

As the effect of 30 μM [CPA] on Ca^{2+} sensitivity of the contractile filaments was negligible (not shown here), we used this concentration of CPA to study the force-[Ca^{2+}]$_i$ relationships during tetani in intact trabeculae. We adjusted muscle length (or initial SL) in such a way that active tetanic force was measured at either SL ~2.15 µm or ~1.85 µm, and studied under these conditions the response of force development to [Ca^{2+}]. We chose these lengths, because at 1.85 µm, the contribution of restoring forces to force output is negligible, while the correction for changes in passive elastic force is small at 2.15 µm. Typically, force during the tetani showed a transient overshoot, followed by the gradual development of a plateau (Figure 4.1). During the transient phase of force development, the sarcomeres shortened initially (throughout the muscle; data not shown), then SL remained constant for the duration of the plateau (Figure 4.1). The initial shortening of the sarcomeres is known to take place at the expense of the series elastic element in the muscle (usually the tricuspid valve attachment in these trabeculae).

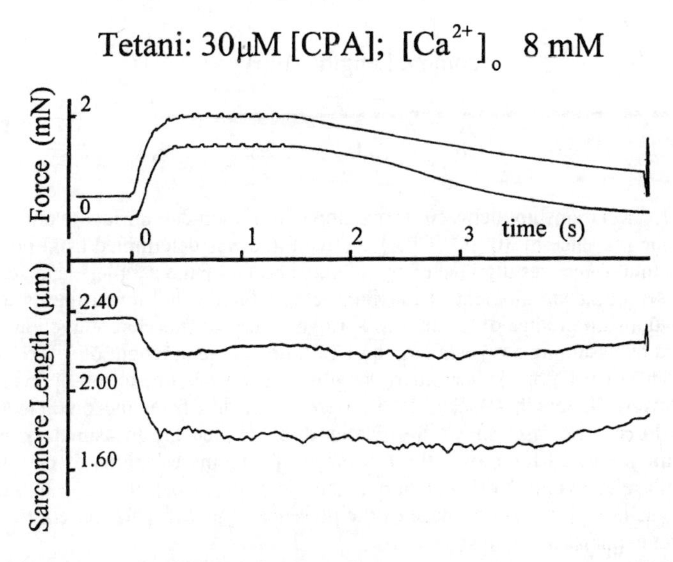

Figure 4.1. Force and sarcomere length tracings obtained at two muscle lengths during 1.5 s lasting tetani. The muscle was stimulated with 12 Hz pulses of 40 ms duration. [Ca^{2+}]$_o$ was 8 mM. Tetani were repeated at varied initial muscle lengths so that SL during the tetani ranged from 1.6 to 2.2 µm.

INFLUENCE OF [Ca^{2+}]$_o$

[Ca^{2+}]$_o$ was varied to modulate force during the tetani. The resultant variation of [Ca^{2+}]$_i$ between stimuli was small at a short length (i.e., resting [Ca^{2+}]$_i$ increased from ~100 to 250 nM; Figure 4.3B) and had no effect on force development in the muscle held at SL = 1.85 µm. Stretch caused a clear increase of resting [Ca^{2+}]$_i$, while a further increase of [Ca^{2+}]$_o$ in the stretched muscle increased both [Ca^{2+}]$_i$ (up to ~600 nM) and caused substantial force development (Figure 4.3).

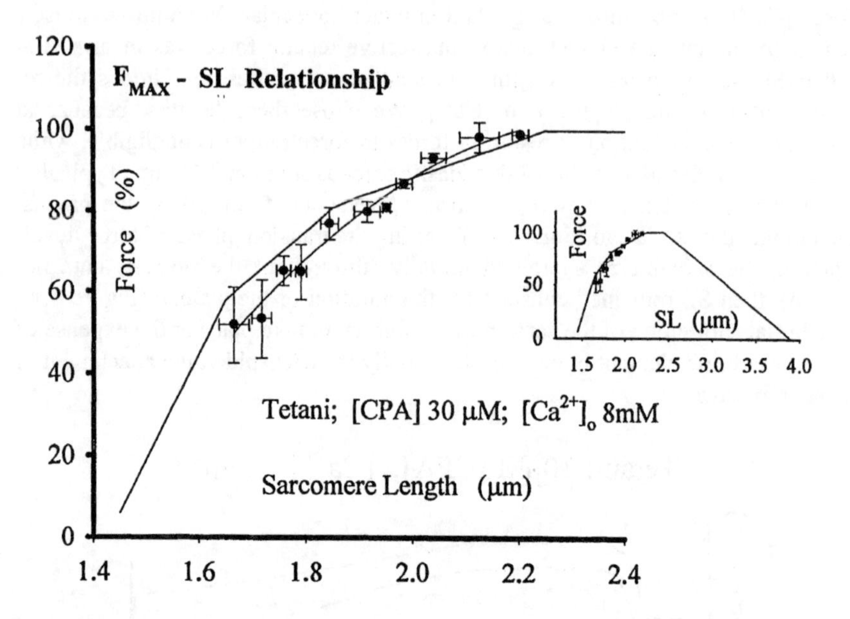

Figure 4.2. Relationship between active force development during tetani at 8 mM $[Ca^{2+}]_o$ in the presence of 30 μM [CPA]. Active force was determined as in ter Keurs et al. (1980): total force was diminished by the force borne by the parallel elastic element at the SL observed at the moment of maximal tetanic force. The drawn line is the force-length relationship predicted by the cross-bridge theory of muscle contraction, and was constructed by assuming a myosin length of 1.65 μm, an actin length of 1.125 μm, and a Z-band width of 0.1 μm. At sarcomere lengths below 1.86 μm, forces opposing shortening below slack length (Backx, 1989) were subtracted from the predicted actively generated force. The inset shows the relationship between the measured experimental data and the predicted force over the full range of sarcomere lengths of the theoretical force-length relationship. Note that cardiac muscle cannot reach the SL range of partial overlap of actin and myosin because of the presence of a stiff, parallel, elastic element. (For further explanation see text.)

During the plateau of the tetani, $[Ca^{2+}]_i$ increased linearly with $\log([Ca^{2+}]_o)$ by 0.72 μM for every tenfold increase of $[Ca^{2+}]_o$ (between 0.2 and 16 mM). SL had no effect on the interrelationship between $[Ca^{2+}]_i$ and $[Ca^{2+}]_o$ (data not shown). The relationship between force and $[Ca^{2+}]_o$ fitted well to a modified Hill equation (Figure 4.4). The steepness of the relationship was independent of length (n was 1.4 and 1.3 at SL = 2.15 and 1.85 μm respectively; n.s.). However, the EC_{50} for $[Ca^{2+}]_o$ decreased substantially with stretch (from 1.85 mM to 0.3 mM; $p<0.0001$), and maximal tetanic force increased by ~30% (from 87±7 mN/mm^2 to 107±10 mN/mm^2 at SL = 1.85 and 2.15 μm respectively; $p<0.0001$). Figure 4.3 shows this effect at $[Ca^{2+}]_o$ of 1.0, 3.0, and 8.0 mM. Maximum tetanic force was reached at $[Ca^{2+}]_o$ >3 mM. Further increase of $[Ca^{2+}]_o$ (up to 16 mM in three experiments; data not shown) did not further increase force at either of the two SL

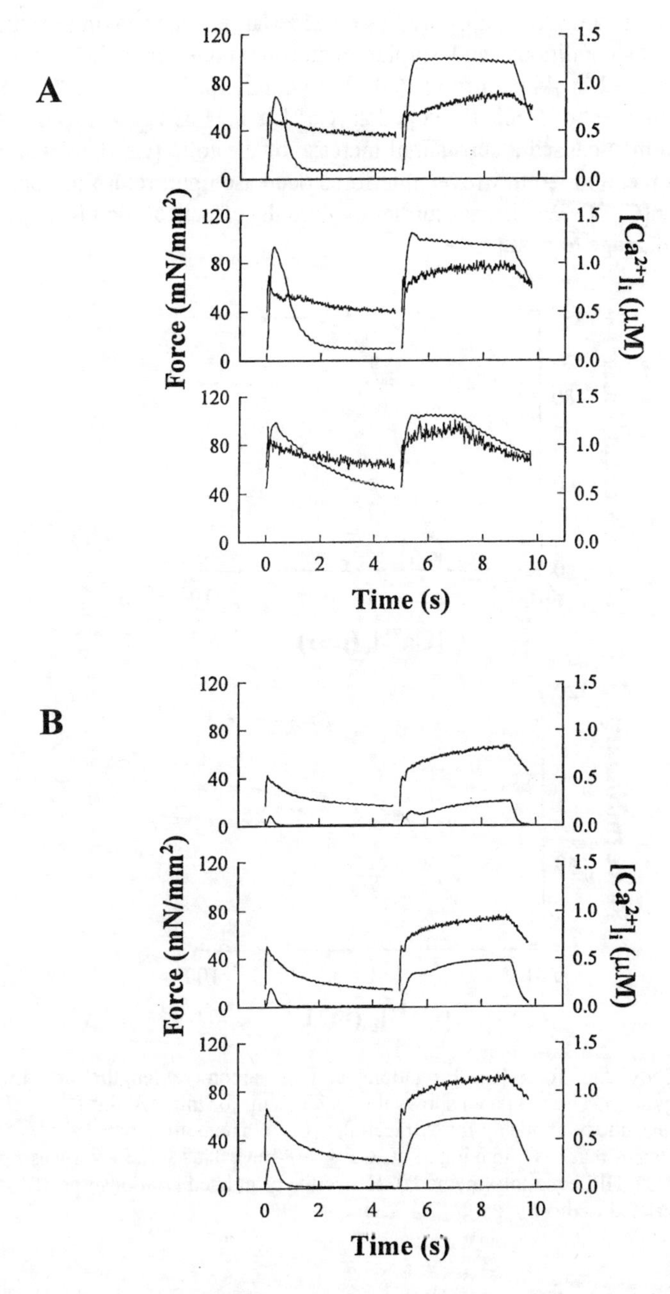

Figure 4.3. Twitch and tetanus force tracings at three $[Ca^{2+}]_o$ (1.0, 3.0, and 8.0 mM, top panels, middle panels, and bottom panels, respectively). A) Measurements made at SL = 2.15 μm. B) Measurements made at SL = 1.85 μm. The smooth tracings reflect force development; the noisier tracings reflect the calculated cytosolic $[Ca^{2+}]$.

studied. Maximal tetanic force appeared to be 25% larger than maximal twitch force in drug-free conditions, and similar to the maximal force that we and others have observed in skinned fibres (~110 mN/mm^2, Backx et al., 1995). So we have used the level of force developed at $[Ca^{2+}]_o$ 8 mM as F_{max}. Elevation of $[Ca^{2+}]_o$ above 8 mM caused a substantial increase of diastolic $[Ca^{2+}]_i$ as well as unstimulated force, and led to irreversible force decrease upon return to control solutions at low $[Ca^{2+}]_o$. We did not further explore the effects of high $[Ca^{2+}]_o$, in order to avoid damage to the muscle.

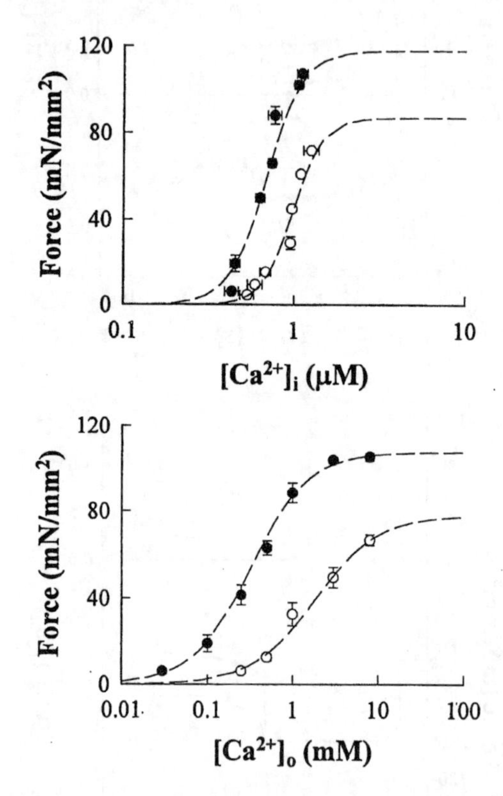

Figure 4.4. Steady-state force-$[Ca^{2+}]$ relations at two sarcomere lengths fit with a modified Hill equation. The top panel shows the relationship for intracellular $[Ca^{2+}]$. The bottom panel shows the relationship for extracellular $[Ca^{2+}]$. The bottom panel shows that the force of the tetani reached a maximum at $[Ca^{2+}]_o > 8$ mM. This force level was used to set the F_{max} of the Hill equation (methods). Hence the calculated relationships of force -$[Ca^{2+}]_i$ were restricted to this F_{max}.

EFFECT OF SL ON MAXIMAL FORCE OF THE TETANI

Figure 4.2 shows the increase of F_{max} with increasing SL, during tetani in [CPA] 30 μM and at $[Ca^{2+}]_o = 8$ mM. This increase in F_{max} with SL paralleled the force increment that is predicted by the assumption that F_{max} generation depends on the mechanical constraints of the interaction between the cross-bridges and the actin filaments. In order to calculate the predicted force-SL relationship, we used the following values for myosin length, actin length, and a Z-band width in the cardiac muscle of the rat: 1.65 μm, 1.125 μm, and 0.1 μm, respectively (ter Keurs et al., 1984; Robinson and Winegrad, 1979). These dimensions predict a force plateau from 2.25 to 2.45 μm. Increasing double overlap would proportionally reduce force at shorter lengths so that when SL = 1.65 μm, ~70% of the maximal force is generated. Myosin distortion would cause a further and steeper decline with an intercept at the SL axis near 1.25 μm. The same assumptions have generated the ascending limb of the classical force-length relationship for skeletal muscle. In addition, it has been shown that cardiac muscle generates a force that opposes shortening below slack length (SL = 1.86 μm, ter Keurs et al., 1980). The opposing force has been shown to increase in cardiac trabeculae in proportion to shortening; at SL = 1.5 μm the opposing force was 25% of the maximal active force (Backx, 1989). The opposing force is generated in part by the titin filament in the sarcomere itself (Granzier and Irving, 1995), with an additional component owing to the collagen meshwork around the myocytes (Robinson et al., 1983). These forces are probably substantially larger than those in skeletal muscle fibres, which are the source of the textbook force-length relationship, because titin in skeletal muscle fibres is more compliant and single skeletal muscle fibres do not contain a collagen meshwork like myocardium. We have, therefore, diminished the force anticipated on the basis of the actin and myosin properties accordingly (see Figure 4.2, line with intercept at SL = 1.44 μm). Figure 4.2 shows that the actually measured maximal force during the tetanus coincided exactly with the force predicted on the basis of these mechanical considerations. Hence, we conclude that there is no significant difference between the factors that dictate maximal force development in skeletal and in cardiac muscle.

FORCE-[Ca^{2+}]$_i$ RELATIONSHIPS DURING TETANI

It is reasonable to conclude from the force-SL relationship at maximal [Ca^{2+}] that force development is influenced only by double overlap of the actin filaments at SL>1.85 μm. In addition, it is known (ter Keurs et al., 1980; de Tombe and ter Keurs, 1992) that these trabeculae generate substantial passive elastic forces at SL>2.2 μm; the presence of the latter complicates the interpretation of force output by the sarcomere. Hence, we studied the relationship between force development and [Ca^{2+}]$_i$ at two SL (1.85 and 2.15 μm) where only double overlap is expected to modulate the properties of the contractile apparatus. The force-[Ca^{2+}]$_i$ relationships derived from plateau force and the plateau of [Ca^{2+}]$_i$ during tetani fitted the Hill equation exactly (Figure 4.4). The effect of sarcomere stretch was to reduce the EC_{50} for [Ca^{2+}]$_i$ (from 0.98±0.09 to 0.65±0.04 μM; $p<0.003$) and to increase F_{max}. The Hill coefficient of the force-

$[Ca^{2+}]_i$ relationships did not change significantly with stretch (n increased from 4±1 to 5±1; n.s.).

The relationships shown in Figure 4.1, Figure 4.2, and Figure 4.4 were obtained using data from the plateau of the tetanus where the force and $[Ca^{2+}]_i$ are in steady state. The steady state provided a spatially uniform distribution of $[Ca^{2+}]_i$ and allowed unambiguous correlation between force development and $[Ca^{2+}]_i$ (Cannell and Allen, 1984).

DISCUSSION

The following sections provide an interpretation and discussion of our results.

EFFECT OF SARCOMERE LENGTH ON MAXIMAL FORCE DEVELOPMENT DURING TETANI

The relationship between $[Ca^{2+}]_i$ during the tetani and $[Ca^{2+}]_o$ was independent of SL, as has been shown previously during the twitch in drug-free conditions (Backx and ter Keurs, 1993). This observation indicates that the Ca^{2+} transport process across the sarcolemma responsible for activation of rat trabeculae is not sensitive to stretch. It follows that any changes in force development during the tetani due to stretch must be secondary to changes in the myofilament sensitivity, and not to effects on Ca^{2+} transport.

F_{max} developed by the trabeculae in the presence of CPA was similar to maximal force that has been reported for skinned fibres, both by others and in our laboratory, and was 25% larger than the maximal twitch force in drug-free conditions. This is consistent with several studies that indicate that twitch force is limited by the release of Ca^{2+} ions from the SR, while the truly maximal force is limited by the properties of the contractile filaments (Schouten et al., 1990; Banijamali et al., 1991). Figure 4.1, Figure 4.2, and Figure 4.4 show that F_{max} increased ~20% with stretch of the sarcomeres from 1.85 to 2.15 μm. This force change with stretch is twofold less than has been reported for skinned fibres (Kentish et al., 1986; Fozzard, 1977). A probable cause for the difference between skinned and intact cardiac muscle is that the dispersion of sarcomere length during forceful contractions as reflected by the width of the first order of the diffraction pattern is, at least in our hands, usually larger in skinned fibres than in intact trabeculae. One likely cause for this observation is that activation of the skinned fibre results from diffusion of Ca^{2+} ions over a distance of ~50 μm (half the thickness of the muscle). The rate of activation is therefore expected to be two orders of magnitude slower than that of the intact muscle, which derives its Ca^{2+} ions from the T-tubuli and the surface membrane at a distance less than 5 μm. As a result of the large dispersion of SL, one would expect lower force development than in muscle in which the sarcomere distribution is uniform. Fabiato's observation that maximally activated skinned single cells exhibit a similar F-SL relationship as the one in Figure 4.1 is consistent with this interpretation (Fabiato and Fabiato, 1975). We obtained previously a similar F-SL relationship in a study of trabeculae tetanized in the presence of high [Sr^{2+}] (Schouten et al., 1990). So, three widely different conditions in which the contractile apparatus was activated rapidly and maximally (i.e., tetani in the presence of Sr^{2+} [Schouten et al., 1990] or Ca^{2+} [this study] and contracture in isolated skinned cells in the presence of Ca^{2+} [Fabiato and Fabiato, 1975]) have yielded the same F-SL relationships. These relationships were nearly identical to the ascending limb of the classical relationship between force and SL for skeletal muscle fibres (Gordon et al., 1966), provided that the dimensions that apply to the sarcomere of rat heart are used. This relationship shows a maximum at SL near 2.25 μm if actin length is 1.125 μm as has been observed in rat cardiac (ter Keurs, 1993) and skeletal muscle (ter Keurs et al., 1984). Previously, it was reported that filament length appears variable (Robinson and Winegrad, 1979). We have assumed that actin filament is constant in the cardiac sarcomeres, because a contribution of titin filaments to the observed variability has not been ruled out in these reports (Robinson and Winegrad, 1979, 1977). Shortening below this length is accompanied by increasing double overlap of the actin filaments. The observed decrease of force could have been caused by the increased double overlap as was originally proposed for skeletal muscle (Gordon et al., 1966) or by increased lattice spacing as has been proposed more recently (Fuchs and Wang, 1996). With further shortening of the cardiac sarcomere below slack length (1.86 μm), force should decrease progressively because of the deformation of three structures. The first is distortion of the myosin filament, as was originally proposed as an explanation of the ascending limb of the F-SL relationship of skeletal muscle at SL<1.65 μm (Gordon et al., 1966). Deformation of titin in the myocyte and of collagen around the myocyte will generate additional opposing forces upon shortening below slack SL. These opposing forces have been estimated on the basis of the elastic properties of activated trabeculae. They have been shown to increase in proportion to shortening to a value of 30% of maximal active force of the twitch (Backx, 1989) (i.e., 25% of the truly maximal force of the contractile apparatus, at SL = 1.65 μm). The combined effects of double overlap and other mechanical constraints predict an intercept of the F-SL relationship at SL ~1.45 μm in the intact trabecula. Figure 4.1 and Figure 4.2 show that the actual data correspond closely to the values of force predicted by this theoretical approach. Hence, these data from intact cardiac muscle provide the first direct support for the hypothesis that SL controls F_{max} in heart muscle in the same way as in skeletal muscle.

THE STEADY-STATE FORCE-[Ca^{2+}]$_i$ RELATIONSHIP IN INTACT MUSCLE

In the above, we showed that SL governs F_{max} in cardiac muscle in a manner comparable to that in skeletal muscle. This implies that SL, between 1.85 and

2.25 μm, dictates force development by variation of both double overlap and filament lattice spacing, and controls, in this way, the number of cross-bridges that interact with actin of the opposite Z-band. At shorter lengths, elastic opposing forces complicate this effect. We have, therefore, evaluated the effect of changing overlap by studying the force-length relationship in intact muscle at SL = 1.85 and 2.15 μm.

Our results show a substantial length-related change of the EC_{50} for $[Ca^{2+}]_i$ (0.98±0.09 to 0.65±0.04 μM for sarcomere stretch from 1.85 to 2.15 μm) of the contractile apparatus (Figure 4.4). The EC_{50} in the intact stretched trabeculae observed here is identical to the value that was reported previously (Gao et al., 1994; Backx et al., 1995). We will discuss in the following the proposal that both the observed EC_{50} and the SL dependence of the EC_{50} are consistent with observations on skinned fibres. By doing so, we will show that there is no need to assume that loss of cytosolic proteins or peptides during skinning causes an essential difference in behaviour between intact and skinned muscle. Several studies have shown that the EC_{50} of the contractile apparatus in skinned fibres varies with sarcomere length (~0.5 pCa unit/μm [Kentish et al; 1986; Fuchs and Wang, 1996; de Beer et al., 1988]). If the change of EC_{50} is also 0.5 pCa unit/μm in intact fibres, shortening of intact fibres from SL of 2.15 μm to 1.85 μm should increase the EC_{50} from 0.65 μM to 0.92 μM, i.e., closely similar to the value experimentally observed in this study (0.98 μM). This similarity is consistent with the assumption that the EC_{50} is length dependent, but it does not clarify the mechanism, which couples length changes to a change in sensitivity of the filaments. Two candidate mechanisms have been proposed. The first is that length changes influence double overlap and thereby modify the number of cross-bridges that can attach to actin. The force generated by the attached cross-bridges could, then, be translated into a change of sensitivity of the contractile system to Ca^{2+} ions (Adelstein and Eisenberg, 1980; Greene and Eisenberg, 1988). The second mechanism that has been proposed is that the length change causes a change of spacing between the actin and the myosin lattice, and thereby modifies the number of cross-bridges that can attach to actin. Cross-bridge force would, then, be translated in the same way into a change of sensitivity of the contractile system to Ca^{2+} ions (Adelstein and Eisenberg, 1980; Greene and Eisenberg, 1988). Fuchs et al. have shown in skinned fibres that when they eliminated stretch-induced changes of lattice spacing by the use of dextran, the change of the EC_{50} was also eliminated, despite SL changes (Fuchs and Wang, 1996). As a result of these studies, these authors have proposed that the lattice spacing determines the sensitivity of cardiac muscle to Ca^{2+} ions. Their intervention also eliminated the difference of force between contractions at short versus long sarcomeres, so that it is possible that lattice spacing dictates the probability of cross-bridge force development, irrespective of the cause of a change of the spacing (double overlap due to SL change or skinning).

This study does not provide information about whether double overlap or the concomitant force change cause the EC_{50} to change, because SL and force change in this range by the same amount (both ~20%; Figure 4.1). However, the change in lattice spacing between intact and skinned fibres is enough to explain the different Ca^{2+} sensitivity of force development in these preparations. Godt and Maughan have shown that muscle fibres swell by 20% due to the skinning procedure (Godt and Maughan, 1981). This would be equivalent to ~30% sarcomere shortening at constant lattice volume, as has been shown in intact cardiac muscle by Matsubara and Millman (1974). If it matters little whether the lattice changes are due to SL changes or due to the skinning procedure, it follows that the EC_{50} (0.65 μM at SL = 2.15 μm) observed in this study should increase to 1.8 μM following skinning at the same length. This value is close to the observed value (3.8 μM) at this SL (Kentish, 1986). Kerrick and collaborators have shown that the EC_{50} for Ca^{2+} increases by 3.4 μM/mM [Mg^{2+}] (Bolitho et al., 1978). Hence, the residual difference between the observed and the predicted value of the EC_{50} for Ca^{2+} is fully explained by the high [Mg^{2+}] that is commonly used in skinned fibre solutions (e.g., Kentish, 1986) compared to the cytosolic level (0.72 mM) (Gao et al.,1994).

Scrutiny of the force-$[Ca^{2+}]_i$ relationships shows that the Hill coefficients at both SL were larger than would be expected on the basis of the binding of one Ca^{2+} ion to the low-affinity site of cardiac Tn-C. This observation is consistent with previous studies (Kentish et al., 1986) and with the concept of co-operativity between force development and Ca^{2+} binding. The notion that it is actually force that determines Ca^{2+} binding to and/or dissociation from Tn-C is supported by a multitude of observations. These observations indicate that Ca^{2+} binding to Tn-C closely follows myosin attachment (Hofmann and Fuchs, 1987a), formation of strongly attached cross-bridges (Brandt et al., 1987; Zot and Potter, 1989; Metzger and Moss, 1991), and force development (Hannon et al., 1991; Hofmann and Fuchs, 1987b; Allen and Kentish, 1988; Sweitzer and Moss, 1990). The effect of the cross-bridges is possibly due to conformational changes in Tn-C as a result of force development (Hannon et al., 1991). In turn, changes in the apparent affinity of Tn-C for Ca^{2+} will affect further force development, thereby closing the feedback loop that is the basis of the co-operativity between force development and Ca^{2+} sensitivity of the thin filament of cardiac muscle (Brandt et al.,1987; Moss, 1992) (for review, see Landesberg and Sideman, 1994b; Zot and Potter, 1987). Figure 4.1 and Figure 4.2 show that predicted force development without the contribution of opposing forces varies with SL along the ascending limb of the F-SL relationship. Hence, a corresponding variation of EC_{50} is expected on the basis of the hypothesis that force generation influences Ca^{2+} binding kinetics of Tn-C.

CONCLUSION

This study describes, for the first time, the force-sarcomere length interrelationship of maximally Ca^{2+}-activated intact cardiac muscle. This relationship is close to identical to the ascending limb of the force-length relationship based on the cross-bridge theory of muscle contraction for skeletal muscle, provided that the dimensions of the filaments for the rat are used and provided that forces opposing shortening below slack length are incorporated in the relationship. The correspondence between the observed data and the theoretical relationship suggests that there is no difference in the mechanisms that dictate force at saturating [Ca^{2+}] between cardiac and skeletal muscle. Secondly, this study confirms in intact muscle that stretch of cardiac sarcomeres over the range of lengths where only double overlap of the actin filaments plays a role leads to an increase of the apparent sensitivity of the contractile system to Ca^{2+} ions. Comparison of the data in this study with the data from skinned fibre experiments shows that the mechanisms underlying the length dependence of Ca^{2+} sensitivity are the same, despite the loss of soluble cytosolic proteins from the skinned fibre. These data are consistent with the hypothesis that the degree of stretch of the sarcomeres determines force development by the cross-bridge owing to physical intra-sarcomeric factors; in turn, developed force affects the sensitivity of the contractile system to Ca^{2+} ions, observed at submaximal [Ca^{2+}]. Taken together, the data provide a rational framework of mechanisms in intact cardiac muscle underlying Frank Starling's Law of the Heart.

Further studies on the mechanism of co-operativity are needed to provide a detailed pathway of feedback from filament spacing, force, or number of attached cross-bridges to Ca^{2+} binding to Troponin C.

ACKNOWLEDGEMENTS

This study was supported by grants from the Medical Research Council of Canada. Dr. H.E.D.J. ter Keurs is a medical scientist with the Alberta Heritage Foundation for Medical Research (AHFMR). Ellen H. Hollander was a recipient of an AHFMR studentship.

REFERENCES

Adelstein, R.S., and Eisenberg, E. (1980). Regulation and kinetics of the actin-myosin-ATP interaction. *Annual Review of Biochemistry* **49**, 921–956.

Allen, D.G., and Kentish, J.C. (1985). The cellular basis of the length-tension relation in cardiac muscle. *Journal of Molecular and Cellular Cardiology* **17**, 821–840.

Allen, D.G., and Kentish, J.C. (1988). Calcium concentration in the myoplasm of skinned ferret ventricular muscle following changes in muscle length. *Journal of Physiology* **407**, 489–503.

Allen, D.G., and Kurihara, S. (1982). The effects of muscle length on intracellular calcium transients in mammalian cardiac muscle. *Journal of Physiology* **327**, 79–94.

Babu, A., Sonnenblick, E.H., and Gulati, J. (1988). Molecular basis for the influence of muscle length on myocardial performance. *Science* **240**, 74–76.

Backx, P.H.M. (1989). *Force sarcomere relation in cardiac myocardium.* Ph.D. thesis, University of Calgary, Calgary, Alberta, Canada.

Backx, P.H.M., Gao, W.D., Azan-Backx, M.D., and Marban, E. (1995). The relationship between contractile force and intracellular [Ca^{++}] in intact rat cardiac trabeculae. *Journal of General Physiology* **105**, 1–19.

Backx, P.H.M., and ter Keurs, H.E.D.J. (1993). Fluorescent properties of rat cardiac trabeculae microinjected with fura-2 salt. *American Journal of Physiology* **264**, H1098–H1110.

Banijamali, H.S., Gao, W.D., MacIntosh, B.R., and ter Keurs, H.E.D.J. (1991). Force-interval relations of twitches and cold contractures in rat cardiac trabeculae: Influence of ryanodine. *Circulation Research* **69**, 937–948.

Bolitho, S., Donaldson, S.K., Best, P.M., and Kerrick, W.G.L. (1978). Characterization of the effects of Mg^{++} on Ca^{++} and Sr^{++} activated tension generation of skinned rat cardiac fibers. *Journal of General Physiology* **71**, 645–655.

Brandt, P.W., Diamond, M.S., and Rutchik, J.S. (1987). Co-operative interactions between troponin-tropomyosin units extend the length of the thin filament in skeletal muscle. *Journal of Molecular Biology* **195**, 885–896.

Burkhoff, D., de Tombe, P.P., Hunter, W.C., and Kass, D.A. (1991). Contractile strength and mechanical efficiency of left ventricle are enhanced by physiological afterload. *American Journal of Physiology* **260**, H569–H578.

Burkhoff, D., Sugiura, S., Yue, D.T., and Sagawa, K. (1987). Contractility-dependent curvilinearity of end-systolic pressure-volume relations. *American Journal of Physiology* **252**, H1218–H1227.

Cannell, M.B., and Allen, D.G. (1984). Model of calcium movements during activation in the sarcomere of frog skeletal muscle. *Biophysical Journal* **45**, 913–925.

de Beer, E.L., Grundeman, R.L.F., Wilhelm, A.J., van den Berg, C., Caljouw, C.J., Klepper, D., and Schiereck, P. (1988). Effect of sarcomere length and filament lattice spacing on force development in skinned cardiac and skeletal muscle preparations from the rabbit. *Basic Research in Cardiology* **83**, 410–423.

de Tombe, P.P., and ter Keurs, H.E.D.J. (1992). An internal viscous element limits unloaded velocity of sarcomere shortening in rat myocardium. *Journal of Physiology* **454**, 619–642.

Fabiato, A., and Fabiato, F. (1975). Dependence of the contractile activation of skinned cardiac cells on the sarcomere length. *Nature* **256**, 54–56.

Fozzard, H.A. (1977). Heart: Excitation-contraction coupling. *Annual Review of Physiology* **39**, 201–220.

Fuchs, F., and Wang, Y-P. (1996). Sarcomere length versus interfilament spacing as determinants of cardiac myofilament Ca^{2+} sensitivity and Ca^{2+} binding. *Journal of Molecular and Cellular Cardiology* **28**, 1375–1383.

Gao, W.D., Backx, P.H.M., Azan-Backx, M., and Marban, E. (1994). Myofilament Ca^{2+} sensitivity in intact versus skinned rat ventricular muscle. *Circulation Research* **74**, 408–415.

Godt, R.E., and Maughan, D.W. (1981). Influence of osmotic compression on calcium activation and tension in skinned muscle fibers of the rabbit. *Pflügers Archiv* **391**, 334–337.

Gordon, A.M., Huxley, A.F., and Julian, F.J. (1966). The variation in isometric tension with sarcomere length in vertebrate muscle fibres. *Journal of Physiology* **184**, 170–192.

Granzier, H.L.M., and Irving, T.C. (1995). Passive tension in cardiac muscle: Contribution of collagen, titin, microtubules, and intermediate filaments. *Biophysical Journal* **68**, 1027–1044.

Greene, L.E., and Eisenberg, E. (1988). Relationship between regulated actomyosin ATPase activity and cooperative binding of myosin to regulated actin. *Cell Biophysics* **12**, 59–71.

Gulati, J., Sonnenblick, E., and Babu, A. (1990). The role of troponin C in the length dependence of Ca^{2+}-sensitive force of mammalian skeletal and cardiac muscles. *Journal of Physiology* **441**, 305–324.

Hannon, J.D., Martyn, D.A., and Gordon, A.M. (1991). Effects of cycling and rigor crossbridges on the conformation of cardiac troponin C. *Circulation Research* **71**, 984–991.

Hibberd, M.G., and Jewell, B.R. (1982). Calcium- and length-dependent force production in rat ventricular muscle. *Journal of Physiology* **329**, 527–540.

Hofmann, P.A., and Fuchs, F. (1987a). Effect of length and cross-bridge attachment on Ca++ binding to cardiac troponin-C. *American Journal of Physiology* **253**, 90–96.

Hofmann, P.A., and Fuchs, F. (1987b). Evidence for a force-dependent component of calcium binding to cardiac troponin-C. *American Journal of Physiology* **253**, 541–546.

Housmans, P.R., Lee, N.K.M., and Blinks, J.R. (1983). Active shortening retards the decline of the intracellular calcium transient in mammalian heart muscle. *Science* **221**, 159–161.

Hunter, W.C. (1989). End-systolic pressure as a balance between opposing effects of ejection. *Circulation Research* **64**, 265–275.

Jewell, B.R. (1977). A re-examination of the influence of muscle length on myocardial performance. *Circulation Research* **40**, 221–230.

Kentish, J.C. (1986). The effects of inorganic phosphate and creatine phosphate on force production in skinned fibres from rat ventricle. *Journal of Physiology* **370**, 585–604.

Kentish, J.C., ter Keurs, H.E.D.J., Ricciardi, L., Bucx, J.J.J., and Noble M.I.M. (1986). Comparison between the sarcomere length-force relations of intact and skinned trabeculae from rat right ventricle. *Circulation Research* **58**, 755–768.

Landesberg, A. (1996). End-systolic pressure-volume relationship and intracellular control of contraction. *The American Physiological Society* **270**, H338–H349.

Landesberg, A., Beyar, R., and Sideman, S. (1994). Left ventricle pressure-volume relationship based on the intracellular control mechanisms: Analysis of the time-varying elastance. *Computers in Cardiology*, Washington, D.C., September 1994, IEEE Supplement, pp. 9–12.

Landesberg, A., and Sideman, S. (1994a). Coupling calcium binding to troponin-C and cross-bridge cycling in skinned cardiac cells. *American Journal of Physiology* **266**, H1260–H1271.

Landesberg, A., and Sideman, S. (1994b). Mechanical regulation in the cardiac muscle by coupling calcium kinetics with crossbridge cycling: A dynamic model. *American Journal of Physiology* **267**, H779–H795.

Le Peuch, C.J., Haiech, J., De Maille, J.G. (1979). Concerted regulation of cardiac sarcoplasmic calcium transport by cyclic adenosine monophosphate dependent and calcium-calmodulin-dependent phosphorylations. *Biochemistry* **18**, 5150–5157.

Matsubara, I., and Millman, B.M. (1974). X-ray diffraction patterns from mammalian heart muscle. *Journal of Molecular Biology* **82**, 527–536.

Maughan, W.L., Sunagawa, K., Burkhoff, D., Graves, W.L., Hunter, W.C., and Sagawa, K. (1985). Effect of heart rate on the canine end-systolic pressure-volume relationship. *Circulation* **72**, 654–659.

McDonald, K.S., Fields, L.J., Parmacek, M.S., Soonpaa, M., Leiden, J.M., and Moss, R.L. (1995). Length dependence of Ca^{2+} sensitivity of tension in mouse cardiac myocytes expressing skeletal troponin C. *Journal of Physiology* **483**, 131–139.

Metzger, J.M., and Moss, R.L. (1991). Kinetics of a Ca^{++}-sensitive crossbridge state transition in skeletal muscle fibers. *Journal of General Physiology* **98**, 233–248.

Moss, R.L. (1992). Ca^{2+} regulation of mechanical properties of striated muscle: Mechanistic studies using extraction and replacement of regulatory proteins. *Circulation Research* **70**(5), 865–884.

Robinson, T.F., Cohen-Gould, L., and Factor, S.M. (1983). Skeletal framework of mammalian heart muscle: Arrangement of inter- and pericellular connective tissue structures. *Laboratory Investigation* **49**, 482–498.

Robinson, T.F., and Winegrad, S. (1977). Variation of thin filament length in heart muscle. *Nature* **267**, 74–75.

Robinson, T.F., and Winegrad, S. (1979). The measurement and dynamic implications of thin filament lengths in heart muscle. *Journal of Physiology* **286**, 607–619.

Rodriguez, E.K., Hunter, W.C., Royce, M.J., Leppo, M.K., Douglas, A.S., and Weisman, H.F. (1992). A method to reconstruct myocardial sarcomere lengths and orientations at transmural sites in beating canine hearts. *American Journal of Physiology* **263**(32), H293–H306.

Sagawa, K., Maughan, L., Suga, H., and Sunagawa, K. (1988). *Cardiac Contraction and the Pressure-volume Relationship.* Oxford University Press, Oxford, England, pp. 171–231.

Sagawa, K., Suga, H., Shoukas, A.A., and Bakalar, K.M. (1977). End-systolic pressure-volume ratio: A new index of ventricular contractility. *American Journal of Cardiology* **40**, 748–753.

Schouten, V.J.A., Bucx, J.J.J., de Tombe, P.P., and ter Keurs, H.E.D.J. (1990). Sarcolemma, sarcoplasmic reticulum, and sarcomeres as limiting factors in force production in rat heart. *Circulation Research* **67**, 913–922.

Suga, H., Yamada, O., Goto, Y., and Igarashi, Y. (1986). Peak isovolumic pressure-volume relation of puppy left ventricle. *American Journal of Physiology* **250**, 167–172.

Sweitzer, N.K., and Moss, R.L. (1990). The effect of altered temperature on Ca^{++}-sensitive force in permeabilized myocardium and skeletal muscle. Evidence for force development of thin filament activation. *Journal of General Physiology* **96**, 1221–1245.

ter Keurs, H.E.D.J. (1993). Regulation of cardiac contraction in the normal and failing heart: Cellular aspects. *Canadian Journal of Cardiology* **9F**, 1–11.

ter Keurs, H.E.D.J., Bucx, J.J.J., de Tombe, P.P., Backx, P.H.M., and Iwazumi, T. (1988). The effects of sarcomere length on force and velocity of shortening in cardiac muscle. In: *Molecular Mechanisms of Muscle Contraction.* Sugi, H., and Pollack, G.H. (eds.), Plenum Press (a division of Plenum Publishing Corporation), New York, pp. 581–593.

ter Keurs, H.E.D.J., Luff, A.R., and Luff, S.E. (1984). Force-sarcomere length relation and filament length in rat extensor digitorum muscle. *Advances in Experimental Medicine and Biology* **170**, 511–525.

ter Keurs, H.E.D.J., Rijnsburger, W.H., van Heuningen, R.V., and Nagelsmit, M.J. (1980). Tension development and sarcomere length in rat cardiac trabeculae. Evidence of length-dependent activation. *Circulation Research* **46**, 703–714.

Wang, Y-P., and Fuchs, F. (1994). Length, force, and Ca^{2+}-troponin C affinity in cardiac and slow skeletal muscle. *American Journal of Physiology* **266**, C1077–C1082.

Wang, Y-P., and Fuchs, F. (1995). Osmotic compression of skinned cardiac and skeletal muscle bundles: Effects on force generation, Ca^{2+} sensitivity and Ca^{2+} binding. *Journal of Molecular and Cellular Cardiology* **27**, 1235–1244.

Wannenburg, T., Schulman, S.P., and Burkhoff, D. (1992). End systolic pressure-volume and MVO2-pressure-volume area relations of isolated rat hearts. *American Journal of Physiology* **262**, H1287–H1293.

Zot, A.S., and Potter, J.D. (1987). Structural aspects of troponin-tropomyosin regulation of skeletal muscle contraction. *Annual Review of Biophysics and Biophysical Chemistry* **16**, 535–559.

Zot, A.S., and Potter, J.D. (1989). Reciprocal coupling between troponin C and myosin crossbridge attachment. *Biochemistry* **28**, 6751–6756.

5 Length Dependence of Force Production and Ca^{2+} Sensitivity in Skeletal Muscle

DILSON E. RASSIER
Health Sciences Center, University of Vale do Rio dos Sinos, São Leopoldo, Brazil

WALTER HERZOG
Human Performance Laboratory, University of Calgary, Calgary, Alberta, Canada

INTRODUCTION

The cross-bridge model of muscle contraction predicts that force is proportional to the number of myosin cross-bridges attached to the actin filaments (Huxley, 1957; Huxley and Simmons, 1971). In this model, myosin-actin interaction is the key event during muscle contraction and force generation. This interaction is possible due to calcium (Ca^{2+}) binding to troponin C (TnC), a step that is directly responsible for muscle activation. Muscle activation is defined here as the amount of TnC with bound Ca^{2+}; when a Ca^{2+}/TnC complex is formed, the actin-active sites are available for myosin attachment, initiating the molecular events of muscle contraction and force production.

Force production in skeletal muscle is length-dependent. In a classic study of muscle physiology, Gordon et al. (1966) investigated the effects of sarcomere length on force production in a mid-section of isolated frog muscle fibres. They observed that there was an optimal sarcomere length for force development, which was related to the optimal overlap of thick and thin filaments. At lengths greater than optimal, force was observed to decrease linearly with increases in sarcomere length. This relation has been referred to as 'the length-tension relation', and it provides strong evidence in favour of the cross-bridge theory of muscle contraction.

Besides influencing the degree of filament overlap, it has been recognized that sarcomere length also affects Ca^{2+} sensitivity of the myofilaments. In this chapter, Ca^{2+} sensitivity will be referred to as the response of the contractile apparatus to a given level of Ca^{2+} concentration. Length-dependent changes in Ca^{2+} sensitivity imply that changes in muscle length alter the effectiveness of the filaments to produce force at a given level of Ca^{2+} concentration.

Skeletal Muscle Mechanics: From Mechanisms to Function. Edited by W. Herzog.

Although length dependence of Ca^{2+} sensitivity in skeletal muscle has been proposed for many years (Endo, 1972; Moisescu and Thieleczek, 1979), the mechanisms underlying this phenomenon are not known. This is probably the reason why this specific feature of muscle contraction has not been considered in most models of force production. Ignoring the effects of length dependence of Ca^{2+} sensitivity may have a profound impact on force predictions at submaximal levels of muscle stimulation.

The length dependence of Ca^{2+} sensitivity will be the focus of this chapter. In the first part, a brief description of concepts associated with Ca^{2+} sensitivity and length dependence of Ca^{2+} sensitivity will be made. In the second part, experimental evidence for the length dependence of Ca^{2+} sensitivity in skinned and intact muscles will be discussed. In the third part, current problems associated with the length dependence of Ca^{2+} sensitivity will be considered, specifically the origin of its intrinsic mechanism will be discussed. In the fourth part, selected conclusions are presented, together with suggestions for future research directions.

THE FORCE/Ca^{2+} RELATION AND INVESTIGATIONS OF Ca^{2+} SENSITIVITY

Muscle activation, directly responsible for force generation, is associated with Ca^{2+} binding to TnC. Therefore, if the intracellular Ca^{2+} concentration ($[Ca^{2+}]_i$) increases, force also increases. This relation between force and Ca^{2+} concentration is well described in the literature and is illustrated in Figure 5.1. The force/pCa^{2+} relation has a sigmoid shape, with a plateau of force development at a given level of Ca^{2+} concentration. The plateau value of $[Ca^{2+}]_i$ has been measured in different preparations, and may vary from 1 to 10 μM during a tetanic contraction (Westerblad and Allen, 1996). The increase in $[Ca^{2+}]_i$ is transient, and is associated with Ca^{2+} release from the sarcoplasmic reticulum after the action potential spreads into the myofibrils.

The force/pCa^{2+} relation has been investigated in preparations that use intact (Balnave and Allen, 1996) and skinned skeletal muscle fibres (Endo, 1972; Stephenson and Williams, 1982). In intact fibres, $[Ca^{2+}]_i$ has typically been measured using fluorescence probes that are placed inside the muscle fibres. When excited with specific wavelengths, these probes respond to different $[Ca^{2+}]_i$ concentrations (Grynkiewicz et al., 1985). Since the force response is graded according to the stimulation frequency, it is possible to measure $[Ca^{2+}]_i$ at different levels of force (Westerblad and Allen, 1991).

The force/pCa^{2+} relation has been studied in much detail in skinned muscle fibre preparations. In this preparation, the muscle fibre membrane is mechanically or chemically removed, but the contractile proteins are maintained intact. Therefore, it is possible to control the Ca^{2+} concentration, induce different levels of force, and correlate Ca^{2+} with force production. Naturally, the actual amount of Ca^{2+} bound to TnC may differ from the media Ca^{2+} concentration, as there are Ca^{2+} buffers

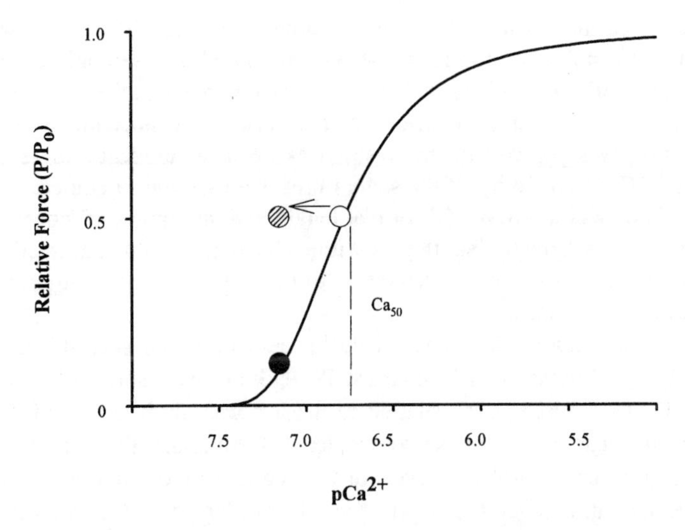

Figure 5.1. The force/pCa^{2+} relation in skeletal muscle. Force (P) is expressed relative to maximal force (P_o), and Ca^{2+} is expressed as pCa^{2+} ($-\log_{10}(Ca^{2+})$). The figure also shows the amount of Ca^{2+} needed to produce half-maximal tension (Ca_{50}). At a low Ca^{2+} concentration (solid circle), the force produced is minimal. When Ca^{2+} concentration increases along the X-axis (from solid to open circle), there is an increase in force. An increased Ca^{2+} sensitivity of the filaments causes a leftward shift on the force/pCa^{2+} relation (from open to striped circle), and a decrease in Ca_{50}.

inside the cell, like parvalbumin, calmodulin, and the myosin light chains, among others, that might affect the relationship between Ca^{2+} in the medium and Ca^{2+} bound to TnC.

The force/pCa^{2+} relation may be used to investigate changes in Ca^{2+} sensitivity. Ca^{2+} sensitivity is defined here by the position of the force/pCa^{2+} relation along the Ca^{2+} axis. Generally, Ca^{2+} sensitivity is characterized by the amount of Ca^{2+} needed to produce half-maximal tension (Ca_{50}). Any situation that increases Ca^{2+} sensitivity causes a leftward shift of the force/pCa^{2+} relation (decreasing the Ca_{50}); therefore less Ca^{2+} is needed to produce a given amount of force. Variations in Ca^{2+} sensitivity can be observed in intact and skinned fibres in a variety of situations, including twitch potentiation (Persechini et al., 1985), fatigue (Westerblad and Allen, 1991), changes in pH (Martyn and Gordon, 1999), and sarcomere/muscle length (Endo, 1972; Martyn and Gordon, 1999; Stephenson and Williams, 1982), among others. In this chapter, the changes induced by sarcomere/muscle length will be considered in detail.

LENGTH DEPENDENCE OF Ca^{2+} SENSITIVITY

The effects of length on Ca^{2+} sensitivity were first reported by Endo (1972; 1973). Using skinned, single muscle cells of the toad *Xenopus laevis*, the author (Endo,

1973) observed that, at low Ca^{2+} concentrations (less than 10^{-6} Ca^{2+}), isometric force was maximal when the response was measured at a sarcomere length of ~2.9 μm. Since this length is associated with a region in which the overlap of thick and thin filaments is decreasing (Gordon et al., 1966), and since the external Ca^{2+} concentration was controlled, this result was taken as evidence of length dependence of Ca^{2+} sensitivity. If the same fibres were set at a sarcomere length of ~2.3 μm, force was not maximal for that length and the same Ca^{2+} concentration as observed at a sarcomere length of 2.9 μm. This observation suggested that the contractile apparatus responds differently to the same Ca^{2+} concentration at different sarcomere lengths.

In several other studies, the same conclusion as Endo's was reached (Martyn and Gordon, 1999; Moisescu and Thieleczek, 1979; Stephenson and Williams, 1982), i.e., a shift of the force/pCa^{2+} relation to the left was observed when the force response was measured at long sarcomere lengths. Naturally, the increase in force was more pronounced at lower levels of Ca^{2+} concentration; in other words, the higher the activation level, the smaller the effect of length on Ca^{2+} sensitivity.

There is also evidence of a length dependence of Ca^{2+} sensitivity in intact skeletal muscles. It has been observed that muscles that contract at submaximal levels of stimulation respond differently to changes in muscle length than muscles that are maximally stimulated (Balnave and Allen, 1996; Close, 1972; Rack and Westbury, 1969; Rassier et al., 1998). At submaximal activation, peak tension is developed at a length longer than the length associated with maximal filament overlap (Close, 1972; Rack and Westbury, 1969; Rassier et al., 1998) (Figure 5.2).

The observation that force during submaximal contractions is higher beyond optimal filament overlap than at optimal filament overlap (in contrast to results for tetanic contractions) has been referred to as length dependence of muscle activation. However, here, the concept of muscle activation refers to muscle stimulation frequency. Since muscle activation was defined above as the Ca^{2+} bound to TnC, we will refer to this property as length dependence of force production during submaximal stimulation.

Using an elegant preparation in which single intact cells were isolated from the mouse flexor digitorum brevis muscle, Balnave and Allen (1996) observed that the force/pCa^{2+} relation is shifted to the left when sarcomere length is increased. Using the Ca^{2+} indicator Indo-2, they observed that $[Ca^{2+}]_i$ did not change as a function of muscle length for a given stimulation frequency. This result agrees with those obtained in studies using skinned fibre preparations cited above (where Ca^{2+} concentration is controlled). Therefore, it appears that there is a length dependence of Ca^{2+} sensitivity in skinned and intact cells, and that this Ca^{2+} sensitivity is independent of the intracellular Ca^{2+} concentration.

At present, it is difficult to investigate the length dependence of Ca^{2+} sensitivity in whole muscle. An approach that may allow this type of study is to use drugs that can enhance or decrease the length dependence of Ca^{2+} sensitivity. We have used caffeine, a drug that increases Ca^{2+} sensitivity at short sarcomere lengths (De Beer

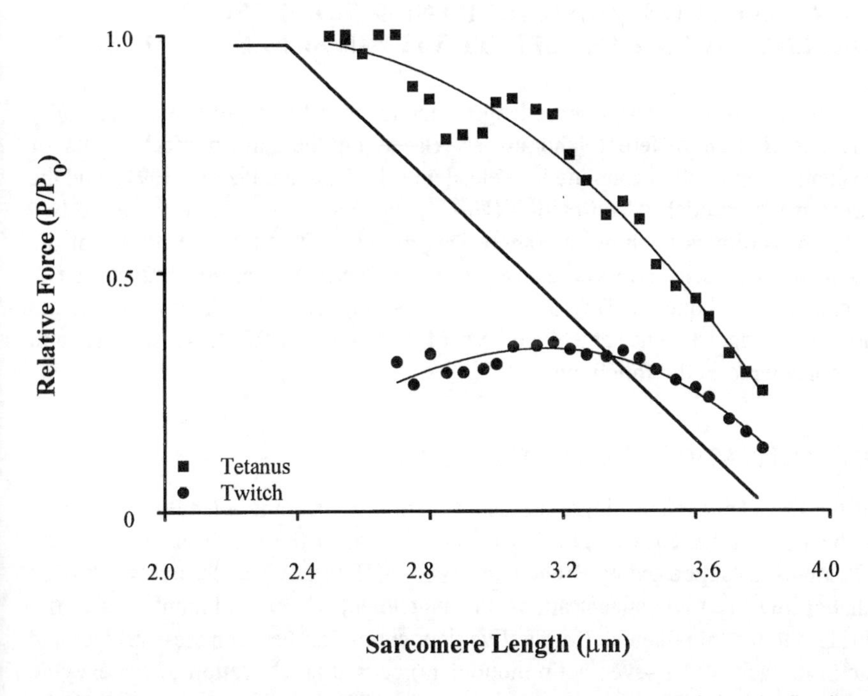

Figure 5.2. The force-sarcomere length relationship for twitch and tetanic fixed-end contractions of an intact fibre bundle of the mouse extensor digitorum muscle. Data are from one representative experiment conducted in our laboratory at 22°C, in which the sarcomere length was measured by laser diffraction. The solid thick line represents the theoretical force-length relation (Gordon et al., 1966), with the length of the filaments taken from the rat, with a plateau extended from 2.26 μm to 2.43 μm, and a zero intercept at 3.79 μm (ter Keurs et al., 1984). The thin lines represent third order polynomial least squares fits. Note that the peak force during twitch contractions is obtained at a sarcomere length that is longer than the length where maximal force is obtained during tetanic contraction.

et al., 1988a), to investigate the force-length relation during twitch contractions measured *in situ* in the whole isolated rat gastrocnemius (Rassier et al., 1998). As expected, before caffeine treatment, the force-length relation obtained using twitch contractions reached a peak force at a length longer than that obtained using tetanic contractions. However, after caffeine treatment, twitch contractions produced the peak force at the same length as the tetanic contractions, indicating that force development at short lengths was increased presumably because Ca^{2+} sensitivity was increased by caffeine. Therefore, it appears that the length dependence of Ca^{2+} sensitivity is also present in whole muscle.

In conclusion, length-dependent Ca^{2+} sensitivity seems to be a phenomenon that occurs in skinned fibres, intact isolated fibres, and whole muscle, and so may have important functional implications, and should be considered in future models of muscle contraction and force production.

CURRENT PROBLEMS: MECHANISMS OF LENGTH DEPENDENCE OF Ca^{2+} SENSITIVITY IN SKELETAL MUSCLE

The mechanisms underlying the length dependence of Ca^{2+} sensitivity are under investigation in different laboratories. Based on the known mechanisms of excitation-contraction coupling in skeletal muscle (Rios and Pizarro, 1991), and the cross-bridge model of contraction (Huxley and Simmons, 1971), there are two main possibilities for how sarcomere length could affect Ca^{2+} sensitivity of the filaments: Ca^{2+} sensitivity could be affected via an increased muscle activation, i.e., increased Ca^{2+} bound to TnC, and/or Ca^{2+} sensitivity could be affected by changes in cross-bridges kinetics at a given level of Ca^{2+} activation. Both possibilities will be considered in the following.

Ca^{2+} ACTIVATION AND TnC AFFINITY

If the Ca^{2+}/TnC affinity increases as a function of sarcomere length, one would expect a simultaneous length-dependent increase in the Ca^{2+} sensitivity of the filaments, as suggested by Bremel and Weber (1972). These researchers showed that at low MgATP concentrations, the attachment of a critical number of cross-bridges to the thin filament (four to five rigor complexes between cross-bridges and actin in each of the seven actin monomers) caused an activation of the enzyme actin-activated ATPase. This enzyme causes ATP hydrolysis, a key step in the cross-bridge cycle (Eisenberg and Hill, 1985). Furthermore, an increase in the TnC affinity for Ca^{2+} was observed under the same condition. Therefore, the binding of some cross-bridges co-operatively activated the thin filament by an increased ATPase activity and an increased TnC affinity for Ca^{2+}. If these results can be extrapolated to physiological situations, one would expect that a fibre on the ascending limb of the force-length relation with a large number of attached cross-bridges may produce an increased ATPase activity and TnC affinity for Ca^{2+} (Grabarek et al., 1983; Güth and Potter, 1987). In turn, this phenomenon could be responsible for the length dependence of Ca^{2+} sensitivity.

Results on cardiac muscle, working only on the ascending limb of the force-length relation, support the hypothesis of an increased TnC affinity for Ca^{2+} at increased sarcomere lengths (Hofmann and Fuchs, 1987a, 1987b). Also, studies that use protein replacement techniques have suggested a strong role for the TnC complex in regulating the length dependence of Ca^{2+} sensitivity in skeletal muscle (Babu et al., 1987; Babu et al., 1988; Gulati et al., 1988). However, these results are controversial, as others have reached opposite conclusions using what appears to be the same replacement techniques (Moss et al., 1991; Moss, 1992).

The hypothesis that TnC affinity for Ca^{2+} changes as a result of cross-bridge attachment has one serious problem for skeletal muscle: on the descending limb of the force-length relation, the number of cross-bridges attached to actin decreases as myosin/actin overlap is reduced, but the length-dependent Ca^{2+} sensitivity is very pronounced at these fibre lengths. In fact, a series of studies has failed to detect length-dependent changes in TnC affinity in skeletal muscle. For example, using markers to detect Ca^{2+} binding to TnC (Fuchs and Wang, 1991; Wang and Fuchs, 1994, 1995), or variations in Ca^{2+} concentrations following flash photolysis of caged Ca^{2+} (Patel et al., 1997), it was shown that the level of Ca^{2+} binding to TnC is unchanged with changes in muscle length.

An interesting study aimed at detecting possible changes in Ca^{2+} binding to TnC as a result of sarcomere length was performed by Patel et al. (1997). These authors measured simultaneously changes in sarcomere length, tension, and $[Ca^{2+}]_i$ signal following flash photolysis of caged Ca^{2+} (which causes Ca^{2+} liberation into the myoplasm) in normal fibres and in fibres that had the TnC complex removed. All measurements were made in skinned rabbit psoas muscles loaded with Ca^{2+} fluorophore Fluo-3. Using this technique, Patel et al. (1997) could 'instantaneously' increase $[Ca^{2+}]_i$. By comparing the amount of $[Ca^{2+}]_i$ in regular fibres with the amount of $[Ca^{2+}]_i$ in fibres that had the TnC complex removed, they could evaluate the amount of Ca^{2+} binding to TnC. They observed that Ca^{2+} binding to TnC was not different whether the response was measured at short (1.97 μm) or long (2.51 μm) sarcomere lengths, even though the tension response was 50% greater at a greater sarcomere length. This result indicates that the affinity of TnC for Ca^{2+} is unchanged with sarcomere length, and that the difference in Ca^{2+} sensitivity is not due to the regulatory protein.

When investigating the Ca^{2+}/TnC affinity of intact muscles, quick release of active muscle fibres (as described by Huxley and Simmons, 1971) has been performed to decrease force quickly. This approach is used to identify Ca^{2+} transients as a result of muscle shortening, assuming that any extra Ca^{2+} observed in the myoplasm as a result of quick shortening has been released from TnC because of a decreased affinity between Ca^{2+} and TnC. Although shortening of barnacle fibres (Gordon and Ridgway, 1987; Ridgway and Gordon, 1984) and cardiac muscles (Allen and Kentish, 1988; Allen and Kurihara, 1982) shows extra Ca^{2+} following quick shortening, investigations performed with skeletal muscle have failed to show extra Ca^{2+} after shortening, or even found a decrease in Ca^{2+} (Allen, 1977; Taylor et al., 1982) following muscle shortening. These latter results are in agreement with the above-mentioned skinned fibre studies in which it was not possible to detect changes in TnC affinity for Ca^{2+} as a function of sarcomere length.

In summary, the length dependence of Ca^{2+} sensitivity in skinned and intact skeletal muscle fibres does not seem to be related to changes in Ca^{2+}/TnC affinity.

INCREASED RATE OF CROSS-BRIDGE CYCLING

If we assume that muscle activation via Ca^{2+} binding to TnC is not affected by sarcomere length, and that there is no length dependence of $[Ca^{2+}]_i$, then the length dependence of Ca^{2+} sensitivity is likely related to changes in cross-bridge kinetics at a given level of Ca^{2+} activation.

In order to investigate cross-bridge kinetics, it is necessary to mention some characteristics of the models used to describe muscle contraction. The detailed steps of cross-bridge attachment and detachment are not fully explained by the original two-state cross-bridge model described by Huxley (1957). Current models of muscle contraction assume many more than two cross-bridge steps involved in force generation (Eisenberg and Greene, 1980; Eisenberg and Hill, 1985).

Many models assume two basic attached states of cross-bridges bound to actin: weakly and strongly attached states (Lymn and Taylor, 1971; Stein et al., 1979). The weak binding state of cross-bridges is characterized by a low affinity (Stein et al., 1979) but rapid kinetics of attachment and detachment from and to actin (Lymn and Taylor, 1971; Stein et al., 1979). These weakly binding cross-bridges have a low ability to regulate actin and generate force (Chalovich et al., 1983). However, the weak binding state precedes the strong binding state, with a high affinity for actin association (Greene and Eisenberg, 1980b), slower actin binding kinetics, but with the ability to regulate actin activation and force generation (Greene and Eisenberg, 1980a) (Figure 5.3).

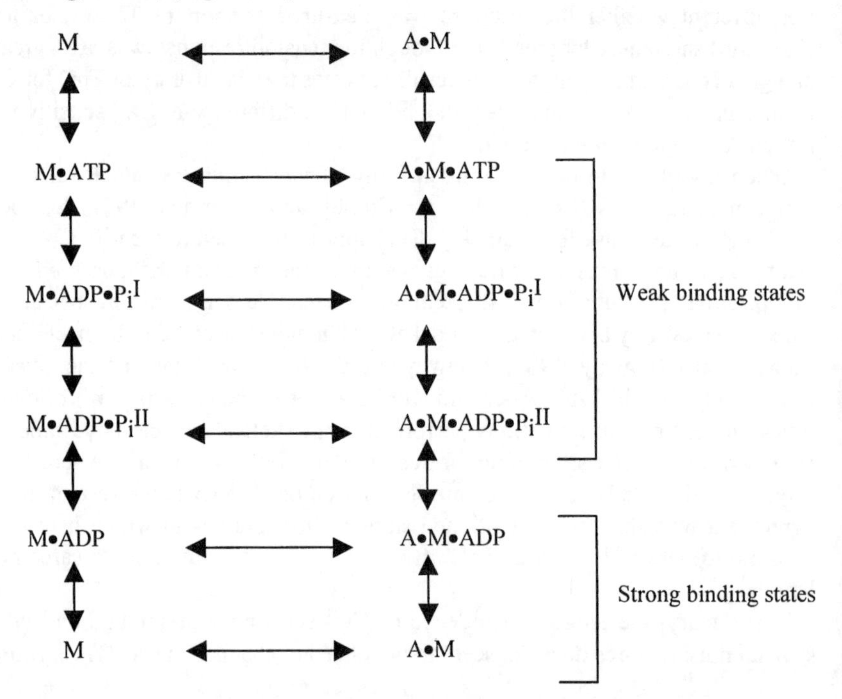

Figure 5.3. An example of a cross-bridge model with multiple steps (redrawn from Stein et al., 1979), where M=myosin, A=actin, ATP=adenosine triphosphate, ADP=adenosine diphosphate, and P_i =inorganic phosphate. The steps involving reactions between P_i^I and P_i^{II} are limited by similar rate constants, and the release of P_i is involved in the strong binding conformation of cross-bridges. Note that the weak-binding cross-bridges necessarily precede the strong-binding cross-bridges.

Based on cross-bridge models with multiple attachment states, an increased muscle force at a given time (and a given Ca^{2+} activation) can be obtained by an increase in the number of attached cross-bridges and/or by an increase in the average force produced by the attached cross-bridges. An increase in the number of attached cross-bridges can be obtained by an increased rate of cross-bridge attachment, and/or a decreased rate of cross-bridge detachment. An increase in the average force produced by the attached cross-bridges can be achieved by an increased transition from weak states to strong binding states. Changes in the potential energy for attachment and the x-distance for attachment along the actin axis (Huxley, 1957) could also change the amount of force produced by the muscle, but these options will not be considered, as they appear less likely to change as a function of sarcomere length.

An increased rate of cross-bridge attachment may be associated with a) an increased displacement of the cross-bridges away from the thick filament (changing the attachment angle between the myosin cross-bridge and the thin filament) (Figure 5.4), and/or b) a decreased distance between actin and myosin (Schoenberg 1980a, 1980b) (Figure 5.5). These possibilities will be considered in the next paragraphs. It is recognized that the angular speed of the attached cross-bridge head could also change the probability of cross-bridge interaction with actin, but we will ignore this possibility and others here.

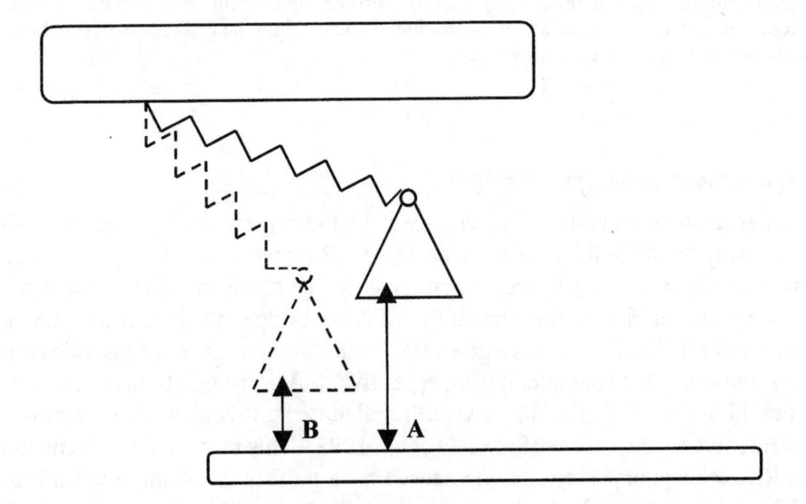

Figure 5.4. Schematic representation of a thick filament, cross-bridge, and thin filament, before (A) and after (B) an increase in sarcomere length. In this model, stretching the sarcomere causes a displacement of the cross-bridge away from the thick filament, decreasing the lateral distance between the myosin cross-bridge and actin, therefore increasing the probability of cross-bridge attachment. For clarity, the actual differences between filament and cross-bridge diameters, and the distance between the filaments, are not represented in this figure.

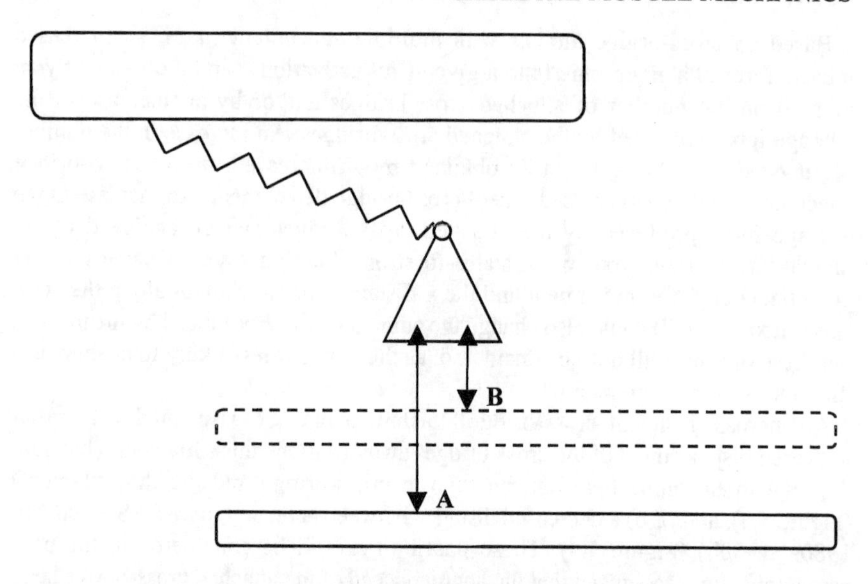

Figure 5.5. Schematic representation of a thick filament, cross-bridge, and thin filament, before (A) and after (B) an increase in sarcomere length. In this model, stretching the sarcomere causes a decrease in the lateral distance between thick and thin filaments, therefore increasing the probability of cross-bridge attachment. For clarity, the actual differences between filament and cross-bridge diameters, and the distance between the filaments, are not represented in this figure.

Displacement of the cross-bridges

In order to engage in a force-generating state, the myosin cross-bridge must first move away from the thick filament backbone (Rayment et al., 1993). If changing sarcomere lengths alters the release of cross-bridge connections from the myosin backbone, it could alter the probability of cross-bridge attachment to actin, and thus, the force development at a given Ca^{2+} concentration. Cross-bridge release has been proposed, for example, during repetitive muscle stimulation, a situation in which phosphorylation of the myosin regulatory light chains (RLC) causes an increase in Ca^{2+} sensitivity (Sweeney et al., 1993). This increased Ca^{2+} sensitivity via RLC phosphorylation may be caused by a movement of the myosin cross-bridge away from the thick filament, as shown in X-ray diffraction studies (Levine et al., 1991; Levine et al., 1996). This movement may be caused by a change in the charge potential present in the RLC/myosin backbone connection.

There is no report in the literature showing a greater (or smaller) displacement of the cross-bridges away from the thick filament as a function of sarcomere length. However, this possibility should be addressed. It may be, for example, that changes in the interfilament space alter the electrostatic potential at the surface of each filament, which in turn may alter the position of the myosin cross-bridges, affecting

Ca^{2+} sensitivity. It has been shown that at physiological pH, the myofilaments have a net negative charge (Elliot and Bartels, 1982; Stephenson et al., 1981), resulting in the development of a Donnan equilibrium. A decrease in lattice space increases the lattice charge density (Aldoroty et al., 1987), and likely alters the degree of superposition of electrostatic fields from adjacent filaments, as suggested by Stephenson and Wendt (1984). That, in turn, could alter cross-bridge kinetics or the interaction between myosin and actin (Eisenberg and Hill, 1985; Krasner and Maughan, 1984).

There are some indications that the above idea may be correct. Martyn and Gordon (1999) performed a series of experiments in which lattice spacing of rabbit skinned fibres was altered by changing sarcomere length and osmotic compression (see details below). Further, myofilament charge and electrostatic interactions were investigated by changing the pH and the ionic strength of the media solution. They observed that the influence of sarcomere length on Ca^{2+} sensitivity decreased as pH was lowered, and increased as ionic strength was lowered. These changes were observed even when sarcomere length was increased, showing that variations in myofilament charges (induced by pH and ionic strength) affect the length-dependent Ca^{2+} sensitivity. This hypothesis needs to be fully addressed.

Proximity of the filaments

A greater proximity of the filaments associated with an increased Ca^{2+} sensitivity at long sarcomere lengths seems a feasible mechanism to explain the length dependence of Ca^{2+} sensitivity. There are studies showing that fibre diameter (Godt and Maughan, 1981; Martyn and Gordon, 1999; Maughan and Godt, 1979) and interfilament spacing (Matsubara and Elliot, 1972; Rome, 1968) decrease when skeletal muscle is stretched. This reduction in interfilament spacing should reduce the lateral distance (not the x-distance, which would reduce average force per cross-bridge [Huxley, 1957]) for myosin cross-bridge attachment to actin.

Godt and Maughan (1981) investigated the effects of a decreased lateral filament spacing at constant sarcomere lengths on Ca^{2+} sensitivity. These authors produced osmotic shrinkage of the lattice spacing by adding the high-molecular-weight polymer (Dextran T-5000, polyvinylpyrrolidone-40) to the bathing solution in which the fibres were stabilized. This substance does not penetrate into the inter-filament space. When using low concentrations of the polymer (5%), the authors observed a large increase in myofibrillar Ca^{2+} sensitivity, as measured by a leftward shift in the force/pCa^{2+} relation, which was accompanied by a decrease in fibre diameter (96±7%). This result suggests that the increased Ca^{2+} sensitivity may be the result of a decrease in filament lattice spacing.

Other investigators (De Beer et al., 1988b; Fuchs and Wang, 1991; Martyn and Gordon, 1999; Moss et al., 1983; Stienen et al., 1985; Wang and Fuchs, 1994, 1995) have observed the same phenomenon. It is important to mention that although osmotic compression increases Ca^{2+} sensitivity and force in skinned skeletal muscle, it does not affect Ca^{2+} binding to TnC (Fuchs and Wang, 1991; Wang and

Fuchs, 1994, 1995); therefore the results are consistent with other studies showing no relation between Ca^{2+} sensitivity and Ca^{2+}/TnC affinity.

It appears reasonable to assume that a reduction in lattice spacing results in an increased probability of cross-bridge attachment. If the probability of cross-bridge attachment is increased, the force is also increased at a given level of Ca^{2+} activation, increasing the Ca^{2+} sensitivity. Of course, this interpretation of the results assumes that the proposed increase in the rate of cross-bridge attachment with decreased lattice spacing is not affected by a corresponding change in the rate of cross-bridge detachment.

The detachment rate may also contribute to the force increase due to length-dependent changes in Ca^{2+} sensitivity. Zhao and Kawai (1993) investigated the effects of lattice compression on Ca^{2+} sensitivity, and analyzed their data using a six-step theoretical cross-bridge cycle. It was assumed that the cycle included a transition from a weak-binding to a strong-binding cross-bridge attachment, and that this transition was necessary for force generation. After compression of skinned fibres with 3.6%–6.3% Dextran, the authors observed that the forward rate of strong-binding cross-bridges did not change, but the detachment rate was reduced. A decreased detachment rate, together with constant attachment rates, results in an increased number of attached cross-bridges in a steady-state configuration (Huxley, 1957; Huxley and Simmons, 1971), therefore an increased force. The authors interpreted their results using a model in which, at low levels of compression, the neck region of the cross-bridges is pressed against the thick filament, and the thermal motion of the myosin head is limited, resulting in a smaller detachment rate.

Overall, the results presented in this section suggest that the length dependence of Ca^{2+} sensitivity is associated with changes in sarcomere length *per se*, and not with changes in the muscle activation process. The changes in sarcomere length are likely associated with changes in the cross-bridge kinetics, increasing the probability of cross-bridge interaction with actin filaments.

CONCLUSIONS AND FUTURE DIRECTIONS

There is evidence that muscle length alters force development by variations in filament overlap and changes in Ca^{2+} sensitivity of the filaments. At submaximal levels of muscle activation, increasing muscle length enhances Ca^{2+} sensitivity of the myofilaments. This result is observed as a leftward shift in the force/pCa^{2+} relationship. This means that at a given level of Ca^{2+} activation, more force is produced when the muscle is stretched. Although not completely solved yet, it appears that length dependence of Ca^{2+} sensitivity may be related to an increased probability of cross-bridge attachment.

These conclusions are based on several investigations in which the effects of increasing muscle length on Ca^{2+} sensitivity, muscle fibre diameter, and force development were studied. However, the effects of myofilament spacing on cross-bridge kinetics should be investigated in more detail. Using X-ray diffraction, it might be possible to detect whether cross-bridges become disordered when sarcomere length is increased. If cross-bridges are moved away from the myosin backbone at increasing sarcomere lengths, they may become more mobile to interact with actin, which in turn could indicate a greater probability of attachment of cross-bridges to actin filaments.

Mathematical modelling of cross-bridge kinetics could be useful to investigate the effects of sarcomere length on cross-bridge attachment/detachment rates. This work is insofar needed, as current models of force production have largely neglected this specific property of muscle contraction. By changing fibre diameter and thus lattice spacing, and leaving the rates of attachment and detachment as a function of the x-distance (Huxley, 1957; Huxley and Simmons, 1971) exclusively, further insights into the mechanisms of length-dependence of Ca^{2+} sensitivity could be obtained.

REFERENCES

Aldoroty, R.A., Garty, N.B., and April, E.W. (1987). Donnan potentials from striated muscle liquid crystals: Lattice spacing dependence. *Biophysical Journal* **51**, 371–381.

Allen, D.G. (1977). Shortening of tetanized skeletal muscle causes a fall of intracellular calcium concentration. *Journal of Physiology* **275**, 63P.

Allen, D.G., and Kentish, J.C. (1988). Calcium concentration in the myoplasm of skinned ferret ventricular muscle following changes in muscle length. *Journal of Physiology* **407**, 489–503.

Allen, D.G., and Kurihara, S. (1982). The effects of muscle length on intracellular calcium transients in mammalian cardiac muscle. *Journal of Physiology* **327**, 79–94.

Babu, A., Scordilis, S., Sonnenblick, E., and Gulati, J. (1987). The control of myocardial contraction with skeletal fast muscle troponin C. *Journal of Biological Chemistry* **262**, 5815–5822.

Babu, A., Sonnenblick, E., and Gulati, J. (1988). Molecular basis for the influence of muscle length on myocardial performance. *Science* **240**, 74–76.

Balnave, C.D., and Allen, D.G. (1996). The effects of muscle length on intracellular calcium and force in single fibres from the mouse skeletal muscle. *Journal of Physiology* **492**, 705–713.

Bremel, R.D., and Weber, A. (1972). Cooperation within actin filament in vertebrate skeletal muscle. *Nature (New Biology)* **238**, 97–101.

Chalovich, J.M., Greene, L.E., and Eisenberg, E. (1983). Crosslinked myosin subfragment 1: A stable analogue of the subfragment ATP complex. *Proceedings of the National Academy of Sciences USA* **80**, 4909–4913.

Close, R.I. (1972). The relations between sarcomere length and characteristics of isometric twitch contractions of frog sartorius muscle. *Journal of Physiology* **220**, 745–762.

De Beer, E.L., Grundeman, R.L.F., Wilhelm, A.J., Caljouw, C.J., Klepper, D., and Schiereck, P. (1988a). Caffeine suppresses length dependence of calcium-activation curves in skinned skeletal and cardiac muscle preparations of the rabbit. *American Journal of Physiology* **254**, C491–C497.

De Beer, E.L., Grundeman, R.L.F., Wilhelm, A.J., Van Den Berg, C., Caljouw, C.J., Klepper, D., and Schiereck, P. (1988b). Effect of sarcomere length and filament lattice spacing on force development in skinned cardiac and skeletal muscle preparations from the rabbit. *Basic Research in Cardiology* **83**, 410–423.

Eisenberg, E., and Greene, L.E. (1980). The relation of muscle biochemistry to muscle physiology. *Annual Review of Physiology* **42**, 293–309.

Eisenberg, E., and Hill, T.L. (1985). Muscle contraction and free energy transduction in biological systems. *Science* **227**, 999–1006.

Elliot, G.F., and Bartels, E.M. (1982). Donnan potential measurements in extended hexagonal polyelectrolyte gels such as muscle. *Biophysical Journal* **38**, 195–199.

Endo, M. (1972). Stretch-induced increase in activation of skinned fibres by calcium. *Nature (New Biology)* **237**, 211–213.

Endo, M. (1973). Length dependence of activation of skinned muscle fibers by calcium. *Cold Spring Harbour Symposia for Quantitative Biology* **37**, 505–510.

Fuchs, F., and Wang, Y. (1991). Force, length, and Ca^{2+}-troponin C affinity in skeletal muscle. *American Journal of Physiology* **261**, C787–C792.

Godt, R.E., and Maughan, D.W. (1981). Influence of osmotic compression on calcium activation and tension in skinned muscle fibers of the rabbit. *Pflügers Archiv* **391**, 334–337.

Gordon, A.M., Huxley, A.F., and Julian, F.J. (1966). The variation in isometric tension with sarcomere length in vertebrate muscle fibres. *Journal of Physiology* **184**, 170–192.

Gordon, A.M., and Ridgway, E.B. (1987). Extra calcium on shortening in barnacle muscle. Is the decrease in calcium binding related to decreased cross-bridge attachment, force, or length? *Journal of General Physiology* **90**, 321–340.

Grabarek, Z., Grabarek, J., Leavis, P.C., and Gergely, J. (1983). Cooperative binding to the Ca^{2+}-specific sites of troponin C in regulated actin and actomyosin. *Journal of Biological Chemistry* **258**, 14098–14102.

Greene, L.E., and Eisenberg, E. (1980a). Cooperative binding of myosin subfragment-1 to the actin-troponin-tropomyosin complex. *Proceedings of the National Academy of Sciences USA* **77**, 2616–2620.

Greene, L.E., and Eisenberg, E. (1980b). Dissociation of the actin subfragment 1 complex by adenyl-5′-yl imidodiphosphate, ADP, and PPi. *Journal of Biological Chemistry* **255**, 543–548.

Grynkiewicz, G., Poenie, M., and Tsien, R.Y. (1985). A new generation of fluorescence calcium indicators with greatly improved fluorescence properties. *Journal of Biological Chemistry* **260**, 3440–3450.

Gulati, J., Scordilis, S., and Babu, A. (1988). Effect of troponin C on the cooperativity in Ca^{2+} activation of cardiac muscle. *FEBS Letters* **236**, 441–444.

Güth, K., and Potter, J.D. (1987). Effect of rigor and cycling cross-bridges on the structure of troponin C and on the Ca^{2+} affinity of the Ca^{2+}-specific regulatory sites in skinned rabbit psoas fibers. *Journal of Biological Chemistry* **262**, 13627–13635.

Hofmann, P.A., and Fuchs, F. (1987a). Effect of length and cross-bridge attachment on Ca^{2+} binding to cardiac troponin C. *American Journal of Physiology* **253**, C90–C96.

Hofmann, P.A., and Fuchs, F. (1987b). Evidence for a force-dependent component of calcium binding to cardiac troponin C. *American Journal of Physiology* **253**, C541–C546.

Huxley, A.F. (1957). Muscle structure and theories of contraction. *Progress in Biophysics and Biophysical Chemistry* **7**, 255–318.

Huxley, A.F., and Simmons, R.M. (1971). Proposed mechanisms of force generation in striated muscle. *Nature* **233**, 533–538.

Krasner, K., and Maughan, D.W. (1984). The relationship between ATP hydrolysis and active force in compressed and swollen skinned muscle fibers of the rabbit. *Pflügers Archiv* **400**, 160–165.

Levine, R.J.C., Chantler, P.D., Kensler, R.W., and Woodhead, J.L. (1991). Effects of phosphorylation by myosin light chain kinase on the structure of limulus thick filaments. *Journal of Cell Biology* **113**, 563–572.

Levine, R.J.C., Kensler, R.W., Yang, Z., Stull, J.T., and Sweeney, H.L. (1996). Myosin light chain phosphorylation affects the structure of rabbit skeletal muscle thick filaments. *Biophysical Journal* **71**, 898–907.

Lymn, R.W., and Taylor, E.W. (1971). Mechanism of adenosine triphosphate hydrolysis by actomyosin. *Biochemistry* **10**, 4617–4624.

Martyn, D.A., and Gordon, A.M. (1999). Length and myofilament spacing-dependent changes in calcium sensitivity of skeletal fibres: Effects of pH and ionic strength. *Journal of Muscle Research and Cell Motility* **9**, 428–445.

Matsubara, I., and Elliot, G.F. (1972). X-ray diffraction studies on skinned single fibers of the frog skeletal muscle. *Journal of Molecular Biology* **72**, 657–669.

Maughan, D.W., and Godt, R.E. (1979). Stretch and radial compression studies on relaxed skinned muscle fibers of the frog. *Biophysical Journal* **28**, 391–402.

Moisescu, D.G., and Thieleczek, R. (1979). Sarcomere length effects on the Sr^{++}- and Ca^{++}- activation curves in skinned frog muscle fibres. *Biochimica et Biophysica Acta* **546**, 64–76.

Moss, R.L. (1992). Ca^{2+} regulation of mechanical properties of striated muscle. Mechanistic studies using extraction and replacement of regulatory proteins. *Circulation Research* **70**, 865–884.

Moss, R.L., Nwoye, L.O., and Greaser, M.L. (1991). Substitution of cardiac troponin C into rabbit muscle does not alter the length dependence of Ca^{2+} sensitivity of tension. *Journal of Physiology* **440**, 273–289.

Moss, R.L., Swinford, A.E., and Greaser, M.L. (1983). Alterations in the Ca^{2+} sensitivity of tension development by single skeletal muscle fibers at stretched lengths. *Biophysical Journal* **43**, 115–119.

Patel, J.R., McDonald, K.S., and Wolff, M.R. (1997). Ca^{2+} binding to troponin C in skinned skeletal muscle fibers assessed with caged Ca^{2+} and a Ca^{2+} fluorophore. Invariance of Ca^{2+} binding as a function of sarcomere length. *Journal of Biological Chemistry* **272**, 6018–6027.

Persechini, A., Stull, J.T., and Cooke, R. (1985). The effect of myosin phosphorylation on the contractile properties of skinned rabbit skeletal muscle fibers. *Journal of Biological Chemistry* **260**, 7951–7954.

Rack, P.M.H., and Westbury, D.R. (1969). The effects of length and stimulation rate on tension in the isometric cat soleus muscle. *Journal of Physiology* **204**, 443–460.

Rassier, D.E., Tubman, L.A., and MacIntosh, B.R. (1998). Caffeine and length dependence of staircase potentiation in skeletal muscle. *Canadian Journal of Physiology and Pharmacology* **76**, 975–982.

Rayment, I., Holden, H.M., Whittaker, M., Yohn, C.B., Lorenz, M., Holmes, K.C., and Milligan, R.A. (1993). Structure of the actin-myosin complex and its interactions for muscle contraction. *Science* **261**, 58–65.

Ridgway, E.B., and Gordon, A.M. (1984). Muscle calcium transient. Effect of post-stimulus length changes in single fibers. *Journal of General Physiology* **83**, 75–103.

Rios, E., and Pizarro, G. (1991). Voltage sensor of excitation-contraction coupling in skeletal muscle. *Physiological Reviews* **71**, 849–908.

Rome, E. (1968). X-ray diffraction studies on the filament lattice of striated muscle in various bathing media. *Journal of Molecular Biology* **37**, 331–344.

Schoenberg, M. (1980a). Geometrical factors influencing muscle force development. I. The effect of filament spacing upon axial forces. *Biophysical Journal* **30**, 51–68.

Schoenberg, M. (1980b). Geometrical factors influencing muscle force development. II. Radial forces. *Biophysical Journal* **30**, 69–78.

Stein, L.A., Schwarz, R.P., Chock, P.B., and Eisenberg, E. (1979). Mechanism of actomyosin adenosine triphosphate. Evidence that adenosine 5′-triphosphate hydrolysis can occur without dissociation of the actomyosin complex. *Biochemistry* **18**, 3895–3909.

Stephenson, D.G., and Wendt, I.R. (1984). Length dependence of changes in sarcoplasmic calcium concentration and myofibrillar calcium sensitivity in striated muscle fibres. *Journal of Muscle Research and Cell Motility* **5**, 243–272.

Stephenson, D.G., Wendt, I.R., and Forrest, Q.G. (1981). Non-uniform ion distributions and electrical potentials in sarcoplasmic regions of skeletal muscle fibres. *Nature* **289**, 690–692.

Stephenson, D.G., and Williams, D.A. (1982). Effects of sarcomere length on the force-pCa relation in fast- and slow-twitch skinned fibres from the rat. *Journal of Physiology* **333**, 637–653.

Stienen, G.J.M., Blangé, T., and Treijtel, B.W. (1985). Tension development and calcium sensitivity in skinned muscle fibres of the frog. *Pflügers Archiv* **405**, 19–23.

Sweeney, H.L., Bowman, B.F., and Stull, J.T. (1993). Myosin light chain phosphorylation in vertebrate striated muscle: Regulation and function. *American Journal of Physiology* **264**, C1085–C1095.

Taylor, S.R., Lopez, J.R., Griffiths, G., Trube, G., and Cecchi, G. (1982). Calcium in excitation-contraction coupling of frog skeletal muscle. *Canadian Journal of Physiology and Pharmacology* **60**, 489–502.

ter Keurs, H.E.D.J., Luff, A.R., and Luff, S.E. (1984). Force-sarcomere-length relation and filament length in rat extensor digitorum muscle. *Advances in Experimental and Medical Biology* **170**, 511–525.

Wang, Y., and Fuchs, F. (1994). Length, force, and Ca^{2+}-troponin C affinity in cardiac and slow skeletal muscle. *American Journal of Physiology* **266**, C1077–C1082.

Wang, Y., and Fuchs, F. (1995). Osmotic compression of skinned cardiac and skeletal muscle bundles: Effects on force generation, Ca^{2+} sensitivity and Ca^{2+} binding. *Journal of Molecular and Cellular Cardiology* **27**, 1235–1244.

Westerblad, H., and Allen, D.G. (1991). Changes on myoplasmic calcium concentration during fatigue in single mouse muscle fibers. *Journal of General Physiology* **98**, 615–635.

Westerblad, H., and Allen, D.G. (1996). Mechanisms underlying changes of tetanic $[Ca^{2+}]_i$ and force in skeletal muscle. *Acta Physiologica Scandinavica* **156**, 407–416.

Zhao, Y., and Kawai, M. (1993). The effect of the lattice spacing change on cross-bridge kinetics in chemically skinned rabbit psoas muscle fibers. *Biophysical Journal* **64**, 197–210.

II Theoretical Modelling of Muscle and Muscle Contraction

6 Considerations on the Theoretical Modelling of Skeletal Muscle Contraction

W. HERZOG
Human Performance Laboratory, University of Calgary, Calgary, Alberta, Canada

Models of skeletal muscle can be developed on different structural levels. Depending on the structural level, and depending on the problem to be solved, theoretical models of muscle differ substantially. In fact, they may differ so much that a mathematician not familiar with models of skeletal muscle would be hard pressed to realize that they all represent the same biological structure. Although a muscle has many structural levels, those of interest are the entire muscle, a single muscle fibre, and the molecular interaction of a cross-bridge head (S1) with the thin (actin) myofilament attachment site.

ENTIRE MUSCLE

In biomechanics, the primary interest in skeletal muscle modelling has been the representation of the forces of entire muscles (or muscle groups). The primary approach for this type of skeletal muscle modelling has been Hill-type models (Epstein and Herzog, 1998). Hill-type models are composed of rheological elements, such as springs and dashpots, arranged in series and/or in parallel. Typically, but not always (Forcinito et al., 1998), Hill-type models contain a contractile element that has force-velocity properties related to the well-known relation between the speed of shortening and the corresponding maximal tetanic force obtained at optimal muscle length (Hill, 1938). In order to account for the force-length relationship of skeletal muscle, Hill's force-velocity properties are often scaled to account for the decrease in maximal isometric force potential when working at muscle lengths below or above optimal length. Such scaling has been done by replacing the maximal isometric force in Hill's equation by the maximal isometric force as a function of length. This particular approach has the disadvantage that the maximal speed of shortening becomes a function of muscle length. However, experimental results suggest that the maximal speed of shortening is largely independent of length within a range including the top part of the ascending limb, the plateau, and the top part of the descending limb of the force-

Skeletal Muscle Mechanics: From Mechanisms to Function. Edited by W. Herzog.

length relationship (Edman, 1979). Other scaling methods exist that avoid this particular problem, e.g., by scaling the entire force-velocity relationship by a normalized force-length value, but none of these approaches is without limitations.

The big disadvantage of Hill-type models is that the rheological elements, arranged in series and in parallel, merely serve to derive analytical expressions for force as a function of the contractile conditions. History-dependent effects are largely ignored in these models and, if they are not ignored, they are implemented as yet another phenomenological function matching some observed experimental data. The rheological elements are not associated with particular structural elements of the muscle, therefore, interpretation of the model results becomes difficult. Furthermore, Hill-type models cannot be used readily to provide insight into the mechanisms of contraction and force production. Relationships between mechanical and biochemical events underlying contraction are lost, and the relationship of total muscle force and total muscle length with the force and length of the various rheological elements of the model is not trivial.

Summarizing, although Hill-type muscle models have dominated the field of biomechanics, they are ill suited for providing scientific insight into the events underlying muscular contraction. They may be viewed as tools that can be used to simulate movements of entire systems.

MUSCLE FIBRE

In muscle physiology, much of the relevant work in the past 50 years has centred around experimental research on isolated muscle fibres. From this work, the cross-bridge theory of muscular contraction has evolved (Huxley, 1957; Huxley, 1969; Huxley and Simmons, 1971). The cross-bridge theory describes the events of muscular contraction on the cross-bridge level. The basic idea underlying the cross-bridge theory is that side-pieces from the thick filaments (the cross-bridges) attach cyclically on specific attachment sites of the thin (actin) filaments, produce force, and pull the two sets of filaments (which are presumed to be rigid) past one another. This set of events then causes shortening of the sarcomeres (assuming that the external resistance to shortening is smaller than the force produced by the cross-bridges within the sarcomeres). These events of force production and contraction are associated with biochemical steps in which adenosine triphosphate (ATP) is hydrolyzed to adenosine diphosphate (ADP) and free phosphate, an event that is known to yield free energy that is then used (partly) to produce the force and contractile steps.

The basic assumptions underlying the classic cross-bridge theory are that the cross-bridges are arranged in a uniform way along the thick filament, that the thin filament attachment sites are arranged uniformly along the thin filament, that each cross-bridge acts as an independent force generator that produces the same steady-state force as the other cross-bridges, that the two sets of filaments are largely rigid, and that the contractile events are directly coupled to the ATP hydrolysis steps. For completeness, it needs to be pointed out that many of the assumptions of the classic cross-bridge model are not considered valid anymore. For example, it has been shown convincingly in independent laboratories that thin (and possibly thick) filaments are not strictly rigid, but contribute 50-70% of the sarcomere compliance (Huxley et al., 1994; Wakabayashi et al., 1994; Goldman and Huxley, 1994; Kojima et al., 1994). This compliance in the myofilaments shows that the speed of a point on the thin filament relative to a point on the thick filament is not only dependent on the relative filament speed, as has been assumed up to a few years ago, but also depends on the forces (and therefore the elongations) acting on the filaments (Mijailovich et al., 1996). Similarly, the traditional notion that the number of attached cross-bridges can be directly estimated from the stiffness of a fibre has been proven incorrect (Forcinito et al., 1997). Also, the assumption of time-independent force generation may not be correct. On the muscle, and the muscle-fibre level, it has been shown convincingly that the isometric forces obtained following shortening (stretching) of the preparation are smaller (larger) than the corresponding forces obtained for purely isometric conditions. However, it has not been elucidated in detail whether or not these time-dependent phenomena are caused by structural non-uniformities (Edman et al., 1993), by time-dependent events on the cross-bridge level (Herzog, 1998), or by other unexplained mechanisms. The suggestion of time-dependent events on the cross-bridge level would be of interest insofar as the cross-bridge model, or any model of S1 interaction with actin, does not account for time or history-dependent effects.

In contrast to the Hill-type models described above, cross-bridge models are aimed to be structural models in which the elements of the model represent distinct structural elements of the biological system. Of course, any model of a biological system breaks down at one level where the structural complexity cannot be modelled. Therefore, the cross-bridge model becomes phenomenological at one point of representation. Depending on the sophistication of the model, this might happen at different levels. For example, some aspects of the rate functions of attachment and detachment are phenomenological and do not have a good structural representation at this time.

Cross-bridge models have not been used extensively in biomechanical applications, primarily because they are considered too complex and numerically too time consuming to be practical. However, in chapter 7, Zahalak provides a suggestion for how cross-bridge models might be used for complex biomechanical analyses. The model discussed by Zahalak is a two-state cross-bridge model with cross-bridges either in an attached or detached state. It is argued that this model, although in no way suited to represent the high non-linearities of fast transient events observed experimentally (Huxley and Simmons, 1971), can be used without much error for representing the transient responses of muscle observed during normal everyday (i.e., slow) movements. In chapter 8, Ait-Haddou presents a general framework for how any cross-bridge model can be extended to a new model without losing the properties of the old model. For example, the two-state model by Zahalak gives good agreement with the steady-state concentric force-velocity properties observed by Hill (1938). However, as pointed out, it cannot predict with any degree of accuracy the fast force transients following a quick-step change in length. If we would like to add the property associated with the quick-length steps, but retain simultaneously the accurate prediction of the concentric force-velocity relationship of the two-state model, then the approach described by Ait-Haddou gives the mathematical framework for how to do just that.

S1 ACTIN INTERACTION

There is one distinctly smaller structural level of muscle modelling than fibre that has been largely ignored. This is the molecular level of single cross-bridge interactions with the corresponding actin attachment site. In 1993, Rayment et al. (1993a; 1993b) published two studies on the three-dimensional atomic structure of the S1 cross-bridge head and the corresponding actin attachment site. With the atomic structure known to a good extent, it becomes possible to study theoretically the interactions between the two molecules, and to derive the possible mechanisms underlying force production, contraction, and the mechanochemical coupling that occurs during force production. At first glance, this approach might seem impossible because of the large number of possible interactions between S1 and the actin attachment site. However, much is known from other studies regarding some of the likely interactions, therefore, many configurations may be excluded from serious analysis, reducing the number of possibilities. Theoretical models of possible S1 interactions with corresponding actin attachment sites are rare, and much exciting theoretical work is awaiting proper solution on this structural level of muscle modelling.

COMPLETE MUSCLE MODEL

Of course, many scientists would argue that all of the above-proposed models of muscle are incomplete, as muscle does not produce force for the sake of producing force. Muscular force production is controlled by the central nervous system, and muscular contraction produces an effect on the skeletal system. Movement control scientists see muscle as the effector organ at the end of a long chain of decisions occurring in the central nervous system. Biomechanists see muscle as the primary organ at the beginning of a long chain producing well-co-ordinated and sophisticated movements. Therefore, one could make a convincing argument that a complete muscle model needs to contain at least three basic parts: a control system, a muscle model, and a musculoskeletal model.

Let's briefly look at these three systems in turn. The control system for all muscle actions is the nervous system. Skeletal muscle activation is primarily under voluntary control of motor centres in the brain. However, activation and force production in skeletal muscle also depends on the input from pattern generators and from peripheral feedback. In order to understand muscular activation and force sharing among synergistic muscles during movement, it is necessary to understand all aspects of control of muscular force arising from these neural pathways. Such a comprehensive model does not exist to date and the exact contributions of brain, pattern generators, and feedback systems to dynamic force production and to force sharing among muscles are not known.

The muscle model needs to contain mechanical, physiological, and structural properties. The mechanical properties relate to the force-length, force-velocity, and force-time properties of maximally and submaximally activated muscle. The physiological properties relate to the fibre-type distribution of the muscle, its fatigue properties, the force-calcium relationship, potentiation properties, and all aspects of the metabolic demands of muscular contraction. Finally, the structural properties relate to the fibre arrangement within the muscle, the aponeuroses and tendon arrangements (and corresponding mechanical properties), and the size of the muscle.

The musculoskeletal model defines the line of action of the muscle relative to the joints that are crossed by the muscle, and thus, by definition, contains the variable moment arms of the muscle about each of the joints it crosses.

As can be appreciated from the above (incomplete) description of a 'complete' muscle model, most of the model parameters are unknown at this time, therefore, a truly complete muscle model integrating control, muscle force production, and effects of the muscular actions on the skeletal movement system awaits solution.

REFERENCES

Edman, K.A.P. (1979). The velocity of unloaded shortening and its relation to sarcomere length and isometric force in vertebrate muscle fibres. *Journal of Physiology* **291**, 143–159.

Edman, K.A.P., Caputo, C., and Lou, F. (1993). Depression of tetanic force induced by loaded shortening of frog muscle fibres. *Journal of Physiology* **466**, 535–552.

Epstein, M., and Herzog, W. (1998). Hill and Huxley-type models: Biological considerations. In: *Theoretical Models of Skeletal Muscle*. John Wiley & Sons, Ltd., New York, pp. 70–84.

Forcinito, M., Epstein, M., and Herzog, W. (1997). Theoretical considerations on myofibril stiffness. *Biophysical Journal* **72**, 1278–1286.

Forcinito, M., Epstein, M., and Herzog, W. (1998). Can a rheological muscle model predict force depression/enhancement? *Journal of Biomechanics* **31**, 1093–1099.

Goldman, Y.E., and Huxley, A.F. (1994). Actin compliance: Are you pulling my chain? *Biophysical Journal* **67**, 2131–2136.

Herzog, W. (1998). History dependence of force production in skeletal muscle: A proposal for mechanisms. *Journal of Electromyography and Kinesiology* **8**, 111–117.

Hill, A.V. (1938). The heat of shortening and the dynamic constants of muscle. *Proceedings of the Royal Society of London B* **126**, 136–195.

Huxley, A.F. (1957). Muscle structure and theories of contraction. *Progress in Biophysics and Biophysical Chemistry* **7**, 255–318.

Huxley, A.F., and Simmons, R.M. (1971). Proposed mechanism of force generation in striated muscle. *Nature* **233**, 533–538.

Huxley, H.E. (1969). The mechanism of muscular contraction. *Science* **164**, 1356–1366.

Huxley, H.E., Stewart, A., Sosa, H., and Irving, T. (1994). X-ray diffraction measurements of the extensibility of actin and myosin filaments in contracting muscles. *Biophysical Journal* **67**, 2411–2421.

Kojima, H., Ishijima, A., and Yanagida, T. (1994). Direct measurement of stiffness of single actin filaments with and without tropomyosin by *in vitro* nanomanipulation. *Proceedings of the National Academy of Sciences USA* **91**, 12962–12966.

Mijailovich, S.M., Fredberg, J.J., and Butler, J.P. (1996). On the theory of muscle contraction: Filament extensibility and the development of isometric force and stiffness. *Biophysical Journal* **71**, 1475–1484.

Rayment, I., Holden, H.M., Whittaker, M., Yohn, C.B., Lorenz, M., Holmes, K.C., and Milligan, R.A. (1993a). Structure of the actin-myosin complex and its implications for muscle contraction. *Science* **261**, 58–65.

Rayment, I., Rypniewski, W.R., Schmidt-Bäse, K., Smith, R., Tomchick, D.R., Benning, M.M., Winkelmann, D.A., Wesenberg, G., and Holden, H.M. (1993b). Three-dimensional structure of myosin subfragment-1: A molecular motor. *Science* **261**, 50–58.

Wakabayashi, K., Sugimoto, Y., Tanaka, H., Ueno, Y., Takezawa, Y., and Amemiya, Y. (1994). X-ray diffraction evidence for the extensibility of actin and myosin filaments during muscle contraction. *Biophysical Journal* **67**, 2422–2435.

7 The Two-state Cross-bridge Model as a Link Between Molecular and Macroscopic Muscle Mechanics

GEORGE I. ZAHALAK
Departments of Biomedical Engineering and Mechanical Engineering, Washington University, St. Louis, Missouri, USA

ABSTRACT

Remarkable progress continues to be made by biophysicists and biochemists in understanding the molecular mechanisms underlying muscle contraction. How can this progress contribute to advancing the macroscopic biomechanics of movement and its control? The two-state cross-bridge model, being both mathematically tractable and containing the essential kernel of molecular contraction dynamics (actin-myosin bonding), provides a convenient vehicle to connect muscle behaviour at the micro- and macro-levels. The development of the two-state cross-bridge model into a state-variable model of whole muscle via the distribution moment (DM) approximation is reviewed. This DM model encompasses not just contraction, but also activation and energetics, and even some aspects of metabolic dynamics. Further, the DM approximation clarifies the relation between biophysical cross-bridge models and the A.V. Hill phenomenological model traditionally employed by biomechanists. The relation between the two-state cross-bridge model and the more current and accurate multi-state models is examined from the mathematical perspective of matched asymptotic expansions. This analysis shows that the two-state rate equation governing the bond distribution is a first approximation, valid on time scales that are slow compared to rapid transitions between attached states (the usual case in macroscopic biomechanics). Consequently, the two-state cross-bridge model may be useful for applications to whole-muscle mechanics, even if it is inadequate to predict some fine details of the molecular contraction mechanism. But asymptotic analysis reveals also that the *effective* cross-bridge force in the two-state model is in general a non-linear, typically saturating, function of cross-bridge strain—which contradicts the usual assumption of two-state cross-bridge theory that the cross-bridge stiffness is constant. Therefore, this assumption must be regarded as a convenient simplifying approximation, the associated

Skeletal Muscle Mechanics: From Mechanisms to Function. Edited by W. Herzog.

errors anticipated, and allowances made for them. Three additional topics are noted briefly that bear on the application of the two-state model in macroscopic biomechanics: (1) cross-bridge interaction with remote actin binding sites, (2) sarcomere inhomogeneity, and (3) the effects of non-axial deformation on axial and non-axial active muscle stress.

INTRODUCTION

I believe that Professor Yanagida's excellent keynote lecture (Yanagida, 1999) has served to remind us of what striking progress has been achieved by biophysicists and biochemists during the past few decades in explaining the molecular mechanisms underlying muscle contraction. We may hope and expect that this impressive increase of fundamental understanding will be reflected in corresponding progress in macroscopic biomechanics—in improved modelling of the mechanics, energetics, and metabolism of skeletal muscles as they produce movement. But it seems to me that the connection between molecular muscle biophysics and macroscopic muscle biomechanics has been tenuous. In most biomechanical research, the muscle representation of choice continues to be some variant of the A.V. Hill visco-elastic analogy (Hill, 1938), which was proposed before most of what we now know about muscle was known. New discoveries about contraction mechanisms at the molecular level have no impact on this model, which is independent of such details. Depending on one's point of view, this may or may not be a good thing.

My view has been that it would indeed be desirable to have a close and clear connection between biophysical and biomechanical muscle models. This would allow macroscopic biomechanics to share, at least to some extent, in the continuing discoveries made by a large community of active and clever muscle researchers. Further, the molecular muscle models imply natural interrelations between activation, mechanics, energetics, and metabolism—factors which must also be related at the macroscopic level. Finally, the molecular models provide the possibility of estimating quantitatively the effects of drugs or disease states on the various components of the muscle machine, be they contractile proteins, calcium channels, or passive elastic structures. These are certainly important considerations for macroscopic biomechanics.

I propose that the two-state cross-bridge model is well suited to bridging the gap between the molecular and whole-muscle levels. This is the model first introduced by A.F. Huxley (1957), and subsequently modified by T.L. Hill et al. (1975) to account for certain thermodynamic constraints. Why the restriction to only two states? Certainly such a simple model is known to be too simple to explain the fine details of the cross-bridge cycle. Indeed, the number of putative states in cross-bridge models has grown concurrently with accumulating experimental data; models have been published containing up to 18 (!) states (Propp, 1986), although the usual number is between five and seven. Huxley has referred to his two-state model as a 'skeleton theory' (Huxley, 1988). But even a skeleton defines a framework that may be very useful for certain purposes. My concern is less with the problem of representing the true and complete molecular mechanism of contraction than with integrating at least some salient features of that mechanism into macroscopic muscle models. And for that purpose, I will argue, the two-state cross-bridge model will serve.

Multiple states were originally introduced into cross-bridge models to account for fast transitions in the cross-bridge cycle that were discovered by precise mechanical and chemical measurements. The classic case is that of the introduction of two, rather than one, attached states by Huxley and Simmons (1971) in order to reproduce fast mechanical transients—the so-called T_1–T_2 phenomenon. But these fast transitions, important as they are for understanding the fundamental mechanisms of contraction, may not be directly relevant to macroscopic muscle mechanics under most conditions. The T_1–T_2 transition has a characteristic time scale of about 1 millisecond. Few, if any, macroscopic movements occur that quickly. Indeed, any macroscopic muscle model that attempted to include transitions that fast would have to treat a muscle as a distributed continuum, because action potentials and elastic waves propagate along a muscle on comparable time scales. Any whole-muscle model that represents the muscle as a lumped, rather than distributed, system must perforce restrict itself to sufficiently slow time scales, thus ignoring fast cross-bridge transitions.

Here is another consideration prompting a restriction to the two-state model. To facilitate interpretation, biophysicists usually choose to deal with precisely controlled isometric or isotonic experiments at maximum activation. While these may be technically difficult experiments, they present relatively simple boundary-value problems for mathematical analysis. But the biomechanist studying movement must consider muscles subject to histories of varying activation and loading, and requires tractable models that can deal with these conditions. As I shall discuss shortly, cross-bridge models can be coupled rationally to models of calcium activation, energetics, and even some aspects of metabolism. If the cross-bridge model is of the simple two-state kind, then a compatibly simple model can be employed for activation (and energetics, and metabolism). But if the cross-bridge model itself grows to multi-state complexity, then the associated activation and other sub-models must also become correspondingly complicated. For example, it has been shown (Inesi, 1979) that the protein pump responsible for sequestering calcium in the sarcoplasmic reticulum (SR) appears to have 12 separate biochemical states. In coupling a model of the SR calcium pump to a simple two-state cross-bridge model, one could perhaps gloss over the complexity of the former and represent it by simple Michaelis-Menten kinetics. But such a crude approximation would hardly be permissible if the cross-bridge model itself had five states, including fast transitions. Another similar consideration is the rate of binding of calcium to (the low-affinity sites of) troponin-C. This is known to be fast (Robertson et al., 1981; Cannell and

Allen, 1984) and can be considered 'instantaneous' on the relatively slow rate-limiting time scale of actin-myosin dynamics, which simplifies modelling greatly. But if an activation-contraction model is to be valid on millisecond time scales, then such a simplifying assumption is not permissible. The moral is: complicated cross-bridge models require complicated associated models for activation, energetics, and metabolism, and tractability will be hard put to survive this piling up of complications.

So, in the remainder of this presentation I will focus on the two-state model. First, I will review how, by a simple mathematical approximation, this model can be converted to a corresponding state-variable model of the macroscopic muscle actuator. In this process, I will try to identify the connection, such as it is, between the two-state cross-bridge model and the traditional A.V. Hill model. I will also indicate briefly how models of activation, energetics, and metabolism can be coupled rationally to the contraction models. Then, I will consider the two-state model in the context of multi-state models, via the mathematical formalism of matched asymptotic expansions, and try to define in what sense the two-state model may be regarded as a valid approximation to multi-state models. Finally, I will touch briefly on a few further issues relating to two-state models and their applications: cross-bridge binding to remote actin sites, sarcomere inhomogeneity, and non-axial stresses and deformations.

MICRO- TO MACRO-TRANSITION VIA THE DISTRIBUTION-MOMENT (DM) APPROXIMATION

The two-state model introduced by A.F. Huxley in 1957 is governed by the well known rate equation on the actin-myosin bond-distribution function, $n(x,t)$, shown in the top part of Figure 7.1.

$$\frac{\partial n}{\partial t} - v(t)\frac{\partial n}{\partial x} = f(x)(1-n) - g(x)n \tag{7.1}$$

Later, T.L. Hill pointed out that to generate work, a cycling cross-bridge needed two separate paths between the attached and detached states (Hill et al., 1975; Hill, 1977), and modified the model and rate equation as shown in the bottom part of Figure 7.1.

$$\frac{\partial n}{\partial t} - v(t)\frac{\partial n}{\partial x} = \left[f(x) + g'(x)\right](1-n) - \left[g(x) + f'(x)\right]n \tag{7.2}$$

If in Hill's equation one simply relabels the net attachment rate, $(f + g')$, as f and the net detachment rate, $(g + f')$, as g, one recovers the Huxley equation, so for most purposes the two models can be regarded as identical. Some time ago, I proposed that the two-state cross-bridge model could be converted to an approximate state-variable model for whole muscle by a simple mathematical

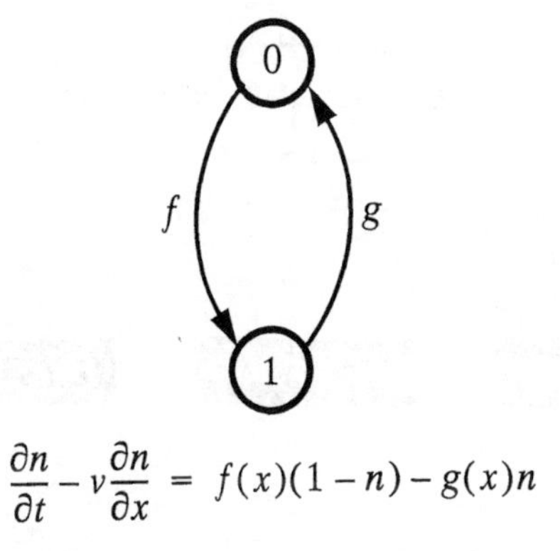

$$\frac{\partial n}{\partial t} - v\frac{\partial n}{\partial x} = f(x)(1-n) - g(x)n$$

$$\frac{\partial n}{\partial t} - v\frac{\partial n}{\partial x} = [f(x) + g'(x)](1-n) - [g(x) + f'(x)]n$$

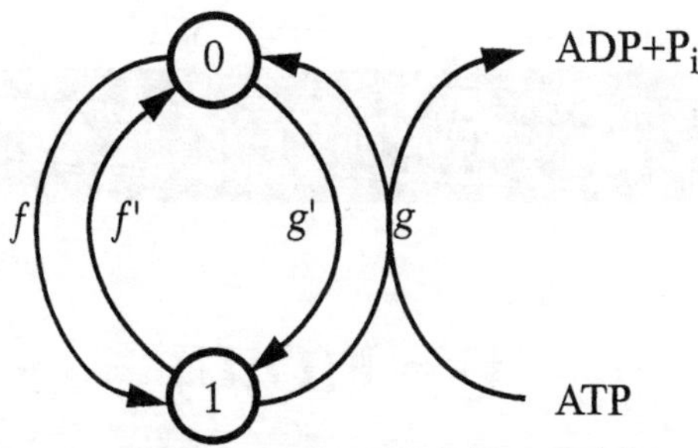

Figure 7.1. Two versions of the two-state cross-bridge model: top, A.F. Huxley (1957), and bottom, T.L. Hill et al. (1975).

approximation (Zahalak, 1981), which I called the distribution-moment (DM) approximation. Mathematically, it corresponds to the first term of a Hermite-function expansion of the bond-distribution function, *n*, and is illustrated in Figure 7.2. The name of the approximation derives from the fact that it focuses on the moments of the bond-distribution, defined in the usual way. The λth moment of $n(\xi,t)$ is:

$$Q_\lambda(t) = \int_{-\infty}^{\infty} \xi^\lambda n(\xi,t)d\xi \tag{7.3}$$

where $\xi = x/h$ is the normalized bond length, and h is Huxley's bond-length scaling parameter.

For the two-state model with constant cross-bridge stiffness, the three lowest order-moments have direct physical significance (Zahalak and Ma, 1990):

$Q_0(t)$ ~ *Stiffness* $\quad$ $Q_1(t)$ ~ *Force* $\quad$ $Q_2(t)$ ~ *Elastic Energy*

$$Q_\lambda = \int_{-\infty}^{\infty} \xi^\lambda n(\xi, t) d\xi$$

Q_0 ~ Stiffness Q_1 ~ Force Q_2 ~ Elastic Energy

$$n(\xi, t) = \frac{Q_0}{\sqrt{2\pi}\ q} e^{-\frac{(\xi - p)^2}{2q^2}}$$

$$\dot{Q}_\lambda = \beta_\lambda - \phi_\lambda(Q_0, Q_1, Q_2) - \lambda u Q_{\lambda - 1}$$

$$\dot{\mathbf{Q}} = \mathbf{F}(\mathbf{Q};\mathbf{U})$$

Figure 7.2. Elements of the distribution-moment (DM) approximation.

The approximation consists of representing $n(\xi,t)$ by an un-normalized Gaussian probability density function (or zero-order Hermite function):

$$n(\xi,t) = \frac{Q_0}{\sqrt{2\pi}q} e^{-\frac{(\xi-p)^2}{2q^2}} \tag{7.4}$$

where $p = Q_1/Q_0$ and $q^2 = (Q_2/Q_0) - (Q_1/Q_0)^2$ are, respectively, the centroid and variance of the distribution. Combining Equation 7.4 with Equation 7.1 yields the state-variable system (Zahalak, 1981):

$$\dot{\boldsymbol{Q}} = \boldsymbol{F}(\boldsymbol{Q}; \boldsymbol{U}) \tag{7.5}$$

where the state vector is $\boldsymbol{Q} = (Q_0, Q_1, Q_2)$, and the control vector in this case has only one component, $\boldsymbol{U} = u = v/h$, the normalized myofilament shortening velocity. (Although u appears formally as a control variable of the system, it is not usually under the experimenter's control.) If the rate parameters of the two-state cross-bridge model are specified, then the system function F of Equation 7.5, displayed in expanded form in Figure 7.2, can be calculated. Then, this system of non-linear ordinary differential equations can be solved, subject to appropriate auxiliary conditions, to yield the time histories of force, stiffness, and elastic energy, if desired. We have simulated many experiments with these equations and found that they usually give quite good approximations to the solutions of the corresponding cross-bridge rate equation (see, for example, Figure 7.3). The advantage of the DM ordinary differential equations over the cross-bridge partial differential equation is that the former are much more tractable computationally and provide direct approximations of force and stiffness, without first having to solve for the bond-distribution function, n.

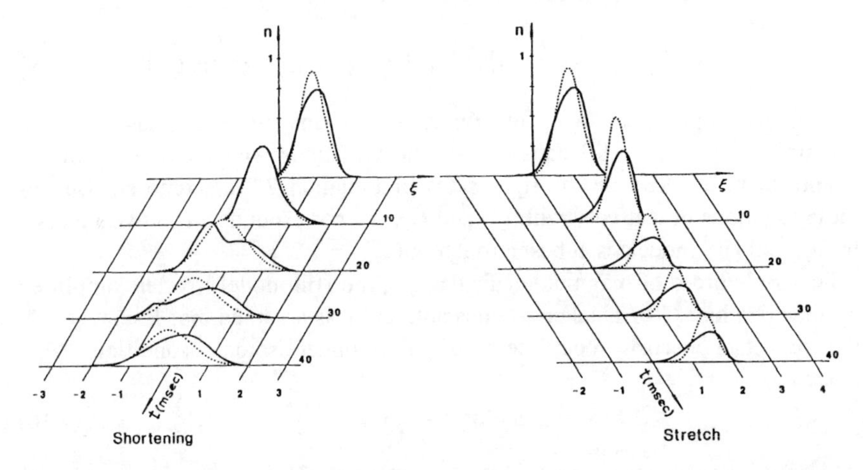

Figure 7.3. A comparison between exact solutions (solid curves) of the two-state rate equation for the bond-distribution function, $n(x,t)$, and the Gaussian approximations (dashed curves) with the moments Q_0, Q_1, and Q_2 (representing stiffness, force, and elastic energy) computed using the DM equations. The moments according to the DM equations are good approximations to the exact moments. (Reproduced from Ma and Zahalak, 1987, with the permission of Elsevier Science.)

The DM equations also suggest a connection between cross-bridge models of contraction and the A.V. Hill phenomenological model of muscle (see Figure 7.4). Ignoring the so-called 'parallel elasticity', the Hill model represents a muscle by the single ordinary differential equation (Hill, 1938):

$$\dot{L} = C(P)\dot{P} - V(P;P_0) \qquad (7.6)$$

where L is the muscle length, P is the force, P_0 is the isometric force (or 'active state'), $C(P)$ is the compliance of the 'series elastic element', and $V(P;P_0)$ is the force-velocity relation of the 'contractile element'. On the other hand, the specific, expanded form of the DM equations (Equation 7.5) is:

$$\dot{Q}_\lambda = \beta_\lambda - \phi_\lambda(Q_\mu) - \lambda u Q_{\lambda-1} \qquad (7.7)$$

for λ, $\mu = 0, 1, 2$ (with $Q_{-1} = 0$). Here β_λ is the λth moment of f, the attachment rate function, $u = v/h$ is the normalized myofilament sliding velocity, and the ϕ_λ are functions of the moments Q_0, Q_1, and Q_2 that can be defined when the parameters of the cross-bridge model are specified (see Figure 7.2). Further, if a passive (dimensionless) compliance $\kappa(Q_1)$ is in series with the contractile tissue, the dimensionless length (stretch ratio) Λ satisfies the equation (Zahalak and Ma, 1990):

$$\dot{\Lambda} = \kappa(Q_1)\dot{Q}_1 - \gamma u \tag{7.8}$$

where γ is a structural parameter. Combining Equation 7.8 with the second of Equation 7.7 (for $\lambda = 1$) we get:

$$\dot{\Lambda} = [(\gamma / Q_0) + \kappa(Q_1)]\,\dot{Q}_1 - [\gamma\{\beta_1 - \phi_1(Q_0, Q_1, Q_2)\}/Q_0] \tag{7.9}$$

As Λ is proportional to L, and Q_1 is proportional to P, this last equation resembles the classic Hill equation—if the variations with Q_0 and Q_2 of the quantities in brackets on the right sides of Equation 7.9 are ignored. But, as these functions in fact vary with Q_0 and Q_2, the correspondence between cross-bridge and Hill models is at best a rough one.

For crude order-of-magnitude calculations, the Hill model is often simplified by linearization: $C(P)$ is taken as constant, and it is assumed that $V = \eta(P_0 - P)$ with constant viscosity coefficient, η^{-1}; thus under isotonic conditions ($P = constant$):

$$\dot{L} = \eta(P - P_0) \tag{7.10}$$

which is a linear force-velocity relation. The DM equations suggest a simple cross-bridge model that yields the linearized Hill model (Figure 7.5). If the attachment and detachment rate functions of a two-state cross-bridge model satisfy the special condition:

$$(f + g) = \tau^{-1} = constant \tag{7.11}$$

everywhere, then the corresponding DM equations become:

$$\dot{Q}_\lambda = \beta_\lambda - \tau^{-1} Q_\lambda - \lambda u Q_{\lambda-1} \qquad \lambda = 0, 1, 2, \ldots \tag{7.12}$$

Indeed, these relations are *exact* for any λ. Under steady-state isotonic conditions (Q_1, u=constant), Equation 7.12 yields:

$$Q_0 = \tau\beta_0 \quad \text{and} \quad u = (\tau^2\beta_0)^{-1}(\tau\beta_1 - Q_1) \tag{7.13}$$

Thus, these special choices for the attachment and detachment rate functions lead to a constant intrinsic stiffness (associated with cross-bridge elasticity) and a linear force-velocity relation, as in the linearized Hill model.

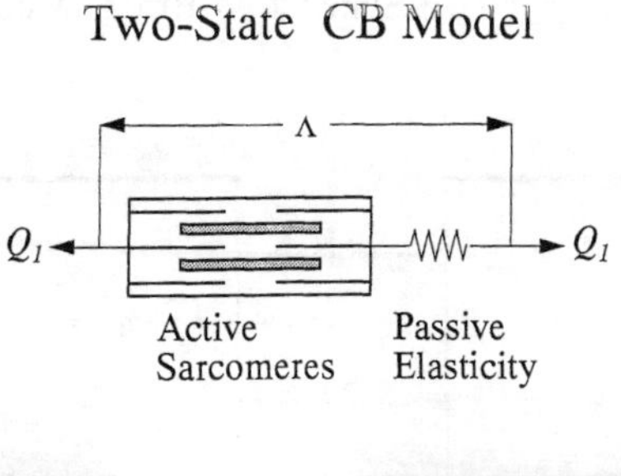

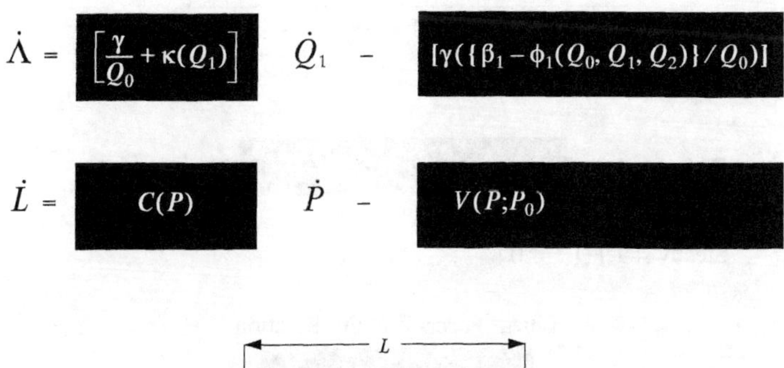

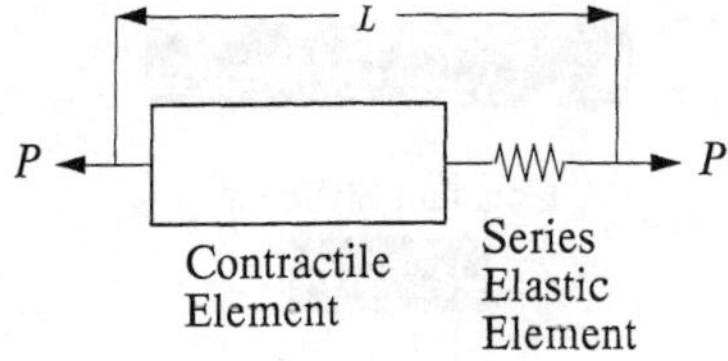

Hill Model

Figure 7.4. The relation established by the DM model between the two-state cross-bridge model and the A.V. Hill phenomenological model of muscle. In the cross-bridge theory, the compliance of the 'series elastic element' depends on the number of attached cross-bridges, Q_0, as well as the actual passive elasticity, and the velocity of the 'contractile element' depends on the moments Q_0 and Q_2 as well as the force, Q_1.

The preceding has been concerned only with contraction dynamics. But cross-bridge models lend themselves readily to rational extensions encompassing activation dynamics, energetics, and even some aspects of metabolism. For example, if it is assumed that two low-affinity, fast-kinetics calcium binding sites on troponin control activation (Figure 7.6) in the sense that both these sites must be sequentially occupied before a cross-bridge can attach, then it is not difficult to show that, if the calcium-troponin rates are fast compared to the actin-myosin rates, the rate equation for n becomes:

$$\frac{\partial n}{\partial t} - u(t)\frac{\partial n}{\partial \xi} = r([Ca])f(1-n) - gn \tag{7.14}$$

A Linear Two-State Cross-Bridge Model

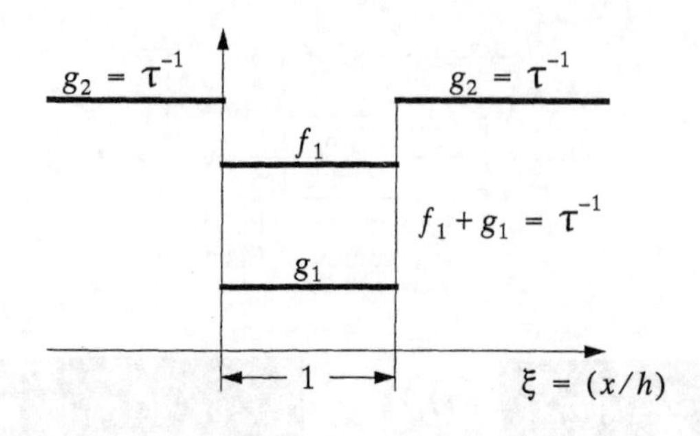

DM equations: $\dot{Q}_\lambda = \beta_\lambda - \tau^{-1} Q_\lambda - \lambda u Q_{\lambda-1}$ (Exact for all λ)

Steady state($\dot{Q}_\lambda = 0$):

Linear Force-Velocity Relation

$$u = (\tau^2 \beta_0)^{-1} (\tau \beta_1 - Q_1)$$

Constant Stiffness

$$Q_0 = \tau \beta_0$$

where $\beta_0 = 2\beta_1 = f_1$

Figure 7.5. A particularly simple two-state cross-bridge model. If the sum of the attachment and detachment rate functions, $f(x)+g(x)$, is a constant for all values of cross-bridge strain, then the DM equations are linear, ordinary, differential equations that are exactly satisfied by the moments of the bond distribution. Under steady-state isotonic conditions, the force-velocity relation is linear and the contractile tissue stiffness is a constant, independent of velocity.

where:

$$r([Ca]) = \frac{k_1^2 [Ca]^2}{k_1^2 [Ca]^2 + k_1 k_{-1} [Ca] + k_{-1}^2} \tag{7.15}$$

is a pure function of sarcoplasmic free calcium concentration, [Ca], which we have dubbed the 'activation factor' (Zahalak and Ma, 1990). Thus, through the activation factor, calcium changes the effective attachment rate from f to *r([Ca])f*, while leaving the detachment rate, *g*, unaffected.

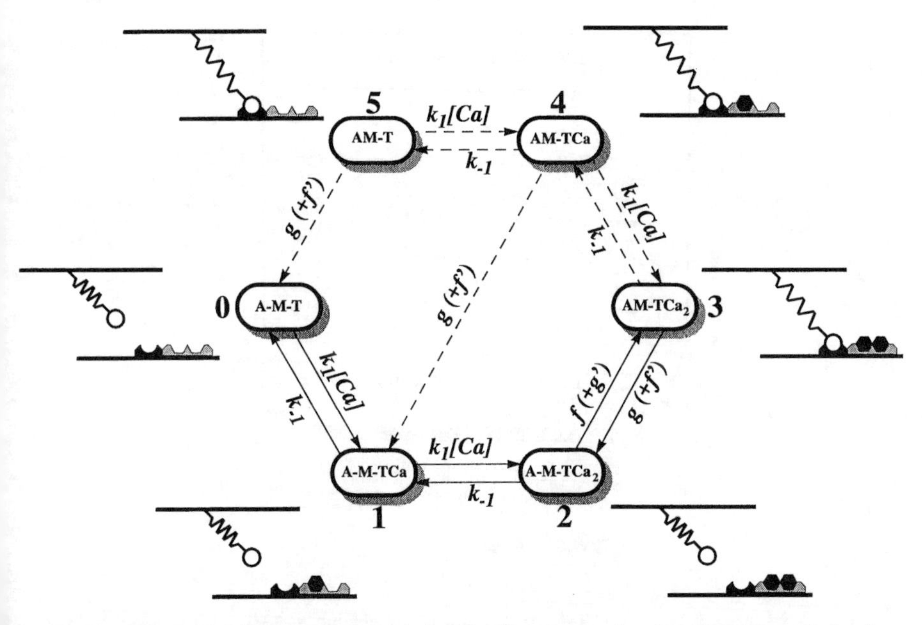

Figure 7.6. State diagram for coupled activation-contraction kinetics. The whole six-state diagram describes a 'loose-coupling' model in which calcium can detach from troponin even if myosin is bound to actin. The lower four states describe a 'tight coupling' model in which release of calcium from troponin is possible only if myosin is detached from actin. (Reproduced from Zahalak and Motabarzadeh, 1997, with the permission of The American Society of Mechanical Engineers.)

The DM approximation can be applied to Equation 7.14, yielding a state-variable model for coupled activation-contraction dynamics in muscle of the form (Zahalak and Ma, 1990; Zahalak and Motabarzadeh, 1997):

$$\dot{\boldsymbol{Q}} = \boldsymbol{F}(\boldsymbol{Q}; \boldsymbol{U}) \tag{7.16}$$

where now the state vector is $\boldsymbol{Q} = (Q_0, Q_1, Q_2, c, \Lambda)$; c is the normalized sarcoplasmic free calcium concentration, and Λ is the normalized muscle length, which has been included on the assumption that there is a passive elastic element in series with the contractile tissue (see Equation 7.8). In addition to the myofilament sliding velocity u, the control vector $\boldsymbol{U} = (u, \mathcal{X})$ now contains a function $\mathcal{X}(t)$ representing a pulse train that models the history of calcium injection by action potentials.

This whole-muscle model, which has been derived directly from a two-state cross-bridge model, does a reasonable job of mimicking several dynamic behaviours of muscle. For example, Figure 7.7 shows the isometric force response to stimulating pulse trains with sinusoidally modulated interpulse intervals (Zahalak and Ma, 1990).

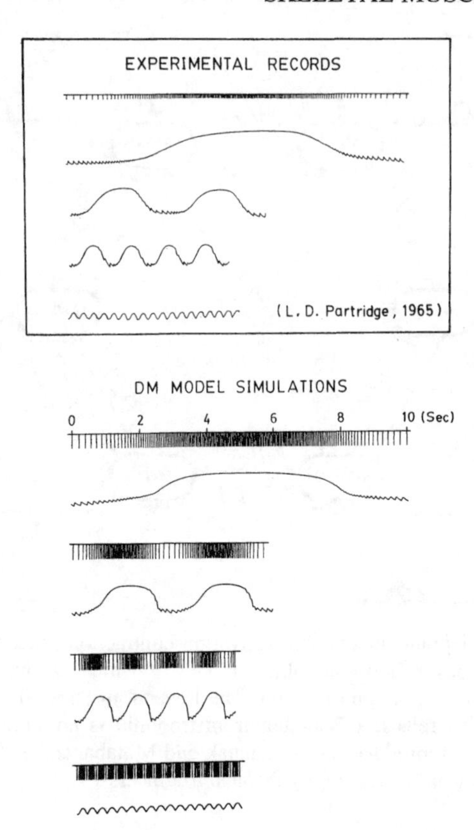

Figure 7.7. DM model simulations of the periodic isometric force produced by sinusoidally modulated pulse stimulation. The inset shows corresponding experimental data. (Reproduced from Zahalak and Ma, 1990, with the permission of The American Society of Mechanical Engineers.)

If a specific assumption is made about the coupling of phosphate compounds to the cross-bridge cycle—for example, that ATP is hydrolyzed in the detachment step—then simple thermodynamic considerations permit the development of expressions for the rate of chemical energy liberation (that is, phosphocreatine hydrolysis), $\dot{E}$, and heat production, $\dot{H}$, as functions of the state variables (Zahalak and Ma, 1990), that is:

$$(\dot{E}/P_0V_m) = \Phi(Q_0, Q_1, Q_2, c, \Lambda) \quad \text{and} \quad (\dot{H}/P_0V_m) = \Psi(Q_0, Q_1, Q_2, c, \Lambda) \qquad (7.17)$$

Figure 7.8 shows a DM model simulation of an isometric twitch with the heat production decomposed into the total heat, H, and the activation heat, H_a (the latter being associated with movement of calcium ions across the SR membrane and binding of calcium to troponin). For comparison, corresponding experimental data obtained by Rall (1982) are included in the figure.

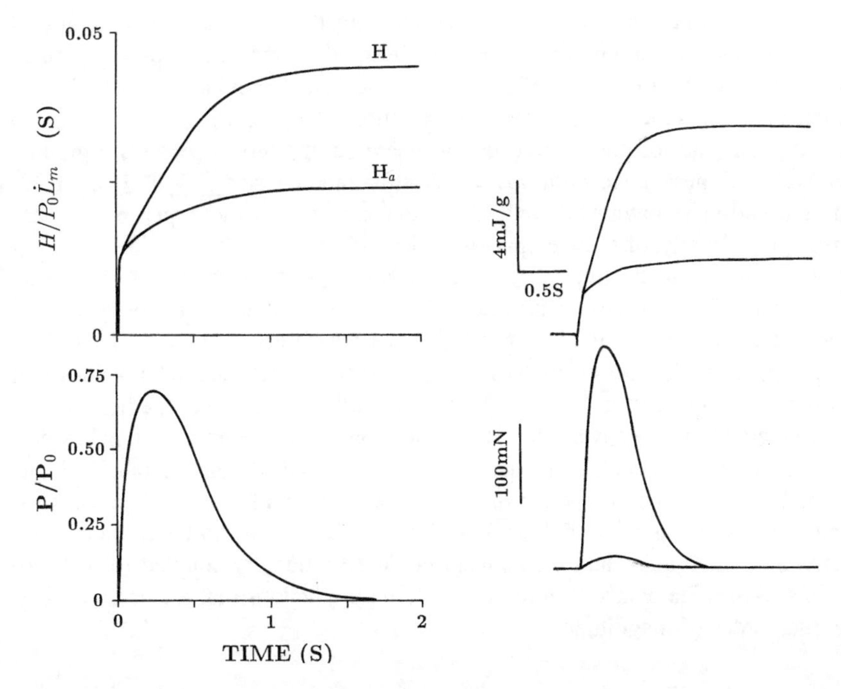

Figure 7.8. DM model predictions of cumulative heat production in an isometric twitch. The upper left panel shows heat production vs. time, decomposed into total heat (upper curve) and activation heat (lower curve). The lower left panel shows twitch force vs. time. The upper and lower right panels show corresponding experimental data. (Reproduced from Ma and Zahalak, 1991, with the permission of Elsevier Science.)

All the preceding models have been for short-term behaviour, defined as contractions sufficiently brief that accumulating metabolic by-products of contraction do not substantially affect the contractility of the muscle itself. In a very ambitious modelling exercise, Rouhaud (1993) combined activation, contraction, energetics, and metabolism, via a two-state cross-bridge model and the DM approximation. The resulting model, while very complicated in detail, again has the form of a state-variable system, this time of ninth order:

$$\dot{\boldsymbol{Q}} = \boldsymbol{F}(\boldsymbol{Q};\boldsymbol{U}) \tag{7.18}$$

where the state vector, $\boldsymbol{Q} = (Q_0, Q_1, Q_2, c, A, c_p, c_{ADP}, c_{Pha}, c_{F6P})$, now contains four additional variables: the concentrations of inorganic phosphate, ADP, phosphorylase-a, and fructose-6-phosphate. The control vector remains $\boldsymbol{U} = (u, \chi)$. This model covers a large range of time scales, from milliseconds to hours. It requires a large number of numerical parameters for its specification (about 60)—but, fortunately, most of these can be estimated from the biochemical literature, at least for frog sartorius. An example of the performance of this model is illustrated in Figure 7.9. Here, an isometric frog sartorius is tetanized for 15 seconds under aerobic conditions. During the tetanus, the force decreases somewhat due to a build-up of inorganic phosphate. Stimulation of the muscle causes, among other things, the concentration of the inorganic phosphate to increase and that of phosphocreatine to decrease; the model predicts a gradual recovery of these concentrations to pre-stimulation resting levels over a 30-minute period following the tetanus, in agreement with the experimental data displayed. Finally, the lower panels show the oxygen consumption and heat production during the recovery period, lasting about an hour. An interesting prediction of the model, displayed in Figure 7.9, is that the recovery heat is about the same as the initial heat, which had been observed by A.V. Hill (1938) many years earlier; no effort was made to select model parameter values that would assure this result. Under anaerobic conditions, the model predicts similar behaviours but the recovery period is much longer. The model also makes predictions of the concentration histories of lactate and various metabolic intermediates. While this theory is probably too complicated for most biomechanical applications, it was nevertheless a worthwhile exercise to show that models for activation, energetics, and metabolism could be rationally coupled to the two-state cross-bridge contraction model, yielding predictions that are certainly better than order-of-magnitude.

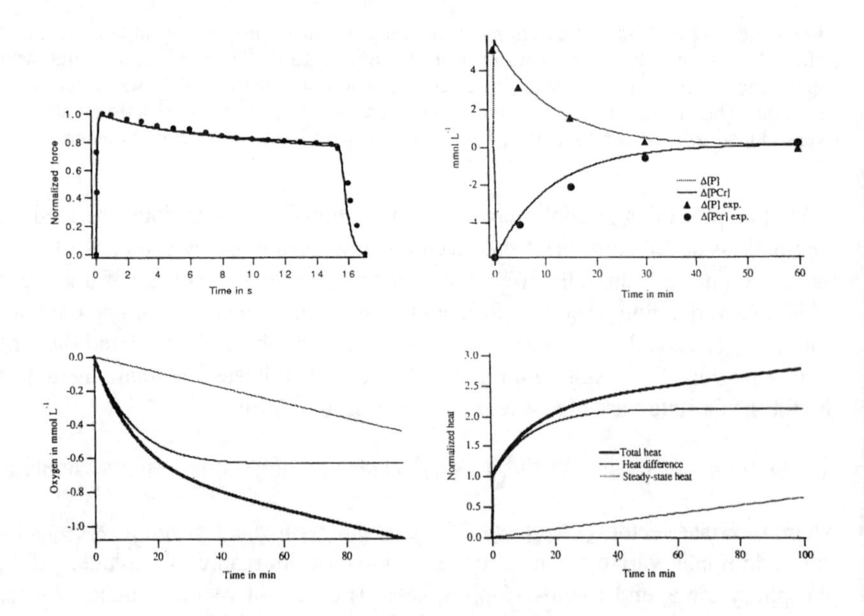

Figure 7.9. Some simulations of an isometric tetanus with the DM metabolic model of Rouhaud (1993). The upper left panel shows isometric force in a 15 s tetanus, with experimental data for a frog sartorius muscle. The upper right panel shows recovery of

phoshocreatine (ΔPCr) and inorganic phosphate (ΔP) concentrations following the end of stimulation. The lower left panel shows oxygen consumption, decomposed into total, resting and incremental components. The lower right panel shows heat production, decomposed into total resting and incremental components; the recovery heat is approximately equal to the initial heat. Note the differences in time scales on the four panels.

THE RELATION BETWEEN TWO-STATE AND MULTI-STATE MODELS

If one accepts that the two-state cross-bridge model provides a useful connection between microscopic and macroscopic muscle mechanics, it becomes important to clarify the relationship between this model (which is regarded by some researchers as inadequate or outdated) and more current and presumably more accurate multi-state models. A detailed asymptotic analysis (Zahalak, 2000) that has instructive implications for the general case can be made of a three-state model of the type first introduced by Huxley and Simmons in 1971 to explain fast mechanical transients (see Figure 7.10). This model has one detached state (0) and two attached states (1 and 2). The normalized populations of these states satisfy the condition $n_0+n_1+n_2=1$ and are governed by the two coupled rate equations:

$$\frac{Dn_1}{D\hat{t}} = \hat{f}_{01} - (\hat{f}_{10} + \hat{f}_{01} + \hat{f}_{12})n_1 + (\hat{f}_{12} - \hat{f}_{01})n_2$$
$$\frac{Dn_2}{D\hat{t}} = \hat{f}_{02} + (\hat{f}_{12} - \hat{f}_{02})n_1 - (\hat{f}_{20} + \hat{f}_{02} + \hat{f}_{21})n_2 \tag{7.19}$$

where the superposed carets denote physical (dimensional) variables and parameters. $(D/D\hat{t}) = (\partial/\partial\hat{t})_x - \hat{v}(\hat{t})(\partial/\partial x)_{\hat{t}}$ is the 'material' derivative based on the filament sliding velocity $\hat{v}(\hat{t})$ (Figure 7.10).

Define $\hat{f}_i$ as a characteristic transition rate between attached states, and $\hat{f}_0$ as a characteristic transition rate between detached and attached states. It is assumed, as experiments suggest, that $(\hat{f}_0/\hat{f}_i)=\varepsilon<<1$, and this opens the possibility of useful asymptotic analysis. For this purpose, the following dimensionless variables and parameters are defined:

$$\xi = x/h_0 \qquad t = \hat{f}_0\hat{t} \qquad T = \hat{f}_i\hat{t}$$
$$(f_{12}, f_{21}) = (\hat{f}_{12}, \hat{f}_{21})/\hat{f}_i \qquad (f_{10}, f_{01}, f_{20}, f_{02}) = (\hat{f}_{10}, \hat{f}_{01}, \hat{f}_{20}, \hat{f}_{02})/\hat{f}_0 \tag{7.20}$$

where h_0 is the classic Huxley bond-length scaling parameter, indicating the range of bond lengths over which attachment can occur with significant probability.

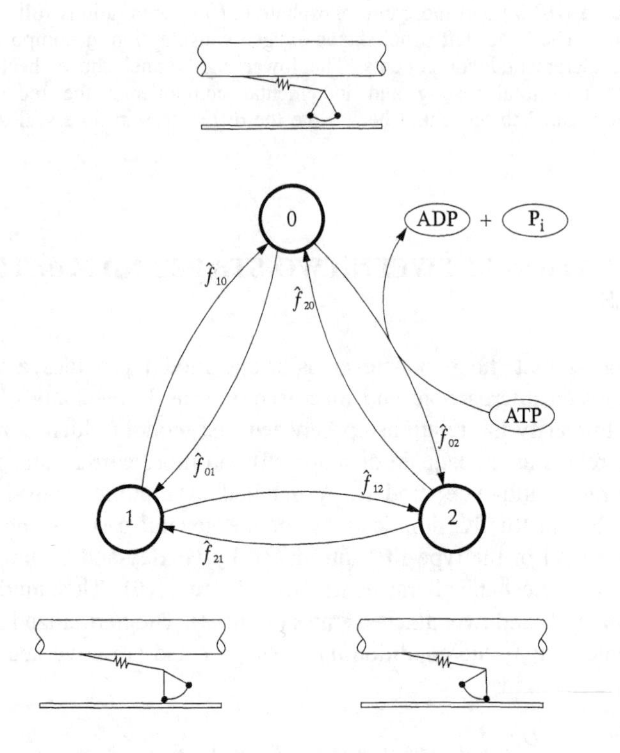

Figure 7.10. The Huxley-Simmons (1971) three-state model.

Note that two time scales have been defined: a fast or 'inner' time, T, and a slow or 'outer' time, t. Further, it is convenient to change the independent variables from $n_1(\hat{x}, \hat{t})$ and $n_2(\hat{x}, \hat{t})$ to:

$$n = n_1 + n_2 \quad \text{and} \quad \theta = n_1 / (n_1 + n_2) \tag{7.21}$$

In terms of the slow outer time the governing equations are transformed into:

$$\frac{Dn}{Dt} = (f_{01} + f_{02})(1-n) - [f_{10}\theta + f_{20}(1-\theta)]n$$

$$\varepsilon \frac{D\theta}{Dt} = [f_{21} - \theta(f_{12} + f_{21})] + \varepsilon[(1-\theta)(f_{01}/n) - \theta(f_{02}/n) + \theta(1-\theta)(f_{20} + f_{02} - f_{10} - f_{01}) - (1-\theta)^2 f_{01} + \theta^2 f_{02}] \tag{7.22}$$

where $(D/Dt) = (\partial/\partial t)_\xi - u(t)(\partial/\partial\xi)_t$, with $u(t) = \hat{v}(t/\hat{f}_0)/h_0\hat{f}_0$.

Clearly, in the 'outer' limit—as $\varepsilon \to 0$ with the outer variables (ξ, t) held fixed—the derivative term on the left of the second of Equation 7.22 vanishes, and it is impossible to satisfy prescribed initial conditions. This is the classic sign of a 'stiff' system of differential equation, which defines singular perturbation problems. Mathematical techniques for analyzing such problems are well known (Van Dyke, 1964; Kevorkian and Cole, 1981) and are based on Prandtl's boundary layer theory of fluid mechanics. 'Inner' and 'outer' expansions are defined, respectively, as follows:

$$n(\xi,t;\varepsilon)=\sum_{k=0}^{\infty}\varepsilon^k n^{(k)}(\xi,t) \quad \text{and} \quad n(\xi,T;\varepsilon)=\sum_{k=0}^{\infty}\varepsilon^k N^{(k)}(\xi,T)$$
$$\theta(\xi,t;\varepsilon)=\sum_{k=0}^{\infty}\varepsilon^k \theta^{(k)}(\xi,t) \quad \text{and} \quad \theta(\xi,T;\varepsilon)=\sum_{k=0}^{\infty}\varepsilon^k \Theta^{(k)}(\xi,T) \tag{7.23}$$

When these expansions are inserted into the governing differential equations and the limits as $\varepsilon \to 0$ are taken with (ξ,t) and (ξ,T) held fixed, respectively, they result in the zero-order outer problem (Zahalak, 2000):

$$\left(\frac{\partial n^{(0)}}{\partial t}\right)_{\xi}-u(t)\left(\frac{\partial n^{(0)}}{\partial \xi}\right)_{t}=(f_{01}+f_{02})(1-n^{(0)})-f_{10}\theta^{(0)}+f_{20}(1-\theta^{(0)})]\,n^{(0)}$$
$$0=f_{21}-(f_{12}+f_{21})\theta^{(0)} \tag{7.24}$$

and the zero-order inner problem:

$$\left(\frac{\partial N^{(0)}}{\partial T}\right)_{\xi}-\lim_{\varepsilon\to 0}\{\varepsilon u(T;\varepsilon)\}\left(\frac{\partial N^{(0)}}{\partial \xi}\right)_{T}=0$$
$$\left(\frac{\partial \Theta^{(0)}}{\partial T}\right)_{\xi}-\lim_{\varepsilon\to 0}\{\varepsilon u(T;\varepsilon)\}\left(\frac{\partial \Theta^{(0)}}{\partial \xi}\right)_{T}=f_{21}-(f_{12}+f_{21})\Theta^{(0)} \tag{7.25}$$

The velocity term in the inner expansion has been left in the form $\lim_{\varepsilon\to 0}\{\varepsilon u(T;\varepsilon)\}$ to provide for the possibility of a step displacement, in which case this limit becomes a Dirac δ-function; if the velocity is bounded, this limit vanishes (Zahalak, 2000). The inner and outer problems can be solved subject to appropriate initial conditions on n and θ, except that the outer solution (the second of Equation 7.24) cannot satisfy arbitrary initial conditions on $\theta^{(0)}$. This loss of initial condition is compensated by connecting the inner and outer solutions by an *asymptotic matching condition* (VanDyke, 1964):

$$\lim_{T\to\infty}\{N^{(0)}(\xi,T)\}=\lim_{t\to 0}\{n^{(0)}(\xi,t)\} \quad \text{and} \quad \lim_{T\to\infty}\{\Theta^{(0)}(\xi,T)=\lim_{T\to\infty}\{\theta^{(0)}(\xi,t)\} \tag{7.26}$$

The asymptotic analysis described above is outlined schematically in Figure 7.11. Once the zero-order approximations to n and θ have been determined by solving the appropriate boundary-value problems, the muscle force, $\hat{P}$, can be calculated by:

$$\hat{P}(t) = \Gamma \int_{-\infty}^{\infty} \left[\xi + (\delta/2)(1-2\theta)\right] n \, d\xi \tag{7.27}$$

where Γ and δ are microstructural parameters. Similarly, if it is assumed that ATP is hydrolyzed in the $2 \rightarrow 0$ transition, then the chemical energy release rate is given by:

$$\left(\frac{d\hat{E}}{d\hat{t}}\right) = \Gamma^* \int_{-\infty}^{\infty} \left[f_{02} + (1-\theta) f_{20}\right] n \, d\xi \tag{7.28}$$

where Γ^* is another microstructural constant.

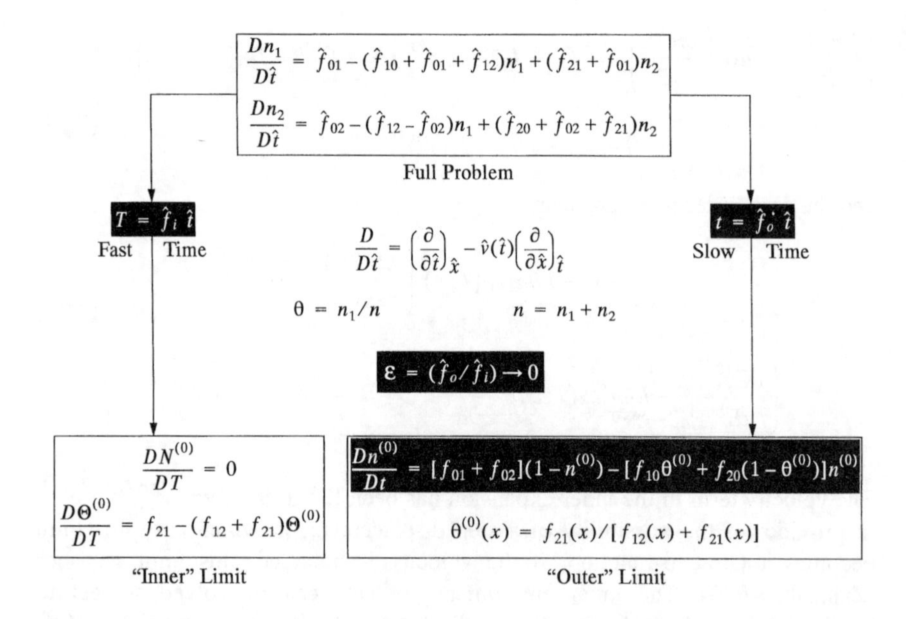

Figure 7.11. Asymptotic analysis of the three-state model of Figure 7.10, based on the assumption that the characteristic rate of transitions between attached states, $\hat{f}_i$, is much greater than the rate of transitions between the detached and attached states, $\hat{f}_0$. The zero-order outer (slow time) rate equation on n is the same as the rate equation of the two-state model (see Figure 7.1), but note that the effective cross-bridge force associated with this equation is non-linear (see text and Figure 7.13).

If this analysis is applied to the case of an instantaneous step length change imposed on an isometrically contracting muscle, the force transients corresponding to the inner and outer solutions are as shown in Figure 7.12. The inner solution (valid for short times) predicts the well-known fast isometric transients, while the outer solution (valid for long times) predicts the gradual return of the force to its pre-perturbation steady state.

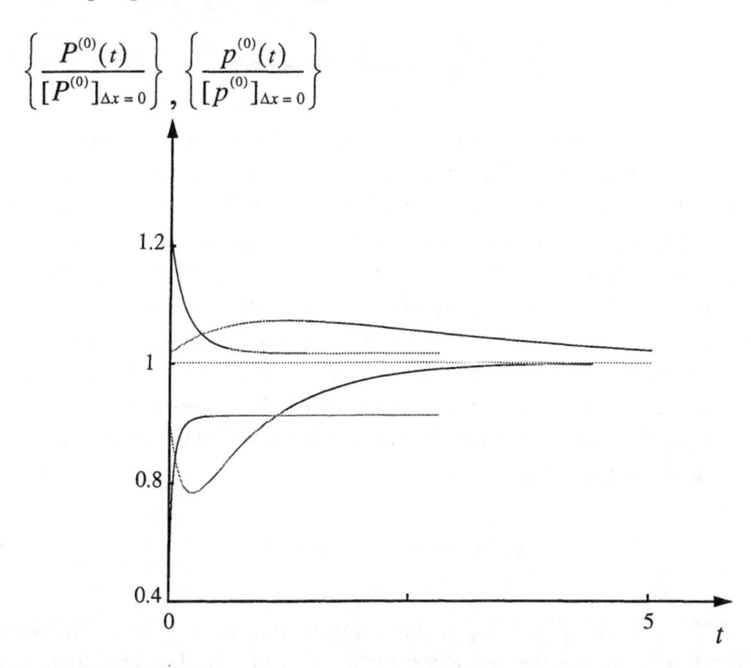

Figure 7.12. Isometric force transients produced by step length changes; one shortening step and one lengthening step are displayed. For each step, the force vs. time is calculated from the inner equations (valid for $t \to 0$) and the outer equations (valid for $t \to \infty$). The grey levels on the inner and outer curves are meant to suggest regions of validity—darker implies increasing validity. (Reproduced from Zahalak, 2000, with the permission of Academic Press.)

The second of the inner equations, Equation 7.25, implies that θ approaches the asymptotic value:

$$\theta^{(0)}(\xi,T) \to \theta^{(0)}(\xi) = f_{21}(\xi) / [f_{12}(\xi) + f_{21}(\xi)] \tag{7.29}$$

with a fast time constant of the order $\hat{f}_i^{-1}$. This asymptotic value is a time-independent function of cross-bridge strain, and is the zero-order approximation to θ in the outer solution *for all times*. For the present purpose, it is instructive to examine the form of the outer equations (Equation 7.24). If we define:

$$f(\xi) = f_{01}(\xi) \quad \text{and} \quad f'(\xi) = f_{10}(\xi)\theta^{(0)}(\xi)$$

$$g'(\xi) = f_{02}(\xi) \quad \text{and} \quad g(\xi) = \left(1-\theta^{(0)}(\xi)\right)f_{20}(\xi)$$

then the first outer equation is identical to Equation 7.2, the T.L. Hill formulation of the A.F. Huxley two-state model. If we redefine $(f+g')$ as f and $(g+f')$ as g, then we recover Equation 7.1, the rate equation of Huxley's 1957 two-state model:

$$\frac{\partial n}{\partial t} - u(t)\frac{\partial n}{\partial \xi} = f(\xi)(1-n) - g(\xi)n \tag{7.30}$$

Thus, formal asymptotic analysis reveals the traditional two-state model to be the zero-order outer limit of a three-state model with two attached states. The two-state model can be expected to produce reasonably accurate results if the strain rate experienced by the muscle is slow compared to the characteristic transition rate between attached states. Then, viewed on a slow outer time scale, cross-bridges rapidly assume an equilibrium distribution among attached states; this distribution is described by $\theta^{(0)}(\xi)$ and depends on cross-bridge strain.

But there are also some substantial differences between the traditional two-state cross-bridge theory and the outer limit of the three-state model. These become apparent if one examines the force, which, following Equation 7.27, in the zero-order outer limit is:

$$(\hat{P}/\Gamma) \rightarrow p^{(0)} = \int_{-\infty}^{\infty} F(\xi)n\,d\xi \tag{7.31}$$

where $F(\xi)=\xi + (\delta/2)(1-2\theta^{(0)}(\xi))$. An analysis similar to that of Huxley and Simmons (1971) shows that, as a consequence of kinetic detailed-balance relations, $\theta^{(0)}(x)$ has the specific form:

$$\theta^{(0)}(\xi) = \frac{1}{2}\left[1 + Tanh\{\frac{\omega}{2}(\xi + \xi_0)\}\right] \tag{7.32}$$

where ω and ξ_0 are microstructural constants that depend on the energy difference between the attached states.

Figure 7.13 displays graphs of F and $\theta^{(0)}$ as functions of bond length. Equation 7.31 states that in the outer limit, the apparent force exerted by an attached cross-bridge, $F(\xi)$, is a non-linear function of cross-bridge strain (this is true even if the actual force in each attached state varies linearly with cross-bridge strain). The reason for the non-linearity is that attached cross-bridges tend to move from a more-strained state to a less-strained state as the strain increases. In the traditional two-state model, it is almost invariably assumed that the cross-bridge force is proportional to cross-bridge strain, so that $F(\xi)= (\xi - \xi^*)$. Equation 7.31 and Equation 7.32 show that this is at best a rough approximation, and this should be borne in mind when interpreting the results of the standard two-state cross-bridge model.

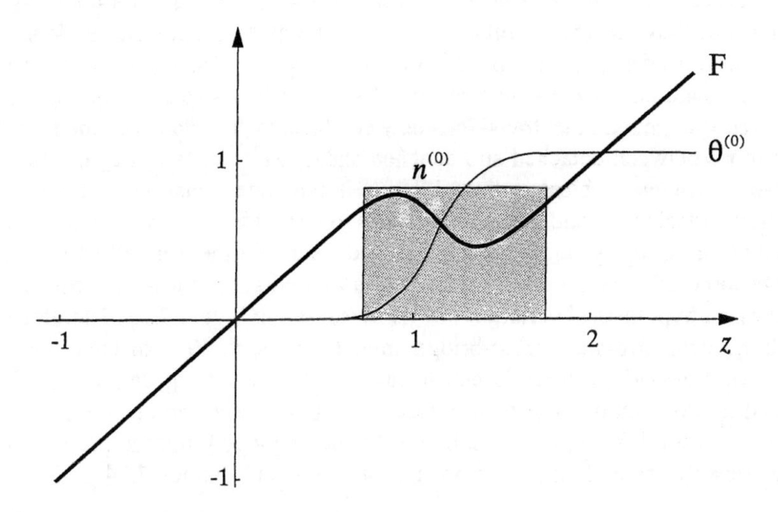

Figure 7.13. Variables associated with the asymptotic outer solution of the three-state model: the effective cross-bridge force, *F(z)*; the strain-dependent equilibrium distribution of cross-bridges among attached states, $\theta^{(0)}(z)$; the steady isometric distribution of attached cross-bridges, $n^{(0)}(z)$. (Reproduced from Zahalak, 2000, with the permission of Academic Press.)

Similar considerations apply in the case of the energy release rate. If we assume that $f_{02} = 0$ (as this transition implies ATP synthesis on cross-bridge attachment), then Equation 7.28 states that the chemical energy release rate is:

$$\left(\Gamma^{*}\right)^{-1}\left(\frac{d\hat{E}}{dt}\right)=\dot{E}=\int_{-\infty}^{\infty}(1-\theta^{(0)})f_{20}\,n\,d\xi \tag{7.33}$$

Qualitatively, it can be seen from this equation that the energy release rate tends to decrease as the stretch velocity increases, in general agreement with experimental observations. This is because as the stretch velocity increases, attached cross-bridges are pulled into regions of larger x, where $\theta^{(0)} \rightarrow 1$ and the integrand of Equation 7.32 decreases (if f_{20} does not increase too rapidly with x). This behaviour resolves a minor paradox in the original Huxley two-state model, which had the energy release rate proportional to the *total* detachment rate, which in the present notation is:

$$\int_{-\infty}^{\infty}\left[f_{10}+(1-\theta^{(0)})f_{20}\right]n\,d\xi$$

It can be shown that, under this assumption, the isotonic energy release rate is an *even* function of velocity, which is not observed experimentally. Equation 7.32, through the asymmetry of $\theta^{(0)}$, provides for an asymmetric energy rate-velocity relation.

The preceding detailed results have been obtained specifically for a three-state model, but they suggest qualitative generalizations to multi-state models. For example, consider the situation illustrated in Figure 7.14, where there is one detached state and several (in this case, four) attached states. Assuming, as in the three-state model, that transitions between the attached states are much faster than those between detached and attached states, we may expect again that the zero-order outer problem will yield the Huxley rate equation, but with the apparent attachment and detachment rates depending on a quasi-equilibrium distribution of cross-bridges among attached states. Further, the effective cross-bridge force, F, depends on this distribution as well as the intrinsic cross-bridge stiffness; a hypothetical variation of F is illustrated in Figure 7.14. Therefore, in working with two-state cross-bridge models, it is to be expected that the effective cross-bridge force is non-linear with the saturating behaviour of the dashed approximation shown in Figure 7.14. If the cross-bridge force is non-linear, then the DM approximation can still be employed, but the muscle force becomes a function of all three moments, as shown in Equation 7.34.

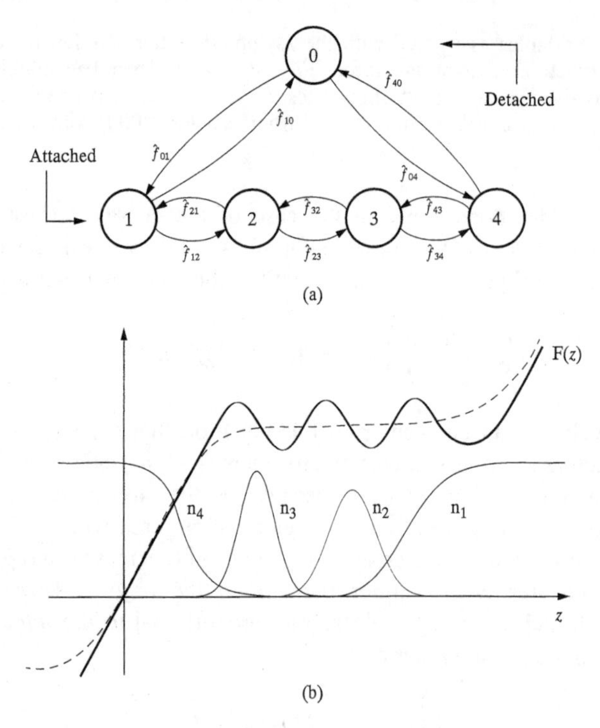

Figure 7.14. The upper part shows a kinetic diagram of a five-state cross-bridge model with one detached and four attached states. The lower part shows qualitative populations of attached cross-bridge states (n_1, n_2, n_3, n_4) as a function of bond length (thin curves). The solid curve represents the qualitative effective cross-bridge force as a function of

cross-bridge length, and its smoothed approximation is represented by a dashed curve. The approximation admits the possibility of cross-bridge buckling under compressive strain. (Reproduced from Zahalak, 2000, with the permission of Academic Press.)

$$\hat{P}(Q_0,Q_1,Q_2) = \Gamma \int_{-\infty}^{\infty} \frac{Q_0}{\sqrt{2\pi}q} e^{-\frac{(\xi-p)^2}{2q^2}} F(\xi)d\xi \tag{7.34}$$

which reduces to $\hat{P} = \Gamma Q_1$ if $F(\xi) = \xi$.

The asymptotic analysis resolves another apparent difficulty associated with the DM model. Cole et al. (1996) compared the predictions of the DM model in isovelocity stretch to published experimental measurements in slow [strain rate = $O(0.1\ s^{-1})$] and fast [strain rate $=O(10\ s^{-1})$] stretches, where the strain rates differed by two orders of magnitude. They found that the force transients could not be predicted accurately in both slow and fast stretch by the DM model (and, by implication, the two-state cross-bridge model) unless the actin-myosin interaction rates were increased by about two orders of magnitude when the speed increased. This need to adjust the rate parameters was regarded as a deficiency of the DM/cross-bridge model. But the asymptotic analysis presented here reveals that the required change in parameters reflects the change from slow 'outer' dynamics to fast 'inner' dynamics produced by the large change in stretch speed.

SOME ADDITIONAL CONSIDERATIONS

Even if we restrict ourselves to the two-state cross-bridge theory as a basis for macroscopic muscle mechanics, several other considerations impinge on the development of the theory. Detailed discussion of these considerations is beyond the scope of this presentation, but some of them should at least be mentioned here.

REMOTE ACTIN BINDING SITES

The simplest two-state cross-bridge model envisions that at every instant a cross-bridge interacts with just one actin binding site—which can for convenience be designated as the 'nearest'. But simulations, particularly those involving fast stretch, often show large strains of attached cross-bridges, suggesting that cross-bridges may be bound to more remote binding sites, not only the nearest. The two-state cross-bridge theory can be modified to include the possibility of remote-site interactions, and such an enhanced theory was published by Yu et al. (1997). In this case, the differential-rate equation of the standard two-state theory becomes a differential-difference equation:

$$\frac{\partial n}{\partial t} - u\frac{\partial n}{\partial \xi} = f(1-n) - gn - f\sum_{i=1}^{\infty}\left\{n\left(\xi + \frac{i}{b}\right) + n\left(\xi - \frac{i}{b}\right)\right\} \tag{7.35}$$

The terms in the summation can be thought of as a correction to the standard rate equation. Typical effects of these remote-site interactions on force-velocity characteristics are illustrated in Figure 7.15. Increasing the number of remote site interactions decreases the force; this is because a cross-bridge has fewer accessible actin binding sites if some of these are already occupied by neighbouring cross-bridges. This is yet another possible mechanism to limit the muscle force in stretch. (The other two are high detachment rates at large stretch and, possibly, a non-linear saturating effective cross-bridge force (see Figure 7.13 and Figure 7.14).)

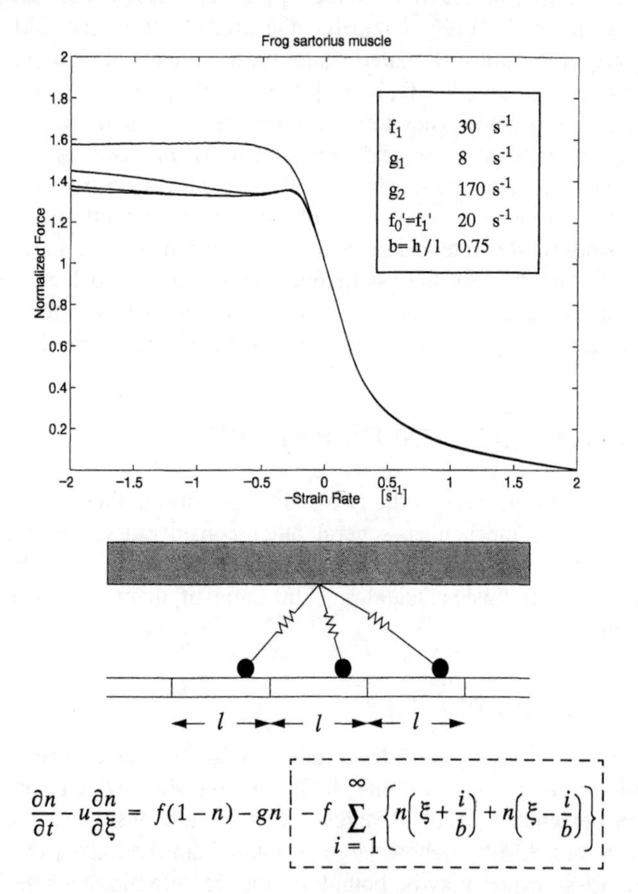

$$\frac{\partial n}{\partial t} - u\frac{\partial n}{\partial \xi} = f(1-n) - gn \boxed{- f\sum_{i=1}^{\infty}\left\{n\left(\xi + \frac{i}{b}\right) + n\left(\xi - \frac{i}{b}\right)\right\}}$$

Figure 7.15. Steady-state force-velocity curves for a two-state cross-bridge model that permits cross-bridge binding to remote actin sites (Yu et al., 1997). The boxed terms in the rate Equation 7.35 represent a correction to the standard two-state rate equation that accounts for binding to sites beyond the nearest. The successive curves in the graph show the reduction of the force as successively more remote interactions are permitted. The force-velocity curve attains a limiting shape after two orders of interaction ($i = 2$). In this calculation, the ratio of the bond-length scaling parameter, h, to the distance between actin binding sites is 3/4. The DM approximation was employed for this calculation.

INHOMOGENEITY

While it is very convenient to assume so for modelling purposes, muscles are not composed of identical, uniform fibres running from origin to insertion. It has long been recognized that sarcomere inhomogeneities may affect the macroscopic mechanical behaviour of muscle (Huxley and Peachey, 1961). Figure 7.16(a) (from Motabarzadeh, 1998) shows a model of fibre consisting of a chain of DM elements with distributed non-linear parallel elasticity. The only inhomogeneity in this model is initial sarcomere length in the unstimulated fibre, determined by the neutral lengths of the parallel elastic elements. The simulation displayed in Figure 7.16(b) corresponds to activating the isometric fibre maximally at $t=0$, and shows that an inhomogeneous fibre model can predict, at least qualitatively, two of the 'anomalous' mechanical behaviours that have been observed in both muscle fibres and whole muscle: force enhancement on stretch, and isometric force creep. Two other observed anomalous behaviours—force depression on shortening and the inconsistency between isometric and isotonic static length-tension curves—can also be reproduced qualitatively by this inhomogeneous fibre model (Motabarzadeh, 1998). But we have not been able to explain all the observed anomalies of muscle mechanics with this model—in particular, the apparent inconsistency between the sarcomere and whole-muscle length-tension relation in the cat soleus (Herzog et al., 1992). The effects of sarcomere inhomogeneity are difficult to incorporate rationally into lumped muscle models, but they should be accounted for in some fashion, if only as error estimates.

NON-AXIAL DEFORMATIONS AND STRESSES

There is a growing interest among investigators in the three-dimensional distributions of stress and strain within actively contracting muscles (Huijing, 1999; Monti et al., 1999). For example, Figure 7.17 shows some stress distributions in an isometric rat gastrocnemius muscle calculated with a finite-element model (Gielen, 1998). The model consists of active DM-model fibres embedded in a passive hyper-elastic matrix. Thus, muscle is treated as a three-dimensional, inhomogeneous contractile continuum. Under these circumstances, muscle fibres are subjected to deformations in directions other than along the fibre axes. This raises two important questions. Do non-axial deformations affect active axial muscle stress? Do non-axial deformations produce active non-axial (e.g., shear) stresses? There is little relevant experimental data in the literature, but I have proposed a tentative generalization of the two-state cross-bridge model (Zahalak, 1996) to address these non-axial effects. Figure 7.18 shows typical steady-state stresses predicted by this theory when the contractile tissue is sheared parallel to the fibres at a constant rate. This theory predicts that non-axial deformations indeed affect axial muscle stress, and this effect may be large. Further, non-axial deformations generate non-axial active muscle stresses

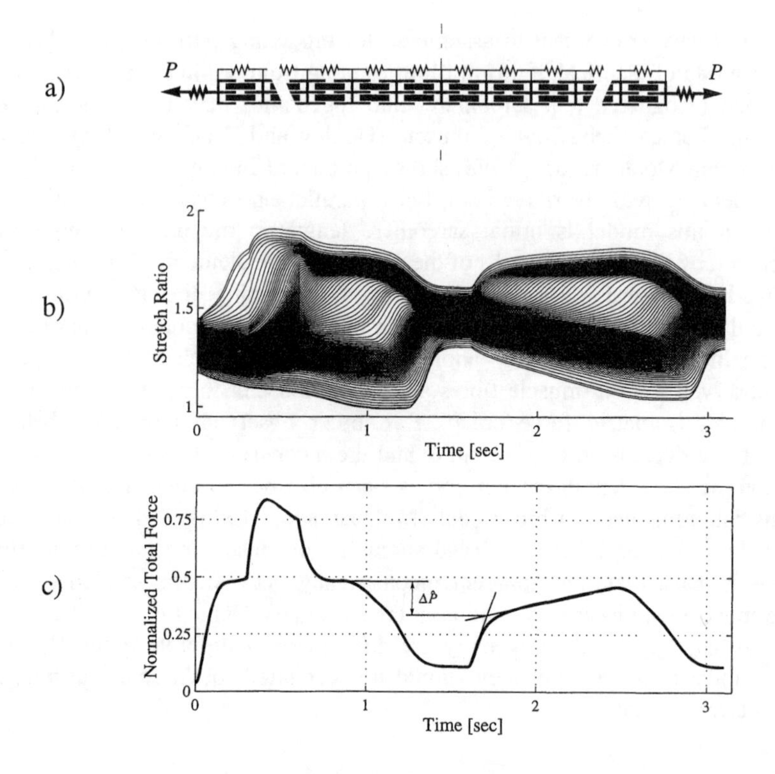

Figure 7.16. The upper panel shows a model of a muscle fibre as a chain of DM models with a distributed variable parallel elasticity. The middle panel shows stretch ratio vs. time of several segments along the muscle fibre. The fibre is first held isometric and activated, then stretched to a longer length, then allowed to relax, and finally re-activated at the longer length. The lower panel shows muscle force vs. time corresponding to the strains shown in the middle panel. This model exhibits both isometric creep and force enhancement on stretch. (From Motabarzadeh, 1998.)

that are absent under purely axial loading, but these stresses appear to be small compared to the axial stress. While these results are suggestive, more experimental and theoretical attention needs to be directed at non-axial effects before they are clearly understood.

CONCLUSIONS

Two-state cross-bridge models can serve as a vehicle to connect the rapidly progressing field of molecular muscle biophysics to macroscopic biomechanics. While inadequate to explain all details of force generation at the molecular level, they are

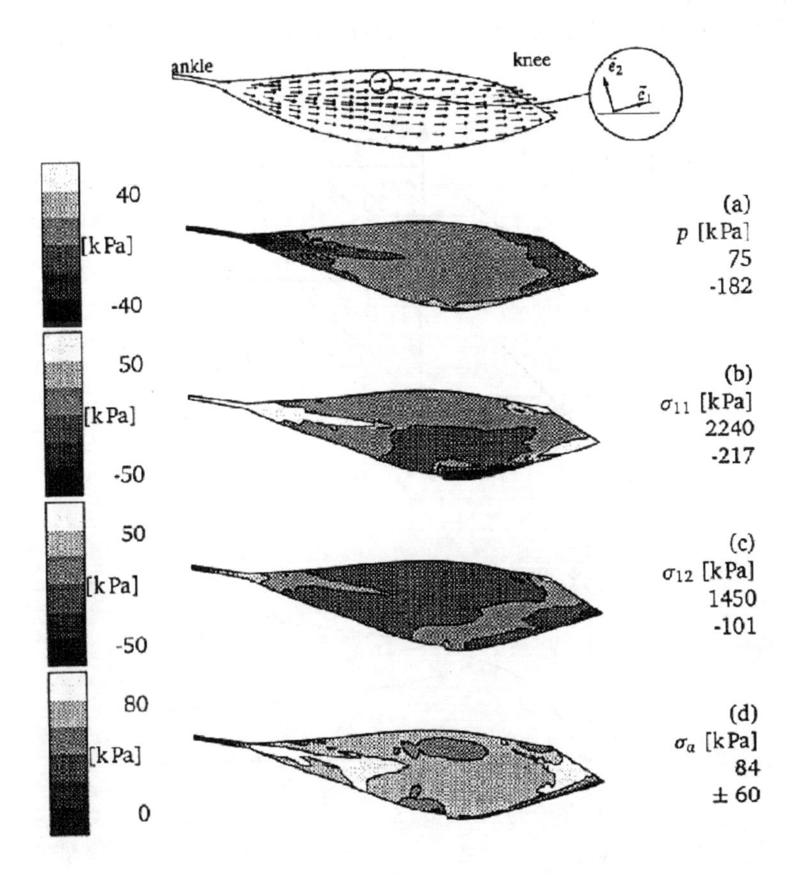

Figure 7.17. Distributed stresses in a rat tibialis anterior muscle predicted by a DM finite-element continuum model: pressure (p), total axial fibre stress (σ_{11}), shear stress (σ_{12}), and active contractile fibre stress (σ_a). (From Gielen, 1998.)

mathematically tractable and contain the essential kernel of molecular contractile dynamics: actin-myosin binding. When subjected to the distribution-moment (DM) approximation, they yield tractable state-variable models of muscle that can encompass not only contraction but also activation and energetics, and even some aspects of metabolic dynamics. The DM approximation also clarifies the connection between cross-bridge models and the well established traditional A.V. Hill phenomenological model of muscle. The rate equations of A.F. Huxley and T.L. Hill for the two-state cross-bridge models can be regarded as the asymptotic, zero-order 'outer' limit of multi-state cross-bridge models, valid on the relatively slow time scales appropriate to whole-body mechanics. But, in contrast to the usual two-state cross-bridge theory, asymptotic analysis indicates that the effective cross-bridge force is a non-linear function of strain, with characteristics affected by a fast-equilibrium distribution of cross-bridges among the attached states.

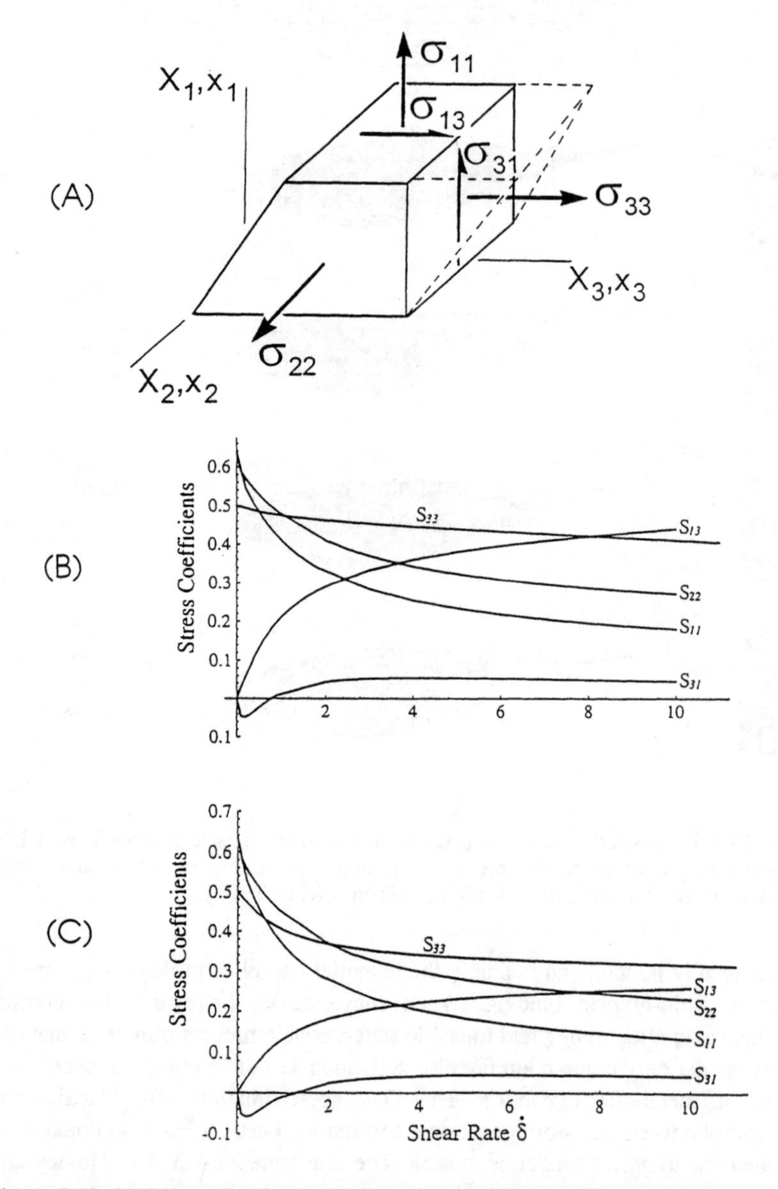

Figure 7.18. Variation of active stress components with shear rate, according to a three-dimensional generalization of the two-state cross-bridge model, for shear at a constant rate parallel to the fibres. The upper panel shows definition of the stress components. The lower two panels show stress vs. shear rate for two different sets of attachment and

detachment rate functions. Note that (1) non-axial deformation reduces the axial stress, $\sigma_{33} \propto S_{33}$, and (2) non-axial deformation produces non-axial active stresses, $\sigma_{13} \propto S_{13}$, etc. (Reproduced from Zahalak, 1996, with the permission of Academic Press.)

ACKNOWLEDGEMENTS

This work was supported by the U.S. National Science Foundation under NSF Grant BES–9626736.

REFERENCES

Cannell, M.B., and Allen, D.G. (1984). Model of calcium movement during activation in the sarcomere of frog skeletal muscle. *Biophysical Journal* **45**, 913–925.

Cole, G.K., van den Bogert, A.J., Herzog, W., and Gerritsen, K.G. (1996). Modeling of force production in skeletal muscle undergoing stretch. *Journal of Biomechanics* **29**(8), 1091–1104.

Gielen, A.W.J. (1998). A continuum approach to the mechanics of contracting skeletal muscle. Doctoral dissertation, Department of Mechanical Engineering, Eindhoven University of Technology, Eindhoven, The Netherlands.

Herzog, W., Kamal, S., and Clarke, H.D. (1992). Myofilament lengths of cat skeletal muscle; theoretical considerations and functional implications. *Journal of Biomechanics* **25**(8), 945–948.

Hill, A.V. (1938). The heat of shortening and the dynamic constants of muscle. *Proceedings of the Royal Society of London B*, **126**, 136–195.

Hill, T.L. (1977). *Free Energy Transduction in Biology*. Academic Press, New York.

Hill, T.L., Eisenberg, E., and Chen, Y. (1975). Some self-consistent two-state sliding filament models of muscle contraction. *Biophysical Journal* **15**, 335–372.

Huijing, P.A. (1999). Muscle as a collagen fiber reinforced composite: A review of force transmission in muscle and whole limb. *Journal of Biomechanics* **32**(4), 329–345.

Huxley, A.F. (1957). Muscle structure and theories of contraction. *Progress in Biophysics and Biophysical Chemistry* **7**, 225–318.

Huxley, A.F. (1988). Muscular contraction. *Annual Review of Physiology* **50**, 1–16.

Huxley, A.F., and Peachey, L.D. (1961). The maximum length of contraction in vertebrate skeletal muscle. *Journal of Physiology* **156**, 150–165.

Huxley, A.F., and Simmons, R.M. (1971). Proposed mechanism of force generation in striated muscle. *Nature* **233**, 533–538.

Inesi, G. (1979). Transport across sarcoplasmic reticulum in skeletal and cardiac muscle. In: *Membrane Transport in Biology*. Giebisch, G., Tosteson, D.C., and Ussing, H.H. (eds.), Springer-Verlag, Berlin.

Kevorkian, J., and Cole, J.D. (1981). *Perturbation Methods in Applied Mathematics*. Springer, New York.

Ma, S.P., and Zahalak, G.I. (1987). A simple self-consistent distribution-moment model for muscle: Energy and heat rates. *Mathematical Biosciences* **24**(1), 211–230.

Ma, S.P., and Zahalak, G.I. (1991). A distribution-moment model of energetics in skeletal muscle. *Journal of Biomechanics* **84**(1), 21–35.

Monti, R.J., Roy, R.R., Hodgson, J.A., and Edgerton, V.R. (1999). Transmission of forces within mammalian skeletal muscles. *Journal of Biomechanics* **32**(4), 371–380.

Motabarzadeh, I. (1998). A distribution-moment model of mechanical instability in non-uniform skeletal muscle fibers. D.Sc. Dissertation, Department of Mechanical Engineering, Washington University, St. Louis, Missouri.

Partridge, L.D. (1965). Modification of neural output signals by muscles: A frequency-response study. *Journal of Applied Physiology* **20**, 150–156.

Propp, M.B. (1986). A model of muscle contraction based upon component studies. *Lectures on Mathematics in the Life Sciences* (American Mathematical Society, Providence, Rhode Island) **16**, 61–119.

Rall, J.A. (1982). Energetics of Ca^{2+} cycling during skeletal muscle contraction. *Federation Proceedings* **41**, 155–160.

Robertson, S.P., Johnson, J.D., and Potter, J.D. (1981). The time course of Ca^{2+} exchange with calmodulin, troponin, parvalbumin, and myosin in response to transient increases in Ca^{2+}. *Biophysical Journal* **34**, 559–569.

Rouhaud, E. (1993). A distribution-moment model for coupled contraction and metabolic dynamics in skeletal muscle, D.Sc. Dissertation, Department of Mechanical Engineering, Washington University, St. Louis, Missouri.

VanDyke, M. (1964). *Perturbation Methods in Fluid Mechanics*. Academic Press, New York.

Yanagida, T. (1999). Simultaneous observation of individual ATPase and mechanical events by a single myosin molecule during interaction with actin. *Canmore Symposium on Skeletal Muscle*, August 6-7, 1999, Canmore, Alberta, Canada.

Yu, S.-N., Crago, P.E., and Chiel, H.J. (1997). A nonisometric kinetic model for smooth muscle. *American Journal of Physiology* **272,** *Cell Physiology* **41**, C1025–C1039.

Zahalak, G.I. (1981). A distribution-moment approximation for kinetic theories of muscular contraction. *Mathematical Biosciences* **55**, 89–114.

Zahalak, G.I. (1996). Non-axial muscle stress and stiffness. *Journal of Theoretical Biology* **182**, 59–84.

Zahalak, G.I. (2000). The two-state cross-bridge model of muscle is an asymptotic limit of multi-state models. *Journal of Theoretical Biology* (in press).

Zahalak, G.I., and Ma, S.P. (1990). Muscle activation and contraction: constitutive relations based directly on cross-bridge kinetics. *Journal of Biomechanical Engineering* **112**, 52–56.

Zahalak, G.I., and Motabarzadeh, I. (1997). A re-examination of calcium activation in the Huxley cross-bridge mode. *Journal of Biomechanical Engineering* **119**, 20–29.

8 A General Mathematical Framework for Cross-bridge Modelling

RACHID AIT-HADDOU AND WALTER HERZOG
Human Performance Laboratory, University of Calgary, Calgary, Alberta, Canada

ABSTRACT

In this chapter, we give a general mathematical framework underlying cross-bridge modelling. For simplicity of illustration, the framework is built on the assumptions that the force generators (cross-bridges) are independent of each other and independent of the attachment history, that the two sets of filaments (actin and myosin) are rigid, and that the cross-bridge can only attach to the nearest actin site. However, these assumptions are not crucial or limiting for our approach, as any new assumption or modification can be incorporated into the proposed formalism.

INTRODUCTION

In the 1940s, muscle contraction was widely believed to be caused by contraction and the corresponding shortening of protein filaments within the sarcomere. However, careful electron microscopic studies and great advances in X-ray diffraction techniques have provided detailed information on the structure and biochemistry of striated muscle in the past few decades. This remarkable progress helped to elucidate, at least to a certain extent, the molecular mechanisms underlying muscular contraction. The principal molecular participants in the process of contraction have been identified as: myosin, actin filaments (G-actin, tropomyosin, troponin), ATP (adenosine triphosphate), water, Mg^{++}, and Ca^{++}. A crucial first step was the discovery of the geometric arrangement of actin and myosin within the sarcomere—a step that has been largely achieved by the work of H. E. Huxley and his collaborators (Huxley, 1953; Huxley and Brown, 1967). The myosin exists as a set of parallel filaments located at the centre of the sarcomere, whereas the actin filaments extend from the Z-bands into the centre of a sarcomere. Myosin contains a long tail consisting of light meromyosin (LMM), and two globular head portions consisting of heavy meromyosin (HMM). Actin is a double-stranded helical fibre formed by a polymerization process from identical actin molecules in the globular form (G-actin). This leads

Skeletal Muscle Mechanics: From Mechanisms to Function. Edited by W. Herzog.

to a regular and periodic structure of the actin filament. Each G-actin is asymmetric, with a definite polarity. When actin monomers assemble to form the actin filament, the sense of polarity is the same in all molecules of the filament, so that the entire chain is endowed with the same polarity (Jülicher, 1999). This polarity determines the direction of motion during contraction. In 1954, H. E. Huxley, J. Hansen, A. F. Huxley, and R. Niedergerke independently refuted any attempts to explain the process of muscular contraction by a change in myosin filament length, by showing that the length of the thick filaments did not change appreciably during active and passive shortening and stretching of single muscle fibre and isolated myofibrils. Based on these findings, the 'sliding filament theory' was proposed, in which the contraction is caused by the relative sliding of the thin and thick filaments (Huxley, 1957). The energy of contraction was associated with the hydrolysis of ATP molecules. More precisely, the myosin head is thought to bind ATP (with Mg^{++} as a co-enzyme), attach to the active site on the actin filament (provided that calcium is bound to the high-affinity sites on troponin C, which causes a removal of inhibition for cross-bridge attachment), break the phosphate group by hydrolysis, liberate an extra amount of free energy, and release the products ADP (adenosine diphosphate) and P (phosphate). The energy liberated in this way is used by the myosin head to pull the actin filament past the myosin filament. During this process, the myosin head undergoes conformational changes and is thought to have hydrolyzed one ATP molecule (Figure 8.1).

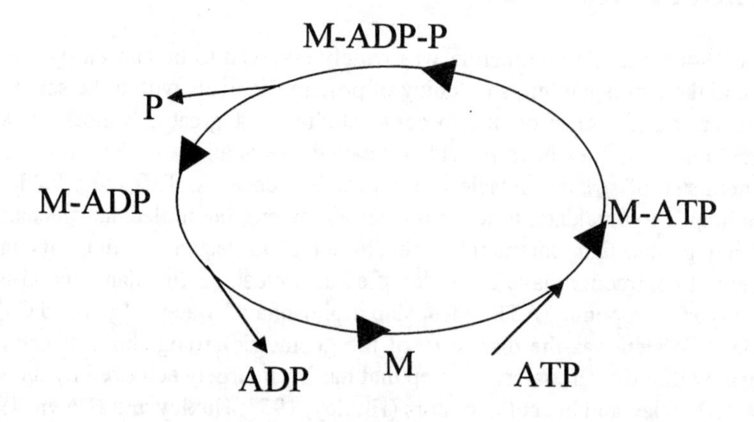

Figure 8.1. Chemical cycle of the myosin head, M.

In order to account for the experimental findings of Ramsey and Street (1940), i.e., that the isomeric force developed by a tetanically stimulated muscle decreases linearly (Gordon et al., 1966) with length when stretched beyond the resting length, it was assumed that the force generators that cause filament sliding are independent of each other. One of the most spectacular predictions of the cross-bridge theory is the shape of the force-length relationship without any analytical considerations, only structural ones (Gordon et al., 1966).

However, despite the remarkable progress made in understanding the molecular mechanisms underlying muscular contraction, it appears that the precise relationship between the elementary steps of the enzymatic ATPase activity of actomyosin, the mechanical interactions between the two sets of filaments, and the structural change of the myosin head during the contractile process have not been completely identified. Therefore, even if the major steps underlying the contractile process are known, other steps have been incorporated into the model to account for unexplained experimental observations. This leads to a growing number of cross-bridge models with an increasing number of biochemical steps (Hill, 1974; Huxley and Simmons, 1971; Propp, 1986). The first such model was introduced by Huxley in 1957. The contractile mechanism was modelled with only two generic states: one attached and one detached cross-bridge state. In order to account for the rapid tension transients after a sudden length step, Huxley and Simmons (1971) proposed a model with multiple attached states that had different properties, and one detached state. In order to account for certain thermodynamic constraints, Hill (1974) modified the original theory by incorporating conditions on the backward and forward rate reactions.

The original cross-bridge models did not account for the level of activation. In order to incorporate the activation process and its effect on contraction, Zahalak and Ma (1990) developed a model with a separate pathway for the calcium transients representing muscle activation.

As stated above, new variations of the cross-bridge model are generated virtually on a daily basis. These models are conceptually all similar, but have their own distinct features to account for specific experimental observation. It almost seems that every new property of muscle is guaranteed with a slightly modified or adapted model of an existing model. However, as people adapt one model to account for a specific experimental observation, they may lose accurate predictions of the original model on another experimental observation. This state of affairs has led to a situation in which many models may predict a newly discovered phenomenon really well, but cannot account adequately any longer for some long-standing, well-accepted result. There is no framework that incorporates 'all' cross-bridge models on a conceptual level.

In this paper, we give a general mathematical framework that conceptually incorporates all these models. For simplicity of illustration, this framework is built on the assumptions that the force generators (cross-bridges) are independent of each other and independent of the attachment history, that the two sets of filaments (actin and myosin) are rigid, and that cross-bridges can only attach to the nearest actin site. However, these assumptions are not crucial for the formulation of our mathematical framework; they are merely convenient and can be changed, deleted, or otherwise modified within the basic approach presented here.

THE DYNAMICS OF A SINGLE MYOSIN HEAD

We assume that the chemistry of one myosin head, or one cross-bridge, can be described by a number m of discrete states or conformations $i=1,2,...,m$. Consider a myosin head in the conformation state i, and denote by x the positive displacement of the cross-bridge to the nearest binding site on actin from the resting position of the myosin head (Figure 8.2).

The conformation of the system is fully characterized by the pair $\{i, x\}$ of internal state and position with respect to the closest actin site.

The chemical reaction between states $\{i, x\}$ and $\{j, x\}$ is assumed to follow the Poisson statistics with reaction rates $\omega_{ij}(x)$:

$$\{i, x\} \underset{\omega_{ji}(x)}{\overset{\omega_{ij}(x)}{\rightleftarrows}} \{j, x\} \tag{8.1}$$

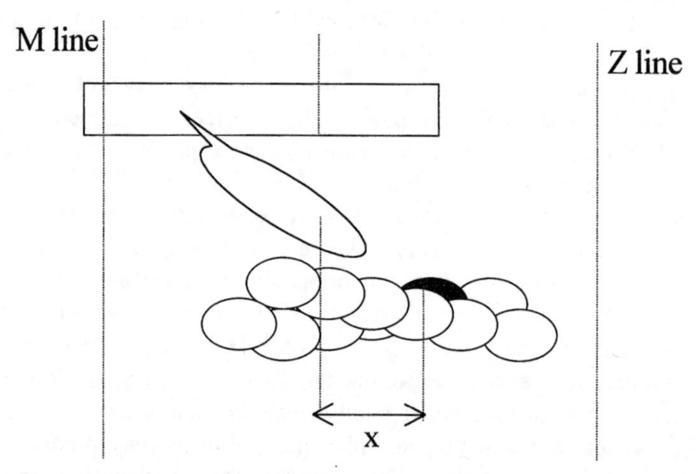

Figure 8.2. A schematic representation of the configuration of a myosin head and the actin filament. The x-value is defined as the positive displacement of the myosin head to the nearest binding site on the actin filament from the resting position of the myosin head.

The reaction rates are, in general, x-dependent since it is assumed that the myosin head changes its shape during the displacement. We also assume that the transition between states happens instantaneously, without a change in the x-value. This assumption may be justified by the fact that the thermal relaxation is very fast compared to the chemical cycle (Jülicher, 1999). The full characterization of the system can be represented by a diagram (Figure 8.3) popularized by the work on muscle by T. L. Hill (1974, 1975).

Denote by $W(x)$ the matrix defined by:

$$W(x) = W_{nn'}(x) = \begin{cases} \omega_{nn'}(x) & for \quad n \neq n' \\ -\sum_{j=1, j\neq n'}^{m} \omega_{nj}(x) & for \quad n = n' \end{cases} \tag{8.2}$$

The matrix $W(x)$ of the cross-bridge model gives a complete characterization of the model and satisfies the two properties:

$$W_{nn'}(x) \geq 0 \quad for \quad n \neq n'; \quad \sum_{n'=1}^{m} W_{nn'}(x) = 0 \tag{8.3}$$

Any matrix that satisfies the relations given in Equation 8.3. is called a *Q*-matrix. From the diagram in Figure 8.3, we recognize that the dynamics of a single myosin head can be described by a continuous-time Markov chain in which the rate constants are x-dependent. Therefore, if we denote by $p_i(x,t)$ the probability distribution for the myosin head to be at position x, at time t, in state i, then the probability distribution obeys the master equations (or the Chapman-Kolmogorov equations) (Haken, 1977):

$$\frac{dp\,(t,x)}{dt} = p(x,t)W(x) \tag{8.4}$$

where $p(x,t)=(p_1(x,t),\dots,p_m(x,t))$. From the relations in Equation 8.3, we deduce that $W(x)$ has a left eigenvector, $\phi(x)$, with zero eigenvalue. Each $\phi(x)$ is a time-independent solution of the master equation. When it is normalized, it represents a stationary probability distribution of the system (the steady-state distribution). Cross-bridge models are influenced by the structure of the matrix $W(x)$. If the matrix $W(x)$ of the model is 'decomposable'—that is, by a suitable simultaneous permutation of rows and columns, it can be written in the form:

$$\begin{pmatrix} A(x) & 0 \\ 0 & B(x) \end{pmatrix}$$

where $A(x)$ and $B(x)$ are two square matrices—then the states of the model can be decomposed into two subsets into which no transition is possible. This merely means that the model can be decomposed into two or more non-interacting models. If the matrix $W(x)$ is 'reducible'—that is, by a suitable simultaneous permutation of rows and columns, it can be written in the form:

$$\begin{pmatrix} A(x) & 0 \\ D(x) & B(x) \end{pmatrix}$$

where $A(x)$ and $B(x)$ are square matrices, but $D(x)$ is not a square matrix in general, then the states of the model can be decomposed into two sets of states, S_1 and S_2, with the following two master equations:

$$\begin{aligned} \frac{dp_{S_1}(x,t)}{dt} &= p_{S_2}(x,t)D(x) + p_{S_1}(x,t)A(x) \\ \frac{dp_{S_2}(x,t)}{dt} &= p_{S_2}(x,t)B(x) \end{aligned} \tag{8.5}$$

The solution of the last equation can be obtained without knowing $p_{S_1}(x,t)$. But $p_{S_2}(x,t)$ loses all probabilities to the states S_1. A solution of the master equation can only be stationary if all the probabilities of the states in S_2 are equal to zero. In this case, states in S_1 are called 'absorbing' states, whereas the states in S_2 are called 'transient'. This last case represents the original Huxley model in which the detached state becomes an absorbing state for a negative *x*-distance, whereas the model has an irreducible matrix for all positive *x*-values. If the matrix *W(x)* is 'splitting', for example, it can be written as:

$$\begin{pmatrix} A(x) & 0 & 0 \\ 0 & B(x) & 0 \\ D(x) & E(x) & C(x) \end{pmatrix}$$

where *A(x)* and *B(x)* are *Q*-matrices, *C(x)* is a square matrix, and at least some elements of *D(x)* and *E(x)* are non-zero. In this case, there are at least two linearly independent eingenvectors with eigenvalue zero. Therefore, the model has more than one stationary distribution. It must be taken into account that the nature of the matrix *W(x)* depends on its *x*-value. A cross-bridge model can have more than one structure for different *x*-values.

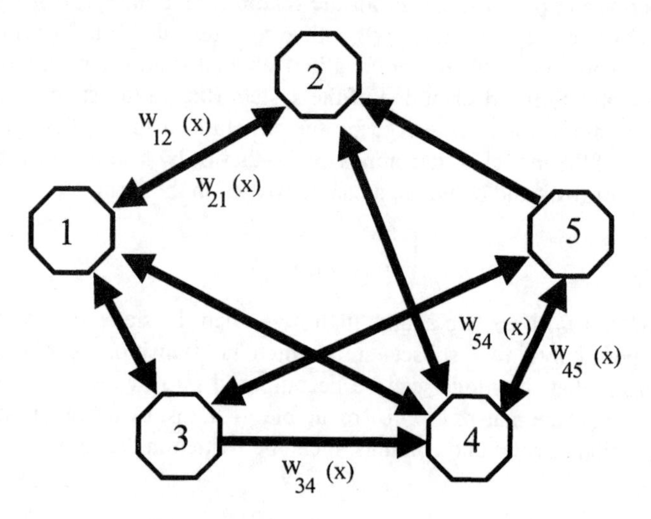

Figure 8.3. Kinetic diagram of a multi-state cross-bridge model.

DYNAMICS OF THE COLLECTIVE SYSTEM

Consider now a system consisting of a collection of *N* similar myosin heads, independent of one another, and each governed by Equation 8.4. The precise state of this collective system is specified by a set $n_1, n_2, \ldots, n_N$, which are the labels of the conformation in which the myosin heads reside. The probability for the collective system to be in this precise conformation is the product of the probabilities of the individual myosin heads:

$$p_{n_1}(x,t)p_{n_2}(x,t)\ldots p_{n_N}(x,t) \tag{8.6}$$

Now suppose that one is merely interested in the number of myosin heads occupying each state, regardless of their identity—that is, the gross state defined by the set of occupation numbers $\{N\}=N_1,N_2,\ldots,N_m$. The probability of the collective system to be in this gross state is:

$$P(\{N\},x,t)=\sum p_{n_1}(x,t)p_{n_2}(x,t)\ldots p_{n_N}(x,t) \tag{8.7}$$

where the sum extends over all values of $n_1,n_2,\ldots,n_N$ that are consistent with the prescribed occupation numbers. The probability of the gross state changes whenever one of the myosin heads jumps from one state, n, to another state, n'. The probability that one of the N_n myosin heads in state n makes such a jump during Δt is $W_{nn'}(x)N_n\,\Delta t$. The probability that two or more make a jump is of order $\Delta t \cdot \Delta t$. Hence, the probability, P, obeys the multivariate master equation:

$$\begin{aligned}\frac{dP(\{N\},x,t)}{dt} &= \sum_{n,n'} W_{nn'}(x)N_{n+1}P(\{N_1,\ldots,N_{n+1},\ldots,N_{n'-1},\ldots,N_m\},x,t) \\ &- \sum_{n,n'} W_{nn'}(x)N_n P(\{N\},x,t)\end{aligned} \tag{8.8}$$

Denote by $<N_n(x,t)>$ the average number of myosin heads in state n, with displacement x at time t. From Equation 8.8, we can deduce that this average number satisfies the same master Equation 8.4, that is:

$$\frac{d<N(x,t)>}{dt} = <N(x,t)>W(x) \tag{8.9}$$

where $<N(x,t)> = (<N_1(x,t)>,<N_2(x,t)>,\ldots,<N_m(x,t)>)$. Therefore, there is an equivalence in the dynamics of the average number of myosin heads in a given state and the probabilistic dynamics of a single myosin head, that is, $<N_n(x,t)> = N\,P_n(x,t)$ for each state n.

SHORTENING AND LENGTHENING CONSIDERATIONS

We now consider the dynamics of the cross-bridge model when the muscle is shortening or lengthening, and we adopt the Eulerian description of the myosin head by accounting for the x-displacement over time. For this case, Equation 8.4 becomes:

$$\frac{dp(x(t),t)}{dt} = p(x,t)W(x) \tag{8.10}$$

By differentiation this leads to:

$$\frac{\partial p(x,t)}{\partial t} - v(t)\frac{\partial p(x,t)}{\partial x} = p(x,t)W(x) \tag{8.11}$$

where $v(t)$ stands for the velocity of shortening when it is positive, and lengthening when it is negative. The partial differential Equation 8.8 can be converted to a time-dependent evolution equation. We can prove that a solution of Equation 8.11 is given by:

$$p(x,t) = \Psi(x + \delta(t), t) \tag{8.12}$$

where $\psi(t)$ is the solution of the first-order differential system:

$$\Psi'(.,t) = \Psi(.,t)W(x - \delta(t)) \quad \text{and} \quad \delta(t) = \int_0^t v(\tau)d\tau \tag{8.13}$$

CONTRACTILE FORCE

In order to derive the contractile force of one thin filament in one half-sarcomere, we label the myosin heads in sequence by $i=1,2,...,n$. Denote by A the set of those states, i, in which the myosin head is attached to the actin site and is able to transmit force, $F_i(x)$, and by D those states in which the myosin head is unable to transmit force. Let $x_i(t)$ be the positive displacement of the nearest actin binding site from the resting position of the myosin head at time t. The force in the filament as a function of time is then given by:

$$F(t) = \sum_j \sum_{i \in A} F_i(x_j(t)) p_i(x_j(t), t) \tag{8.14}$$

Myosin heads with the same orientation occur every $a = 42.9\ nm$ along the thick filament, whereas the repeat distance of the binding site on the actin filament is $b = 38.5\ nm$. Therefore, no periodicity occurs as i ranges from one to 100. The average spacing between adjacent values of x_i can be approximated by a uniformly dense distribution over the interval $[-b/2, b/2]$. As a consequence, the first sum in Equation 8.14 can be approximated by an average integral, that is:

$$F(t) = \frac{n}{b} \sum_{i \in A} \int_{-b/2}^{b/2} F_i(x) p_i(x,t) dx \tag{8.15}$$

STEADY-STATE TRANSITION

A fundamental property of the master equation is the so-called 'ergodic theorem' (Haken, 1977) in continuous-time Markov-chains, which asserts: *All solutions of the master equation tend to the same stationary solution when* $t \to \infty$

independent of the initial distribution, provided that the matrix W(x) is irreducible. In the case where the matrix *W(x)* is decomposable or splitting, the master equation converges to one of the stationary solutions. This property shows that any cross-bridge model in which the rate constants depend only on the x-value cannot account for history-dependent force production, such as force depression following muscle shortening, or force enhancement following stretch within the half-sarcomere.

The kinetic diagram in Figure 8.3 gives an elegant way of deriving the stationary distribution, and is given by the so-called 'Kirchhoff Theorem' (Haken, 1977). To state the theorem, we define a directional diagram D of a state i, any path in the diagram of the cross-bridge model such that 1) it possesses the maximum possible number of edges that can be included in the diagram without forming any cycle, 2) for any state $j \neq i$, there is exactly one path from j to i, and 3) no edge of D starts at the state i. Now let D be a diagram of a cross-bridge model and we set:

$$\Pi_D = \Pi \omega_{ij}(x) \quad with \quad (i,j) \in D \tag{8.16}$$

For any state, i, we set $\Sigma_i = \Sigma \Pi_D$, where the sum extends over all directional diagrams of the state i. Then, we can prove that the steady-state distribution at a state i is given by:

$$\pi_i(x) = -\frac{\Sigma_i}{Tr(W(x))} \tag{8.17}$$

where Tr stands for the sum of the diagonal elements in matrix *W(x)*.

CONCLUSIONS

In this paper, we have considered the general mathematical framework underlying the cross-bridge theory. We have chosen to work with the original cross-bridge theory (Huxley, 1957), which contains limiting assumptions. We did not take into consideration the thermodynamic constraints suggested by T. L. Hill because, although consistent, it would not change the generic mathematical approach shown here. Energetic considerations, as well as Nyquist analysis for small oscillations and cross-bridge co-operativity, could be incorporated into the mathematical framework provided here, that is, we can define all thermodynamic and energetic quantities, incorporate oscillation analysis, and model the contractile filaments with elastic properties analytically from the matrix *W(x)* and the force at each state.

REFERENCES

Gordon, A.M, Huxley, A.F., and Simmons, R.M. (1966). The variation in isometric tension with sarcomere length in vertebrate muscle fibres. *Journal of Physiology (London)* **197**, 405–419.

Haken, H. (1977). *Synergetics. Non-equilibrium Phase Transitions and Self-organization in Physics, Chemistry, and Biology*. Springer-Verlag, Berlin and Heidelberg.

Hill, T.L. (1974). Theoretical formalism for the sliding filament model of contraction of striated muscle. Part I. *Progress in Biophysics and Molecular Biology* **28**, 267–340.

Hill, T.L. (1975). Theoretical formalism for the sliding filament model of contraction of striated muscle. Part II. *Progress in Biophysics and Molecular Biology* **29**, 105–159.

Huxley, A.F. (1957). Muscle structure and theories of contraction. *Progress in Biophysics and Biophysical Chemistry* **7**, 255–318.

Huxley, A.F., and Niedergerke, R. (1954). Interference microscopy of living muscle fibres. *Nature* **173**, 971–973.

Huxley, A.F., and Simmons, R.M. (1971). Proposed mechanism of force generation in striated muscle. *Nature* **233**, 533–538.

Huxley, H.E. (1953). Electron microscope studies of the organisation of the filaments in striated muscle. *Biochimica et Biophysica Acta* **12**, 387–394.

Huxley, H.E., and Brown, W. (1967). The low-angle X-ray diagram of vertebrate striated muscle and its behaviour during contraction and rigor. *Journal of Molecular Biology* **30**, 383–434.

Huxley, H.E., and Hanson, J. (1954). Change in the cross-striations of muscle during contraction and stretch and their structural interpretation. *Nature* **173**, 973–976.

Jülicher, F. (1999). Force and motion generation of molecular motors: A generic description. Chapter 3 In: *Transport and Structure in Biophysical and Chemical Phenomena.* Muller, S.C., Parisi, J., and Zimmermann, W. (eds.), Lecture Notes in Physics, Springer, Berlin.

Propp, M.B. (1986). A model of muscle contraction based upon component studies. *Lectures on Mathematics in Life Science* (American Mathematical Society, Providence, Rhode Island) **16**, 61–119.

Ramsey, R.W., and Street, S.F. (1940). The isometric length-tension diagram of isolated skeletal muscle fibres of the frog. *Journal of Cellular Composition* **15**, 11–34.

Zahalak, G.I., and Ma, S.P. (1990). Muscle activation and contraction: Constitutive relations based directly on cross-bridge kinetics. *Journal of Biomechanical Engineering* **112**, 52–62.

9 Parameter Identification of a Distribution-moment Approximated Two-state Huxley Model of the Rat Tibialis Anterior Muscle

M. MAENHOUT
Department of Materials Technology, Faculty of Mechanical Engineering, Eindhoven University of Technology, Eindhoven, The Netherlands; Department of Movement Science, Faculty of Health Science, University of Maastricht, Maastricht, The Netherlands

M. K. C. HESSELINK
Department of Movement Science, Faculty of Health Science, University of Maastricht, Maastricht, The Netherlands

C. W. J. OOMENS
Department of Materials Technology, Faculty of Mechanical Engineering, Eindhoven University of Technology, Eindhoven, The Netherlands

M. R. DROST
Department of Movement Science, Faculty of Health Science, University of Maastricht, Maastricht, The Netherlands

INTRODUCTION

Mechanical loading of muscle tissue can lead to regional changes in the tissue, such as change of fibre diameter, number of sarcomeres in series in muscle fibres, or even regional damage (Ebbeling and Clarkson, 1989; van der Meulen et al., 1993; Lieber et al., 1994; Koh and Herzog, 1998a). External loads of the muscle (e.g., active muscle force) result in regional loads (stresses and strains) within the tissue, which are supposed to be responsible for the regional changes (Ebbeling and Clarkson, 1989; Watson, 1991). The process of changes is referred to as adaptation. Knowledge of the mechanisms of damage and adaptation may have important implications for the clinical fields of myoplasty and tendon or muscle transfer (Brunner, 1995; Koh and Herzog, 1998b). An example of myoplasty is a cardiac assist device, in which transformed skeletal

Skeletal Muscle Mechanics: From Mechanisms to Function. Edited by W. Herzog.

muscle is used to power an assist device (Gealow et al., 1993; Badhwar et al., 1997; Trumble et al., 1997; Mizahara et al., 1999). Tendon and muscle transfers can be considered as therapy in restoring the mobility of joints affected by spinal cord injury or in increasing a decreased range of knee motion during gait (Bliss and Menelaus, 1986; Illert et al., 1986; Delp et al., 1994; Keith et al., 1996).

To study the influence of mechanical load on muscle tissue, a continuum model of the tibialis anterior (TA) of the rat has been developed by Gielen (1998). The geometry of the continuum model is based on an MRI image of a longitudinal cross-section of the hindlimb of a rat. The local fibre direction is determined by diffusion-weighted MRI. A distributed moment (DM) approximated two-state Huxley model (Zahalak, 1981; Ma, 1988; Zahalak and Ma, 1990; Zahalak and Motabarzadeh, 1997) is used to describe the contractile properties. The passive tissue is described by a 3D, non-linear, anisotropic, elastic model. Although 3D hexahedral elements are used, the continuum model represents a 'flat slice' in the median plane of the TA. The equations covering the continuum model are solved with the finite-element method.

In order to determine whether or not the model describes the real stresses and strains in the muscle tissue, unknown parameter values need to be identified and the model needs to be validated. This paper will discuss the first step that was focused on the contraction model or, more specifically, the distribution-moment approximated two-state Huxley cross-bridge model, including calcium activation dynamics (Zahalak, 1981; Ma, 1988; Zahalak and Ma, 1990; Zahalak and Motabarzadeh, 1997). A big advantage of this model compared to the Hill-type models is the integration of mechanical, structural, and biochemical features of muscle tissue in one model. The parameters can be interpreted physically and the model is suitable for an inclusion of metabolic processes. The objective of the present research is to identify the unknown DM parameter values for the rat tibialis anterior (TA) muscle.

THE MODEL

Because model simulations will be compared to macroscopic behaviour like muscle torque, a 1D model of the muscle-tendon complex was used to identify the unknown parameters. This 1D model consists of a force-generating contractile element, described by the DM-approximated two-state Huxley model, in series with a non-linear spring representing the tendon (Figure 9.1).

A parallel elastic element representing the passive behaviour of the muscle tissue was omitted because, for an actively contracting muscle in the physiological range of muscle lengths, the passive force is negligible compared to the active force (Hill, 1938, 1950).

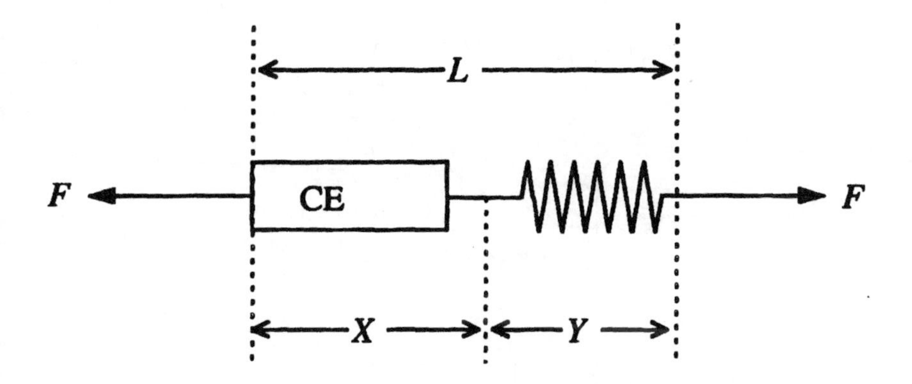

Figure 9.1. The 1D muscle model consisting of a contractile element (CE) generating force F and an elastic spring element. L represents the length of the whole muscle tendon complex, X is the length of the contractile tissue, and Y represents the length of the tendon.

THE HUXLEY RATE EQUATION WITH CALCIUM ACTIVATION

The contractile property of the muscle tissue is described by a two-state Huxley cross-bridge model. The basic Huxley theory focuses on an ensemble of myosin heads, which are assumed to be capable of binding to an actin binding site to form a so-called cross-bridge. Since the original equation was postulated by Huxley (1957), many modifications were applied in order to improve the agreement of the model with experimental results. To include the calcium activation and the filament overlap, Ma and Zahalak modified the original equation to the so-called 'generalized Huxley equation including calcium activation' (Ma, 1988; Zahalak and Ma, 1990; Zahalak and Motabarzadeh, 1997). The equation reads:

$$\frac{dn(\xi,t)}{dt} = \frac{\partial n(\xi,t)}{\partial t} - u(t)\frac{\partial n(\xi,t)}{\partial \xi} = r(t)f(\xi)[\alpha(l_s) - n(\xi,t)] - g(\xi)n(\xi,t) \tag{9.1}$$

where ξ represents the bond length with respect to the scaling factor h, which is defined as the maximum displacement of the myosin head at which attachment can occur. $n(\xi,t)$ represents the fraction of attached cross-bridges with scaled bond-length ξ, which can be interpreted as the actin-myosin bond distribution function; $u(t)$ is the scaled shortening velocity of a half sarcomere; $r(t)$ is the activation factor depending on the calcium present in the muscle fibre; and α is the overlap factor of the filaments. $f(\xi)$ and $g(\xi)$ represent the rate parameters for attachment and detachment of myosin to actin, respectively. They are postulated to vary with the bond length ξ as shown in Figure 9.2.

$$f(\xi) = \begin{cases} 0 & -\infty < \xi < 0 \\ f_1\xi & 0 \le \xi \le 1 \\ 0 & 1 < \xi < \infty \end{cases} \tag{9.2}$$

$$g(\xi) = \begin{cases} g_2 & -\infty < \xi < 0 \\ g_1\xi & 0 \le \xi \le 1 \\ g_1\xi + g_3(\xi - 1) & 1 < \xi < \infty \end{cases} \tag{9.3}$$

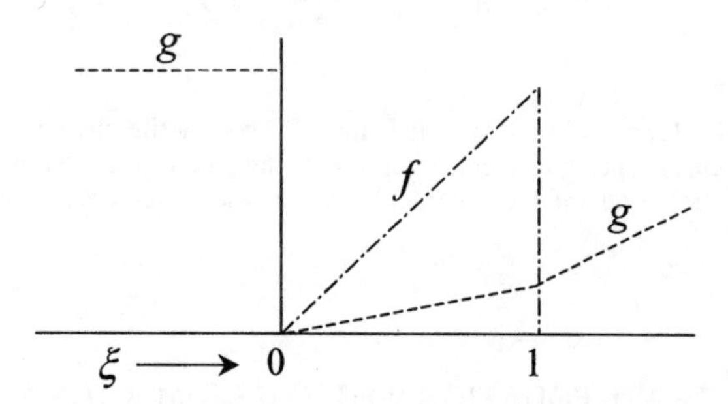

Figure 9.2. Rate constants for attachment (*f*) and detachment (*g*) of myosin to actin. x=1 is the maximum displacement of the myosin head at which binding can occur.

In contrast to Ma and Zahalak (Ma, 1988; Zahalak and Ma, 1990; Zahalak and Motabarzadeh, 1997), we used simplified linear relations for $f(\xi)$ and $g(\xi)$ instead of the exponential relations that satisfy the thermodynamic equilibrium. From a mechanical point of view, the differences are marginal (Gielen, 1998).

The overlap factor α is described (Herzog et al., 1992) as:

$$\alpha(l) = \begin{cases} 1 - 6.25(l-1)^2 & l \le 1 \\ 1 - 1.25(l-1) & l > 1 \end{cases} \qquad l = l_s / l_{s,opt} \tag{9.4}$$

where l_s and $l_{s,opt}$ are the current and the optimal sarcomere length, respectively.

The activation factor $r(t)$ describes the calcium activation in the Huxley cross-bridge model. A myosin head can only interact with an actin binding site if two calcium ions are bound to the specific receptor sites on the troponin molecule of that actin site (Zahalak and Ma, 1990). The fraction of activated actin is defined as $r(t)$ and depends on the calcium present in the muscle (Zahalak and Ma, 1990; Zahalak and Motabarzadeh, 1997):

$$r(t) = \frac{[Ca]^2}{[Ca]^2 + \mu[Ca] + \mu^2} \tag{9.5}$$

$[Ca]$ represents the free calcium concentration in the myofibrilar space and μ the troponin-calcium reaction ratio constant. In order to determine the activation factor $r(t)$, the free calcium concentration $[Ca]$ needs to be known. The net rate of the total calcium concentration $[Ca]_t$ in the myofibrilar space equals the injection rate of calcium from the sarcoplasmatic reticulum (SR), minus the uptake rate of calcium into the SR:

$$\frac{d}{dt}[Ca]_t = [\dot{Ca}]_{in} - [\dot{Ca}]_{up} \tag{9.6}$$

It is assumed (Zahalak and Ma, 1990) that each muscle action potential releases a fixed quantity of calcium with a fixed stereotyped time course. The calcium injection rate may be written as:

$$[\dot{Ca}]_{in} = R_0\left(1 - \frac{[Ca]}{[Ca]^*}\right)\chi(t) \tag{9.7}$$

where R_0 is the increase in total calcium concentration due to one action potential, $[Ca]^*$ is the average calcium concentration in the muscle, and $\chi(t)$ is defined as the sum of normalized impulse functions. If the stimulation of the muscle is started at $t=0$, then $\chi(t)$ can be represented as:

$$\chi(t) = \sum_i \hat{\chi}(t - t_i) \tag{9.8}$$

in which t_i is the time period between successive stimulation pulses,

$$\hat{\chi}(t) = 0 \text{ for } t < 0 \quad \text{and} \quad \int_0^\infty \hat{\chi}(t)dt = 1 .$$

The following analytical representation satisfies the conditions mentioned above:

$$\hat{\chi}(t) = \frac{t}{\tau^2} e^{-t/\tau} \tag{9.9}$$

where τ is a time constant determining the time course of the calcium release. The calcium injection rate has now been defined. The rate of calcium uptake into the SR is controlled by membrane-bound pumping proteins. According to Zahalak and Ma (1990), the classic Michaelis-Menten enzyme kinetics can be used to define the rate of calcium uptake as:

$$[\dot{Ca}]_{up} = V_m \frac{[Ca]}{[Ca] + K_m} \tag{9.10}$$

where V_m and K_m are the two Michaelis-Menten parameters of the reaction.

Finally, the rate of the total calcium concentration in the myofibrilar space can be written as:

$$[\dot{Ca}]_t = R_0\left(1-\frac{[Ca]}{[Ca]^*}\right)\chi(t) - V_m\frac{[Ca]}{[Ca]+K_m} \tag{9.11}$$

The total calcium concentration in the myofibrilar space can now be determined at any time, t, by solving the differential Equation 9.11. However, in order to determine the activation factor $r(t)$, the concentration of free calcium needs to be known. From the assumption of tight coupling and fast calcium-troponin equilibrium, Zahalak and Ma (1990) found an algebraic equation to relate the total calcium to the free calcium. $r(t)$ and the free calcium can be determined by using this additional algebraic equation:

$$[Ca]_t = [Ca] + m^* b\left(2Q_0 + r(t)\left(2+\tfrac{\mu}{[Ca]}\right)\left(\tfrac{1}{b}-Q_0\right)\right) \tag{9.12}$$

where m^* is the concentration of cross-bridges, b is referred to as the cross-bridge structural parameter (which is defined as h/l_a, where l_a is the distance between two successive actin binding sites), and Q_0 is the zeroth moment of the actin-myosin bond distribution function $n(\xi,t)$, defined as:

$$Q_0 = \int_{-\infty}^{\infty} n(\xi,t)\,d\xi \tag{9.13}$$

THE DISTRIBUTION-MOMENT (DM) APPROXIMATION

Since we are interested in the contractile behaviour at the regional rather than the cross-bridge level, it suffices to approximate the solution of the Huxley equation by the so-called 'distribution-moment (DM) approximation method'. It was shown by Zahalak (1981) that by multiplying both sides of the Huxley equation by ξ^λ (λ=0, 1, 2, ...) and by integrating this equation with respect to the scaled bond length variable ξ (=x/h), the Huxley equation becomes equivalent to an infinite system of coupled equations. Furthermore it was proposed by Zahalak (1981) to approximate $n(\xi,t)$ by a Gaussian distribution, which is determined by the first three moments $Q_0(t)$, $Q_1(t)$, and $Q_2(t)$ of $n(\xi,t)$. The calculations reported in Zahalak (1981) result in a set of three coupled, non-linear ordinary differential equations for the first three moments of n:

$$\dot{Q}_\lambda = \alpha r \beta_\lambda - r\phi_{1\lambda}(Q_0,Q_1,Q_2) - \phi_{2\lambda}(Q_0,Q_1,Q_2) - \lambda u(t) Q_{\lambda-1}, \quad \lambda = 0,1,2,... \tag{9.14}$$

where the superimposed dot denotes time differentiation, and Q^λ represents the λth normalized moment of the bond distribution function n defined by:

$$Q_\lambda = \int_{-\infty}^{\infty} \xi^\lambda n(\xi,t)\,d\xi, \qquad \lambda = 0,1,2,... \tag{9.15}$$

The terms β_λ, $\phi_{1\lambda}$, and $\phi_{2\lambda}$ have been defined by Zahalak (1981). The three differential equations (Equation 9.14) and the differential equation describing the net calcium rate (Equation 9.11) define the complete DM contraction model, including calcium activation. An additional advantage of the DM model is that the first three moments of the bond distribution $n(\xi,t)$ have a physical meaning (Zahalak and Motabarzadeh, 1997; Zahalak and Ma, 1990; Zahalak, 1986). The first moment Q_1, for example, is proportional to the instantaneous active muscle force F_m:

$$F_m = \Gamma Q_1 \tag{9.16}$$

where the constant Γ is referred to as the force-scaling parameter.

DISTRIBUTION-MOMENT MODEL FOR A MUSCLE-TENDON COMPLEX

It is known from experimental data (Hatze, 1981; Hawkins and Bey, 1997) that the force-strain curves for tendons show an exponential course. Therefore, the tendon is described by a non-linear exponential elastic spring:

$$F_t = \frac{K}{s}\left(e^{(s\varepsilon_t)} - 1\right) \tag{9.17}$$

where K can be interpreted as the initial resistance of the tendon against deformation, and s determines the speed at which the resistance increases with increasing strain of the tendon. This strain is defined as:

$$\varepsilon_t = \frac{(Y - Y_0)}{Y_0} \tag{9.18}$$

in which Y is the length of the tendon, and Y_0 represents the initial length. During rest, at optimal length, the muscle-tendon complex is already pre-strained. If the pre-strain of the tendon of the TA of the rat is defined as ε_0, the tendon force F_t^0 at rest equals:

$$F_t^0 = \frac{K}{s}(e^{(s\varepsilon_0)} - 1) \tag{9.19}$$

So, the tendon force of the actively contracting muscle equals:

$$F_t = F_t^0 + F_m \tag{9.20}$$

Now let's consider the muscle as represented in Figure 9.1 as a cylinder of active contractile tissue with length X, in series with a passive tendon of length Y. The total length of the muscle-tendon complex is defined as $L = X+Y$. To account for a finite tendon compliance, one more differential equation must be added to the DM equations, Equation 9.11 and Equation 9.14. This fifth differential equation may be written as:

$$\dot{L} = \dot{X} + \dot{Y} \tag{9.21}$$

The time derivative of X and Y are written (Zahalak, 1986) as:

$$\dot{X} = -\left(\frac{2hX_0}{l_{s0}}\right)u(t) \tag{9.22}$$

$$\dot{Y} = \frac{dY}{dQ_1}\frac{dQ_1}{dt} \tag{9.23}$$

Using Equation 9.16, Equation 9.17, Equation 9.18, and Equation 9.20 we may write:

$$\frac{dY}{dQ_1} = \left(\frac{s}{Y_0}\left(Q_1 + \frac{F_t^0}{\Gamma}\right) + \frac{K}{Y_0\Gamma}\right)^{-1} \tag{9.24}$$

Now, the muscle stretch ratio, with respect to the initial length L_0, is defined as:

$$\Lambda = \frac{L}{L_0} \tag{9.25}$$

Further, two dimensionless geometric parameters θ and η and a dimensionless compliance function $\kappa(Q_1)$ are defined as:

$$\theta = \frac{X_0}{L_0}, \quad \eta = \frac{2h}{l_{s0}}, \quad \kappa(Q_1) = \frac{1}{L_0}\left(\frac{s}{Y_0}\left(Q_1 + \frac{F_t^0}{\Gamma}\right) + \frac{K}{Y_0\Gamma}\right)^{-1} \tag{9.26}$$

The fifth differential, Equation 9.21, that must be added to the DM equations (Equation 9.11 and Equation 9.14) to account for the tendon compliance can finally be written as:

$$\dot{\Lambda} = \kappa(Q_1)\dot{Q}_1 - \theta\eta u(t) \tag{9.27}$$

SUMMARY

The 1D model of the muscle-tendon complex consisting of a force-generating contractile element (CE) and a series non-linear elastic element (Figure 9.1) has been explained. The five first-order coupled differential equations that embody this 1D model can be collected and are summarized below. By utilizing the normalized quantities:

$$C = [Ca]_t / m^* \qquad c = [Ca]/m^* \qquad C^* = [Ca]^* / m^* \qquad \rho = R_0 / m^*$$

$$\tau_0 = m^* / V_m \qquad k_m = K_m / m^* \qquad \mu = k_- / k_+ m^* \qquad b = h / l_a$$

where m^* represents the concentration of myosin heads and l_a the distance between successive actin binding sites, the model may be written as:

$$\dot{Q}_\lambda = \alpha r \beta_\lambda - r\phi_{1\lambda}(Q_0, Q_1, Q_2) - \phi_{2\lambda}(Q_0, Q_1, Q_2) - \lambda u(t) Q_{\lambda-1}, \quad \lambda = 0,1,2,\ldots \tag{9.28}$$

$$\dot{C} = \rho\left(1 - \frac{c}{c^*}\right)\chi(t) - \tau_0^{-1}\left(\frac{c}{c + k_m}\right) \tag{9.29}$$

$$\dot{\Lambda} = \kappa(Q_1)\dot{Q}_1 - \theta\eta u(t) \tag{9.30}$$

These equations were implemented in MATLAB, which enables the solving of the system differential equations by numerical integration using a Runge-Kutta integration scheme. By choosing the right initial values for the state variables Q_0, Q_1, Q_2, C, and Λ, and input values for the collection of model parameters, muscle contractions can be simulated.

PARAMETER IDENTIFICATION

Most of the parameter values that appear in the five coupled differential equations embodying the 1D muscle model of the TA could be measured or taken from the literature (Ma, 1988; Hawkins and Bey, 1997), and are given in Table 9.1.

Table 9.1. Parameter values of the DM model that could be measured or taken from the literature.

Model Parameters DM Model								
μ	k_m	b	c^*	K	s	ε_0	θ	η
0.2_a	0.006_a	0.77_a	1.0_a	0.32_b	60_b	0.07_b	0.6_c	0.02_d
[-]	[-]	[-]	[-]	[N]	[-]	[-]	[-]	[-]

a Taken from Ma (1988).
b Fitted to data from Hawkins and Bey (1997).
c Measured at dissected hindlimb of a rat.
d Determined from Ma (1988).

However, the values of the rate constants of attachment and detachment of the cross-bridges, f_1, g_1, g_2, and g_3, are unknown for the TA of the rat and cannot be measured directly. Three parameters associated with calcium activation are also unknown for the rat TA. These parameters are ρ, the calcium injection magnitude; τ, the time constant determining the duration of the calcium injection pulse; and τ_0, the inverse maximum calcium uptake rate. Unlike the rate constants, these activation parameter values are constrained (Ma, 1988; Lehman, 1982). The values of the calcium injection magnitude ρ found in the literature lie between 2 and 5. The reported values of the inverse maximum calcium uptake rate τ_0 in different species vary from $0.0023 s^{-1}$ to $2.3 s^{-1}$. The calcium injection pulse should last longer than 3 ms (Lehman, 1982) and shorter than 15 ms (Zahalak and Ma, 1990). Based on Equation 9.9, this means a value for τ between 0.5 ms and 3.5 ms.

The parameter Γ can be estimated from the maximum, isometric, fused tetanic force generated by the muscle $F_m^{(0)}$ at optimal muscle length via the relation:

$$\Gamma = \frac{F_m^{(0)}}{Q_1^{(0)}} \tag{9.31}$$

The value of $Q_1^{(0)}$ is the maximum value of Q_1 of a fused tetanus (high stimulation frequency) when the isometric constraint is imposed ($\Lambda = 0$) and the overlap alpha is assumed to be 1. However, the value of $Q_1^{(0)}$ depends on the parameter Γ via the fifth differential equation (Equation 9.27) of the 1D muscle-tendon model. A sensitivity analysis indicated that while the rest of the parameters remained unchanged, a change of Γ from 40 N to 70 N (i.e., a 75% higher Γ) resulted in a 1% higher value for $Q_1^{(0)}$. Because the effect of Γ on $Q_1^{(0)}$ during isometric contractions was very small, it was feasible to estimate Γ from a calculated value of $Q_1^{(0)}$ using a first estimate for Γ and the other unknown parameters and the maximum isometric force $F_m^{(0)}$. This force was estimated from the experimentally determined muscle torques T of seven rats as described in the next section and the moment arm R of the TA. The moment arm R was determined from a free dissected hindlimb of a rat and was approximately 2.8 mm. This resulted in an estimate for the maximum isometric muscle force via the relation $F_m^{(0)} = T/R$ of 14.6±3.4 N. From Equation 9.31, the value of Γ was estimated to be 50 N. This estimate will be evaluated in the discussion.

Identification of the unknown parameters requires experimental data of macroscopic muscle behaviour (like muscle torque) to which model simulations can be compared. The parameter values can then be estimated by minimization of a least-squares objective function:

$$J(\underset{\sim}{\theta}) = [\underset{\sim}{m} - \underset{\sim}{s}(\underset{\sim}{\theta})]^T [\underset{\sim}{m} - \underset{\sim}{s}(\underset{\sim}{\theta})] \tag{9.32}$$

with respect to the column $\underset{\sim}{\theta}$ in which the unknown parameters of the 1D muscle model are stored. The measurements (muscle torques) are stored in a column $\underset{\sim}{m}$

and the model simulations using a parameter set $\underset{\sim}{\theta}$ are stored in $\underset{\sim}{s}(\underset{\sim}{\theta})$. The Gauss-Newton algorithm (Meuwissen, 1998) was used to minimize the objective function. A necessary condition at a (local) minimum of $J(\underset{\sim}{\theta})$ is given by:

$$\left[\frac{\partial J(\underset{\sim}{\theta})}{\partial \underset{\sim}{\theta}}\right]^T = \underset{\sim}{0} \tag{9.33}$$

This results in a set of seven non-linear equations for the seven unknown parameters:

$$\underline{H}^T(\underset{\sim}{\theta})[\underset{\sim}{m} - \underset{\sim}{s}(\underset{\sim}{\theta})] = \underset{\sim}{0} \tag{9.34}$$

where the sensitivity matrix $\underline{H}(\underset{\sim}{\theta})$ is given by:

$$\underline{H}(\underset{\sim}{\theta}) = \frac{\partial \underset{\sim}{s}(\underset{\sim}{\theta})}{\partial \underset{\sim}{\theta}} \tag{9.35}$$

Equation 9.34 has to be solved iteratively because the model response $\underset{\sim}{s}(\underset{\sim}{\theta})$ is a non-linear function of the parameters. An estimate of the optimal parameter set $\underset{\sim}{\theta}^{(i)}$, where the index i is the iteration counter, needs to be available. If the optimal parameter set is defined as $\underset{\sim}{\theta}_{LS}$ then:

$$\underset{\sim}{\theta}_{LS} = \underset{\sim}{\theta}^{(i)} + \delta \underset{\sim}{\theta}^{(i)}_{LS} \tag{9.36}$$

Substitution of Equation 9.36 into Equation 9.34 for the optimum $\underset{\sim}{\theta}_{LS}$ leads to:

$$\underline{H}^T(\underset{\sim}{\theta}^{(i)} + \delta \underset{\sim}{\theta}^{(i)}_{LS})\left[\underset{\sim}{m} - \underset{\sim}{s}(\underset{\sim}{\theta}^{(i)} + \delta \underset{\sim}{\theta}^{(i)}_{LS})\right] = \underset{\sim}{0} \tag{9.37}$$

On the assumption that $\delta \underset{\sim}{\theta}^{(i)}_{LS}$ is small, the following two approximations can be made:

$$\begin{aligned} \underset{\sim}{s}(\underset{\sim}{\theta}^{(i)} + \delta \underset{\sim}{\theta}^{(i)}_{LS}) &\approx \underset{\sim}{s}(\underset{\sim}{\theta}^{(i)}) + \underline{H}(\underset{\sim}{\theta}^{(i)})\delta \underset{\sim}{\theta}^{(i)}_{LS} \\ \underline{H}(\underset{\sim}{\theta}^{(i)} + \delta \underset{\sim}{\theta}^{(i)}_{LS}) &\approx \underline{H}(\underset{\sim}{\theta}^{(i)}) \end{aligned} \tag{9.38}$$

Substitution of these approximations into Equation 9.37 results in an iterative scheme for updating the parameters $\underset{\sim}{\theta}^{(i)}$:

$$\underset{\sim}{\theta}^{(i+1)} = \underset{\sim}{\theta}^{(i)} + \delta\,\underset{\sim}{\theta}^{(i)}$$

$$\delta\,\underset{\sim}{\theta}^{(i)} = \underline{K}^{i^{-1}} \left[\underline{H}(\underset{\sim}{\theta}^{(i)})^T [\underset{\sim}{m} - \underset{\sim}{s}(\underset{\sim}{\theta}^{(i)})] \right] \tag{9.39}$$

$$\underline{K}^i = \underline{H}(\underset{\sim}{\theta}^{(i)})^T\, \underline{H}(\underset{\sim}{\theta}^{(i)})$$

where $\delta\underset{\sim}{\theta}^{(i)}$ is an estimate of $\delta\underset{\sim}{\theta}^{(i)}_{LS}$. In the neighbourhood of the optimal solution $\underset{\sim}{\theta}_{LS}$, this so-called Gauss-Newton scheme has quadratic convergence.

EXPERIMENTS

The experimental data were obtained by inducing isometric contractions of the TA muscle of seven male Wistar rats, during which the muscle torque T was measured using a rat isometric dynamometer (RIDM) (Figure 9.3). A comparable set-up was also used to measure muscle torques of hindlimb flexor muscle complexes of intact mice. Gorselink et al. (2000) described this set-up in more detail. Isometric contractions are defined here as contractions where the total length of the muscle-tendon complex remains constant.

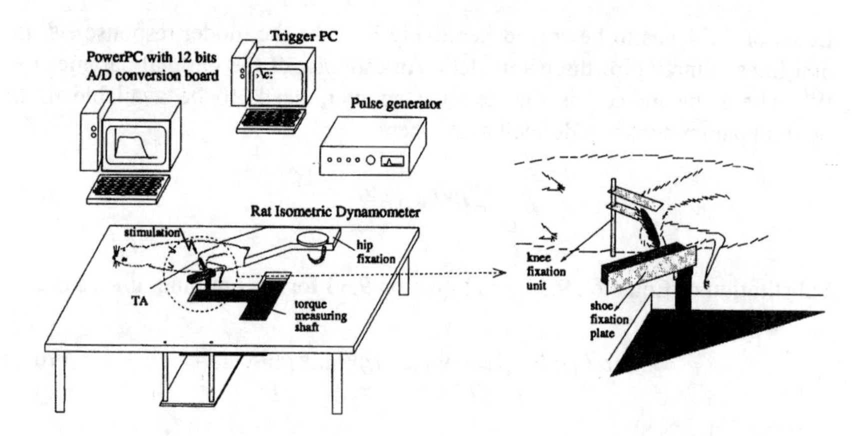

Figure 9.3. Experimental set-up of the rat isometric dynamometer (RIDM) for measuring muscle torques. The left part shows a rat in the RIDM and the four units of the set-up. The right part shows the fixation unit of the knee and the way the foot is fixed to the torque-measuring axis in more detail.

APPARATUS DESCRIPTION

The experimental set-up consists of four units (Figure 9.3): a custom-built rat isometric dynamometer (RIDM); an Apple Macintosh 7100 PowerPC® with an

8-channel, 12-bit Lab-NB analogue-to-digital conversion board (National Instruments®, Mopac Expwy, Austin, USA) programmed with Lab-VIEW® 3.1; a pulse generator for the purpose of electrical stimulation (HSE 215/IZ, Freiburg, Germany); and a PC to trigger the stimulation and the data acquisition. During the experiment, the foot was rigidly connected to the shaft of the torque transducer. The torque measurement shaft was connected to two perpendicularly placed custom-built leaf springs and a perpendicularly placed beam. Muscle torque resulted in a minor rotation of the shaft, resulting in a slight displacement of the beam. The displacement of the beam was monitored by a displacement transducer. Assessments of muscle torque using the device presented here ranged from –220 Nmm to +220 Nmm, with an accuracy of ±0.03 Nmm. The resonance frequency of the measurement device is 165 Hz. The axis angle can be varied and was monitored with an optical encoder. Data acquisition was performed at 1000 Hz. The measurement unit was calibrated with an external torque calibration unit connected to the thermostatic platform (38.5±0.1°C). The calibration curve was acquired using precision weights at a well-defined distance from the torque unit axis. The calibration range was between –220 Nmm and +220 Nmm, and calibration was performed with an increasing and decreasing series of weights at a well-defined distance (7.1 cm) from the torque unit axis. Calibration showed an excellent linear fit ($r = 0.999$) with a scaling factor of 250 Nmm/V. The measuring noise during dynamic measurements appeared to be ±0.5 Nmm.

EXPERIMENTAL PROCEDURE

All experiments (n = 7) were performed on 12-week-old male Wistar rats (368±36 grams, mean±SD), positioned on an aluminium platform maintained at 38.5±0.1°C. Initially the rats were anaesthetized by a subcutaneous injection of 0.15 ml ketamine. From this moment on, halothane (Fluothane(r), Zeneca, Ridderkerk, The Netherlands) (1.5-2.0%) in O_2 and N_2O (2:1, 3.0 l min^{-1}) was delivered via a face mask through a flow-meter system (Medec, Montvalle, New Jersey, USA). All experimental procedures were approved by the Institutional Animal Care and Use Committee of the University of Maastricht, and complied with the principles of proper laboratory animal care. After depilating the skin of the hindlimb of the anaesthetized rat, a small incision in the lateral part of the knee was made to obtain access to the peroneal nerve. A bipolar platinum hook electrode, inter-pole distance 0.8 mm, was carefully attached to the nerve, and the dorsal muscle complex could be stimulated using a pulse generator (HSE 215/IZ, Freiburg, Germany). In order to assure an isometric contraction of the TA, movement of the rat's knee and foot was prevented. The foot was cast in a two-component cement (paladur, Heraus Kulzer GmbH, Wehrheim, Germany) shoe using a Teflon mould that was glued with cryano-acrylate (Bison, Goes, The Netherlands) to the shoe-fixation plate. This was done in order to suppress the foot's internal degrees of freedom. This plate was connected to the apparatus shaft. The proximal part of the hindlimb was fixed to the external world through a hip-fixation unit (Figure 9.3). The rat was positioned sidelong to the hip-fixation unit and the tail was led through a small opening in the fixation platform. The hip was fixed to a vertical platform by winding the tail on a pulley. Tail friction, obtained by an external force (F_{tail} = 3 N) on the pulley, fixed the hip onto the vertical platform. The knee was supported by two aluminium plates to prevent lateral movement while permitting sagittal plane movements. Rigid knee fixation is not allowed, because an extra fixation force, as a result of preventing sagittal knee movements, is transmitted via the tibia as an eccentric force, introducing a non-muscular torque to the torque transducer. In a series of contractions, the angle of the rat's ankle was optimized for maximal fused tetanic torque. This angle appeared to be about 90°. After this optimization, the stimulation current was optimized for maximal fused tetanic torque. The pulse duration of the electrical stimulation current was 0.5 ms and the optimal stimulation current was about 1 mA. After these preparations, the exercise protocol was carried out.

THE EXERCISE PROTOCOL

The protocol consisted of contractions which were induced by 300 ms electrical stimulation at different stimulation frequencies: 40, 50, 67, and 125 Hz. Besides these contractions, a twitch stimulation was also given. A series of measurements was started with a twitch, followed by the different stimulation frequencies, starting with the lowest frequency. For each rat, three such measuring series (m = 3) were carried out in order to test the reproducibility of the muscle torques. Between contractions, a resting period of three minutes was maintained in order to avoid fatigue. During the contractions, the muscle torques were measured and, in order to reduce the signal-to-noise ratio, the measurements were filtered using a 5.0 lag moving-average.

RESULTS

For each rat the measured torques T are considered relative to the maximum isometric torque T_m at 125 Hz. Also, the simulated first moments Q_1 of the distribution function $n(\xi,t)$, which are proportional to the muscle force and thus the muscle torque, are considered as a fraction of the maximum isometric first moment $Q_{1,m}$ at 125 Hz. This normalization enables a direct comparison of the normalized measured torques to the normalized simulated first moments. Also, the normalized torques enable a comparison of measurements between rats without having to be concerned about scaling differences.

For each rat and each stimulation frequency, the torque measurements were performed three times (m = 3). The average measured muscle torque at a stimulation frequency of 125 Hz was 40.9 Nmm±9.5 Nmm. The maximal difference relative to the maximum torque at 125 Hz between the three repeat torque measurements on the same rat were small (5%) compared to the maximal differences between measurements on different rats (40%). Therefore, in Figure 9.4, only one set of torque measurements (black lines) is displayed for each rat. These measurements will also be used for parameter identification. Note that on the first rat, the 40 Hz torque measurement was not performed.

The so-called optimal DM parameter values were determined by minimization of the objective function (Equation 9.32) as described in the section titled *Parameter Identification*. By using the optimal parameter values, the isometric torques were simulated for each rat. In Figure 9.4, these simulations are compared to the experimental results for each rat. The optimal parameter values for each rat are presented in Table 9.2.

Figure 9.4 shows a fairly good agreement between model and experiment at the different stimulation frequencies. However, some differences can be found. The most pronounced difference appears at the relaxation of the isometric contractions at a stimulation frequency of 125 Hz and, to a lesser degree, at 67 Hz. The simulated relaxation at these frequencies is much faster than that measured experimentally, while the simulated relaxation of the twitch is generally slower than that obtained experimentally.

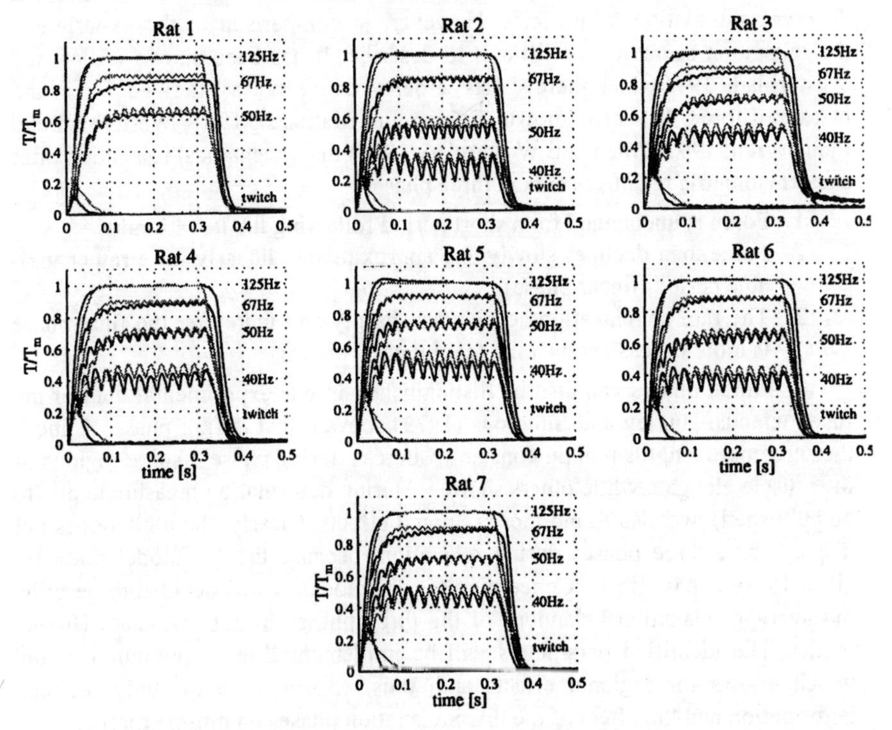

Figure 9.4. Measured (black lines) and simulated (grey lines) normalized isometric torques for each rat. The stimulation frequencies are indicated on the right. The optimal parameter values for each rat are displayed in Table 9.2.

Table 9.2. Optimal parameter values for each rat, and mean values and their standard deviation for the seven rats.

	Rat 1	Rat 2	Rat 3	Rat 4	Rat 5	Rat 6	Rat 7	Mean±SD
$f_1[s^{-1}]$	203	142	137	136	200	169	151	163±29
$g_1[s^{-1}]$	70	79	51	58	66	69	55	64±10
$\rho[-]$	1.8	2.5	3.4	2.7	3.3	2.8	3.5	2.9±0.6
$\tau[ms]$	2.6	2.7	3.1	2.5	3.1	3.0	3.2	2.9±0.3
$\tau_0[ms]$	13.8	9.4	7.5	9.6	7.9	8.8	7.2	9.2±2.2

DISCUSSION

The standard deviation for all parameters presented in Table 9.2 is in the order of 20% of the mean value, which seems reasonable for a biological experiment.

The results in Figure 9.4 indicate that the five identified parameters of the DM model enable a good description of the isometric muscle torques of the rat TA at different stimulation frequencies. However, in comparison to the experiment, the relaxation seems to be difficult to describe. It has been shown (Allen and Westerblad, 1995) that there exists a delay between Ca^{2+}-derived force and measured force. It is also known (Edman and Mattiazzi, 1981; Westerblad and Lännergren, 1991; Allen and Westerblad, 1995) that mechanical relaxation after a short isometric tetanus occurs in three phases:

1. Force is unchanged for a short period following the last stimulus.
2. Force then declines slowly and approximately linearly for a rather variable period (linear phase).
3. The rate of relaxation accelerates sharply and thereafter the final phase is more or less exponential (exponential phase).

These three phases can also be distinguished in our experimental data of the fused tetanus. Huxley and Simmons (1973) showed that during phases 1 and 2 the sarcomere lengths remain constant, whereas during phase 3 some regions of the muscle elongate while others shorten. During deformation measurements (to be published), we also found these regional effects. Clearly, the model does not display these three phases during relaxation, because the 1D model does not allow for regional effects. Concerning the 1D model, it was decided to describe the average relaxation behaviour of the three phases by the two-state Huxley model. The identified parameters will be implemented in a continuum model which allows for regional effects and, thus, can be used to study regional deformation and the effect of the three relaxation phases on muscle torque.

The values found for the DM parameters seem realistic in the sense that the values found for the activation parameters ρ and τ_0 are within the range of constraints for these parameters given in Ma (1988). Furthermore, the values for the rate constants of the TA appear to be higher than the values for the rat soleus identified by Ma (1988). For f_1 and g_1, Ma found 90 s^{-1} and 30 s^{-1}, respectively, while for TA these values are 163 s^{-1} and 64 s^{-1}, respectively. This difference can be expected because soleus consists mainly of slow Type 1 muscle fibres while TA consists mainly of fast Type 2b fibres. Furthermore, as a rule of thumb, based on the maintenance heat rate associated with actin-myosin interaction, g_1 equals about 0.4 to 0.15 times f_1 (Ma and Zahalak, 1988; Zahalak, 1999). Therefore, the values found for f_1 and g_1 for TA seem acceptable.

Once the unknown model parameters are identified, the reliability of the force scaling parameter Γ can be verified, as explained in the section titled *Parameter Identification.* The mean value of $Q_1^{(0)}$ of the seven rats, by using the optimal parameter values and the value of Γ of 50 N, was 0.338±0.027. Using these values, the maximum isometric force $F_m^{(0)}$ is 16.9±1.4 N. From the experiments, an estimate for the maximum isometric muscle force $F_m^{(0)}$ of 14.6±3.4 N was obtained. Therefore, the value of 50 N for Γ seems reasonable.

During the exercise protocol, both hindlimb dorsiflexors (TA and EDL) are stimulated. However, because the mass of the EDL is small compared to the TA (±20%), the contribution of the EDL to the measured muscle torque is expected to be small as well. Because EDL and TA consist primarily of the same Type 2b muscle fibres, it appears feasible to assume that the contribution of the EDL to the measured muscle torque is proportional to the torque at the various stimulation frequencies. Therefore, the net contribution to the normalized torques is expected to be close to zero.

FUTURE RESEARCH

Because the experimental protocol consisted only of isometric contractions, the parameter values of g_2 and g_3 could not be identified due to the fact that these parameters hardly play a role during isometric contractions. They do influence the muscle torque development during shortening and lengthening contractions. Because isometric contractions were defined here as contractions during which the total length of the muscle-tendon complex remains constant, one might expect the muscle tissue to shorten during these contractions. However, during deformation measurements (to be published), it appeared that the maximal global strain of the muscle tissue was less than 4%, which confirms the results of Hawkins and Bey (1997). In the 1D model, the strain of the contractile element at maximal fused tetanus was about 4% when using the identified parameters. By varying the parameters g_2 and g_3, the sensitivity of the parameters was tested during isometric contractions. It appeared that a change of +50% or −50% for each individual parameter g_2 and g_3 from its previous optimal value gave a difference of less than 2.5% of the maximum fused torque for g_2, and less than 0.5% for g_3. So, for a proper identification of the rate constants g_2 and g_3, experimental data on shortening and lengthening contractions are needed. In the future, we will collect experimental data under several experimental conditions. By comparing model simulations for different experimental conditions, the identified parameter values will be validated, and g_2 and g_3 will be identified.

The 1D model can be used to study the macroscopic behaviour of the rat TA for isometric contractions and, after identification of g_2 and g_3, also for shortening and lengthening contractions. Furthermore, the identified parameters will be implemented in a 3D continuum model of TA. For the time being, it will be assumed that the DM parameters are distributed homogeneously through the muscle tissue. In order to determine whether or not the model describes the real stresses and strains, the model needs to be validated. Due to the complexity of the 3D model, the numerical results are difficult to interpret. Therefore, a stepwise approach is chosen with an increasing level of complexity incorporating different measuring techniques. The first validation experiments were focused on the DM contraction model of the muscle. This validation step was described here. To measure deformation at the muscle surface, 3D marker displacements will be measured during contraction by 3D video analysis (van Donkelaar et al., 1999; van Bavel et al., 1996). Validation of the internal tissue deformation will be carried out by an MRI-tagging technique (Kerckhoffs, 1998; Ossevoort, 1997). During the validation steps, it will be investigated whether the assumption that the DM parameters are distributed homogeneously through the tissue is realistic. Special attention will be paid to the sarcomere length distribution, the properties of the passive tissues, and the ability of the 3D model to describe the three relaxation phases of the fused tetanic muscle torque.

REFERENCES

Allen, D.G., and Westerblad, J.H. (1995). Muscle cell function during prolonged activity: Cellular mechanisms of fatigue. *Experimental Physiology* **80**, 497–527.

Badhwar, V., Badhwar, R., Oh, J., and Chui, R., (1997). Power generation from four skeletal muscle configurations: Design implications for a muscle powered cardiac assist device. *American Society for Artificial Internal Organs Journal* **43**(5), M651–M657.

Bliss, D., and Menelaus, M. (1986). The results of transfer of the tibialis anterior to the heel in patients who have a myelomeningocele. *Journal of Bone and Joint Surgery* **68**(8), 1258–1264.

Brunner, R. (1995). Changes on muscle power following tendon lengthening and tendon transfer. *Orthopäde* **24**(3), 246–251.

Delp, S., Ringwelski, D., and Carroll, N. (1994). Transfer of the rectus femoris: Effects of transfer site on moment arms about the knee and hip. *Journal of Biomechanics* **27**(10), 1201–1211.

Ebbeling, C.B., and Clarkson, P.M. (1989). Exercise-induced muscle damage and adaptation. *Sports Medicine* **7**, 207–234.

Edman, K., and Mattiazzi, A. (1981). Effects of fatigue and altered pH on isometric force and velocity of shortening at zero load in frog muscle fibres. *Journal of Muscle Research and Cell Motility* **2**, 321–334.

Gealow, K., Solien, E., Bianco, R., Chui, R., and Shumway, S. (1993). Conformational adaptation of muscle: Implications in cardiomyoplasty and skeletal muscle ventricles. *Annals of Thoracic Surgery* **56**(3), 520–526.

Gielen, A.W.J. (1998). A continuum approach to the mechanics of contracting skeletal muscle. Ph.D. thesis, Eindhoven University of Technology, Eindhoven, The Netherlands.

Gorselink, M., Drost, M.R., de Louw, J., Willems, P.J.B., Rosielle, N., Janssen, J.D., van de Vusse, G.J. (2000). Accurate assessment of *in situ* isometric contractile properties of hind limb plantar and dorsal flexor muscle complex of intact mice. *European Journal of Physiology* **439**, 665–670.

Hatze, H. (1981). *Myocybernetic Control Models of Skeletal Muscle: Characteristics and Applications.* University of South Africa, Muckleneuk, Pretoria, South Africa.

Hawkins, D., and Bey, M. (1997). Muscle and tendon force-length properties and their interactions *in vivo*. *Journal of Biomechanics* **30**(1), 63–70.

Herzog, W., Kamal, S., and Clarke, H.D. (1992). Myofilament lengths of cat skeletal muscle: Theoretical considerations and functional implications. *Journal of Biomechanics* **25**, 945–948.

Hill, A.V. (1938). The heat of shortening and the dynamic constants in muscle. *Proceedings of the Royal Society of London B* **126**, 136–165.

Hill, A.V. (1950). The series elastic component of muscle. *Proceedings of the Royal Society of London* **137**, 273–280.

Huxley, A.F. (1957). Muscle structure and theories of contraction. *Progress in Biophysics and Biophysical Chemistry* **7**, 257–318.

Huxley, A. F., and Simmons, R. (1973). Mechanical transients and the origin of muscle force. *Cold Spring Harbor Symposia on Quantitative Biology* **37**, 669–680.

Illert, M., Trauner, M., Weller, E., and Wiedemann, E. (1986). Forearm muscles of man can reverse their function after tendon transfers: An electromyographic study. *Neuroscience Letters* **67**, 129–134.

Keith, M., Kilgore, K., Peckham, P., Wuolle, K., Creasey, G., and Lemay, M. (1996). Tendon transfers and functional electrical stimulation for restoration of hand function in spinal cord injury. *Hand Surgery* **21**(1), 89–99.

Kerckhoffs, R.C. (1998). *In vivo* determination of a 2D strain field in skeletal muscle using MR tagging. Master's thesis, Eindhoven University of Technology, Eindhoven, The Netherlands. WFW-report 98.023.

Koh, T.J., and Herzog, W. (1998a). Excursion is important in regulating sarcomere number in the growing rabbit tibialis anterior. *Journal of Physiology* **508**, 267–280.

Koh, T. J., and Herzog, W. (1998b). Increasing the moment arm of the tibialis anterior induces structural and functional adaptation. *Journal of Biomechanics* **31**, 593–599.

Lehman, S.L. (1982). A detailed biophysical model of human extraocular muscle. Ph.D. thesis, University of California (Berkeley), Berkeley, California, USA.

Lieber, R.L., Schmitz, M.C., Mishra, D.K., and Fridén, J. (1994). Contractile and cellular remodelling in rabbit skeletal muscle after cyclic eccentric contractions. *Journal of Applied Physiology* **77**, 1926–1934.

Ma, S.P. (1988). Activation dynamics for a distribution-moment model of muscle. Ph.D. thesis, Washington University, St. Louis, Missouri, USA.

Ma, S.P., and Zahalak, G.I. (1991). A distribution-moment model of energetics in skeletal muscle. *Journal of Biomechanics* **24**, 21–35.

Meuwissen, M.H.H. (1998). An inverse method for the mechanical characterisation of metals. Ph.D. thesis, Eindhoven University of Technology, Eindhoven, the Netherlands.

Mizuhara, H., Koshiji, T., Nishimura, K., Nomoto, S., Matsuda, K., and Ban, T. (1999). Evaluation of a compressive-type skeletal muscle pump for cardiac assistance. *Annals of Thoracic Surgery* **67**(1), 105–111.

Ossevoort, L. (1997). Measuring deformation in skeletal muscle using NMR imaging techniques. Master's thesis, Eindhoven University of Technology, Eindhoven, The Netherlands. WFW-report 97.034.

Trumble, D., LaFramboise, W., Duan, C., and Magovern, J. (1997). Functional properties of conditioned skeletal muscle: Implications for muscle-powered cardiac assist. *American Journal of Physiology* **273**(2:1), C588–C597.

van Bavel, H., Drost, M.R., Wielders, J.D.L., Huyghe, J.M., Huson, A., and Janssen, J.D. (1996). Strain distribution on rat medial gastrocnemius (mg) during passive stretch. *Journal of Biomechanics* **29**, 1069–1074.

van der Meulen, J.H., Kuipers, H., van der Wal, J.C., and Drukker, J. (1993). Quantitative and spatial aspects of degenerative changes in rat soleus muscle after exercise of different durations. *Journal of Anatomy* **182**(3), 349–353.

van Donkelaar, C.C., Willems, P.J.B., Muytjens, A., and Drost, M. (1999). Skeletal muscle transverse strain during isometric contraction at different lengths. *Journal of Biomechanics* **32**, 755–762.

Watson, P.A. (1991). Function follows form: Generation of intracellular signals by cell deformation. *Federation of American Societies for Experimental Biology* **5**, 2013–2019.

Westerblad, H., and Lännergren, J. (1991). Slowing of relaxation during fatigue in single mouse muscle fibres. *Journal of Physiology* **434**, 323–336.

Zahalak, G.I. (1981). A distribution-moment approximation for kinetic theories of muscular contraction. *Mathematical Biosciences* **55**, 89–114.

Zahalak, G.I. (1986). A comparison of the mechanical behavior of the cat soleus muscle with a distribution-moment model. *Journal of Biomechanical Engineering* **108**, 131–140.

Zahalak, G.I. (1999). f_1, g_1 ratio. Personal communication.

Zahalak, G.I., and Ma, S.P. (1990). Muscle activation and contraction: Constitutive relations based directly on cross-bridge kinetics. *Journal of Biomechanical Engineering* **112**, 52–62.

Zahalak, G.I., and Motabarzadeh, I. (1997). A re-examination of calcium activation in the Huxley cross-bridge model. *Journal of Biomechanical Engineering* **119**, 20–29.

10 The Effect of Sarcomere Shortening Velocity on Force Generation, Analysis, and Verification of Models for Cross-bridge Dynamics

AMIR LANDESBERG
Department of Biomedical Engineering, Technion - Israel Institute of Technology, Haifa, Israel

LEONID LIVSHITZ
Department of Biomedical Engineering, Technion - Israel Institute of Technology, Haifa, Israel

HENK E. D. J. TER KEURS
Departments of Physiology and Biophysics, Faculty of Medicine, University of Calgary, Alberta, Canada

ABSTRACT

A quick release of the length of contracting cardiac muscle rapidly reduces force to a new load. If the load is then maintained, the muscle shortens at constant velocity. The hyperbolic relationship between velocity and load following this intervention reflects a quasi-steady state. The effect of sarcomere shortening on force development without the quick release is less well known. In this study, we tested the *hypothesis* that the transition rate (G) of the cardiac cross-bridge (XB) from the strong force-generating state to the weak state can be described by: $G = G_0 + G_1 \cdot V_{SL}$, where G_0 is the rate during contraction at constant sarcomere length and G_1 relates G to the rate of shortening of the sarcomeres (V_{SL}).

Methods: Force (F) was measured with a silicon strain gauge in six trabeculae from the right ventricle of rat heart in K-H solution at $[Ca]_o = 1.5$ mM and 25°C. Sarcomere length (SL) was measured with laser diffraction techniques, using both a lateral effect photo diode and a CCD array. Data were sampled at 5 kHz. Twitch F at constant SL, and the F response to shortening at constant V_{SL} (0-8 μm/s; ΔSL 50-100 nm) were measured at different times during the twitch.

An analytical model was used to evaluate XB mechanics. The model included the kinetics of: 1) XB cycling between a strong force-generating state and a weak

Skeletal Muscle Mechanics: From Mechanisms to Function. Edited by W. Herzog.

non-force generating state; 2) the physical interaction between the heads of the myosin and the actin; and 3) the effect of V_{SL} on G. The energy consumption by the sarcomere was calculated from the number of XB cycles, assuming hydrolysis of one ATP per XB cycle.

Results: The response to shortening consisted of an initial exponential decline of F (τ=2 ms) followed by a slow decrease of F. The instantaneous difference (ΔF) between the isometric F (F_M) and F during the second phase of sarcomere shortening depended on the duration of shortening (Δt); the instantaneous F_M and V_{SL}. $\Delta F/F_M$ was independent of the time during the twitch. The relationship between ΔF, F_M, Δt, and V_{SL} followed: $\Delta F = G_1 \cdot F_M \cdot \Delta t \cdot V_{SL} \cdot (1 - V_{SL} / V_{MAX})$, where V_{MAX} is the unloaded V_{SL}, and G_1 was 6.15±2.12 μm^{-1} (mean±SD, n=6). The linear interrelation between ΔF and V_{SL} is consistent with the hypothesis that the transition of strong XBs to weak XBs occurs at a rate that depends linearly on V_{SL}. Using the assumption that V_{SL} determines G, the model predicts a hyperbolic load-velocity relationship. The model also predicts that the mechanical energy output by the sarcomere equals the sum of the external work (W); the pseudo potential energy (E_{PP}), which dissipates as heat but is partially available for external work; and the pseudo visco-elastic energy dissipation by the XBs (Q_η), which is significant at high V_{SL}. The XB derives the mechanical energy it produces from the biochemical energy of ATP, E. The efficiency of biochemical to mechanical energy conversion (ρ) is inversely proportional to the magnitude of feedback of G_1: $\rho \cdot E = W + E_{PP} + Q_\eta$.

Conclusion: The results suggest that XB dynamics are determined by two kinetic components. The fast phase of the F response reflects the visco-elastic property of the XB; the slow phase reflects XB cycling between weak and strong biochemical states. Our observations also suggest feedback of shortening to the transition rate (G) of the strong XB to the weak XB: $G = G_0 + G_1 \cdot V_{SL}$. Including this feedback in the XB model provides a more universal description of the interrelation between shortening and force as well as a basis for the observed linear relationship between biochemical energy consumption and the generated mechanical energy.

BACKGROUND

A most notable paper in physiology published by O.W. Fenn (1923) showed that energy liberation, as work and heat, increases as work production increases. Later, Mommaerts et al. (1962) showed that cardiac and skeletal muscle mobilize energy over and above that needed for activation and the equivalent isometric contraction; this accounts for the work and dissipation of energy accompanying the work process, a phenomenon denoted as the Fenn Effect. Cardiac and skeletal muscles possess a property that enables them to adjust their energy cost to the prevailing mechanical constraints after stimulation (Mommaerts et al., 1962; Rall, 1982). Rall (1982) suggested that there is some internal feedback in active muscle whereby the total energy liberation is regulated by the loading conditions. What is the feedback mechanism? This question is still unresolved.

Energy consumption in the cardiac muscle is characterized by two basic phenomena: 1) the well-known linear relationship between energy consumption by the sarcomere and the generated mechanical energy (Hisano and Cooper, 1987; Sagawa et al., 1988; Suga, 1990), and 2) the ability to modulate the generated mechanical energy and energy consumption to the various loading conditions, as is manifested by the Frank Starling Law (Allen and Kentish, 1985) and the Fenn Effect (Fenn, 1923; Mommaerts et al., 1962; Rall, 1982). These phenomena are analyzed here based on the intracellular control of contraction.

COUPLING CALCIUM BINDING WITH CROSS-BRIDGE CYCLING

The model (Landesberg, 1996; Landesberg and Sideman, 1994a, 1994b) describes the intracellular control mechanisms of contractile filament contraction based on the structural and biochemical model of XB cycling (Brenner and Eisenberg, 1986; Chalovich and Eisenberg, 1986; Eisenberg and Hill, 1985; Lehrer, 1994), and couples the kinetics of calcium binding to troponin with the regulation of XB cycling. The basic assumptions underlying the model are detailed elsewhere (Landesberg, 1996; Landesberg and Sideman, 1994a, 1994b, 1999) but are briefly listed here:

1. The regulatory unit (regulated actin) consists of a single regulatory troponin protein complex with 14 adjacent actin molecules (Lehrer, 1994).
2. The XB cycle is a repeated oar-like cycle between weak and strong conformations that differ in their structure (Eisenberg and Hill, 1985). The XB transitions between weak and strong conformations are described by the biochemical rate limiting steps, which are related to nucleotide binding and release (Eisenberg and Hill, 1985). Force is produced only by the strong conformation. XB turnover from the weak to the strong conformation relates to ATP hydrolysis and phosphate release (Brenner and Eisenberg, 1986; Eisenberg and Hill, 1985). Thus, energy consumption is proportional to the total amount of XB turnover from the weak to the strong conformation.
3. The individual XB acts like a Newtonian pseudo visco-elastic element: the average force-velocity relation of a single XB is linear (de Tombe and ter Keurs, 1992).
4. Calcium binding to troponin low affinity sites regulates the activity of the actomyosin ATPase (Chalovich and Eisenberg, 1986), which is required for XB transition from the weak to the strong conformation (Eisenberg and Hill, 1985). Thus, calcium binding to troponin regulates XB recruitment and the energy consumption by the sarcomere.

5. Calcium can dissociate from the troponin before the XB turnover to the weak conformation; XBs can generate force without having bound calcium on the adjacent troponin (Guth and Potter, 1987; Peterson et al., 1991).

The controls of sarcomere function include two dominant feedback mechanisms, a positive and a negative one. These feedback loops were established and substantiated based on the analysis of intact (Landesberg and Sideman, 1994b) and skinned (Landesberg and Sideman, 1994a) cardiac fibres: a) the positive feedback mechanism, denoted as the co-operativity mechanism, whereby the affinity of the troponin for calcium, and hence the rate of XB recruitment, depends on the number of force producing XBs, and b) the negative feedback mechanism, denoted as the mechanical feedback, whereby the filaments' shortening velocity, or the XB strain rate, determines the rate of XB weakening.

The co-operativity mechanism derives from the interaction between the cross-bridges (XBs) and the troponin, the sarcomere regulatory proteins. The co-operativity mechanism regulates XB recruitment and plays a key role in the regulation of the force-length relationship (FLR) in cardiac muscle. It explains the 'length-dependent calcium sensitivity of the sarcomere' and the Frank Starling Law (Landesberg and Sideman, 1994a, 1999). It also provides the basis for the regulation of energy consumption and the ability of the muscle to adapt its energy consumption to the loading conditions (Landesberg and Sideman, 1999, 2000). The co-operativity mechanism may also exist in slow skeletal muscle, which has a similar troponin T as that found in cardiac muscle (T.L. Hill, 1983). However, the existence of this co-operativity mechanism in skeletal muscle is controversial (Fuchs and Wang, 1991). The co-operativity mechanism, which stems from the analysis of skinned cardiac fibre data (Landesberg and Sideman, 1994a), is substantiated by simulation of intact fibre function (Landesberg, 1996; Landesberg and Sideman, 1994b) and the analysis of energy consumption in cardiac muscle (Landesberg and Sideman, 1999, 2000).

The negative mechanical feedback relates to the physical property of the XBs. It is based on the biochemical model of XB cycling of Eisenberg and Hill (1985), who suggested that the filament shortening velocity affects the rate of XB weakening. The rate of XB weakening is a linear function of the sarcomere shortening velocity, which is substantiated (Landesberg and Sideman, 1994b) by the analytical derivation of the force-velocity relationship (FVR) in agreement with the well-established, experimentally derived, Hill's equation (A.V. Hill, 1964). Moreover, the mechanical feedback regulates the generated power and provides the analytical explanation for the regulation of biochemical-to-mechanical energy conversion by the sarcomere, and for the linear relationship between energy consumption and the generated mechanical energy (Landesberg and Sideman, 2000).

The combined effect of these two feedback loops regulates sarcomere dynamics and explains a wide spectrum of phenomena, such as the effect of loading conditions on free calcium transients (Allen and Kentish, 1988; Landesberg and Sideman, 1994b), and the effect of loading conditions on the end-systolic-pressure-volume relationship in the left ventricle (Landesberg, 1996). Their combined effect explains the above-mentioned two basic phenomena relating to the energy conversion in muscle: 1) the linear relationship between energy consumption and the generated mechanical energy, and 2) the ability of muscle to modulate the generated mechanical energy and energy consumption in response to various loading conditions.

These assumptions allow for the construction of a kinetic model that corresponds to the various possible states of the regulated actin and the associated cross-bridges, and allows calcium binding to troponin with the regulation of XB cycling (Landesberg, 1996; Landesberg and Sideman, 1994a, 1994b, 1999, 2000). The present study focuses on the description of XB dynamics and the role of the two kinetics involved in XB cycling.

CROSS-BRIDGE DYNAMICS

Results from theoretical studies (Landesberg and Sideman, 1994b, 2000) suggested that XB dynamics is determined by two kinetics, a slow and a fast kinetics, and that the shortening velocity (V_{SL}) has two different effects on these kinetics and on force generation by the XBs (Figure 10.1).

The fast kinetics relates to the physical interactions between the myosin and the actin and is denoted as XB attachment-detachment. This kinetics determines the unitary force, i.e., the average force per XB, (F_{XB}). The shortening velocity decreases the unitary force as was shown by de Tombe and ter Keurs (1992). Hence, the XB is described as a pseudo visco-elastic Newtonian element:

$$F_{XB} = \bar{F} - \eta \cdot V_{SL} \tag{10.1}$$

where $\bar{F}$ is the isometric unitary force and η describes the pseudo-viscous property of the XBs (de Tombe and ter Keurs, 1992). Note that this equation applies for steady-state shortening, since it was measured experimentally during steady shortening (de Tombe and ter Keurs, 1992). However, since the kinetics of XB attachment-detachment is very fast—in the order of hundred and thousand per second (Brenner, 1991; Lombardi et al., 1992; Stein et al., 1984) and has a time constant of several milliseconds (see below)—it reaches a steady-state within a couple of milliseconds.

The second kinetics is related to XB cycling between two biochemical conformations: the weak and strong conformations. This kinetics is related to the ATPase cycle and to the kinetics of nucleotide binding and release (Eisenberg and Hill, 1985; Greene et al., 1987; T.L. Hill, 1983; Stein et al., 1984). The shortening velocity determines the rate of XB weakening (G), i.e., the rate of XB turnover from the strong to the weak state:

$$G = G_0 + G_1 \cdot V_{SL} \tag{10.2}$$

where G_0 is the rate of XB weakening in the isometric regime, and G_1 is the mechanical feedback coefficient that describes the effect of filament shortening velocity on the rate of XB weakening, and has the units of [1/m] (Landesberg, 1996; Landesberg and Sideman, 1994b, 2000).

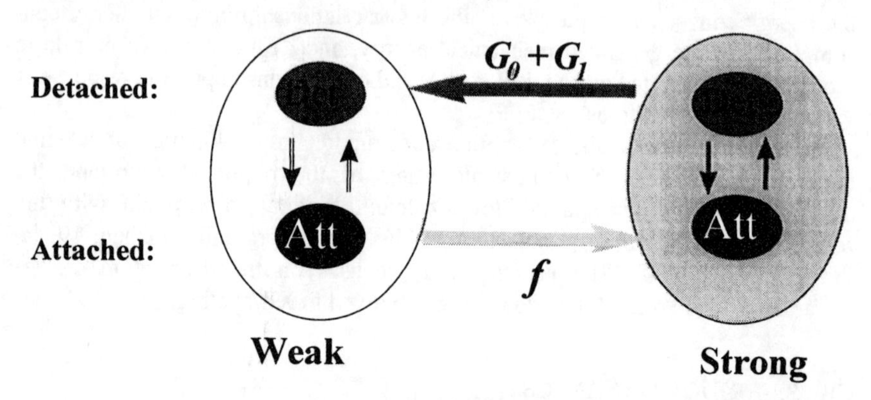

Figure 10.1. Cross-bridge dynamics is determined by two kinetics: the kinetics of cross-bridge cycling between the weak and strong conformations, and the kinetics of cross-bridge attachment-detachment.

The shortening velocity has additional effects on the regulation of force generation by the sarcomere, which were discussed elsewhere (Landesberg and Sideman, 1994b, 2000). This includes the effect on the single overlap length between the actin and myosin filaments, and additional positive and negative effects on the activation level, i.e., on the rate of new XB recruitment during shortening. These effects relate to the phenomenon of 'shortening deactivation' and the 'positive effect of shortening' (Landesberg, 1996; Landesberg and Sideman, 1994b). The present study relates only to the short-term effect of shortening on XB mechanics, and not to its effect on the activation level.

The dominant effect of the shortening velocity on force development is attributed to its effect on the rate of XB weakening (Equation 10.2) at moderate shortening velocities. At large shortening velocities, the effect of the pseudo visco-elastic property of the XB becomes dominant (Equation 10.1).

THE FORCE-VELOCITY RELATIONSHIP

We have shown (Landesberg and Sideman, 1994b, 2000) that the mechanical feedback provides the analytical solution, based on XB dynamics, for the experimentally established Hill's equation for the FVR:

$$V_{HILL} = b_H \frac{F_M - F}{F + a_H} = \frac{G_0(F_M - F)}{\left\{\left(G_1 + \frac{1}{L_S}\right)F + \frac{G_0 F_M}{V_{MAX}}\right\}} \tag{10.3}$$

where a_H and b_H denote Hill's coefficients, F is the isotonic force, F_M is the peak isometric force, and V_{MAX} is the unloaded shortening velocity, which is determined by the pseudo visco-elastic properties of the XBs ($V_{MAX} = \bar{F}/\eta$). Hence, Hill's parameters are given by:

$$a_H = \frac{G_0}{G_1 + L_S^{-1}} \cdot \frac{F_M}{V_{MAX}} = b_H \cdot \frac{F_M}{V_{MAX}} \qquad b_H = \frac{G_0}{G_1 + L_S^{-1}} \cong \frac{G_0}{G_1} \tag{10.4}$$

Equation 10.4 provides the physiological meaning for the experimentally derived Hill's constants a_H and b_H. The dependencies of the shortening velocity, as well as the unloaded shortening velocity on sarcomere length, free calcium, the time of contraction, the rate of XB turnover, and the internal load are in agreement (Landesberg and Sideman, 1994b) with published experimental observation (Daniels et al., 1984; de Tombe and ter Keurs, 1992). Since power is the product of force and shortening velocity, the mechanical feedback regulates the FVR and power. Moreover, the mechanical feedback accurately describes the dependence of the force deficit (Leach et al., 1980) on the instantaneous force, shortening velocity, and the duration of shortening. The force-deficit is defined as the decrease in force during shortening, compared with the isometric force, at the same instant of the twitch (see below).

CONTROL OF ENERGY CONVERSION

Experimental studies have established the existence of a linear relationship between oxygen consumption ($\dot{V}_{O_2}$) and the mechanical energy generated, both at the whole-heart level (Sagawa et al., 1988; Suga, 1990) and at the isolated-fibre level (Hisano and Cooper, 1987). The observed linear correlation between the mechanical energy and energy consumption by the XBs is conveniently and convincingly explained by the mechanical feedback (Landesberg, 1997; Landesberg and Sideman, 2000). Electrical stimulation of cardiac muscle causes a transient elevation of the intracellular calcium, which binds to the regulatory troponin proteins, activates the actomyosin ATPase, and enables XB cycling. However, this one-directional open-loop excitation-contraction coupling from the electrical stimulation to the mechanical output cannot explain the linear relationship between energy consumption and mechanical energy production, and also cannot explain how the loading conditions affect energy consumption.

The mechanical feedback provides and maintains the linear relationship between the biochemical energy consumption and the generated mechanical energy. The energy consumption is calculated from the number of XBs that turn from the weak to the strong state in the XB cycle during contraction, since each cycle from the weak to the strong conformation is assumed to require the hydrolysis of one ATP. The general equation for energy conversion is given by:

$$\rho E = W + E_{PP} + Q_\eta \tag{10.5}$$

where ρ is the efficiency of the biochemical to mechanical energy conversion, E is the energy (ATP) consumption during the twitch, W is the external work, E_{PP} is the pseudo-potential energy which is proportional to the force-time integral (as for the isometric contraction), and Q_η represents energy dissipation due to the pseudo visco-elastic property of the XB. Q_η represents the viscous component since it is the integral over the force multiplied by the square of the velocity and higher orders of the velocity:

$$\rho = \frac{\overline{F} \cdot V_{MAX}}{E_{ATP}} \cdot \frac{1}{G_0 + G_1 V_{MAX}} \qquad E_{PP} = \frac{G_0 V_{MAX}}{\left(G_0 + G_1 V_{MAX}\right)} \int_0^T F(t)dt$$

$$W = \int_0^T F(t) V_{SL}(t)dt \qquad Q_\eta = \eta \int_0^T \frac{F(t) V_{SL}(t)^2}{\overline{F} - nV_{SL}} dt \tag{10.6}$$

where E_{ATP} is the free energy of ATP hydrolysis.

Note that the efficiency index of muscle contraction, which describes the utility of the muscle as a generator of external work, depends on G_1.

Equation 10.5 and Equation 10.6 provide the theoretical explanation of energy conversion, including the Fenn Effect (Fenn, 1923; Rall, 1982) and the linear relationship between energy consumption and the generated mechanical energy described by Suga and Sagawa for cardiac muscle (Sagawa et al., 1988; Suga, 1990). The pseudo-potential energy (Landesberg and Sideman, 1999), E_{pp}, is equal to Suga's (1990) potential energy for isometric contractions, and also dissipates as heat. The detailed derivation of the equations is described elsewhere (Landesberg, 1997; Landesberg and Sideman, 1999, 2000).

The co-operativity mechanism provides the molecular basis for the regulation of cardiac muscle energy consumption. It provides the necessary closed-loop feedback mechanism, which quantifies the bound calcium and the activity of the actomyosin ATPase (Chalovich and Eisenberg, 1986), and explains the ability of the muscle to adapt the energy consumption to the loading conditions. An increase in the afterload increases the affinity of troponin for calcium and increases the bound calcium. An increase in the bound calcium increases the rate of XB recruitment and ATP hydrolysis. Afterload reduction will reduce energy expenditure through the same mechanism (Landesberg and Sideman, 1999, 2000).

The mechanical feedback controls the force-velocity relationship and the power generation (Landesberg, 1996; Landesberg and Sideman, 1994b). The mechanical feedback determines how the biochemical energy, stored in the XBs as biochemical potential energy, is converted to external work (Landesberg and Sideman, 2000). Hence, the co-operativity mechanism determines the rate of ATP consumption, and the mechanical feedback determines the amount that will be converted to external mechanical energy. The balance between the maximal rate of energy consumption and the maximal power (efficiency) are related to the interplay between these two feedback loops.

CURRENT PROBLEMS

The FVR describes only a pseudo steady-state, where the load and the shortening velocity are 'constant'. Hence, the FVR depends on the experimental approach and on the time interval. Figure 10.2 presents the force response of an isolated rat trabecula to sarcomere shortening at various, *but steady*, velocities. As is shown in Figure 10.2, a pseudo steady-state of constant force and shortening velocity can be defined only for one of the force responses. Hence, the FVR describes only a specific relationship between the force and the shortening velocity, which holds only for a specific instant in time and a specific experimental protocol. Moreover, muscle generally does not work under steady-state conditions (constant activation and constant length) and cardiac muscle never works in a steady-state, since activation (free calcium transient) is transient. Hence, the FVR is far from explaining the force control for *in vivo* contractions. As is evident from Figure 10.2, force varies with time even at constant shortening velocities, and no 'steady' FVR that can describe the general contractile behaviour can be defined.

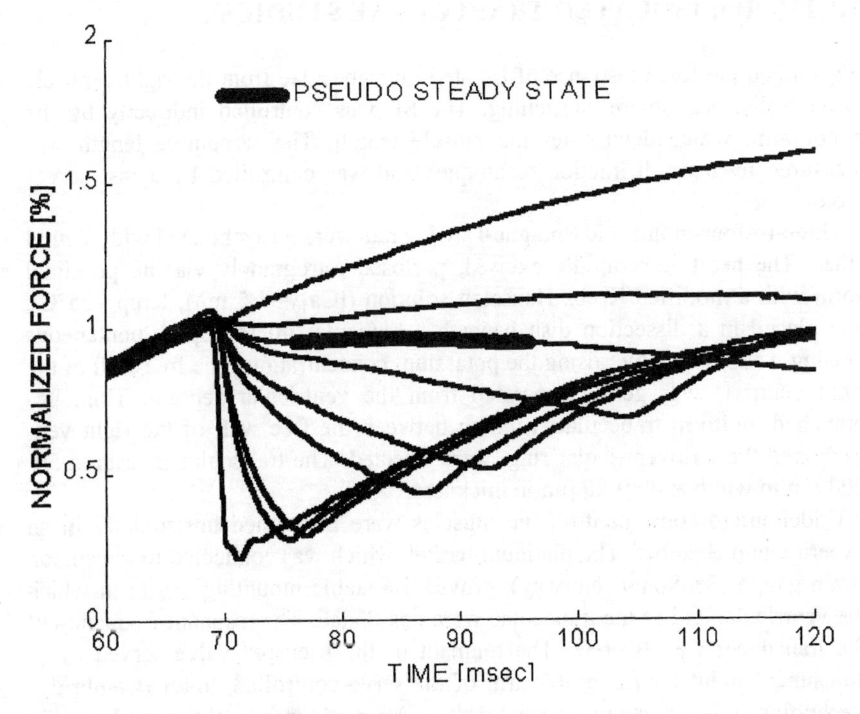

Figure 10.2. The effect of filament shortening at constant velocity on force. Each force response relates to a different, but constant, shortening velocity. Where can we define the 'force-velocity relationship'?

The goal of the present work is to develop a general approach to describe the effect of shortening velocity on generated force, based on intracellular biochemical processes and the regulation of XB dynamics. The following questions are addressed:

- What is the biophysical basis for the force-velocity relationship and the regulation of power generation in cardiac muscle?
- Can the effect of shortening on force be explained by a biochemical model of XB cycling?
- Can we establish and validate the mechanisms underlying the regulation of energy conversion from biochemical to mechanical energy, i.e., the mechanical feedback?
- Can we quantify the regulation of force development by the sarcomere shortening velocity in isolated rat trabeculae?
- Do XB dynamics consist of a single kinetics (attachment-detachment) or two interrelated processes (weak and strong versus attachment-detachment)?

METHODS: ISOLATED TRABECULAE STUDIES

We studied the force response of isolated rat trabeculae from the right ventricle to controlled sarcomere shortening. The SL was controlled indirectly by the motor arm, which determines the muscle length. The sarcomere length was measured by laser diffraction techniques and was controlled by a fast servomotor.

Three-to-four-month-old Sprague-Dawley rats were anaesthetized with diethyl ether. The heart was rapidly excised, perfused retrogradely via the proximal aorta with a modified Krebs-Henseleit solution ($[Ca]_0 = 1.5$ mM, temp = 25°C), and placed in a dissection dish beneath a binocular microscope. Spontaneous beating was stopped by raising the potassium concentration. The free wall of the right ventricle was gently separated from the ventricular septum. Thin, unbranched, uniform trabeculae, running between the free wall of the right ventricle and the atrioventricular ring, were selected. The trabeculae measured 50-200 μm in width and 40-80 μm in thickness.

Under microscopic control, the muscles were positioned horizontally in an experimental chamber. The platinum basket, which was connected to the silicon strain gauge (SenSonor, Norway), provided a stable mounting cradle in which the ventricular end of the trabeculae were positioned. The resonant frequency of the transducer was 10 kHz. The remnant of the tricuspid valve served as an attachment point for the motor arm of the servo-controlled motor (Cambridge Technology) via a stainless steel hook. After mounting, the muscles were stretched to a sarcomere length of 2.1 μm and left to equilibrate for 1 hour at 25°C and 1.5 mM $[Ca^{2+}]_o$, while being stimulated at 0.5 Hz. The solutions were equilibrated with 95% O_2 and 5% CO_2.

The striations of cardiac muscle acted as an optical grating to incident laser light (17-mW helium-neon laser, 632.8 nm). Two types of detectors were used for measurement of the angle of the first order diffraction band: 1) scanning, 512-element photodiode array (Reticon); photodiode arrays were scanned every 0.5 ms; spatial resolution of the diffractometer is about 4 nm; and 2) photo-optical position detection (Schottky barrier detector), which has a lower spatial resolution but a higher frequency response (25 kHz).

EXPERIMENTAL PROTOCOL

To evaluate the effect of shortening on force, the force generated at constant sarcomere length (sarcomere isometric contraction) was measured, and was used as a baseline for the evaluation of any deterioration in fibre function with time. Isometric contractions were achieved by stretching the fibre with the servo-motor, as shown in Figure 10.3. Usually, an exponential stretch of the muscle, which started 22 ms after the electrical stimulation and had a time constant of 33 ms provided constant SL. However, some adjustment of the stretch function was required for the various trabeculae.

Two types of protocols were used: 1) sarcomere shortening at constant velocity with various amplitudes was imposed at a given instant during the twitch, and 2) sarcomere shortening at constant velocity was imposed at various instants during the twitch (Figure 10.3).

The mechanical response of the sarcomere to length changes was analyzed; the input and output signals of this lumped system were the SL and force. The effect of shortening velocity on force was compared to the isometric force (Figure 10.3).

RESULTS

The force response to shortening at various velocities is characterized by two phases: an initial fast response that has a time constant of about 3 ms, followed by a slow response. The two phases are easily defined when the force response is plotted against sarcomere length. The first phase is characterized by a rapid fall in force. The magnitude of force decline increases with increasing shortening velocities. The second phase appears after sliding less than 20 nm per sarcomere. The normalized force, defined as the instantaneous shortening force ($F_V(t)$) divided by the isometric force ($F_M(t)$), $F_V(t)/F_M(t)$, during the second slow phase is linearly proportional to the length change, (ΔL), for shortening at constant velocity. The slope of the slow phase is a linear function of the sarcomere shortening velocity, while the coefficient G_1 describes this linear relationship.

The effect of shortening duration is demonstrated by imposing shortening at a constant velocity at various instants during the twitch (Figure 10.3). The decrease in force is quantified by measuring the force-deficit, defined as the instantaneous force difference between the isometric force ($F_M(t)$) and the shortening force ($F_V(t)$): $\Delta F(t) = F_M(t) - F_V(t)$.

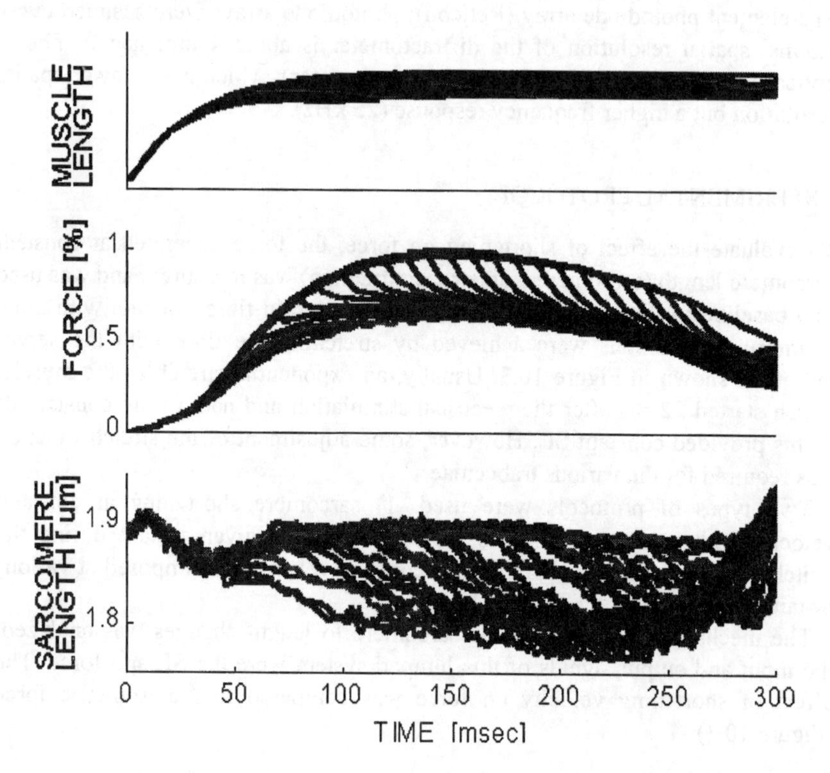

Figure 10.3. Identical sarcomere shortening at a constant velocity was imposed at various instants during the twitch. Sarcomere length (SL) was controlled by changing muscle length while the force response was observed.

The normalized force deficit, $\Delta F_N(t) = (F_M(t) - F_V(t)) / F_M(t)$ is calculated for all shortenings at different times during the twitch. An identical curve was obtained for $\Delta F_N(t)$ versus time by shifting the curves along the time axis so that $t=0$ denotes the time onset of shortening at the different twitches (Figure 10.4). Two phases of the $\Delta F_N(t)$ are well defined in Figure 10.4: an initial fast rise of $\Delta F_N(t)$, and a second moderate linear increase with time.

As seen, $\Delta F_N(t)$ increases linearly with time in the second phase for constant shortening velocity (V_{SL}) and is independent of the instant of onset of shortening. Thus, the dependence of the force deficit on the length change is due to the product of velocity and time:

$$\Delta F = F_M(t) - F_V(t) = G_1 \cdot V_{SL} \cdot t \cdot \left(1 - \frac{V_{SL}}{V_{MAX}}\right) \cdot F_M(t) \tag{10.7}$$

where V_{MAX} is the maximal unloaded velocity, ($V_{MAX} = \bar{F}/\eta$), which is determined by the pseudo visco-elastic properties of the XB (de Tombe and ter Keurs, 1992; Landesberg and Sideman, 1994b).

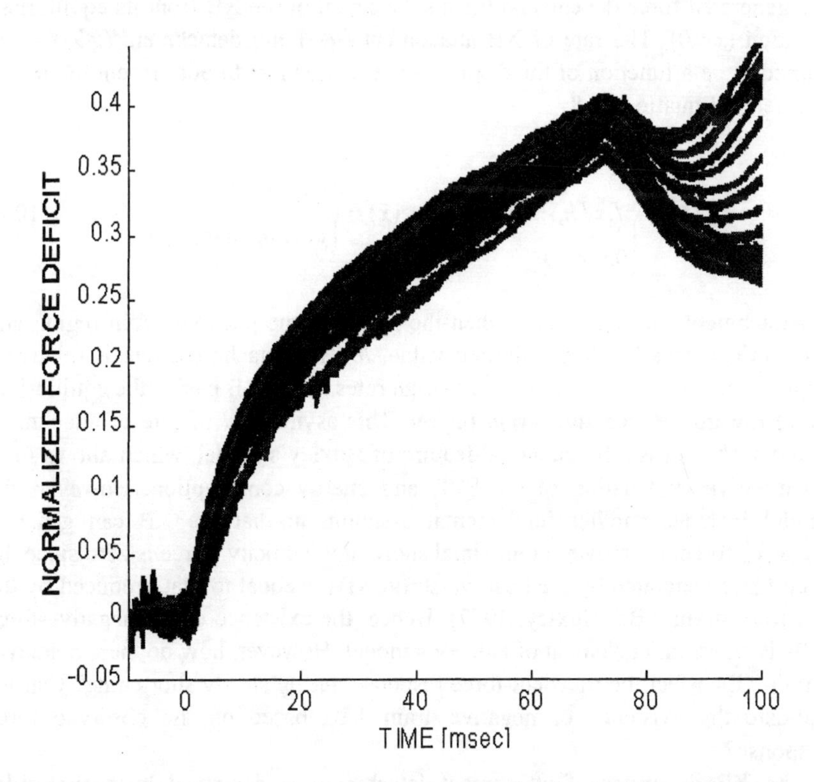

Figure 10.4. The dependence of the normalized force deficit, $\Delta F_N(t)$, on the duration of shortening. The observed effect of shortening is independent of the activation level.

Moreover, the slope of the second phase is independent of the onset time of shortening. Hence, the effect of shortening is independent of the activation level. The slope of the second phase does not depend on the number of strong XBs but only on the intrinsic property of the single XB. Moreover, Figure 10.4 substantiated the validity of the experimental methods and the data analysis that we have used. Although the activation level is not constant during the twitch, as is evident from the isometric contraction, the measurements of the normalized force deficit at various instants during the twitch are independent of the number of XBs. Hence, for all perturbations during a twitch (Figure 10.3), the same result is obtained (Figure 10.4).

The effect of shortening on force was tested on six trabeculae and the value of the mechanical feedback coefficient was: $G_l = 6.15 \pm 2.12$ [1/μm].

COMPARISON WITH HUXLEY'S MODEL OF CROSS-BRIDGE ATTACHMENT-DETACHMENT

According to Huxley's model, the XB is described as a Newtonian spring, and the generated force depends on the displacement of the XB from its equilibrium position ($x=0$). The rate of XB attachment ($f(x)$) and detachment ($g(x)$) is assumed to be a function of the displacement (in contrast to our present model) as defined in Equation 10.8:

$$f(x)=\begin{cases}0, & x<0\\ f_1x/h, & 0<x<h,\\ 0, & x>h\end{cases}\quad g(x)=\begin{cases}g_2, & x<0\\ g_1x/h, & 0<x<\infty\end{cases}\tag{10.8}$$

Attachment can only occur when the XB is in the positive strain region and the strain is smaller than a limited value, h. The detachment rate is relatively small at positive strains but jumps to high rates as the XB passes the equilibrium point toward the negative strain region. This asymmetry of rate of attachment and detachment is a fundamental feature of Huxley's model, which allows for a quantitative explanation of the FVR and energy consumption. However, the model includes another fundamental assumption: that the XB can generate negative force. Moreover, at maximal shortening velocity, force is zero since the total force generated by the positive-strain XBs is equal to that produced by the negative-strain XBs (Huxley, 1957). Hence, the existence of the negative-force XBs is an essential element of Huxley's model. However, how do these negative-strain XBs affect the transient force response during steady shortening? Can we validate the existence of negative-strain XBs based on the observed force response?

The XB dynamics of attachment-detachment is described by a first-order kinetics, based on Huxley's model, where $n(x,t)$ is the distribution function representing the fraction of attached cross-bridges (XBs) with displacement x at time t. Thus, the distribution of attached XBs is given by the differential equation (Huxley, 1957, 1974):

$$\frac{dn(x,t)}{dt}=[1-n(x,t)]f(x)-n(x,t)g(x)\tag{10.9}$$

and, in the full differential form for filament shortening, Equation 10.9 is represented as:

$$\frac{\partial n(x,t)}{\partial t}-V(t)\frac{\partial n(x,t)}{\partial x}=[1-n(x,t)]f(x)-n(x,t)g(x)\tag{10.10}$$

where $V(t)$ represents the speed of shortening of a half sarcomere ($V_{SL}(t)=2V(t)$).

This first-order partial differential equation can be solved by the method of characteristics. In accordance with Zahalak (1981), the solution can be represented as follows:

$$N(\xi,t) = N_0(\xi)e^{-\int_0^t H(\xi-\delta(\tau))d\tau} + \int_0^t f(\xi-\delta(\tau))e^{-\int_\tau^t H(\xi-\delta(\beta))d\beta} d\tau \qquad (10.11)$$

where $N_0(\xi)$ is an integration constant representing the value of $N(\xi,t)$ at t=0.

Zahalak (1981, 1990) solved this equation by assuming that $n(x,t)$ is a normally distributed function with respect to x. We will solve Equation 10.11 without any *a priori* assumptions, and thus are able to investigate the reaction of the model to sarcomere shortening and are able to test the exact transient response of the model. Let us define V as a constant shortening velocity. For constant velocities, Equation 10.11 reduces to:

$$N(\xi,t) = N_0(\xi)e^{-\int_0^t H(\xi-V\tau)d\tau} + \int_0^t f(\xi-V\tau)e^{-\int_\tau^t H(\xi-V\beta)d\beta} d\tau \qquad (10.12)$$

The first term on the right-hand side of Equation 10.12 is the transient response, representing the influence of the initial conditions, which decays with time. The second term represents the steady-state response. Returning to the original parameters x and t for the purpose of further analysis gives:

$$n(x,t) = n_0(x,t)e^{-\frac{1}{V}\int_x^{x+Vt} H(y)dy} - \frac{1}{V}\int_{x+Vt}^{x} f(\eta)e^{-\frac{1}{V}\int_x^{\eta} H(y)dy} d\eta \qquad (10.13)$$

As t tends to go to infinity, Equation 10.13 reduces to the steady-state (time-independent) relation:

$$n_\infty(x) = -\frac{1}{V}\int_{-\infty}^{x} f(\eta)e^{-\frac{1}{V}\int_x^{\eta} H(y)dy} d\eta \qquad (10.14)$$

Using the initial condition, n_0, which represents the distribution of the XBs in the isometric state, as given by Equation 10.15:

$$n_0(x) = \begin{cases} \dfrac{f_1}{f_1+g_1}, & 0 < x < h \\ 0 & \text{,otherwise} \end{cases} \qquad (10.15)$$

provides the steady-state solution of the equation, in agreement with the well-known solution of Huxley (1957):

$$n_{\infty}(x) = \begin{cases} \dfrac{f_1}{f_1+g_1}\left(1-e^{-\frac{f_1+g_1}{2hV}(h^2-x^2)}\right), & \text{for } 0 \le x \le h, \\ \dfrac{f_1}{f_1+g_1}\left(1-e^{-\frac{(f_1+g_1)h}{2V}}\right)e^{\frac{g_2 x}{V}}, & \text{for } x < 0 \end{cases} \tag{10.16}$$

Using the same initial conditions as in Equation 10.15 and the rate constants f and g (Equation 10.8), we obtain the full solution of Equation 10.10 for constant shortening velocity (V):

$$n(x,t) = \begin{cases} \dfrac{f_1}{f_1+g_1}e^{g_2 x/V} & \text{, for } x < 0, \text{ and } 0 < x+Vt < h \\ \dfrac{f_1}{f_1+g_1} & \text{, for } x > 0, \text{ and } 0 < x+Vt < h \\ \dfrac{f_1}{f_1+g_1}\left(1-e^{-\frac{f_1+g_1}{2hV}(h^2-x^2)}\right) & \text{, for } 0 \le x \le h, \text{ and } x+Vt > h \\ \dfrac{f_1}{f_1+g_1}\left(1-e^{-\frac{(f_1+g_1)h}{2V}}\right)e^{\frac{g_2 x}{V}} & \text{, for } x < 0, \text{ and } x+Vt > h \end{cases} \tag{10.17}$$

Note the transient changes in XB distribution (marked by the bold areas in Figure 10.5), which are due to the transition of XBs from the positive-strain region into the negative-strain region during shortening (first 20 ms). This transient shift in the distribution of the XBs does not resemble a normal distribution function. The transient changes in the XB distribution are a consequence of the precise solution of Huxley's model. It implies that there is a transient increase in the number of XBs in the negative strain region during the initial phase of sarcomere shortening. Consequently, a transient undershoot in force response is obtained (Figure 10.6). The undershoot is a feature of Huxley's model that is obtained at any shortening velocity and for any value of XB attachment and detachment rates. An increase in sarcomere shortening velocity enhances the undershoot and, for shortening velocities close to maximal, a negative force is predicted (Figure 10.6).

The predicted undershoot was not observed in any of our experiments, although we observed and recorded thousands of force responses to constant sarcomere shortening up to the maximal shortening velocity. Moreover, we are not aware of any results that have shown this phenomenon. Note that the predicted undershoot, which should appear about 20 ms after the onset of shortening, does not relate to the T_1 plot (Huxley, 1974) observed immediately (within one millisecond) upon imposed length perturbation.

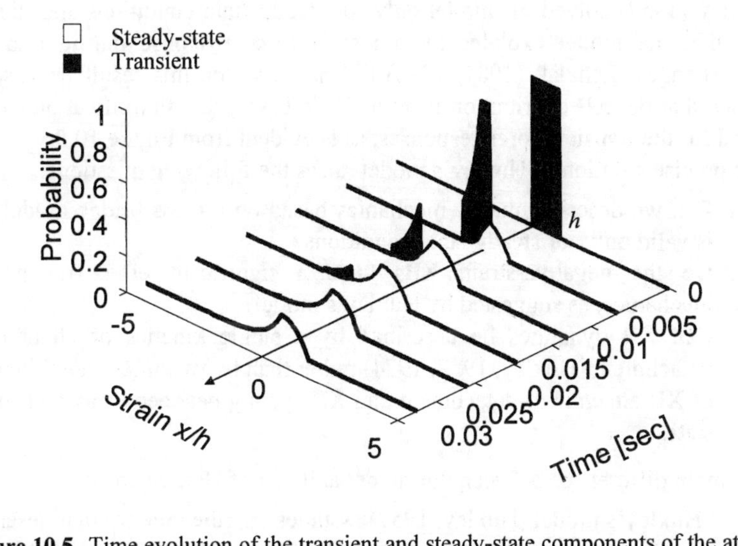

Figure 10.5. Time evolution of the transient and steady-state components of the attached distribution function *n(x,t)* during constant velocity shortening, starting from an isometric regime. Huxley's model (Equation 10.17) with: $g_1 = 10\ s^{-1}$, $g_2 = 209\ s^{-1}$, $f_1 = 43.3\ s^{-1}$, $V/h = 110\ s^{-1}$.

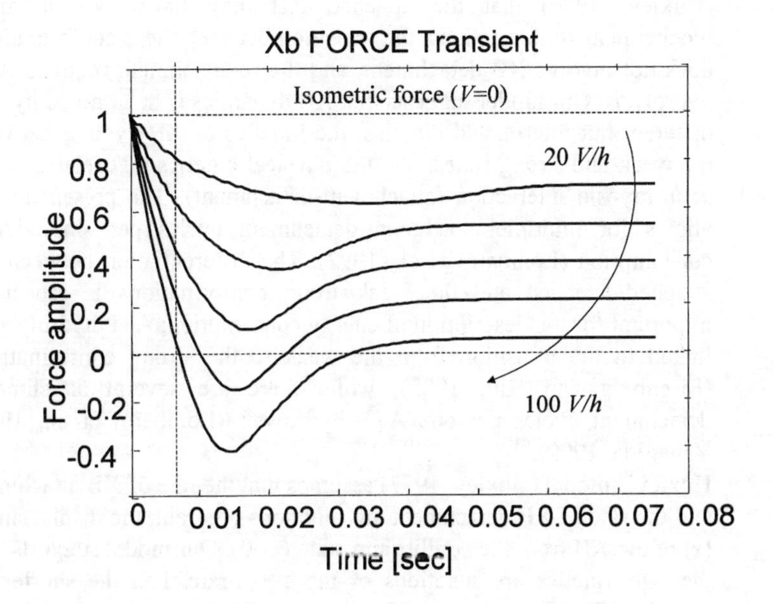

Figure 10.6. The precise transient force responses of Huxley's model to sarcomere shortening at various steady velocities. Note that the expected undershoot increases with increases in the sarcomere shortening velocity.

Huxley (1957) solved his model only for steady-state conditions and did not imply that the model explains the transient force response during transient length changes. Zahalak (1981, 1990) did not describe this result because he assumed that the XB distribution is normal. However, this simplification cannot be used for the transient force responses, as is evident from Figure 10.5.

The precise solution of Huxley's model raises the following questions:

1. Can we describe muscle mechanics based on a cross-bridge model that is valid only for steady-state conditions?
2. Do the negative-strain XBs have a significant effect on muscle mechanics, as suggested by Huxley's model?
3. Can XB dynamics be described by a single kinetics of attachment-detachment (Huxley, 1957, 1974) rather than by two interrelated kinetics of XB attachment-detachment and XB cycling between weak and strong states?

The main differences between our model and that of Huxley are:

1. Huxley's model (Huxley, 1957) assumes that the kinetics that describes the dynamics of XB attachment-detachment is the same as the kinetics of nucleotide (ATP, ADP, P) association and release. Consequently, there is a 1:1 relationship between the number of ATP molecule consumptions and the XB stroke steps. Although Huxley has suggested (Huxley, 1974) that the attached XB may have two or more biochemical conformations, the transition between these conformations does not involve XB detachment, and the re-attachment requires ATP hydrolysis. Our model suggests that XB dynamics is determined by two different but interrelated kinetics: the kinetics of XB cycling between the weak and strong states, and the physical kinetics that relates to the actin-myosin interaction (attachment-detachment). The present model allows for multiple attachment-detachment cycles per single ATP consumption (Lombardi et al., 1992). The differentiation between the attached/detached and the weak/strong conformations is especially important for the description of energy consumption; ATP hydrolysis is linked to the transition from the weak to the strong conformations (Eisenberg and Hill, 1985), while there are several attachment-detachment cycles per one ATP hydrolysis (Lombardi et al., 1992; Yanagida, 1999).
2. Huxley's model (Huxley, 1957) assumes that the rate of XB attachment and the rate of XB detachment are functions of strain, the displacement (x) of the XB from the equilibrium state (x=0). Our model suggests that the rate kinetics are functions of the strain-rate, i.e., the shortening velocity. The dependence of the rate constants on strain rate is simple: The rate of XB turnover from the weak to the strong conformation is constant. The rate of XB weakening (G) is a linear function of filament sliding velocity (Equation 10.2).

3. Huxley's model requires the solution of partial differential equations of time and displacement. In addition, the rate constants of Huxley's model are non-smooth functions of the strain. Consequently, it is impossible to obtain a straightforward analytical solution for general contractions (Zahalak, 1981). The attached XB distribution function is not a smooth function of the strain, and is far from being a normal Gaussian function (Figure 10.5). Therefore, elaborate models are required in order to solve Huxley's model (Zahalak, 1981, 1990). In contrast, the present model is simple and allows for the derivation of the equations for the FVR (Equation 10.3; Landesberg and Sideman, 1994b). Furthermore, it conveniently and convincingly provides the equation for energy conversion by the sarcomeres (Equation 10.5; Landesberg and Sideman, 2000). As far as we know, there is no general solution equivalent to Equation 10.5 based on Huxley's model. The equations for energy conversion using Huxley's model are only given by lumped equations, as those developed by Zahalak (1981).
4. Huxley's model assumes full and steady activation. However, the beating cardiac muscle is never in a steady state. Our model couples XB dynamics with the regulation of XB recruitment by calcium binding to the troponin regulatory proteins.
5. Huxley's model assumes the existence of negative-force-generating XBs during normal muscle function, which are essential for the explanation of the FVR.
6. Huxley's model suggests that the average force per XB decreases slowly (within 20 ms) during sarcomere shortening, and that the kinetics that determines the average force per XB and the number of XBs is the same (if only the attachment-detachment kinetics is considered). Our model and experimental data suggest that the average force per XB reaches its steady-state level within a few milliseconds (Figure 10.4), while the kinetics that determines the number of strong XBs is slower.

CONCLUSION

This study provides new insight into the mechanisms affecting muscle function, and provides a new quantitative approach for evaluating muscle mechanics and energetics based on cross-bridge dynamics, thereby extending our understanding of the intracellular control of contraction.

The results of this study suggest the following model of XB dynamics:

1. XB dynamics is described by two kinetic mechanisms:
    - a fast, physical kinetics of XB attachment-detachment,
    - a slow, biochemical kinetics of XB cycling between the strong force-generating conformation and the weak conformation.

2. The fast kinetics relates to the visco-elastic property of the XB described by de Tombe and ter Keurs (1992) for steady shortening. It also relates to Huxley's T_1–T_2 plots derived from quick-release experiments.
3. The slow kinetics relates to the kinetics of nucleotide binding and release. This kinetics determines the number of strong XBs and the duty cycle of the strong XBs.
4. The mechanical feedback relates to the effect of the shortening velocity on the biochemical rate of XB weakening. This effect of shortening on the rate kinetics is described by the parameter G_1. This feedback loop provides an analytical solution for the force-velocity relationship and for energy conversion by the XBs. A new approach for the quantification of muscle power generation is suggested, based on the magnitude of the mechanical feedback coefficient.

FUTURE RESEARCH

1. Develop a quantitative description of the intracellular control of energy consumption and biochemical to mechanical energy conversion. To fulfil this aim, we will substantiate the reproducibility and the predictability of the magnitude of the mechanical feedback coefficient (G_1) as the general index for the power capability of muscle function and, specifically, as a measurable index for cardiac muscle contractility.
2. Study the effect of various drugs that affect calcium kinetics and XB cycling on energy conversion, economy, and efficiency.
3. Study the balance between the regulation of muscle economy and efficiency. These two parameters, economy and efficiency (Landesberg and Sideman, 1999, 2000), impose contradicting demands on muscle function and are especially important for cardiac muscle, since cardiac adaptation and optimization to loading conditions and physiological requirements are of utmost importance for pathological conditions. We have already shown theoretically that the economy is determined by G_0, the rate of XB weakening at isometric contraction, while the efficiency is determined by the magnitude of the mechanical feedback, G_1. These two parameters can be measured experimentally using various drugs and normal physiological conditions.
4. Develop a new measurable index that will enable the quantification of the mechanical function of the left ventricle, based on the basic property of XB dynamics. A significant step toward the realization of this goal was achieved recently by utilizing a similar approach to that presented here for the analysis of the *in situ* swine heart mechanics.
5. Develop a more precise model for cross-bridge dynamics that will be based on the current biochemical studies and recent studies utilizing motility assays and laser trap techniques.

The advantage and significance of mechanical feedback as an index of the muscle's ability to generate power, and the importance of mechanical feedback for the quantification of the cardiac muscle contractile state are under investigation.

ACKNOWLEDGEMENTS

This study was financed by a grant from the Medical Research Council (MRC, Canada) and from the distinguished Yigal Allon Grant (to A.L., Israel). This research was also supported by the fund for promotion of research at the Technion.

REFERENCES

Allen, D.G., and Kentish, J.C. (1985). The cellular basis of the length tension regulation in cardiac muscle. *Journal of Molecular and Cellular Biology* **17**, 821–840.

Allen, D.G., and Kentish, J.C. (1988). Calcium concentration in the myoplasm of skinned ferret ventricle, following changes in muscle length. *Journal of Physiology* **407**, 489–503.

Brenner, B. (1991). Rapid dissociation and reassociation of actomyosin crossbridge during force generation: A newly observed facet of crossbridges action in muscle. *Proceedings of the National Academy of Science USA* **88**, 10490–10494.

Brenner, B., and Eisenberg, E. (1986). Rate of force generation in muscle: Correlation with actomyosin ATPase activity in solution. *Proceedings of the National Academy of Science USA* **83**, 3542–3546.

Chalovich, J.M., and Eisenberg, E. (1986). The effect of troponin-tropomyosin on the binding of heavy meromyosin to actin in the presence of ATP. *Journal of Biological Chemistry* **261**, 5088–5093.

Daniels, N., Noble, M.I., ter Keurs, H.E.D.J., and Vohlfart, B. (1984). Velocity of sarcomere shortening in rat cardiac muscle: Relationship to force, sarcomere length, calcium and time. *Journal of Physiology* **355**, 367–381.

de Tombe, P.P., and ter Keurs, H.E.D.J. (1992). An internal viscous element limits unloaded velocity of sarcomere shortening in rat myocardium. *Journal of Physiology* **454**, 619–642.

Eisenberg, E., and Hill, T.L. (1985). Muscle contraction and free energy transduction in biological systems. *Science* **227**, 999–1006.

Fenn, O.W. (1923). A quantitative comparison between the energy liberation and work performed by the isolated sartorious muscle of the frog. *Journal of Physiology* **58**, 175–203.

Fuchs, F., and Wang, Y.P. (1991). Force, length and calcium troponin-C in skeletal muscle. *American Journal of Physiology* **261**, C787–C792.

Greene, L.E., Williams, D.L., and Eisenberg, E. (1987) Regulation of actomyosin ATPase by troponin-tropomyosin: Effect of the binding of the myosin subfragment-1·ATP complex. *Proceedings of the National Academy of Science USA* **84**, 3102–3106.

Guth, K., and Potter, J.D. (1987). Effect of rigor and cycling XBs on the structure of troponin-C and on the Ca^{2+} affinity of the Ca^{2+}-specific regulatory sites in skinned rabbit psoas fibers. *Journal of Biological Chemistry* **262**, 15883–15890.

Hill, A.V. (1964). The effect of load on the heat of shortening. *Proceedings of the Royal Society of London B* **159**, 297–318.

Hill, T.L. (1983). Two-element models for the regulation of skeletal muscle contraction by calcium. *Biophysical Journal* **44**, 383–396.

Hisano, G., and Cooper, I.V. (1987). Correlation of force-length area with oxygen consumption in ferret papillary muscle. *Circulation Research* **61**, 318–328.

Hunter, W.C. (1989). ESP as a balance between opposing effects of ejection. *Circulation Research* **64**, 265–275.

Huxley, A.F. (1957). Muscle structure and theories of contraction. *Progress in Biophysics and Biophysical Chemistry* **7**, 255–318.

Huxley, A.F. (1974). Muscle contraction. *Journal of Physiology* **243**, 1–43.

Landesberg, A. (1996). End systolic pressure-volume relation based on the intracellular control of contraction. *American Journal of Physiology* **270**, H338–H349.

Landesberg, A. (1997). Intracellular mechanism in control of myocardial mechanics and energetics. In: *Advances in Experimental Medicine and Biology, Vol. XXX. Analytical and Quantitative Cardiology: From Genetics to Function.* Sideman, S., and Beyar, R. (eds.), Plenum Publishing Corporation, New York, pp. 137–153.

Landesberg, A., and Sideman, S. (1994a). Coupling calcium binding to troponin-C and crossbridge cycling in skinned cardiac cells. *American Journal of Physiology* **266**, H1260–H1266.

Landesberg, A., and Sideman, S. (1994b). Mechanical regulation in the cardiac muscle by coupling calcium binding to troponin with crossbridge cycling: A dynamic model. *American Journal of Physiology* **267**, H779–H795.

Landesberg, A., and Sideman, S. (1999). Regulation of energy consumption in the cardiac muscle: Analysis of isometric contractions. *American Journal of Physiology* **276**, H998–H1011.

Landesberg, A., and Sideman, S. (2000). Force-velocity relationship and biochemical to mechanical energy conversion by the sarcomere. *American Journal of Physiology* (in press).

Leach, J.K., Brady, A.J., Skipper, B.J., and Millar, D.L. (1980). Effect of active shortening on tension development of rabbit papillary muscle. *American Journal of Physiology* **238**, H8–H13.

Lehrer, S.S. (1994). The regulatory switch of the muscle thin filament: Calcium or myosin heads? *Journal of Muscle Research and Cell Motility* **15**, 232–236.

Lombardi, V., Piazzesi, G., and Linari, M. (1992). Rapid regeneration of actin-myosin power stroke in contracting muscle. *Nature* **355**, 638–641.

Mommaerts, M.F.H.M., Seraydarian, I., and Marechal, G. (1962). Work and mechanical change in isotonic muscular contractions. *Biochimica et Biophysica Acta* **57**, 1–12.

Peterson, J.N., Hunter, W.C., and Berman, M.R. (1991). Estimated time course of calcium bound to troponin-c during relaxation in isolated cardiac muscle. *American Journal of Physiology* **260**, H1013–H1024.

Rall, J.A. (1982). Sense and nonsense about the Fenn Effect. *American Journal of Physiology* **242**, H1–H6.

Sagawa, K., Maughan, L., Suga, H., and Sanagawa, K. (1988). *Cardiac Contraction and the Pressure-Volume Relationship*. Oxford University Press, New York.

Stein, L.A., Chock, P.B., and Eisenberg, E. (1984). The rate limiting step in the actomyosin adenosine triphosphatase cycle. *Biochemistry* **23**, 1555–1563.

Suga, H. (1990). Ventricular energetics. *Physiological Reviews* **70**, 247–277.

Yanagida, T. (1999). Simultaneous observation of individual ATPase and mechanical events by a single myosin molecule during interaction with actin. *Canmore Symposium on Skeletal Muscle*, August 6-7, 1999, Canmore, Alberta, Canada.

Zahalak, G.I. (1981). A distribution-moment approximation for kinetic theories of muscular contraction. *Mathematical Biosciences* **55**, 89–114.

Zahalak, G.I. (1990). Modeling muscle mechanics (and energetics). In: *Multiple Muscle System: Biomechanics and Movement Organization*. Winters, J.M., and Woo, S.L-Y. (eds.), Springer-Verlag, New York, pp. 1–23.

11 Three-dimensional Geometric Model of Skeletal Muscle

R. LEMOS
Department of Computer Science, Faculty of Science, University of Calgary, Calgary, Alberta, Canada

M. EPSTEIN
Department of Mechanical Engineering, Faculty of Engineering, University of Calgary, Calgary, Alberta, Canada

W. HERZOG
Human Performance Laboratory, University of Calgary, Calgary, Alberta, Canada

B. WYVILL
Department of Computer Science, Faculty of Science, University of Calgary, Calgary, Alberta, Canada

The main goal presented in this chapter is to investigate function and deformation in whole muscle by determining how force influences muscle deformation and how deformation affects the contractile properties of muscle. A muscle model that predicts fibre forces based on the principle of virtual work, along with appropriate geometric constraints, is proposed and a three-dimensional, structural model of unipennate muscle to assess changes in muscle geometry during contraction is developed as an example. The muscle model consists of an assembly of straight lines (SL) organized in brick-like elements to represent muscle architecture in three-dimensional space. No limitations are imposed on the magnitude of the deformations or rotations of the muscle, which can be measured from an arbitrary reference geometry. The initial muscle architecture was obtained from the measured lengths of muscle fibres, aponeurosis, and tendon in the mid-sagittal plane of the cat medial gastrocnemius. Relaxed and activated fibre lengths, angle of pinnation, and external force were measured for cat medial gastrocnemius muscle (n=8) for isometric contractions at different muscle length. Comparisons of fibre lengths and angles of pinnation were made between theoretically predicted and experimentally measured values. Fibre lengths became shorter and angles of pinnation increased when going from the relaxed to the activated state in both the experiment and the model. Therefore, the conceptual predictions of the model appear correct.

Skeletal Muscle Mechanics: From Mechanisms to Function. Edited by W. Herzog.

INTRODUCTION

Muscle models have been developed to analyze the characteristics of force production (Delp, 1990), and to determine muscle deformation based on architecture (Otten, 1988; van der Linden et al., 1998). The correct representation of muscle architecture is important for gaining an understanding of muscle function. Using a geometric model to represent the muscle architecture, some effects, such as deformation of the muscle, can be observed during muscle contraction. An understanding of these changes in muscle geometry is important, because these changes in geometry will directly affect muscle force production.

Previous papers on muscle models proposed geometric models to represent muscle architecture. These papers can be classified into complex and simple models. The complex models adopt sophisticated techniques to model muscle contraction, such as the finite-element techniques (Anton, 1992; Chen, 1991; Chen and Zeltzer, 1992). The simple models use simple geometric shapes to model muscle contraction, such as straight-line models (Delp, 1990; Delp and Loan, 1995; Epstein and Herzog, 1998; Maurel et al., 1996; Otten, 1988; van der Linden, 1998). In this chapter, we will pay attention to the main features of simple models that use straight lines to represent the tendon-aponeurosis-fibre complex, and we will propose a simple model based on the principle of virtual work.

We would like to investigate muscle function and deformation in whole muscle by determining how force influences muscle deformation and how deformation affects the contractile properties of muscle. Therefore, we developed a three-dimensional, structural model of unipennate muscle in order to assess changes in muscle geometry during contraction.

In the following, some biological and mechanical considerations about muscle modelling are presented, which will be taken into account in the proposed muscle model. Then we will present the architecture of a muscle and discuss how the biological and mechanical considerations were included in the model. Finally, we will present selected results of theoretically predicted and experimentally measured muscle forces, and present some thoughts for future work.

MUSCLE MODELLING

We will develop a geometric model based on straight-line geometry, hereafter referred to as straight-line models (SLM). Straight-line models are constructed by an assembly of straight lines that remain straight during deformation, and elongate or contract in response to internally developed axial forces (Figure 11.1).

The biological considerations along with SLM are related to the shape of skeletal muscle, the active and passive force-producing elements, the angle of pinnation during muscle contraction, and the control of muscle force (Epstein and Herzog, 1998).

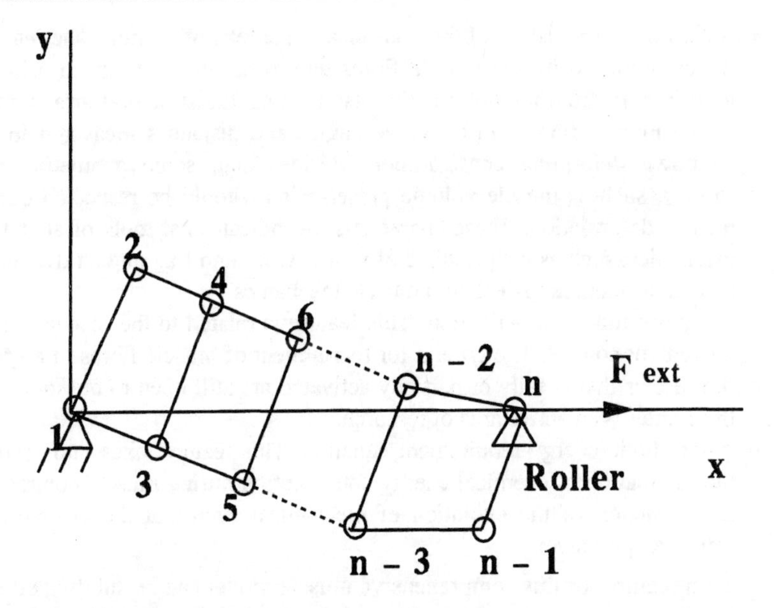

Figure 11.1. Possible level of representation of a muscle.

Experimental results in muscle contraction have been explained in part by the Hill phenomenological model (Hill, 1938) and by the Huxley structural model (Huxley, 1957). However, it is important to establish which features a muscle model should represent, as well as which levels of detail should be used to simulate a muscle model. According to Epstein (1994), a comprehensive muscle model should present the following mechanical considerations:

- An experimentally consistent mathematical representation of the constitutive behaviour of the active and passive force-producing elements. According to Epstein and Herzog (1998), the literature does not give a consistent mathematical model to accommodate all accepted experimental results in the proper representation of the material behaviour of the active and passive force-producing elements. However, the essential features of the experimentally observed behaviour can be approximated. For example, a force-length-velocity relationship based on Hill's model, scaled with a force-length relationship and an activation coefficient, may be used to approximate the active force-producing elements. In addition, the elastic properties of the passive force-producing elements can be used to approximate their material behaviour.
- An accurate geometrical representation of the muscle. The muscle architecture should reflect the dimensions as well as the cross-sectional areas of the tendon-aponeurosis-fibre complex of a given skeletal muscle, the geometric constraints, the angles of pinnation, the aponeuroses thickness, and the details of the tendon insertions.

- A rigorous formulation of the non-linear equations of motion. The internal forces produced by the muscle fibres due to muscle contraction arise, at least in part, from the deformation itself. Thus, the structural arrangement of the muscle fibres must involve angles and distances measured in the unknown deforming configuration. Besides that, some geometric constraints, such as muscle volume preservation, should be respected during muscle deformation. These characteristics indicate that tools of analytical mechanics, such as the principle of virtual work and Lagrangian dynamics, should be adopted to deal with muscle mechanics.
- A control model of activation. This feature is related to the mechanism of movement control. The criteria for recruitment of muscle fibres in a group of muscles that is fully or partially activated are still open to question, and this matter is an active area of research.
- A metabolic energy requirement equation. This feature takes into account the calculation of chemical energy consumption during muscle contraction as a function of the evolution of mechanical events at the cross-bridge (microscopic) level.

The main features of this comprehensive muscle model can be subdivided into three main aspects: mechanics, control, and thermochemistry. In this chapter, a metabolic energy equation is not taken into account. The basic mechanical concepts at the macroscopic level for the constitutive behaviour, for the formulation of the non-linear equations of motion, and for the geometric constraints along with SLM will be considered, using the principle of virtual work in the presence of geometric constraints.

According to Nigg and Herzog (1994), the principle of virtual work states that the equilibrium of a mechanical system is equivalent to the identical vanishing of the total virtual work performed by external and internal forces through all virtual displacements of the system.

DISPLACEMENT

The tendon-aponeurosis-fibre complex is represented using SLM. Therefore, straight lines can represent any of these members (tendons, aponeuroses, or fibres). These straight-line members remain straight during deformation, however they elongate or contract in response to internally developed axial forces. Consider a simple case: just one straight-line member, in which P and Q represent the end-points of the straight-line member, and L is its original length (Figure 11.2).

As a result of the displacement vectors $\vec{u}_p$ and $\vec{u}_q$, new positions of the end-points p and q, as well as a new length l, will be established. Using vectors that provide a link between geometric reasoning and arithmetic calculation, it is possible to describe the elongation produced by this displacement as follows:

$$e = l - L \tag{11.1}$$

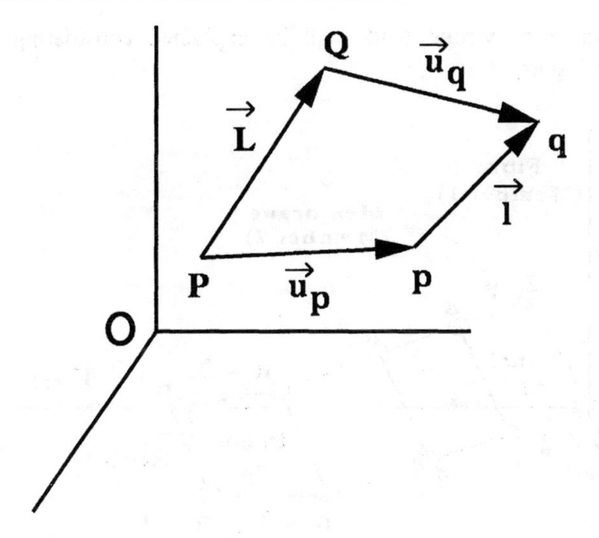

Figure 11.2. Kinematics of a straight line.

From Figure 11.2 and using vectors the following equivalence can be established:

$$\vec{L} + \vec{u}_Q = \vec{u}_p + \vec{l} \tag{11.2}$$

where L and l are the 'length' vectors shown in Figure 11.2.

Taking the dot product of each side (Equation 11.2) with itself gives:

$$l^2 = L^2 + 2\vec{L}\cdot(\vec{u}_Q - \vec{u}_p) + (\vec{u}_Q - \vec{u}_p)\cdot(\vec{u}_Q - \vec{u}_p) \tag{11.3}$$

from which the elongation can be expressed as:

$$e = [L^2 + 2\vec{L}\cdot(\vec{u}_Q - \vec{u}_p) + (\vec{u}_Q - \vec{u}_p)\cdot(\vec{u}_Q - \vec{u}_p)]^{1/2} - L \tag{11.4}$$

THE PRINCIPLE OF VIRTUAL WORK

The principle of virtual work is based on the notion of virtual displacement. A virtual displacement of a system is a set of small variations, δu, of its displacements, compatible with the geometrical constraints of the system. These virtual displacements naturally produce small variations, δe, in the elongations, which can be obtained by differentiating Equation 11.4 as follows:

$$\delta e = (\vec{L} + \vec{u}_Q - \vec{u}_p)\cdot(\delta\,\vec{u}_Q - \delta\,\vec{u}_p)/\,l \tag{11.5}$$

where the virtual displacements $\delta\vec{u}_Q$ and $\delta\vec{u}_P$ are small variations imposed on the actual displacements $\vec{u}_P$ and $\vec{u}_Q$.

The principle of virtual work will be explained considering the system presented in Figure 11.3.

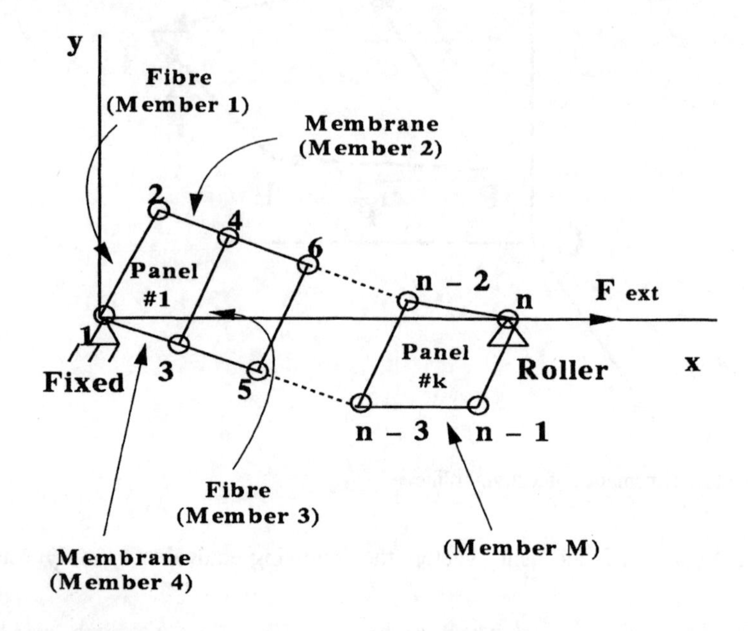

Figure 11.3. Planar assembly with n nodes, external supports (fixed and roller), one external force (F), K panels, and M straight-line members.

The system is composed of M straight-line members, connected at a number of n nodes. Some of these nodes can be prevented from moving by means of supports, and some of them can be subjected to external forces (F). The system is a planar assembly of straight-hinged members so that each member can only develop axial forces, N. When the nodes of this system establish a new configuration in response to internally developed axial forces, this new configuration can be represented by a set of current displacements, $u_i(i=1,\ldots,n)$. As a result of these displacements, the straight-line members of the system will undergo certain elongations, $e_I(I=1,\ldots,M)$, and sustain some internal forces. These internal forces, $N_I\,(I=1,\ldots,M)$, are defined by the constitutive law of the material and they will be positive if in tension and negative if in compression.

Under a superimposed set of virtual displacements, $\delta u_i(i = 1,\ldots,n)$, virtual elongations can be calculated from Equation 11.5. The total internal virtual work IVW of the internal forces N_I, is given by:

$$IVW = \sum_{I=1}^{M} N_I \delta e_I \tag{11.6}$$

and the total external virtual work, EVW, of the external forces is given by:

$$EVW = \sum_{i=1}^{n} F_i \delta u_i \tag{11.7}$$

In addition, according to the definition of the principle of virtual work, the equations of motion of the system are equivalent to the identical satisfaction of the single scalar equation:

$$IVW - EVW \equiv 0 \tag{11.8}$$

for all possible values of the virtual displacements (Epstein, 1994).

GEOMETRIC CONSTRAINTS

During muscle deformation, geometric constraints must be satisfied. The geometric constraints in SLM can be classified into three main groups (Epstein and Herzog, 1998): constraints of assembly, external supports, and other geometric restrictions.

First, the constraints of assembly state that the topology of the muscle architecture must remain fixed during the deformation process. For instance, the geometric representation of *Panel #1* (Figure 11.3) must always be represented by the same four straight-line members, and it should always be quadrilateral.

Second, the external supports state that a node can be completely fixed, or that a node can be partially fixed or free.

Third, other geometric constraints represent aspects that are not directly incorporated in the model. For instance, it was shown by Baskin and Paolini (1967) that muscle volume remains nearly constant during contraction. Since the system presented in Figure 11.3 is a planar system, this constraint may be approximated by the preservation of the area of the panels during the deformation process, assuming that deformations in the out-of-plane direction are small, as they seem to be (Carvalho et al., 1999). Volume preservation is an important constraint and is essential to guarantee the stability of the system.

On the other hand, volume preservation cannot be directly incorporated into the model because it establishes an interdependence between several degrees of freedom. These constraints should be represented by prescribing a functional dependence as follows:

$$\phi(u_1,\ldots,u_n) = 0 \tag{11.9}$$

where $u_i(i=1,\ldots,n)$ are the set of displacements for the n nodes of the system.

Up to now, according to Equation 11.8, the principle of virtual work was used as a method for deriving the equations of motion of an unconstrained system. The principle of virtual work can be modified to take into consideration geometric constraints by the addition of a new variable ('Lagrange multiplier'), λ, to the list of degrees of freedom. The principle of virtual work for a constrained system is:

$$IVW - EVW \equiv \delta(\lambda\phi) \tag{11.10}$$

which expresses mathematically the principle of virtual work in the presence of a geometric constraint. It states that the deformable system is in equilibrium if, and only if, the difference between the internal and external virtual work is identically equal to the variation of the product $\lambda\phi$ for all independent variations of $\delta u_1, \ldots, \delta u_m$, $\delta\lambda$ (Epstein and Herzog, 1998).

If more constraints need to be included, Equation 11.10 can be modified to include additional functional dependencies and additional Lagrange multipliers as follows:

$$IVW - EVW \equiv \sum_{\alpha=1}^{p} \delta(\lambda_\alpha \phi_\alpha) \tag{11.11}$$

where p is the number of constraints. The physical meaning of the Lagrange multipliers is that they are proportional to the 'forces' necessary to maintain the constraints. It should be pointed out here that it is dangerous to derive (or assume) the internal constraint forces intuitively. Such approaches have typically been wrong (e.g., Otten, 1988), and, unfortunately, have had a great influence on the theoretical approaches to muscle modelling.

THREE-DIMENSIONAL GEOMETRIC MUSCLE MODEL

A three-dimensional geometric model of skeletal muscle is introduced. The essential biological and mechanical considerations previously discussed for muscle contraction are included in this model. First, the general characteristics of the model are presented. Second, the methods adopted in the model are described according to the following features: geometric representation of the model, mathematical representation of the constitutive behaviour, and formulation of the non-linear equations of motion, and numerical considerations.

GENERAL CHARACTERISTICS OF THE MODEL

The three-dimensional geometric muscle model exhibits the following characteristics:

- SLM are used to represent the tendon-aponeurosis-fibre complex.
- The approach adopted to describe the equations of motion is the principle of virtual work in the presence of geometric constraints. These equations are exact.
- Geometric constraints included in the model are the following: con-

straints of topology, constraints of support, and constraints of the preservation of muscle volume.

- Different constitutive equations can be used to simulate the behaviour of the force-producing elements during muscle contraction. The present implementation uses a quadratic force-length relationship and an activation coefficient to approximate the active force-producing elements. It uses a quadratic constitutive law to approximate the passive elements in series with the muscle fibres (such as tendons and aponeuroses).
- Force-velocity and history-dependent properties are not considered, but could be included.
- No limitations are imposed on the magnitude of the deformations or the rotations of the muscle; deformations can be measured from an arbitrary reference geometry.

GEOMETRIC REPRESENTATION OF THE MODEL: AN EXAMPLE

A specific muscle architecture for a unipennate muscle is adopted as an example to explain the methods used in the geometric representation of the model; however, the generalization to other muscles and muscle shapes (e.g., parallel-fibred or pennate-fibred muscles) is straightforward.

The present method supports skeletal muscles with other structural arrangements, dimensions, and cross-sectional areas of the tendon-aponeuroses-fibre complex, angles of pinnation, aponeuroses thickness, and positions of tendon insertions than considered here.

Representation of the initial muscle architecture using SLM

The muscle chosen was the unipennate cat medial gastrocnemius. Its initial muscle architecture was obtained from the measured lengths of muscle fibres, aponeurosis, and tendon in the mid-sagittal plane of the skeletal muscle (Carvalho et al., 1999) (Figure 11.4).

During measurement, the muscle was in the relaxed state with both ends fixed at a muscle reference length that we arbitrarily denote as 0 mm. In addition, approximate dimensions of the lateral and medial aponeurosis according to the top view of the skeletal muscle (Figure 11.5) were used to describe the muscle architecture in three-dimensional space.

Based on the measured lengths, obtained with sonomicrometry through crystals placed at specific locations on the muscle, it was possible to calculate the angles of pinnation (Figure 11.4). In addition, nodes were included in the proximal part of the muscle to model the whole muscle. The resultant initial muscle architecture is presented in Figure 11.6, and the resultant three-dimensional co-ordinates for the initial muscle architecture are presented in Table 11.1.

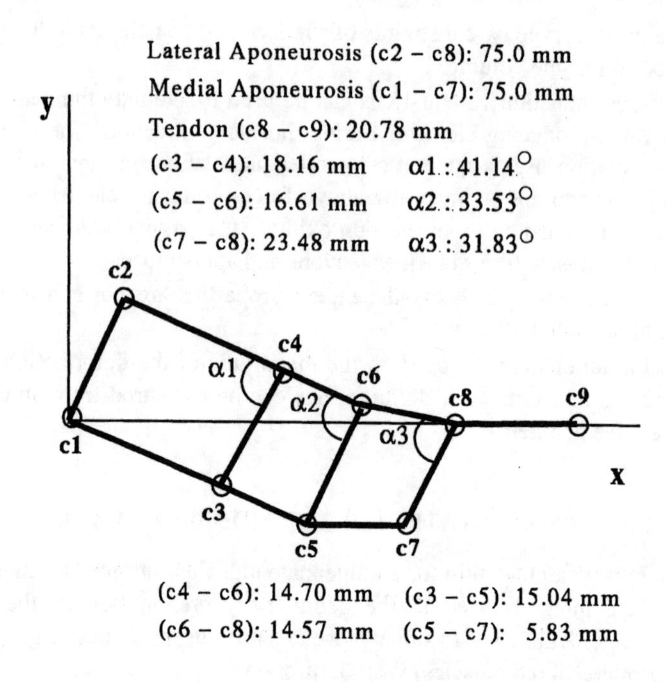

Figure 11.4. Mid-sagittal plane of the cat medial gastrocnemius.

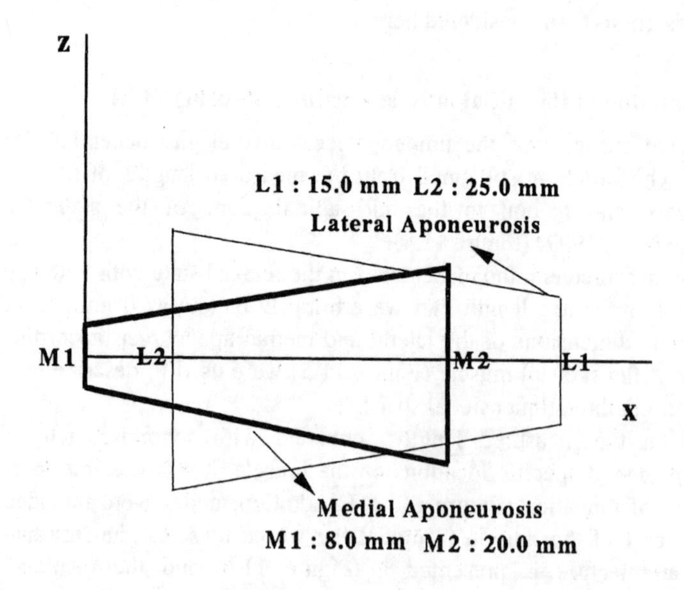

Figure 11.5. Top view of the cat medial gastrocnemius.

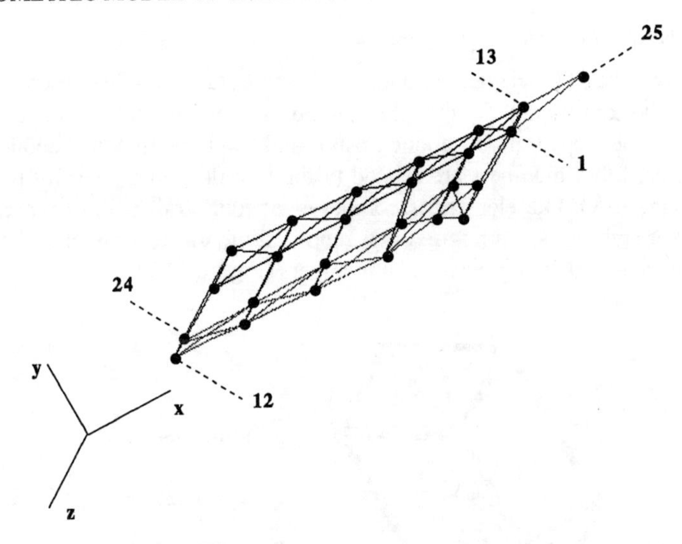

Figure 11.6. Three-dimensional representation of the initial muscle architecture.

Table 11.1. Three-dimensional co-ordinates of the initial muscle architecture.

Node	X (mm)	Y (mm)	Z (mm)	Node	X (mm)	Y (mm)	Z (mm)
1	94.39	0.10	−4.90	14	73.05	−9.24	−22.68
2	73.05	−9.24	−2.31	15	83.18	1.52	−19.67
3	83.18	1.52	−5.0	16	68.30	−5.82	−22.68
4	68.30	−5.82	−2.31	17	68.48	3.38	−20.0
5	68.48	3.38	−5.32	18	53.46	−6.76	−22.68
6	53.46	−6.76	−2.31	19	52.46	5.40	−21.44
7	52.46	5.40	−3.55	20	35.18	−4.45	−20.68
8	35.18	−4.45	−4.32	21	36.45	7.43	−23.22
9	36.45	7.43	−1.77	22	17.39	−2.20	−18.69
10	17.39	−2.20	−6.30	23	20.43	9.45	−25.0
11	20.43	9.45	0.0	24	0.0	0.0	−16.69
12	0.0	0.0	−8.30	25	113.03	0.0	−12.5
13	94.39	0.10	−20.09				

Organization in brick-like elements and biological representation of the skeletal muscle

The SLM was organized in brick-like elements to be used along with the principle of virtual work to derive finite-element-like approximations during muscle contraction. The biological representation of the muscle architecture was obtained through the association of each member of the brick-like element with the biological information that it represents.

General brick-like element structure

The general brick-like element structure was subdivided into three members: top members, bottom members, and fibre members. Top and bottom members are related to the top (lateral aponeurosis) and bottom (medial aponeurosis) aponeurosis. Fibre members are related to the muscle fibres. The full representation of the brick-like element is composed of four straight lines representing the fibre members, six straight lines representing the top members, and six straight lines representing the bottom members (Figure 11.7).

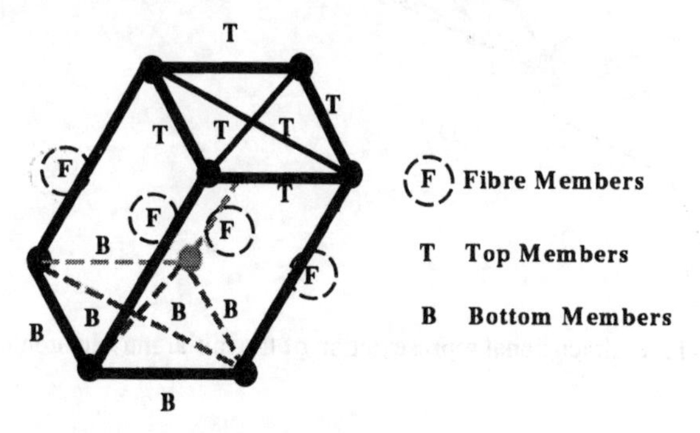

Figure 11.7. General brick-like element structure.

In order to use the principle of virtual work in the formulation of the non-linear equations of motion, the geometric model was divided into these brick-like elements. The model can have several brick-like elements next to one another. The cat medial gastrocnemius was divided into five brick-like elements (Figure 11.8). A larger number of brick-like elements would result in an increase in the number of equations. As a consequence, the calculations would be more extensive. On the other hand, a larger number of brick-like elements would result in a better discretization of the geometric model to derive the deformations of the muscle during contraction.

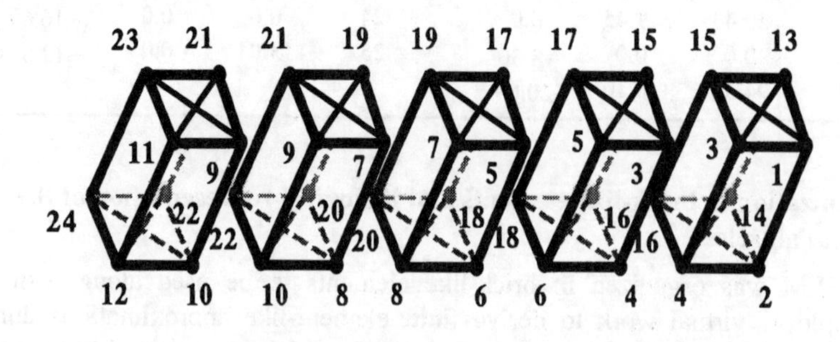

Figure 11.8. Division of the muscle model into brick-like elements.

Biological information associated with each brick-like element

The fibre members, top members, and bottom members of each brick-like element are representative of a bundle of fibres, the top aponeurosis, and the bottom aponeurosis, respectively.

The cross-sectional area of each fibre member, A, is calculated as follows:

$$A = \frac{1}{4} \cdot \frac{VolumeBrick}{AveFibreLength} \tag{11.12}$$

where *VolumeBrick* is the volume of the associated brick-like element, and *AveFibreLength* is the average fibre length.

The cross-sectional area of the members, *Ap*, representing aponeuroses were assumed to be equal. This corresponds roughly to the assumption of incompressibility. The common value of these areas can be calculated as:

$$Ap = \frac{AveThick \cdot AveEdge}{3} \tag{11.13}$$

where *AveThick* is the average thickness of the aponeurosis, and *AveEdge* is the average length of the edge.

In addition, the calculation of the volume of each brick is obtained by decomposing the brick-like element into six tetrahedra (Bloomenthal, 1997), and then calculating the volume of each tetrahedron (Figure 11.9). For example, the three main vectors ($\vec{a}$, $\vec{b}$, and $\vec{c}$) describing the base and height of each tetrahedron are used to calculate its volume, v, as follows (Figure 11.9):

$$v = \frac{1}{6} \cdot (\vec{a} \times \vec{b}) \cdot \vec{c} \tag{11.14}$$

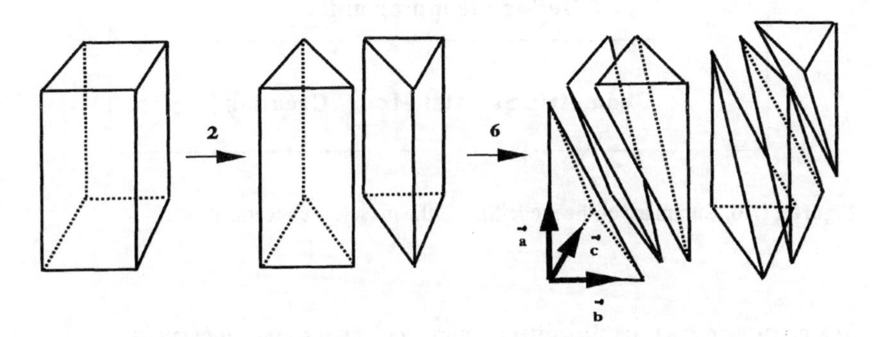

Figure 11.9. Tetrahedral decomposition.

The virtual change of volume, δv, of each tetrahedron is obtained by differentiating the previous equation, which gives:

$$\delta v = \frac{1}{6}\left((d\vec{a}\times\vec{b})\cdot\vec{c} + (\vec{a}\times d\vec{b})\cdot\vec{c} + (\vec{a}\times\vec{b})\cdot d\vec{c}\right) \tag{11.15}$$

Finally, the fibre members of each brick-like element are organized into groups to represent motor units. Slow motor units contain one, two, or three fibre members, and fast motor units contain four fibre members. Thus, the size principle of motor-unit recruitment can be simulated during muscle contraction.

A summary of the modelling of the initial muscle architecture, including the biological considerations, is presented in Figure 11.10.

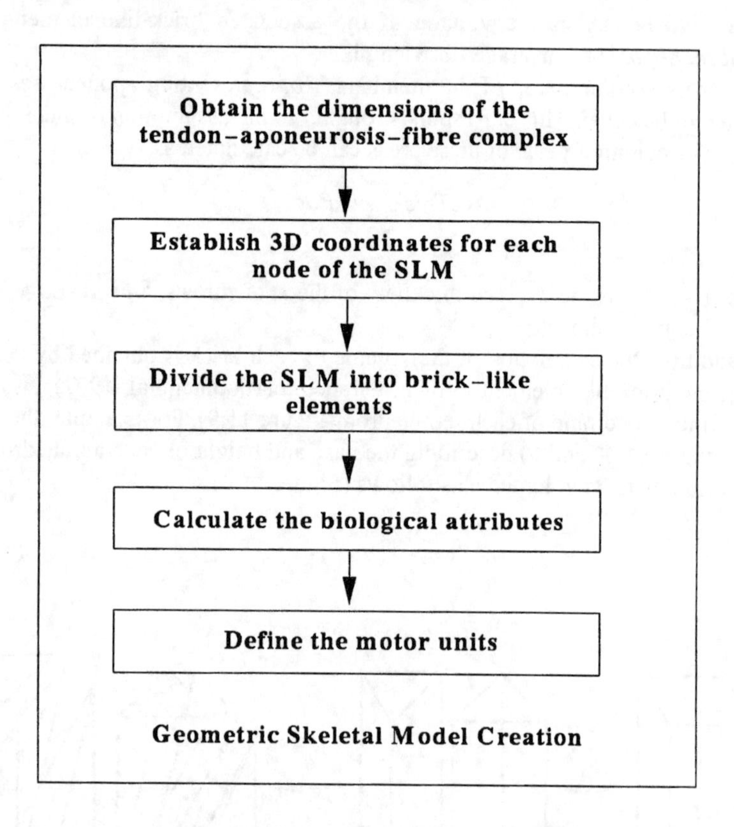

Figure 11.10. Summary of the modelling of the muscle architecture.

MATHEMATICAL REPRESENTATION OF THE CONSTITUTIVE BEHAVIOUR

The representation of the material behaviour of the active force-producing elements (muscle fibres) and the passive force-producing elements (tendons and aponeuroses) is presented.

For the active force-producing elements, a quadratic force-length relationship was adopted:

$$f(r) = -0.772\,r^2 + 1.544\,r - 0.494 \qquad (11.16)$$

where $f(r)$ is the normal stress (N/mm^2) in a fully activated fibre as a function of the ratio between the current fibre length, L, and the optimal fibre length, $L0$:

$$r = \frac{L}{L0} \qquad (11.17)$$

In addition, the range of physical validity of Equation 11.16 is ($0.4 < r < 1.6$) (Figure 11.11).

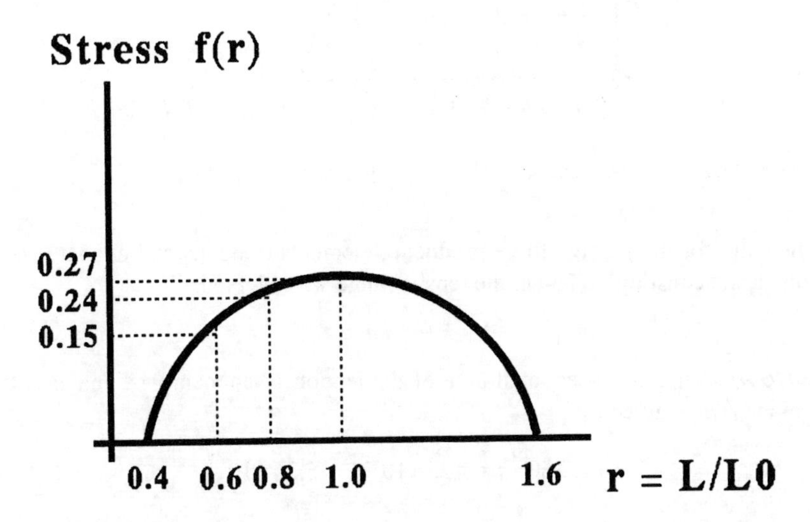

Figure 11.11. Normal stress *f(r)* as a function of the ratio *(r)*.

Using this quadratic force-length relationship, the constitutive equation for the internal force of an isometrically activated muscle becomes:

$$N = A\,a\,f(r) \qquad (11.18)$$

where A is the cross-sectional area of the muscle fibre, and a is the activation coefficient ($0 \le a \le 1$). When the muscle fibre is fully activated ($a=1$) and $r=1.0$, N corresponds to the maximal isometric force.

Additionally, we adopt the following constitutive equation which captures the features of a muscle fibre that is stretched or shortened:

$$N = A\,a\,[f(r0) + k(r0)\,(r - r0)] \qquad (11.19)$$

where $r0$ is the non-dimensionalized fibre length upon first activation, and $k(r0)$ is a length-dependent positive slope for active elongations beyond the initial length (Edman, 1982). *k(r0)* was assumed constant in this example, with a value of 0.9264 that corresponds to the initial slope of the ascending limb (Figure 11.12).

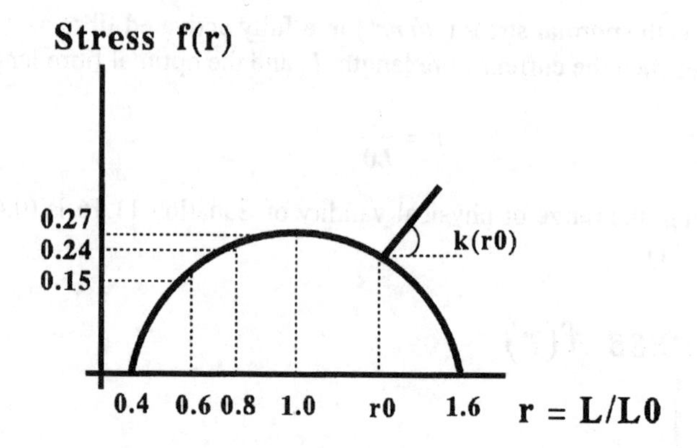

Figure 11.12. Force response to a stretch upon full isometric activation.

Second, for the passive force-producing elements (tendons and aponeuroses), a quadratic constitutive law in the tensile range was adopted:

$$N_p = \sigma \cdot A_p \tag{11.20}$$

where *Ap* is the cross-sectional area of the tendon or aponeurosis, and σ is the stress (N/mm^2) given by:

$$\sigma = 15160\left[\left(\varepsilon + 7.25\times10^{-4}\right)^2 - 5.26\times10^{-7}\right] \tag{11.21}$$

where ε is tendon/aponeurosis strain.

FORMULATION OF THE NON-LINEAR EQUATIONS OF MOTION, AND NUMERICAL CONSIDERATIONS

The formulation of the non-linear equations of motion for the three-dimensional muscle model is presented. In addition, numerical considerations are presented to allow for the development of a computational framework.

CONFIGURATION OF THE VARIABLES OF THE SYSTEM

The above-presented geometrical model of the cat medial gastrocnemius is used as an example. This model contains n ($n=25$) nodes in three-dimensional space (unknown displacements in the x, y, and z directions) and is divided into m ($m=5$) bricks (Figure 11.13).

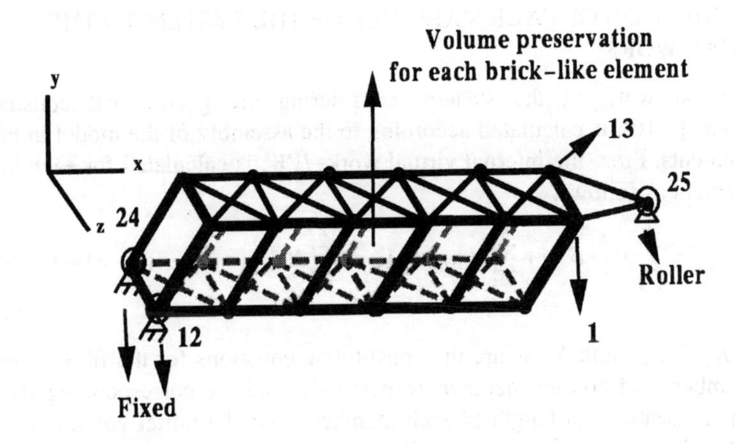

Figure 11.13. Geometric representation of the muscle model with geometric constraints.

In addition, the model contains three external support constraints: two fixed nodes, 12 and 24, and one not fully fixed node, 25, which is allowed to displace in the x direction. The constraints for volume preservation are enforced for each brick. Therefore, each brick will have one Lagrange multiplier, λ. Thus, the total number of unknowns (k) is $(n * 3) + m = 80$ unknowns. The variables of the system are represented by a set of displacements and by the Lagrange multipliers $u_i(i = 1,...,k)$. In addition, the unknowns for the small variations of the displacements and λ_1, λ_2, λ_3, λ_4, and λ_5 are represented by the set $du_i(i = 1,...,k)$ (Figure 11.14).

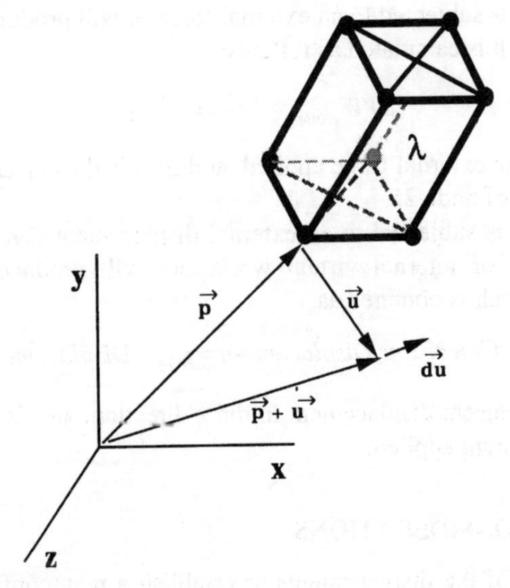

Figure 11.14. Graphical representation of the unknown variables (end-point $\vec{p}$).

CONTRIBUTION OF EACH VARIABLE OF THE SYSTEM TO THE VIRTUAL WORK

The virtual work of the system, considering the geometrical constraints (Equation 11.10), is calculated according to the assembly of the model in brick-like elements. First, the internal virtual work, *IVW*, is calculated for each brick-like element as follows:

$$IVW_{Brick} = \sum_{I=1}^{4} N_I * \delta\, e_I + \sum_{I=1}^{6} N\, top_I * \delta\, e_I + \sum_{I=1}^{6} N\, bot_I * \delta\, e_I + (v - v0)\delta\lambda + \lambda\delta v \qquad (11.22)$$

where N_I, $N\,top_I$, and $N\,bot_I$ are the constitutive equations for the fibre member, top member, and bottom member, respectively, and the corresponding δ_{eI} are the small variations in length of each member. $v0$ is the initial volume, v is the current volume, and δv is the small variation in volume. λ is the Lagrange multiplier, and $\delta\lambda$ is the small variation of the Lagrange multiplier.

In addition, when the tendon members (e.g., composed by the nodes 1-25 and 13-25) produce *IVW*, this is calculated as:

$$IVW_{Tendon} = \sum_{I=1}^{2} N\, ten_I * \delta\, e_I \qquad (11.23)$$

where $N\,ten_I$ is the constitutive equation for the tendon member, and δ_{eI} is the small variation in length.

Second, node 25 can only have displacements in the x direction. This node was subjected to an external force or to an external displacement constraint. For node 25, the only non-vanishing variables are u_{73} and du_{73}.

When node 25 is subjected to an external force, it will produce external virtual work, *EVW*, which is calculated as follows:

$$EVW_{Node} = Force * du_{73} \qquad (11.24)$$

where *Force* is the external force applied, and du_{73} is the current small variation in the x direction of node 25.

When node 25 is subjected to an external displacement constraint, it will not produce external or internal virtual work, but will produce a constraint of displacement, which is obtained as:

$$Constraint\ Displacement = u_{73} - Displacement \qquad (11.25)$$

where u_{73} is the current displacement in the x direction, and *Displacement* is the external displacement applied.

NUMERICAL CONSIDERATIONS

The calculations of the displacements to establish a new configuration involve the equilibrium equations, the constitutive relations, and the constraints to conform a system of non-linear equations, for which there is in general no analytical solution. The approach used to solve these non-linear equations was numerical.

First, a scheme of numerical approximations for a system of non-linear equations will be explained. Second, the calculations for the equations of motion will be presented on the basis of the principle of virtual work and considerations about the geometric constraints.

Assuming that we have a system of algebraic equations that is represented in the following geometric form:

$$f_i(u_1,...,u_n) = 0 \qquad i = 1,...,n \tag{11.26}$$

where $u_1,...,u_n$ represent n unknowns (such as displacements components and Lagrange multipliers), and $f_1,...,f_n$ denote arbitrary functions (such as functions that might be obtained from equilibrium equations, constitutive equations, and/or constraints equations). Assuming that at least some functions (f_i) are not linear, an iterative technique of approximation is needed.

The technique adopted to solve the problems numerically was the Newton-Raphson method. The Newton-Raphson iteration process is started with an initial guess for the unknowns. When this initial guess is entered into the equations (Equation 11.26), these equations fail (in general) to be satisfied, i.e., the residuals do not vanish. The residuals can be represented as:

$$r_i^0 = f_i(u_1^0,...,u_n^0) \qquad i = 1,...,n \tag{11.27}$$

where $u_1^0,...,u_n^0$ represent the n initial guesses.

If, for instance, $f_i = 0$ represents an equilibrium state, then the residual r_1^0 is an unbalanced force that will cause the structure to move to a new configuration (Epstein, 1994). This search for the new configuration is based on the differential of the functions f_i at the initial guess. The values of f_i at points $u_1,...,u_n$ in the neighbourhood of $u_1^0,...,u_n^0$ can be approximated as:

$$f_i(u_1,...,u_n) \cong f_i(u_1^0,...,u_n^0) + \left[\frac{\partial f_i}{\partial u_1}\right]_0 (u_1 - u_1^0) + ... + \left[\frac{\partial f_i}{\partial u_n}\right]_0 (u_n - u_n^0) \tag{11.28}$$

where $[\partial f_i / \partial u_j]_0$ denotes the partial derivative of $f_i(i = 1,...,n)$ with respect to $u_j(j=1,...,n)$, evaluated at the initial guess $u_1^0,...,u_n^0$. This expression becomes:

$$f_i(u_1,...,u_n) \cong r_1^0 + \sum_j \left[\frac{\partial f_i}{\partial u_j}\right]_0 (u_j - u_j^0) \qquad j = 1,...,n \tag{11.29}$$

And, setting this estimate to zero, the following system of linear equations is obtained:

$$\sum_j \left[\frac{\partial f_i}{\partial u_j}\right]_0 \Delta u_j = -r_i^0 \qquad i, j = 1,...,n \tag{11.30}$$

where $\Delta u_j = (u_j - u_j^0)$.

A linear solver, such as the Gauss elimination method, can be used to solve this system of equations. When this system is solved, the new values:

$$u_j^1 = u_j^0 + \Delta u_j \qquad j = 1,\dots,n \tag{11.31}$$

can be used as the new guesses. Thus, iterations can be performed until the residuals become smaller than a specified error.

Using this scheme of numerical approximation along with the principle of virtual work, the residuals are obtained as:

$$VW_i^0(u_j, du_j) = r_i^0 \qquad i, j = 1,\dots,n \tag{11.32}$$

where VW_i^0 is the total virtual work for the unknown i and can be expressed as:

$$IVW_i^0(u_j, du_j) - EVW_i^0(u_j, du_j) = VW_i^0(u_j, du_j) \qquad i, j = 1,\dots,n \tag{11.33}$$

where IVW_i^0 is the total IVW for the unknown i, EVW_i^0 is the total EVW for the unknown i, u_j represent the unknowns $u_j(j = 1,\dots,n)$ that contain the current guesses, and du_j are the small variations of the unknowns $du_j(j = 1,\dots,n)$ that represent the following values: $du_j=1$ if $i=j$ and $du_j=0$ if $i \neq j$.

For instance, the calculation of IVW_i^0 when $i=1$ and $j=1$ will give the following values for the vectors $u_j(j=1,\dots,n)$ and $du_j(j=1,\dots,n)$:

$$IVW_1^0\begin{Bmatrix} u_1 & \leftarrow (j=1) \\ u_j & \\ \vdots & \\ u_n & \\ - & \\ 1 & \leftarrow (i=j=1) \\ 0 & \\ \vdots & \\ 0 & \end{Bmatrix} - EVW_1^0\begin{Bmatrix} u_1 & \leftarrow (j=1) \\ u_j & \\ \vdots & \\ u_n & \\ - & \\ 1 & \leftarrow (i=j=1) \\ 0 & \\ \vdots & \\ 0 & \end{Bmatrix} = VW_1^0\begin{Bmatrix} u_1 & \leftarrow (j=1) \\ u_j & \\ \vdots & \\ u_n & \\ - & \\ 1 & \leftarrow (i=j=1) \\ 0 & \\ \vdots & \\ 0 & \end{Bmatrix}$$

The $VW_i^0(u_j, du_j)$ is calculated according to the contribution of each variable of the system, $u_j(j = 1,\dots,n)$, for the virtual work using Equation 11.22, Equation 11.23, Equation 11.24, and Equation 11.25.

In addition, the partial derivatives are expressed as:

$$\left[\frac{\partial f_i}{\partial u_j}\right]_0 = \frac{1}{h}\left(VWb_i(u_j, du_j) - VWa_i(u_j, du_j)\right) \qquad i, j = 1,\dots,n \tag{11.34}$$

where the partial derivatives $[\partial f_i / \partial u_j]_0$ are the coefficients of the Jacobian matrix, and h is a small increment ($h=0.001$). For VWb_i, $u_j(j=1,\dots,n)$ contains the values u_j as before (current guess) except that the current u_j will be $u_j = u_j + h$, as well as $du_j(j=1,\ldots,n)$ contains the values $du_j=1$ if $i=j$ and $du_j=0$ if $i \neq j$. For VWa_i, $u_j(j=1,\ldots,n)$ contains the values u_j as before (current guess), and $du_j(j=1,\ldots,n)$ contains the values $du_j=1$ if $i=j$ and $du_j=0$ if $i \neq j$.

For instance, the calculation of $[\partial f_i / \partial u_j]_0$ when $i=1$ and $j=1$ gives the following values for the vectors $u_j(j=1,\ldots,n)$ and $du_j(j=1,\ldots,n)$:

$$\left[\frac{\partial f_i}{\partial u_j}\right]_0 = \frac{1}{h}\left(VWb_1 \begin{Bmatrix} u_1 + h & \leftarrow (j=1) \\ u_j & \\ \vdots & \\ u_n & \\ - & \\ 1 & \leftarrow (i=j=1) \\ 0 & \\ \vdots & \\ 0 & \end{Bmatrix} - VWa_1 \begin{Bmatrix} u_1 & \leftarrow (j=1) \\ u_j & \\ \vdots & \\ u_n & \\ - & \\ 1 & \leftarrow (i=j=1) \\ 0 & \\ \vdots & \\ 0 & \end{Bmatrix} \right)$$

Thus, Equation 11.30 can be expressed in matrix form as follows:

$$[J]_{i \times j}[\Delta u]_{j \times 1} = -[r_i]_{i \times 1} \qquad i, j = 1,\ldots,n \tag{11.35}$$

The basic architecture of the computational framework for the calculation of the displacements to establish a new configuration of the skeletal muscle, based on the principle of virtual work, is shown in Figure 11.15.

RESULTS

The computational framework presented above was used to simulate muscle contraction of the cat medial gastrocnemius. The simulation used the same initial conditions as used by Carvalho et al. (1999) to produce an experimental sarcomere force-length relationship for the cat medial gastrocnemius. In the experiment, relaxed and activated fibre lengths, angle of pinnation, and external forces were measured continuously for eight muscles during isometric contractions at different muscle length. The isometric contractions were done at 2, 4, 6, 8, and 10 mm shorter (–) than muscle reference length of 0 mm, and at 2, 4, and 6 mm longer (+) than the reference length (Figure 11.16).

In the experiment, deformations of the muscle from an arbitrary reference geometry were obtained as a function of muscle length and activation. Comparisons of fibre lengths and pennation angles of three muscle fibres from each muscle were made between theoretically predicted and experimentally measured values. All comparisons shown here were made at the muscle reference length of 0 mm, with the muscle maximally activated (Figure 11.17).

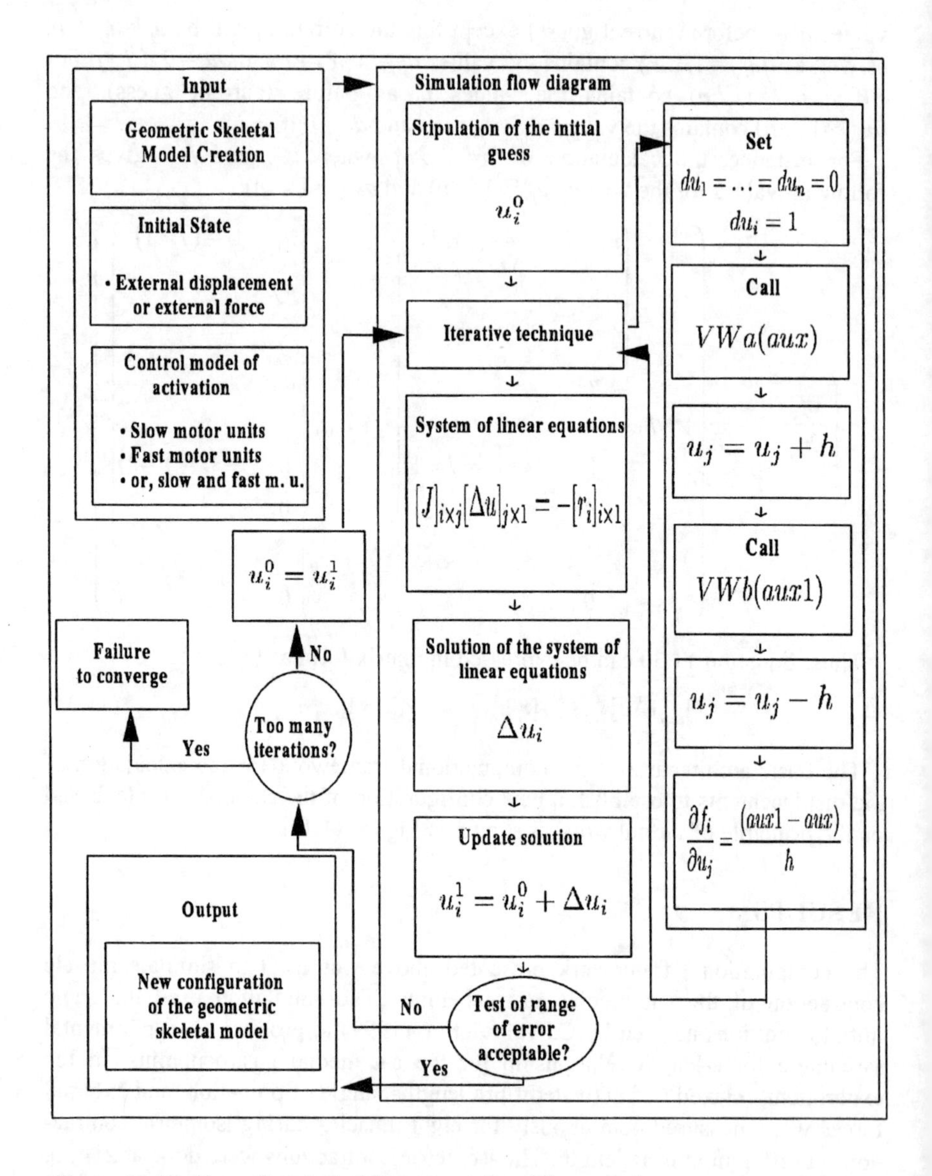

Figure 11.15. Flow of control of the computational framework.

Second, comparisons of external muscle force were made between theoretically predicted and experimentally measured values (Figure 11.18).

In addition, using the three-dimensional model of skeletal muscle, whole muscle deformation from the relaxed to the fully activated state were simulated (Figure 11.19 and Figure 11.20).

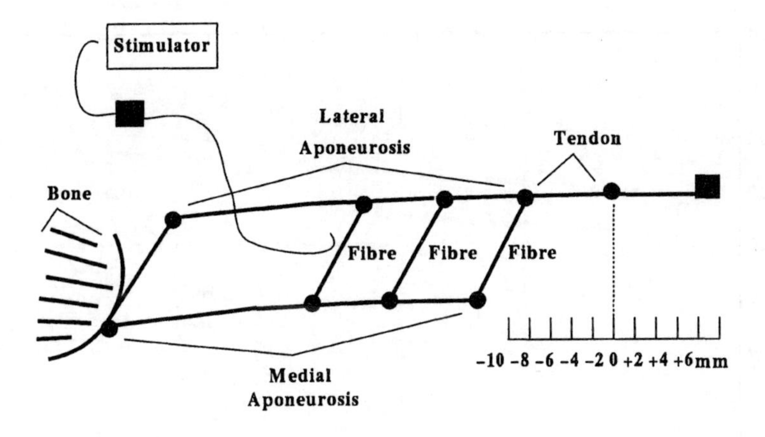

Figure 11.16. Experimental configuration.

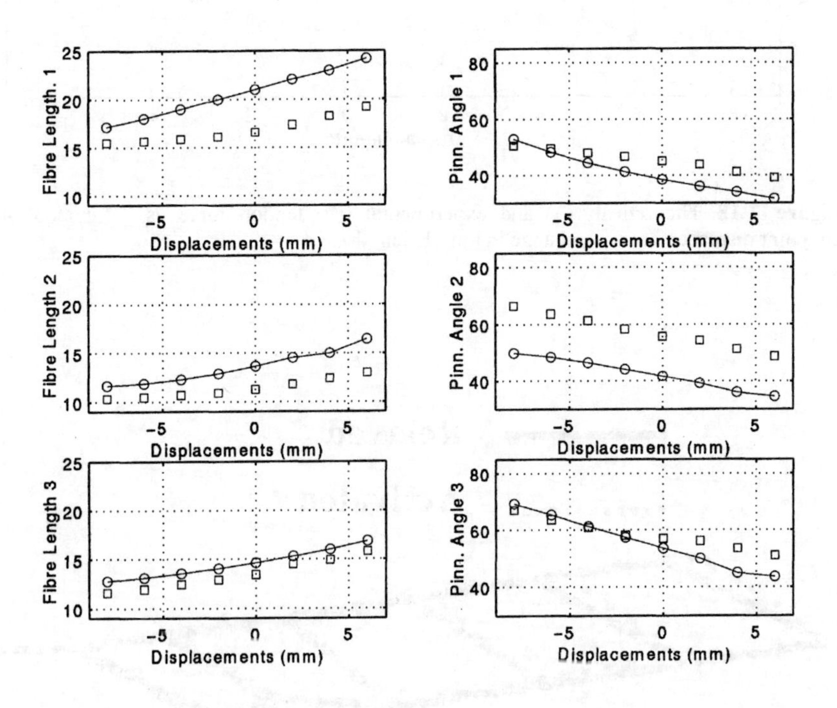

Figure 11.17. Theoretical (○) and experimental (□) fibre lengths and angles of pinnation in a maximally activated cat medial gastrocnemius at reference length.

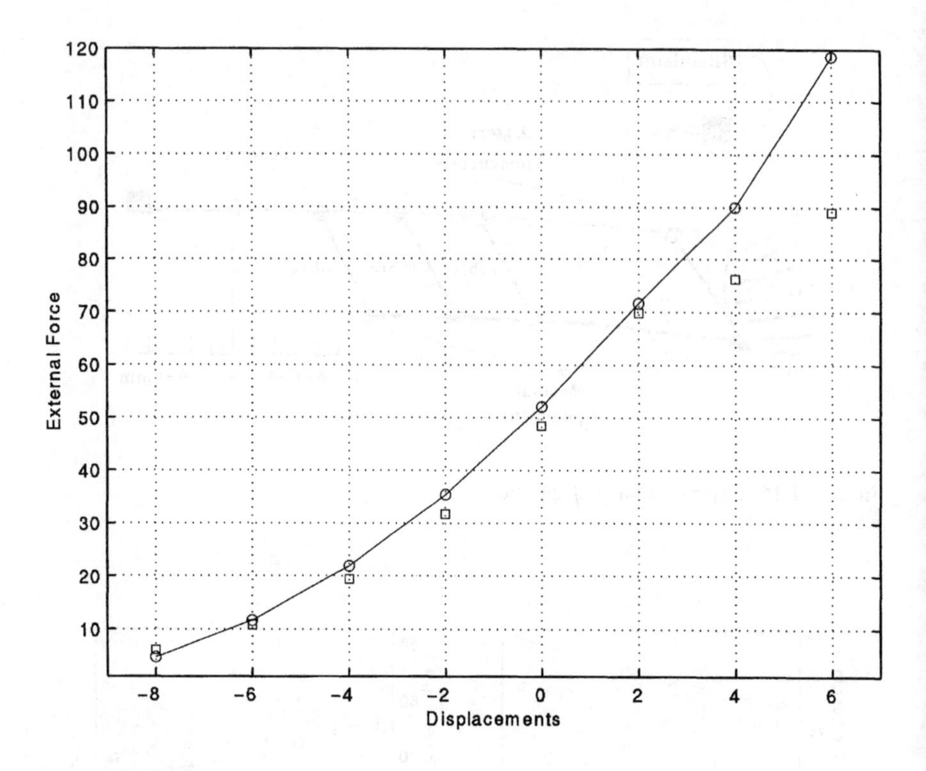

Figure 11.18. Theoretical (O) and experimental (□) tendon force as a function of endpoint displacement (i.e., change in muscle length).

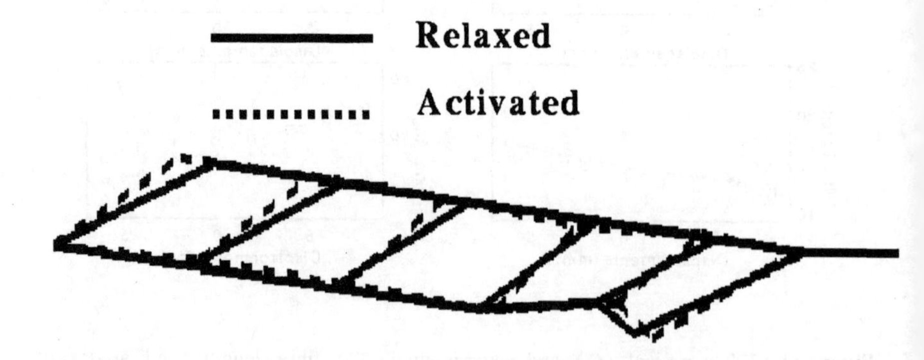

Figure 11.19. Mid-sagittal plane of the 3D model (relaxed and activated).

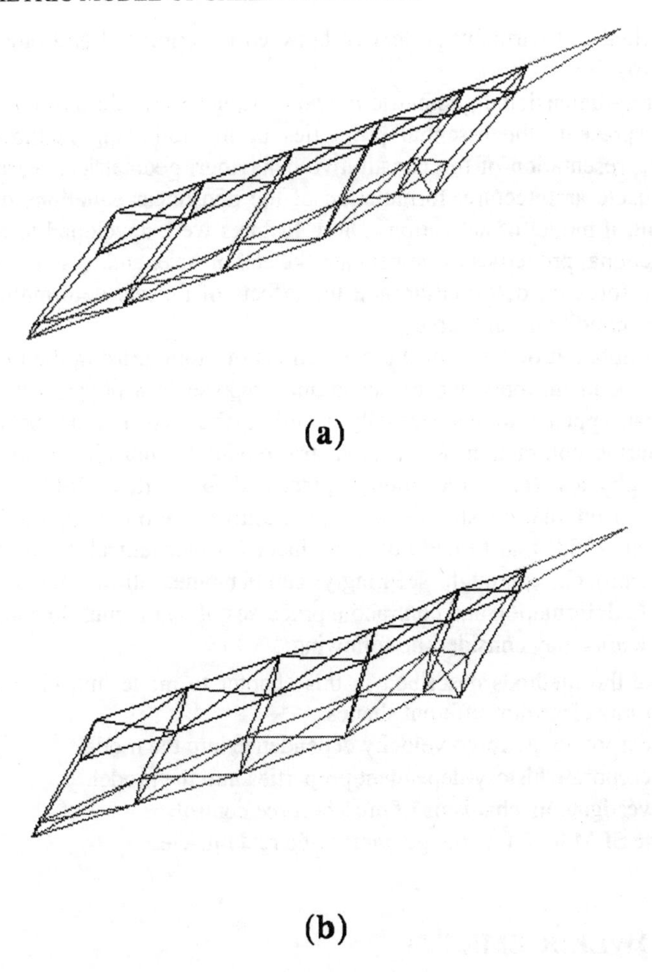

Figure 11.20. 3D model relaxed (a) and activated (b).

DISCUSSION

The results of this study showed that fibre lengths became smaller and angles of pinnation increased when going from the relaxed to the activated state in the experiment and the model. Therefore, the conceptual predictions of the model appear correct. However, the decrease in fibre length and increase in angle of pinnation were not as pronounced in the model as they were in the experiment (Figure 11.17). Several possibilities exist to explain these differences. The most obvious explanation appears to be that the series elastic elements (tendon and aponeuroses) were modelled too stiffly. Decreasing the stiffness of these structures resulted in a virtually perfect fit between experimental and model results (not shown).

The three-dimensional geometric model of skeletal muscle introduced in this chapter represents the essential properties of the following features: mathematical representation of the constitutive behaviour, geometrical representation of the muscle architecture, formulation of the non-linear equations of motion, and a control model of activation. These features were developed to reflect the basic functional properties of unipennate skeletal muscle, and to study the effects of muscle force on deformation and the effects of muscle deformation on the contractile conditions and force.

The formulation of the non-linear equations of motion, using the principle of virtual work in the presence of constraints, represents a physically consistent model that appears to conceptually simulate the deformation behaviour of muscle during contraction. Most geometric models of muscle do not obey the laws of physics (e.g., the model presented by Otten, 1988) and have incorporated internal constraints using an intuitive free-body diagram approach (e.g., Otten, 1988) that turned out to be incorrect mechanically. Therefore, the current approach, although seemingly cumbersome, allows for meaningful insight into deformation and contractile processes of whole muscle contraction.

Future works may consider the following:

- Use the methods described in this chapter to model muscle contraction for muscles with different shapes.
- Incorporate the force-velocity dependence into the model.
- Incorporate history-dependent properties into the model.
- Investigate mechanisms of muscle force control.
- Use SLM to define the geometry of a real muscle.

ACKNOWLEDGEMENTS

CAPES–Brazil, UCS, and NSERC of Canada.

REFERENCES

Anton, M. (1992). Mechanical models of skeletal muscle. Ph.D. dissertation, University of Calgary, Calgary, Alberta, Canada.

Baskin, R., and Paolini, P. (1967). Volume change and pressure development in muscle during contraction. *American Journal of Physiology* **213**, 1025–1030.

Bloomenthal, J. (1997). *Introduction to Implicit Surfaces*. Morgan Kaufmann, San Francisco, California, USA.

Carvalho, W., Leonard, T., and Herzog, W. (1999). The influence of pinnation and series elasticity on the sarcomere force-length behavior of cat skeletal muscle. *Canmore Symposium on Skeletal Muscle*, August 6-7, 1999, Canmore, Alberta, Canada, p. 42.

Chen, D. (1991). Pump it up: Computer animation of a biomechanically based model of muscle using the finite element method. Ph.D. dissertation, Massachusetts Institute of Technology, Media Arts and Science Section, Cambridge, Massachusetts, USA.

Chen, D., and Zeltzer, D. (1992). Pump it up: Computer animation of a biomechanically based model of muscle using the finite element method. *Computer Graphics* **26**(2), 89–98.

Delp, S. (1990). Surgery simulation: A computer graphics system to analyze and design musculoskeletal reconstructions of the lower limb. Ph.D. dissertation, Stanford University, Stanford, California, USA.

Delp, S., and Loan, P. (1995). A graphics-based software system to develop and analyze models of musculoskeletal structures. *Computers in Biology and Medicine* **25**(1), 21–34.

Edman, K.A.P., Elzinga, G., and Noble, M.I.M. (1982). Residual force enhancement after stretch of contracting frog single muscle fibers. *Journal of General Physiology* **80**, 769–784.

Epstein, M. (1994). Muscle modelling: Thermomechanical considerations. In *Proceedings of the VIII Biennial Conference of the Canadian Society for Biomechanics*, Calgary, Alberta, Canada, pp. 18–20.

Epstein, M., and Herzog, W. (1998). *Theoretical Models of Skeletal Muscle*. John Wiley & Sons, Chichester, England.

Hill, A. (1938). The heat of shortening and the dynamic constants of muscle. *Proceedings of the Royal Society of London B* **126**, 136–195.

Huxley, A. (1957). Muscle structure and theories of contraction. *Progress in Biophysics and Biophysical Chemistry* **7**, 255–318.

Maurel, W., Thalmann, D., Hoffmeyer, P., Beylot, P., Gingins, P., Kalra, P., and Magnenat-Thalmann, N. (1996). A biomechanical musculoskeletal model of human upper limb for dynamic simulation. *Proceedings of Computer Animation and Simulation* **96**, 121–136. Boulic, R., and Hegron, G. (eds.), Springer Computer Science, Poitiers, France.

Nigg, B., and Herzog, W. (eds.) (1994). *Biomechanics of the Musculo-Skeletal System*. John Wiley & Sons, Chichester, England.

Otten, E. (1988). Concepts and models of functional architecture in skeletal muscle. *Exercise and Sport Sciences Reviews* **16**, 89–139.

van der Linden, B., Koopman, H., Grootenboer, H., and Huijing, P. (1998). Modelling functional effects of muscle geometry. *Journal of Electromyography and Kinesiology* **8**, 101–109.

12 FEM-Simulation of Skeletal Muscle: The Influence of Inertia During Activation and Deactivation

P. MEIER AND R. BLICKHAN
Institute of Sport Science, Friedrich-Schiller-University, Jena, Germany

INTRODUCTION

Starting with the work of Hill (1922), models have been used to describe the mechanical behaviour of a contracting muscle. In order to simplify the calculation algorithms, lumped parameter models were preferred. The power of a finite-element muscle model rests in the fact that a muscle is subdivided into discrete units. Proper formulation of a model allows for consideration of distributed mass and pressure. In the future, it might be possible to combine finite-element models with multi-body models describing skeletal dynamics. These models are of relevance in studying many biomechanical and medical problems.

The muscle model described by Hatze (1981) consists of a combination of many lumped parameter models, and it might be considered a predecessor of a finite-element muscle model with discrete elements. Bovendeerd (1990) introduced a finite-element continuum model of the left ventricle of the heart. Huyghe et al. (1991) introduced a model for the heart muscle. Van Leeuwen (1991) described the stress in muscle as the sum of an active and a passive stress, and prepared the possibility to include active behaviour into existing theories for materials with passive behaviour similar to those of muscle. Otten and Hulliger (1995) proposed a model with a discrete arrangement of elements. Another formulation of a discrete finite-element code is given by van Leeuwen and Kier (1997) who proposed a 2D-model for the tentacle of a squid. Kojic et al. (1998) published a finite-element muscle model for the software package PAK.

The muscle model used in this study was designed to investigate muscle recruitment and the influence of external forces on the forces produced by activated muscle. The model is shown to be a useful tool in gaining an understanding of the influence of inertia on muscle movement. An important result of our calculations was that the inertia of the muscle limits the speed of contraction. Quick muscle activation has no effect in a contraction dominated by inertia. On the other hand, deactivation should be sufficiently slow to ensure that

Skeletal Muscle Mechanics: From Mechanisms to Function. Edited by W. Herzog.

the properties of active muscle help to dampen unwanted oscillations. In experimental approaches aimed at estimating the velocity of contraction, inertia must be taken into account, especially for large muscles.

FEM-MODEL

The finite-element method (FEM) is a well-known tool for the mechanical analysis of stress-, strain-, or pressure-distributions in structures and tissues. Using a continuum element approach has advantages with respect to formulating isovolumetricity constraints during contraction. Any model of muscle must allow for the implementation of non-linear properties, large deformations, and dynamic active properties that characterize muscle tissue. In the present study, a finite muscle element based on the HYPER58 element of the package ANSYS® was used that had been extended to include the active muscle behaviour (Johansson et al., submitted). This model is described in this chapter.

The behaviour of a muscle can be separated into a passive and an active state. Van Leeuwen (1992) introduced a muscle model with the property,

$$\sigma = \sigma_{passive} + \sigma_{active} \tag{12.1}$$

where σ is the total stress in the muscle, $\sigma_{passive}$ is the passive stress due to the deformation of the muscle, and σ_{active} is the active stress due to the activation of the muscle.

Depending on the input parameters, one element of the finite-element model represents sarcomeres, a part of a single fibre, or a bundle of a few thousand fibres describing the muscle tissue. At present, the mechanical input parameters for a whole muscle are much better known than those for single fibres. Therefore, we would expect much better model results for large muscle compartments than for single fibres. To minimize the calculation time, a simple formulation for the active stress was chosen.

THE PASSIVE MUSCLE BEHAVIOR

The characteristic behaviours of passive muscle are large deflections, a non-linear stress-strain relation, and incompressibility. These properties are the same for rubber-like materials described by Sussmann and Bathe (1987). This is the basis of the hyperelastic HYPER58 element of ANSYS®. The input parameters for the passive properties are muscle density, Poisson's ratio, and the stress-strain relation (see Table 12.1 and Figure 12.1). The stress-strain relation can be described with a user-defined function or the Mooney Rivlin constants defining the strain-energy function (for details see appendix).

Table 12.1. Typical input parameters for passive muscle tissue.

Density ρ [kg/m^3]	1 000
Poisson's ratio ν	0.4999
Mooney Rivlin constant a_{10}	10 000
Mooney Rivlin constant a_{01}	0
Mooney Rivlin constant a_{20}	10 000
Mooney Rivlin constant a_{11}	0
Mooney Rivlin constant a_{02}	0
Mooney Rivlin constant a_{30}	6 666.7
Mooney Rivlin constant a_{21}	0
Mooney Rivlin constant a_{12}	0
Mooney Rivlin constant a_{03}	0

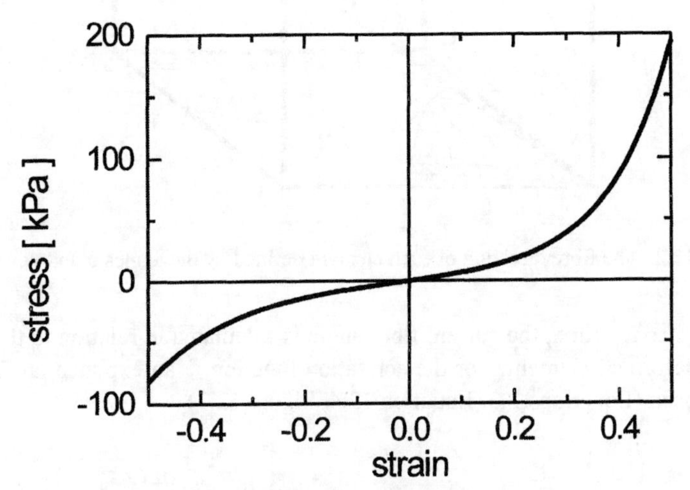

Figure 12.1. Stress-strain relation of passive muscle tissue.

ACTIVE MUSCLE BEHAVIOR

The behaviour of muscle does not only depend on the non-linear stress-strain relation, but also on the large deflection of the tissue, the so-called geometrical non-linearity, and in addition, on the non linear time dependence of the activation function. This threefold non-linearity results in long calculation times for the finite muscle element compared to the calculation times for structural analysis of solid materials. The topic of this study is to create a tool to investigate the effects of mass distribution on muscle properties. The description for the active muscle behaviour was chosen to be as simple as possible. The active stress in the muscle fibre direction is described as the product of the isometric stress $\sigma_{isometric}$, an activation function f_t, a velocity function f_v, and a length function f_l.

$$\sigma_{active} = \sigma_{isometric} \cdot f_t(t) \cdot f_v(v) \cdot f_l(l) \tag{12.2}$$

The fibre direction of each element is defined by two input angles (Figure 12.2).

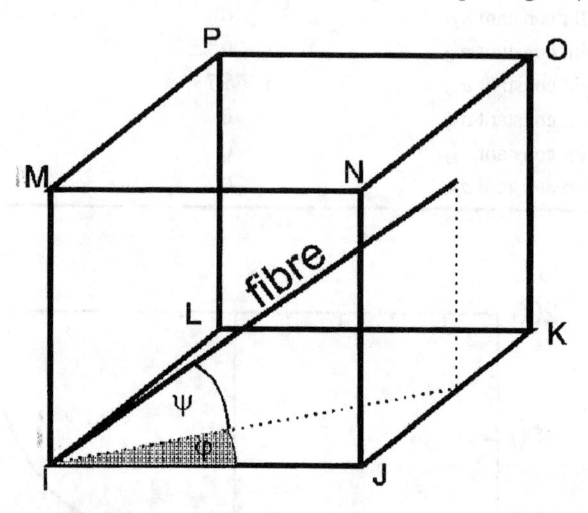

Figure 12.2. The fibre direction of each element defined by the angles ϕ and ψ.

During a contraction, the current fibre angle is calculated in relation to the edges of the deformed element. For the activation function f_t, an exponential function was chosen (Equation 12.3, Equation 12.4, Figure 12.3).

$$f_t(t) = \begin{cases} n_1 & \text{, if } t < t_0 \\ n_1 + (n_2 - n_1) \cdot h_t(t, t_0) & \text{, if } t_0 < t < t_1 \\ n_1 + (n_2 - n_1) \cdot h_t(t_1, t_0) - [(n_2 - n_1) \cdot h_t(t_1, t_0)] \cdot h_t(t, t_1) & \text{, if } t > t_1 \end{cases} \tag{12.3}$$

with

$$h_t(t_i, t_b) = \{1 - \exp[-S \cdot (t_i - t_b)]\} \tag{12.4}$$

where n_1 is the activation level before and after contraction, n_2 is the activation level during contraction, t_0 is the activation time, t_1 is the deactivation time, and S is the exponential factor. In models of single muscle fibres, the magnitude of parameter S should be related to the rate of the chemical processes. For large muscle compartments, S represents the time-dependent recruitment of different motor units.

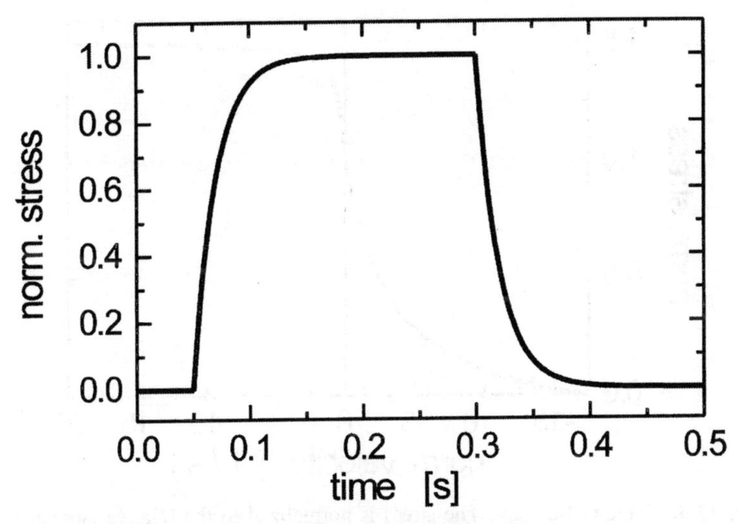

Figure 12.3. Activation function. The stress is normalized relative to the isometric stress $\sigma_{isometric}$.

Because of the force generation process of the cross-bridges, muscle produces forces that depend on the velocity of contraction or extension. In our approach, a hyperbolic force-velocity relationship was chosen for muscle shortening (Hill, 1938) and lengthening (van Leeuwen, 1991) (Equation 12.5, Figure 12.4).

$$f_v(v) = \begin{cases} \dfrac{1 - \dfrac{v}{v_{\min}}}{1 + k_c \cdot \dfrac{v}{v_{\min}}} & \text{, if } v \leq 0 \text{ (concentric)} \\ d - (d-1) \cdot \dfrac{1 + \dfrac{v}{v_{\min}}}{1 - k_c \cdot k_e \cdot \dfrac{v}{v_{\min}}} & \text{, if } v > 0 \text{ (eccentric)} \end{cases} \tag{12.5}$$

The shape parameters of the hyperbolic curves are k_c and k_e. The velocity is normalized relative to the initial length of the element, l_0. That means the velocity v represents a strain rate. The unit for the normalized velocity is $1/s$. Hence, v_{min} is the minimal contraction velocity ($v_{min} < 0$). The maximum eccentric force is dominated by the parameter d:

$$\lim_{v \to \infty} f_v(v) = d + \frac{d-1}{k_c \cdot k_e} \tag{12.6}$$

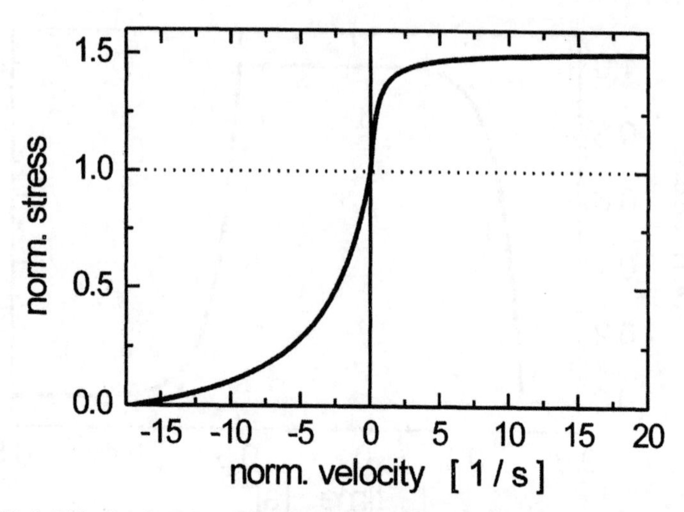

Figure 12.4. Velocity function. The stress is normalized to the relative isometric stress $\sigma_{isometric}$. The velocity is normalized relative to the initial muscle length.

Depending on the overlap of actin and myosin filaments, muscle produces different forces. To include this behaviour, which was first described for frog sarcomeres by Gordon et al. (1966), a piecewise linear length-force function (Equation 12.7, Figure 12.5) was used, where the characteristic fibre lengths are l_1, l_2, ..., l_5; the initial resting length is l_0; and $f(l_2)$ is the stress value at length l_2 normalized to the isometric stress. Typical input parameters for the active muscle behaviour are listed in Table 12.2.

$$f_l(l) = \begin{cases} 0 & , \text{if } l < l_1 \\ f(l_2) \cdot \dfrac{\dfrac{l}{l_0} - \dfrac{l_1}{l_0}}{\dfrac{l_2}{l_0} - \dfrac{l_1}{l_0}} & , \text{if } l_1 \leq l < l_2 \\ (1 - f(l_2)) \cdot \dfrac{\dfrac{l}{l_0} - \dfrac{l_2}{l_0}}{\dfrac{l_3}{l_0} - \dfrac{l_2}{l_0}} + f(l_2) & , \text{if } l_2 \leq l < l_3 \\ 1 & , \text{if } l_3 \leq l < l_4 \\ 1 - \dfrac{\dfrac{l}{l_0} - \dfrac{l_4}{l_0}}{\dfrac{l_4}{l_0} - \dfrac{l_5}{l_0}} & , \text{if } l_4 \leq l < l_5 \\ 0 & , \text{if } l \geq l_5 \end{cases} \tag{12.7}$$

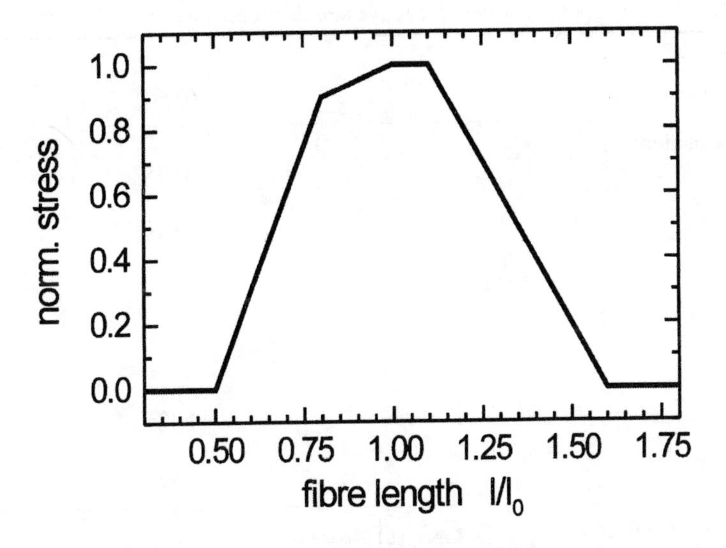

Figure 12.5. Length function. The stress is normalized relative to the isometric stress $\sigma_{isometric}$. The fibre length is normalized relative to the initial fibre length.

RESULTS

The results of selected calculations with this finite-element muscle model are shown in this section. A square block with fibres parallel to the longest edge is used as the geometrical muscle model (Figure 12.6). If the parameters used in these examples are not shown explicitly, then we used the values listed in Table 12.1 and Table 12.2. The purpose for this specific calculation was to study the influence of the mass on the activation and deactivation process. Therefore, geometrically similar muscles with different mass were used.

Figure 12.6. FEM-simulation of a muscle contraction.

Table 12.2. Typical input parameters for active muscle properties.

Fibre angle ϕ		0
Fibre angle ψ		0
Length Function	$\frac{l_1}{l_0}$	0.635
	$\frac{l_2}{l_0}$	0.835
	$f\left(\frac{l_2}{l_0}\right)$	0.83
	$\frac{l_3}{l_0}$	1
	$\frac{l_4}{l_0}$	1.125
	$\frac{l_5}{l_0}$	1.825
Velocity Function	$\dot{\varepsilon}_{\min} = \frac{v_{\min}}{l_0}$ $\left[\frac{1}{s}\right]$	−17
	k_c	5
	k_e	5
	d	1.5
Activation Function	t_0 [s]	0.05
	n_1	0
	t_1 [s]	0.3
	n_2	1
	S $\left[\frac{1}{s}\right]$	50
Isometric Stress	$\sigma_{isometric}$ [Pa]	200 000

ACTIVATION

A muscle (size: $0.3 \cdot 0.05 \cdot 0.05$ m^3) is activated with different parameters S of the activation function. The muscle is fixed at one end and free at the other end.

The displacement of the free end is shown in the top part of Figure 12.7. The displacement-time dependence is shown in the middle panel, and the velocity of the free end is shown in the bottom part of Figure 12.7. The highest contraction velocity of the input velocity function is $v = -17$ s^{-1}. The maximal contraction velocity in the simulation for physiological activation functions ($S = 30 - 50$ s^{-1}) is $v = -9$ s^{-1}, $v = -10$ s^{-1} respectively. The motion is described by Newton's law:

$$F_{isometric} \cdot f_t(t) \cdot f_v(v) \cdot f_l(l) + F_{passive} = m \cdot a \qquad (12.8)$$

where force $F_i = \sigma_i \cdot A$, with muscle cross-section A, muscle mass m, and acceleration a.

Obviously, the shortening velocity of the muscle is only limited by the inertia of the muscle. To verify this result, much higher exponents of the activation function have been used. Even for $S = 1000\ s^{-1}$ (full activation in one step), the maximal contraction velocity is only $v \approx -12\ s^{-1}$. The quick-release set-up (described in the *Quick Release* section of this chapter) leads to the same velocity-time dependency.

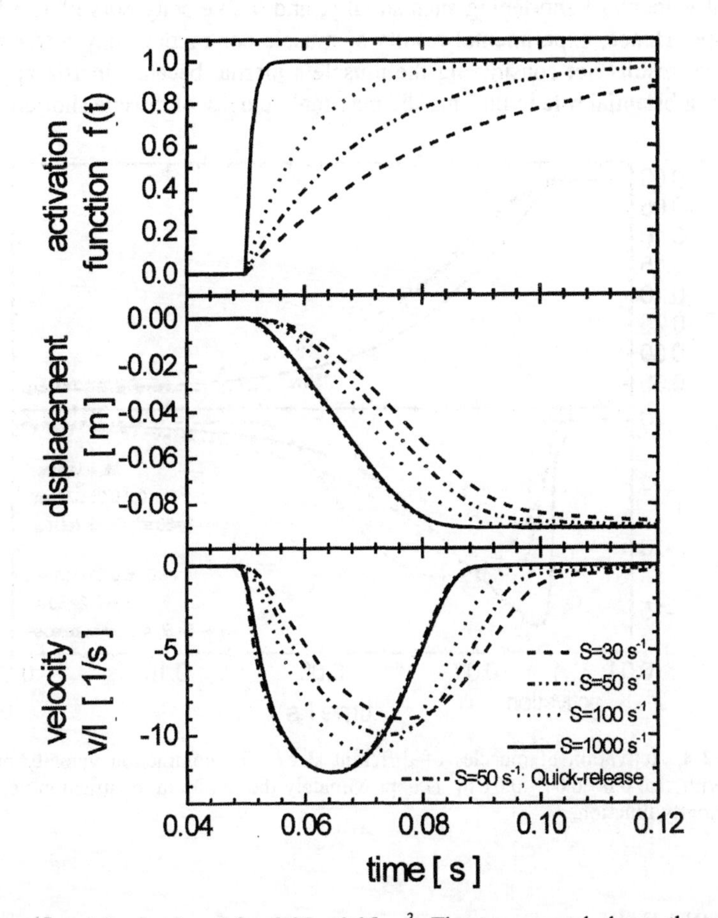

Figure 12.7. Muscle size: $0.3 \cdot 0.05 \cdot 0.05\ m^3$. The upper panel shows the activation functions for different parameters S. The middle panel shows the corresponding displacement-time curves. The lower panel shows the derived velocity-time curves. The velocity is normalized relative to the fibre length.

For each muscle, the absolute maximal shortening velocity (for completely unloaded shortening) cannot be reached because of the inertia of the muscle that provides a load, and so limits the maximal velocity of shortening that can be achieved physiologically. This limit should depend on the size and shape of the muscle. For a small muscle (size of $0.6 \cdot 0.01 \cdot 0.01$ m^3), the influence of inertia is strongly reduced compared to the muscle discussed above (Figure 12.8). The maximum contraction velocity reached, $v = -16$ s^{-1}, is much closer to the absolute maximal velocity (-17 s^{-1}) of unloaded shortening. The shape of the displacement-time curves of the free end of the muscles is quite different too. For the small muscle, the movement starts rapidly and the maximum contraction velocity is reached within a short time. For the large muscle, inertia limits the maximal velocity of shortening substantially, and the velocity curve has a bell-like shape. Hence, experimental results of muscle contraction may differ from theoretical results not considering the muscle's inertia, because inertia appears to play a substantial role in limiting the maximal velocity of muscle shortening.

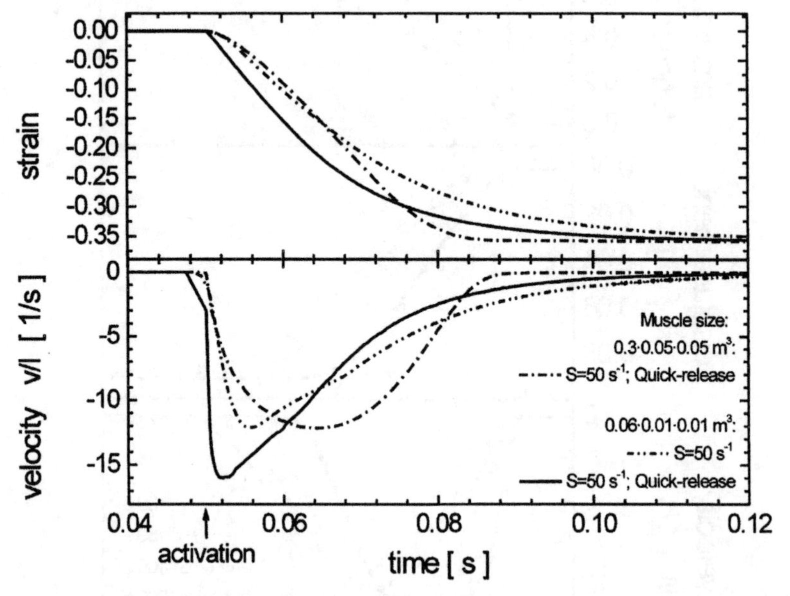

Figure 12.8. Activation of muscles of different sizes. The contraction velocity of the muscle with size $0.6 \cdot 0.01 \cdot 0.01$ m^3 is approximately the minimum contraction velocity of the velocity function.

QUICK RELEASE

Quick release experiments are often performed to get the parameters for the velocity-force function in skeletal muscle. Simulating a quick release experiment allows for separation of the different parts of muscle stress and to compare the input and output parameters. For this next simulation, we chose that the muscle contracts isometrically from $t = 0.05$ s to $t = 0.5$ s; at $t = 0.5$ s, the muscle is released at one end. In addition, the muscle is loaded at the free end with different weights, M. So, the right side of Equation 12.8 must be expanded to $(m + M) \cdot a$.

During the isometric contraction, the stress in the muscle increases according to the activation function f_t; there is no influence of the velocity and the length function on muscle stress ($f_v = f_l = 1$). After release, the muscle shortens and the stress decreases due to the decrease in velocity and muscle length (lower part of Figure 12.9). The maximum acceleration and the maximum velocity can be determined from these curves. If the mass of the muscle ($m = 0$), the change in length, and the passive stress are neglected ($f_t = 1$; $f_l = 1 \Rightarrow F_{passive} = 0$), then Equation 12.8 is reduced to:

$$F_{isometric} \cdot f_v(v) = M \cdot a \tag{12.9}$$

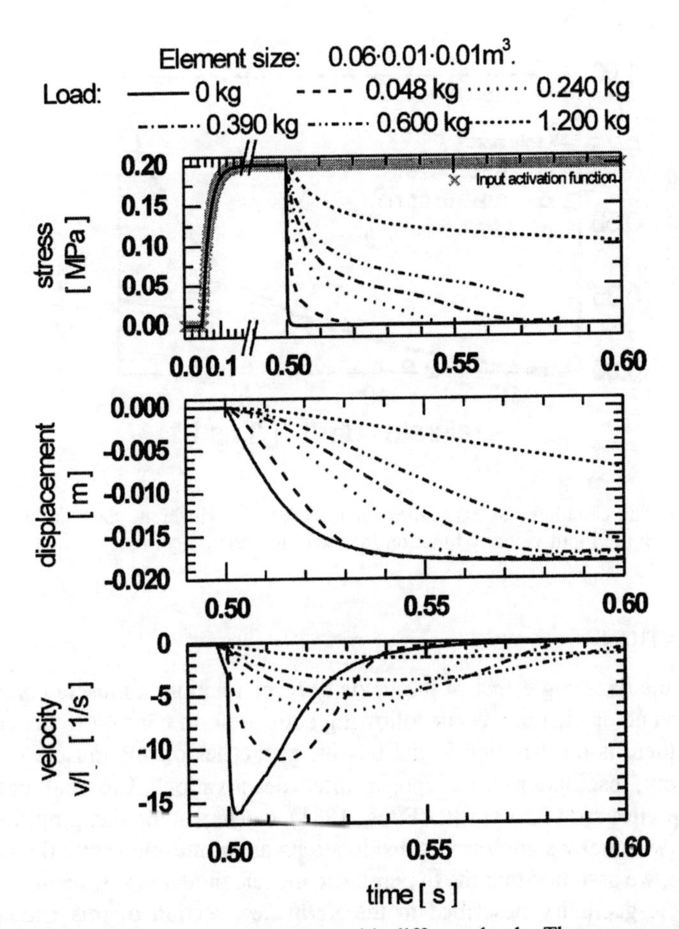

Figure 12.9. Quick release experiment with different loads. The upper panel shows the stress created in the element, the middle panel shows the displacement of the free end after release ($t = 0.5$ s), and the lower panel shows the derived velocity of the muscle.

This is one way to evaluate the experimental results. In Figure 12.10, the input velocity function is compared to the calculated results. Especially for large muscles, the maximal contraction velocity is underestimated. Here, the inertia of the muscle was neglected ($m = 0$). The underestimation of the maximal contraction velocity becomes smaller for increasing external loads, but then the neglecting of length changes and passive stretch become more important. When the maximum contraction velocity is reached, there is no acceleration ($a = 0$), and using Equation 12.8 that means:

$$F_{isometric} \cdot f_t(t) \cdot f_v(v) \cdot f_l(l) = F_{active} = -F_{passive}; \quad \text{at maximum velocity} \quad (12.10)$$

For large external loads, the muscle shortens significantly, and f_l and $F_{passive}$ should not be neglected.

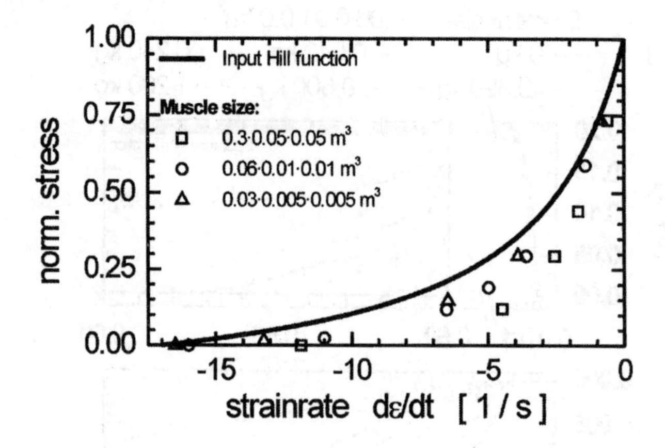

Figure 12.10. The calculated stress values for muscles with different sizes and loads in comparison with the input velocity function for the quick-release set-up.

DEACTIVATION

Because of the damping effect of the velocity-force relation, a muscle contracts smoothly and no oscillations occur following activation. In contrast to the active properties, there is no damping in the passive properties of this muscle model, and, therefore, oscillations can appear after deactivation. The real passive muscle properties are visco-elastic (Fung, 1981), causing some damping. In this section, we will make some basic considerations about muscle oscillations. For this purpose, we assumed that the hyperelastic muscle model is sufficient.

The muscle geometry described in the *Activation* section of this chapter is used. The muscle is deactivated at $t = 0.3$ s, with different exponential factors S. Figure 12.11 shows the deactivation function and the displacement for a muscle of size $0.3 \cdot 0.05 \cdot 0.05$ m^3. The amplitude of the oscillations depends strongly on the exponent S. The smaller the exponent S, the longer the muscle remains active, i.e., the damping of the velocity-force function persists, which in turn reduces the magnitude of the oscillations. In addition, the amplitude of the oscillations depends on the size of the muscle (Table 12.3). Also, following deactivation, a small muscle (with small inertia) is more effectively decelerated than a large muscle. Therefore, large muscles have larger oscillation amplitudes than small muscles.

Table 12.3. Oscillation parameters after deactivation.

Muscle size (m^3)	S (s^{-1})	Period (s)	Amplitude (m)	E (kPa)
$0.3 \cdot 0.05 \cdot 0.05$	30	0.141	$5.9 \cdot 10^{-3}$	90
$0.3 \cdot 0.05 \cdot 0.05$	50	0.139	$2.0 \cdot 10^{-2}$	92
$0.3 \cdot 0.05 \cdot 0.05$	100	0.132	$4.9 \cdot 10^{-2}$	102
$0.3 \cdot 0.05 \cdot 0.05$	1000	0.108	$9.9 \cdot 10^{-2}$	151
$0.06 \cdot 0.01 \cdot 0.01$	50	0.0283	$2.4 \cdot 10^{-6}$	89
$0.03 \cdot 0.005 \cdot 0.005$	50	0.0139	$4 \cdot 10^{-10}$	92

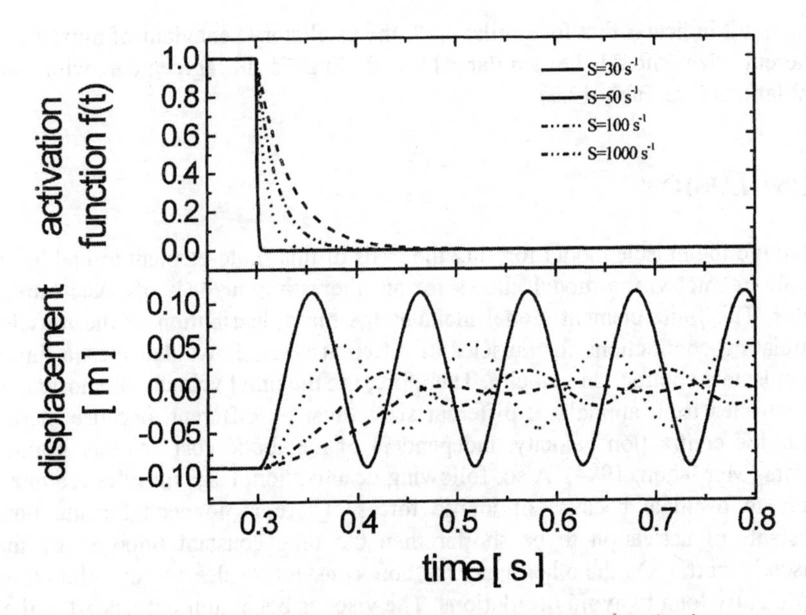

Figure 12.11. Oscillation of the hyperelastic muscle model after deactivation.

On the other hand, the period of oscillation depends only weakly on the exponent S. The characteristic oscillation period depends on the stiffness and the mass of the muscle. Considering the muscle as a spring fixed at one end, with a mass $m^* = ½\, m$ at each end of the spring, leads to the characteristic period T of a spring with an elastic constant D:

$$T = 2 \cdot \pi \cdot \sqrt{\frac{m^*}{D}}; \tag{12.11}$$

Transforming Hooke's law,

$$F = D \cdot (l - l_0) \tag{12.12}$$

to contain a linear Young's modulus (E), and linearizing the strain (ε) leads to:

$$D = \frac{F}{(l - l_0)} = \frac{\sigma \cdot A}{\varepsilon \cdot l_0} = E \cdot \frac{A}{l_0}; \tag{12.13}$$

Substituting Equation 12.11 into Equation 12.13, and inserting $m = V \cdot \rho$, with volume V and density ρ, results in:

$$E = \frac{l_0}{A} \cdot D = \frac{l_0}{A} \cdot \frac{4 \cdot \pi^2 \cdot m^*}{T^2} = \frac{l_0 \cdot V \cdot \rho}{2 \cdot A} \cdot \frac{4 \cdot \pi^2}{T^2} = l_0^{\,2} \cdot \rho \cdot \frac{2 \cdot \pi^2}{T^2}; \tag{12.14}$$

This result indicates that for similar l_0 / T, the oscillatory behaviour of muscles of different size should be similar. This finding is in agreement with our simulations (see Table 12.3).

CONCLUSION

Although the muscle model forming the basis of this finite-element model has a simple geometry, the model allows for an interesting analysis of basic movements. The finite-element model includes the mass distribution of the muscle. Simulating contractions for muscles of different sizes shows that inertia limits the muscle's contraction velocity. Therefore, the maximal velocity of shortening for muscles from animals of different sizes must be different, because inertia limits the contraction velocity, independent of metabolic cost or other scaling factors (McMahon, 1984). Also, following deactivation, large muscles are more likely to oscillate because of inertia forces. There is no need for the time constants of activation to be shorter than the time constant imposed by the muscle's inertia. On the other hand, the time constant for deactivation should be sufficiently long to avoid oscillations. The viscous behaviour embedded in the force-velocity function of the active muscle is sufficient to prevent muscular oscillations, and thus, there is no need for viscous behaviour of the tissue. This strategy may save energy during muscle contraction. Our model suggests that there is a size-dependent error in the calculation of maximal contraction velocities from quick-release experiments that should be corrected.

OUTLOOK

The strength of the finite-element model presented here lies in the description of the passive muscle properties. The influence of mass and incompressibility on the contractile behaviour of muscle can be evaluated. Therefore, the model offers itself to examine the influence of muscle geometry and fibre recruitment. For recruitment studies, it would be helpful to include visco-elastic properties in the model. Also, the model offers itself to study the passive anisotropic properties of muscle and the way by which active fibre forces are transmitted within the active contractile part of the muscle to the passive series elastic elements.

ACKNOWLEDGEMENTS

The authors wish to thank the DFG for support of the work within the INNOVATIONSKOLLEG 'MOTION SYSTEMS' (DFG INK22/A1 project A3). Also, the FEM-model would not exist without the financial support of CADFEM (German distributor of ANSYS®).

APPENDIX

The relationship between stress and strain in a hyperelastic material can be expressed by an elastic potential function W (strain energy function). This function is a scalar, whose derivative with respect to the Lagrangian strain E_{ij} or the Cauchy-Green deformation tensor C_{ij} produces the second Piola-Kirchhoff stress tensor S_{ij}:

$$S_{ij} = \frac{\partial W}{\partial E_{ij}} = 2 \cdot \frac{\partial W}{\partial C_{ij}} \tag{12.15}$$

The relation between the Lagrangian strain E_{ij} and the Cauchy-Green deformation tensor C_{ij} is:

$$E_{ij} = \frac{1}{2}\left(C_{ij} - \delta_{ij}\right)$$
$$\delta_{ij} = \begin{cases} 1 & i = j \\ 0 & i \neq j \end{cases} \tag{12.16}$$

The deformation tensor C_{ij} can be calculated by:

$$C_{ij} = F_{ij} F_{ij} \tag{12.17}$$

F_{ij} are the deformation gradients,

$$F_{ij} = \frac{\partial x_i}{\partial X_j} \tag{12.18}$$

where X_i is the undeformed position of a point in direction i, $x_i = X_i + u_i$ is the deformed position, and u_i is the displacement of a point.

The invariants $I_{1,C}$, $I_{2,C}$, $I_{3,C}$ of the deformation tensor C_{ij} are:

$$I_{1,C} = \sum_i C_{ii} \tag{12.19}$$

$$I_{2,C} = \frac{1}{2} \cdot \left(I_{1,C}^2 - \sum_{i,j} C_{ij} C_{ij} \right) \tag{12.20}$$

$$I_{3,C} = \det \mathbf{C} \tag{12.21}$$

$I_{3,C}$ is the volume change ratio.

Defining the reduced invariants:

$$I_{1,M} = I_{1,C} \left(I_{3,C} \right)^{-1/3} \tag{12.22}$$

$$I_{2,M} = I_{2,C} \left(I_{3,C} \right)^{-2/3} \tag{12.23}$$

$$I_{3,M} = \left(I_{3,C} \right)^{1/2} \tag{12.24}$$

The strain energy function can be written (Rivlin (1984), Mooney (1940)):

$$W = \sum_{i+j=1}^{n} a_{ij} \cdot \left(I_{1,M} - 3 \right)^i \cdot \left(I_{2,M} - 3 \right)^j + \frac{\kappa}{2} \left(I_{3,M} - 1 \right)^2 \tag{12.25}$$

a_{ij} are the Mooney Rivlin constants.

$$\overline{p} = -\kappa \cdot \left(I_{3,M} - 1 \right) \tag{12.26}$$

$\overline{p}$ is the hydrostatic pressure, and κ is the bulk modulus.

REFERENCES

Bovendeerd, P.H.M. (1990). The mechanics of the normal and ischemic left ventricle during the cardiac cycle. Ph.D. thesis, University of Limburg, Maastricht, The Netherlands.

Fung, Y.C. (1981). *Biomechanics: Mechanical Properties of Living Tissues*. Springer-Verlag, New York.

Gordon, A.M., Huxley, A.F., and Julian, F.J. (1966). The variation in isometric tension with sarcomere length in vertebrate muscle fibres. *Journal of Physiology* **184**, 170–192.

Hatze, H. (1981). Myocybernetic control models of skeletal muscle: Characteristics and applications. University of South Africa, Pretoria.

Hill, A.V. (1922). The maximum work and mechanical efficiency of human muscles and their economical speed. *Journal of Physiology* **56**, 19–41.

Hill, A.V. (1938). The heat of shortening and the dynamic constants of muscle. *Proceedings of the Royal Society of London B* **126**, 136–195.

Huyghe, J.M., Van Campen, D.H., Arts, T., and Heethaar, R.M. (1991). The constitutive behavior of passive heart muscle tissue: A quasi-linear viscoelastic formulation. *Journal of Biomechanics* **24**, 7, 841–848.

Johansson, T., Meier, P., and Blickhan R. A finite element model for the mechanical analysis of skeletal muscle. Submitted for publication in March 1999.

Kojic, M., Mijailovic, S., and Zdravkovic, N. (1998). Modeling of muscle behavior by the finite element method using Hill's three-element model. *International Journal for Numerical Methods in Engineering* **43**, 941–953.

McMahon, T.A. (1984). *Muscles, Reflexes and Locomotion*. Princeton University Press, Princeton, New Jersey.

Mooney, M. (1940). A theory of large elastic deformation. *Journal of Applied Physics* **6**, 582–592.

Otten, E. and Hulliger, M. (1995). A finite-element approach to the study of functional architecture in skeletal muscle. *Zoology* **98**, 233–242.

Rivlin, R.S. (1984). Forty years of nonlinear continuum mechanics. In *Proceedings of the IX International Congress on Rheology*, Mexico, pp. 1–29.

Sussmann, T. and Bathe, K.J. (1987). A finite element formulation for nonlinear incompressible elastic and inelastic analysis. *Composite Structures* **26**, 357–409.

van Leeuwen, J.L. (1991). Optimum power output and structural design of sarcomeres. *Journal of Theoretical Biology* **149**, 229–256.

van Leeuwen, J.L. (1992). Muscle function in locomotion. In: *Mechanics of Animal Locomotion*. Volume 11 of Advances in Comparative and Environmental Physiology. Alexander, R. McN. (ed.), Springer-Verlag, Heidelberg, pp. 191–250.

van Leeuwen, J.L., and Kier, W.M. (1997). Functional design of tentacles in squid: Linking sarcomere ultrastructure to gross morphological dynamics. *Philosophical Transactions of the Royal Society of London B: Biological Sciences* **352**, 551–571.

13 Rise and Relaxation Times of Twitches and Tetani in Submaximally Recruited, Mixed Muscle: A Computer Model

H. H. C. M. SAVELBERG
Faculty of Health Sciences, Maastricht University, Maastricht, The Netherlands

INTRODUCTION

Contractile characteristics of twitch contractions in different fibre types are known to differ. Twitches of slow, oxidative fibres have longer rise times and relaxation times than twitches of fast, glycolytic fibres (Burke et al., 1971). Also, the maximal twitch force of slow fibres is known to be less than that of fast fibres. Therefore, it is reasonable to assume that the rise and relaxation times of twitches of muscles can be used to asses the fibre composition of muscle (McComas and Thomas, 1968). For isometric contractions, Gonyea et al. (1981) showed that the rise and relaxation times are indeed correlated to the amount of slow, oxidative fibres.

For recruitment of motor units, the size principle (Henneman et al., 1965; Zajac and Faden, 1985) is the accepted paradigm of motor unit recruitment order. Small and slow motor units are recruited at low force levels and, with increasing force, progressively larger and faster motor units are recruited. Based on the size principle and the correlation between fibre characteristics and contractile type, rise and relaxation times of twitch and tetanic contractions are used as indicators of motor unit recruitment, assuming that long rise and relaxation times are indicators for small recruitment, and short rise and relaxation times are indicators of large recruitment (Fang and Mortimer, 1991b).

CURRENT PROBLEMS

Experimentally, in animal models, muscles are typically maximally stimulated. In experiments, submaximal contractions can be evoked either by modulating the firing rate (Roszek, 1996), or by affecting the recruitment of motor units by varying the electric current applied. The disadvantage of the latter procedure is that it implies a recruitment order that is opposite to the size principle, since the

Skeletal Muscle Mechanics: From Mechanisms to Function. Edited by W. Herzog.

largest and fastest motor units have the lowest excitation threshold for electrical stimulation. For understanding muscle properties under submaximal physiological conditions, the properties derived from maximal tests may not be appropriate. In muscles with a mixed composition of fibre types, electrical nerve stimulation leads to a reversal of the physiologically relevant size principle, i.e., fast motor units are recruited first; slow motor units last. This results in poor force gradation and rapid fatiguing of the muscle (Mortimer, 1981). Besides, it has been suggested that different fibre types in a muscle reach their optimal sarcomere length at different muscle lengths (De Ruiter et al., 1995). This result implies that force-length and force-velocity relationships depend on recruitment and recruitment order. A few studies have addressed the development of procedures to recruit motor units experimentally in an animal model according to the size principle (Petrofsky, 1978; Zhou et al., 1987; Baratta et al., 1989; Fang and Mortimer, 1991a; Huijing and Baan, 1992). In a recent study, we intended to implement a direct current block to selectively block large motor units (Savelberg and Van Bolhuis, 1999) in a rat model. To evaluate the success of this procedure, the change in rise and relaxation times as a function of recruitment level was proposed (Fang and Mortimer, 1991b).

Most muscles are composed of slow and fast fibres and a wide range of intermediate fibres. Consequently, at any recruitment level, a mixture of fibre types will be selected. Therefore, the rise and relaxation times depend on the temporal characteristics of all recruited fibres. However, the corresponding rise and relaxation time will not be a linear average of the rise and relaxation times of the involved fibres, but will depend on a kind of weakest-link-in-a-chain principle: the onset of relaxation is determined by the fastest fibres in the recruited population, and the end of the relaxation and rise periods by the slowest fibres. Thus, the relaxation and rise times of a mixed population might appear to be longer than those of a population of merely slow fibres. Therefore, it is hypothesized that these changes do not only depend on the recruitment level, but may also be affected by composition of the muscle and the mutual relationships between the force-time characteristics of the slow and fast fibres in a muscle.

In experimental studies, it has been shown for different recruitment levels in one muscle that temporal characteristics of twitches increase with decreasing recruitment levels (Fang and Mortimer, 1991b). For muscles of different compositions, it was found that at maximal activation, the twitch characteristics increase with the amount of slow fibres in the muscle (Gonyea et al., 1981). The interaction of muscle composition and recruitment at submaximal recruitment levels is not known. Therefore, a computer simulation model was developed to gain a better understanding of the effects of recruitment on contractile characteristics and to search for possibilities to assess recruitment in submaximal contractions.

METHODS

In the experimental literature, the rise time of force is not strictly defined. Different studies use different definitions; mostly time periods for force to rise from about 20% to about 80% of the maximal force are considered. Therefore, the experimental definitions typically take into consideration the linear part of the rising force curve only. The initial part of the curve might be non-linear due to compliance in series elasticity and the experimental set-up. The non-linearity at the top of the twitch force curve is an inherent feature of force development. In this study, the initial value defined as twitch onset was taken as 0%, since the factors responsible for the initial non-linearity were assumed to be absent in this simulation. As a measure of relaxation time, the time elapsed when the force drops from 50% to 25% of the maximal value of the actual contraction is used in this study; this definition agrees with most of those used in the literature.

SIMULATION MODEL

In the simulation model, a muscle was composed of one hundred muscle fibres that were assumed to be arranged in parallel to each other without the possibility of lateral force-sharing between the fibres. The muscle fibres were characterized by their twitch characteristics. To determine the twitch characteristics of the simulated muscle, the forces of the individual fibres were summed.

The natural range of fibre types was represented by two or three differing twitch force profiles: a force profile typical for fast-fatiguing (FF) fibres, a force profile for fast fatigue-resistant (FR) fibres, and one for slow (S) fibres. Although in the following these abbreviations will be used, no metabolic or structural significance should be attributed to them. The force profiles only represent the mechanical characteristics of these fibre types. The twitch-force characteristics of these fibres were obtained from the literature. In the literature, many different relationships between FF, FR, and S fibres appear. Fibres can differ in amplitude, timing of maximal force, and duration of decay of force. To investigate the effect of these differences between FF, FR, and S fibres, three patterns of force profiles were considered. The first pattern of FF, FR, and S twitch profiles (Burke et al., 1973, cited in McComas, 1996), cat, m. gastrocnemius, (Figure 13.1A) is characterized by FF and FR, mainly differing in force amplitude, and S differing from FF and FR in force amplitude and timing. The ratio of amplitudes of FF and S fibres in this pattern is approximately 3.5.

The second pattern (Gordon et al., 1990, cat, m. tibialis posterior, Figure 13.1B) displays a gradual increase of time-to-peak force from FF to FR to S, and an exponential increase in force amplitude from S to FR to FF. In this pattern, the difference in amplitude between FF and S fibres is maximal, the ratio is 6.

In the third pattern (Latash, 1998—it is not clear whether these patterns result from any experiment—Figure 13.1C), both peak amplitude and time-to-peak force change gradually across the fibre types. In this case, the ratio of amplitudes between FF and S fibres is small, only 2.

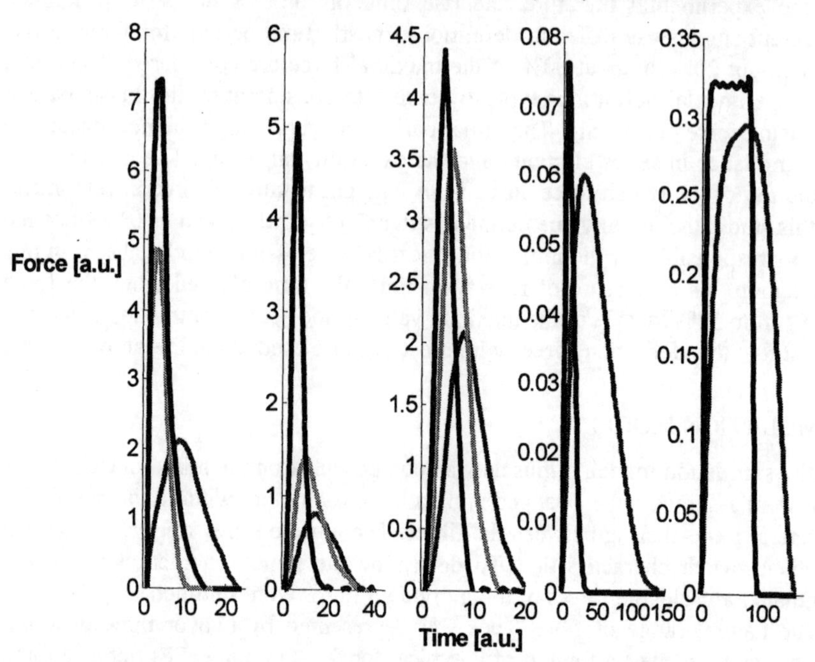

Figure 13.1. Force-time profiles A) through E) from left to right, of twitches for fast-fatiguing, fast fatigue-resistant, and slow fibre types that were used in the simulations: A) from McComas (1996), B) from Gordon et al. (1990), and C) from Latash (1998). D) The figure shows the twitches of a slow- and a fast-fatiguing fibre as obtained from a one-dimensional, Huxley-based simulation model (Maenhout et al., 2000). With this model and using experimentally derived parameters, profiles of tetanic contractions were created (E).

When twitches fuse to tetani, a non-linear summation of force takes place. Given this non-linearity, results found for twitches do not necessarily hold for tetanic contractions. To investigate the changes in relaxation and rise time that occur due to accumulation of twitches to a tetanic contraction, a Huxley-based one-dimensional simulation model (Maenhout et al., 2000) was used to create twitches and tetanic profiles of slow and fast fibres. Based on experimental data from Ma and Zahalak (1991), and Maenhout et al. (2000), model parameters for the m. soleus of the rat, a muscle predominantly consisting of slow fibres, and the m. tibialis anterior of the rat, a muscle predominantly composed of fast fibres, were fitted (Maenhout et al., 2000). With these parameters and the Huxley-based simulation model, twitches (Figure 13.1D) and tetanic contractions (Figure 13.1E) could be simulated. Compared to the three other patterns shown in Figure 13.1A, Figure 13.1B, and Figure 13.1C, this twitch pattern is characterized by differences in the timing of the peak force and the relaxation of force. The ratio between the force amplitudes is only 1.25 (Figure 13.1D).

In the simulations, a muscle was either composed of FF and S fibres, or of FF, S, and FR fibres. By comparing the simulation results obtained using a model of two or three fibre types, insight can be gained about the effect of the number of fibre types on muscle force profiles. In real muscles, fibre properties differ in a continuous way. Sixty-six muscle models were composed with FF, S, and FR fibres. These 66 muscle models, with the corresponding fibre-type distributions, are shown in Table 13.1. Also, 11 muscle models were considered that contained FF and S fibres (Table 13.2).

Table 13.1. With three fibre types, 66 different muscles were created and used in the simulation.

Muscle Composition	1	2	3	4	5	6	....	65	66
# of FF fibres	0	0	10	0	10	20		90	100
# of FR fibres	0	10	0	20	10	0		10	0
# of S fibres	100	90	90	80	80	80		0	0

Table 13.2. With two fibre types, 11 different muscles were created and used in the simulation.

Muscle Composition	1	2	3	4	5	6	7	8	9	10	11
# of FF fibres	100	90	80	70	60	50	40	30	20	10	0
# of S fibres	0	10	20	30	40	50	60	70	80	90	100

In this study, the recruitment level was defined as the maximal twitch or tetanic force a given selection of muscle fibres could generate as a fraction of the maximal twitch or tetanic force respectively when all fibres were recruited. In the simulation model, motor units were recruited either according to or opposite from the size principle. When recruited according to the size principle, only slow fibres were recruited at low forces. With increasing recruitment levels (higher forces), fast-fatigue resistant fibres were selected only if all slow fibres had been recruited, and fast-fatiguing fibres were only recruited if all fast fatigue-resistant fibres had been recruited. When recruited opposite to the size principle, as occurs in artificial electrical stimulation, fast fibres were recruited at low force. Slow fibres were only recruited when all fast fibres had been recruited.

RESULTS

In muscles composed of only FF and S fibres, the effect of recruitment on relaxation and rise time of twitches was found to differ between muscles of different compositions (Figure 13.2A and Figure 13.2B). In a relatively slow muscle, the relaxation time (Figure 13.2A) was found to increase with recruitment level, while it decreased with recruitment level in relatively fast muscle compositions. In mixed muscles containing 50% slow and 50% fast fibres, shortest relaxation times occurred at extreme recruitment levels, while at intermediate recruitment levels, the longest relaxation times were found. For the rise times (Figure 13.2B), this kind of interaction between muscle composition and recruitment level was not found; rather a summation of both effects was found. The rise times were longest in relatively slow muscles and at the lowest recruitment levels.

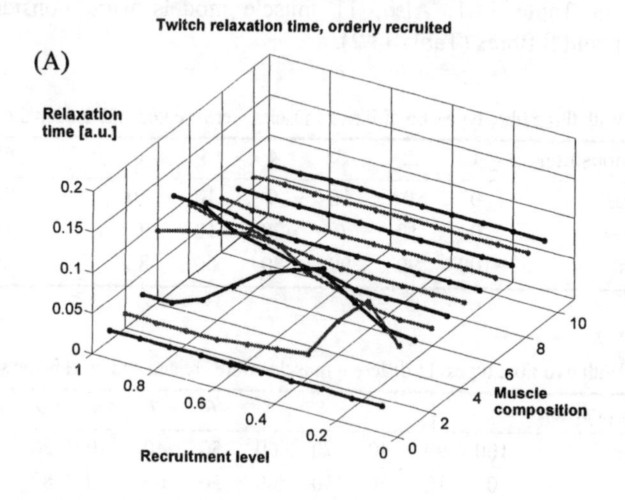

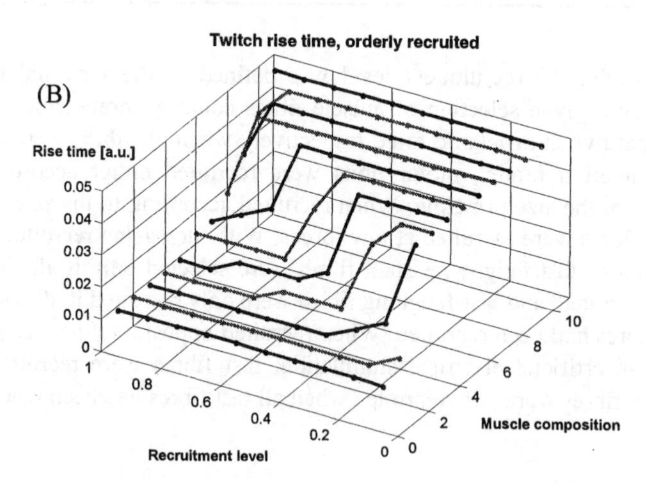

Figure 13.2. (A) Relaxation times of orderly recruited muscle as a function of recruitment level for muscles composed of different combinations of FF and S fibres. (B) Rise times as a function of recruitment level for muscles composed of different combinations

of FF and S fibres for orderly recruited muscles. The twitch profiles for the fibres were obtained from a one-dimensional, Huxley-based simulation model (Maenhout et al., 2000; Figure 13.1D). Different lines represent different muscle compositions. Grey shading of the lines is only meant to improve legibility of the figure. The muscle compositions are ordered from muscles containing 100% FF fibres (0) to muscles containing 100% S fibres (10). A recruitment level of 1 denotes full recruitment, 0.1 denotes the recruitment of the number of motor units to generate 10% of the maximal twitch force. Times are given in arbitrary units (a.u.) to allow comparison among Figure 13.2, Figure 13.5, and Figure 13.6.

The long relaxation times occur at recruitment levels where both fast and slow fibres make a considerable contribution. In these cases, the start of the relaxation is the result of FF fibres being switched off, while the end of the relaxation is determined by the slow relaxation of the S fibres. Consequently, the relaxation period is elongated (Figure 13.3B). In muscles with a large amount of slow fibres, this occurs at relatively high recruitment levels; at lower recruitment levels only slow fibres are selected (Figure 13.3C). In muscles with a considerable amount of fast fibres, this situation is present at relatively low recruitment levels (Figure 13.3A). In these muscle compositions, the contributions of fast fibres overwhelms that of slow fibres at higher recruitment levels. Slow fibres determine only the force-time curve at values lower than 25% of the force level.

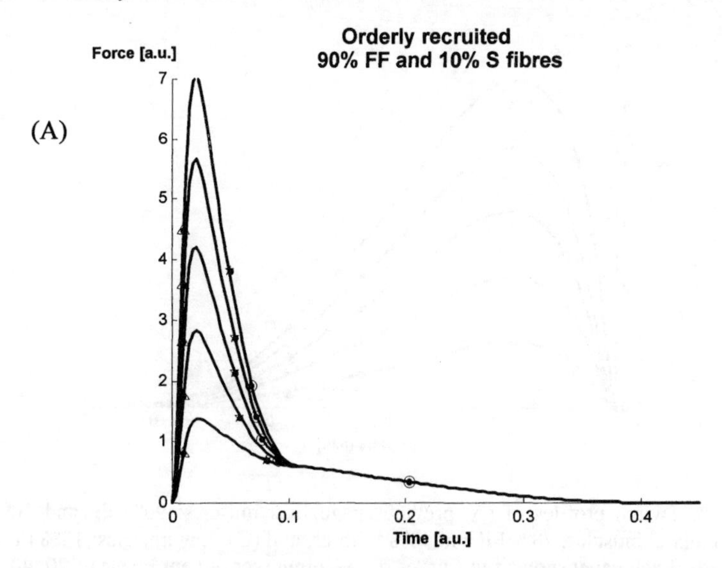

Figure 13.3 (A) Twitch profile of fast muscles: 90% FF and 10% S fibres. The five curves represent recruitment levels of 20, 40, 60, 80 and 100%. The triangles indicate the point from which the rise time was determined (80% of the maximal force). The pentagrams and circles point to 50% and 25% of the maximal force, respectively. The time from a pentagram to the corresponding circle was used as a measure of relaxation time. See also Figure 13.3 (B) and (C), next page.

(B)

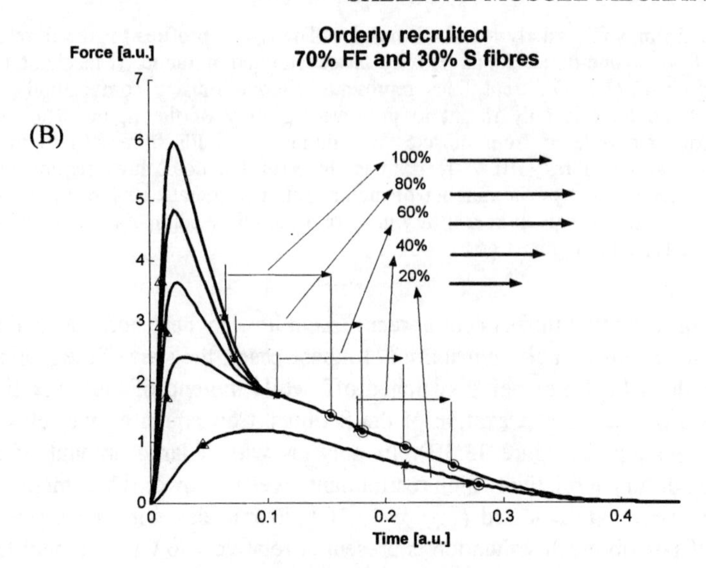

(C)

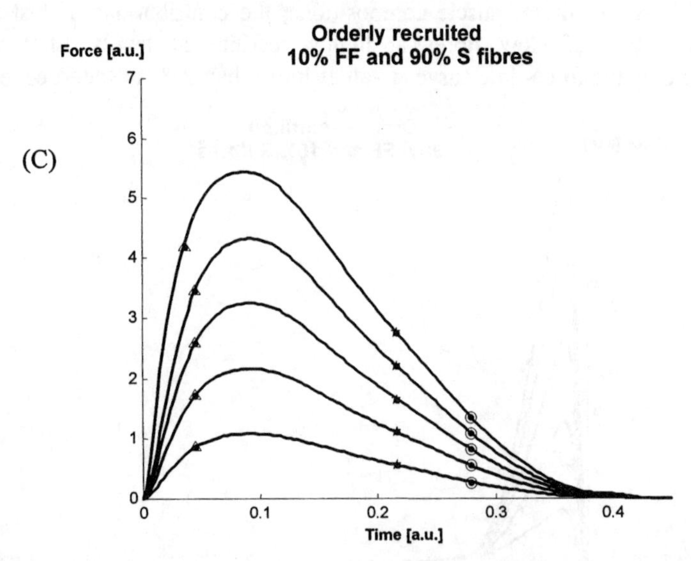

Figure 13.3. Twitch profiles of (A, previous page) fast muscles: 90% FF and 10% S fibres, (B) mixed muscles: 70% FF and 30% S fibres, and (C) slow muscles: 10% FF and 90% S fibres. Each panel shows five curves representing recruitment levels of 20, 40, 60, 80 and 100%. The triangles indicate the point from which the rise time was determined (80% of the maximal force). The pentagrams and circles point to 50% and 25% of the maximal force, respectively. The time from a pentagram to the corresponding circle was used as a measure of relaxation time. In (B), the vertical lines indicate corresponding pentagrams and circles, and the length of the corresponding arrows corresponds to the relaxation time for the five recruitment levels.

This phenomenon—interaction of recruitment level and muscle composition determining the relaxation time—was found to be robust over different patterns of mutual relationship of patterns of FF, FR, and S fibres. Specific relaxation times were affected by changes in these patterns. When the maximal twitch force of the S fibres is small compared to that of the FF fibres (Figure 13.1A or Figure 13.1B), the curves of the relaxation time as a function of recruitment level shift to the right (Figure 13.4 versus Figure 13.2A), compared to when the maximal twitch force of the S fibres is smaller compared to that of the FF fibres (Figure 13.1D). For the rise time, similar shifts were seen.

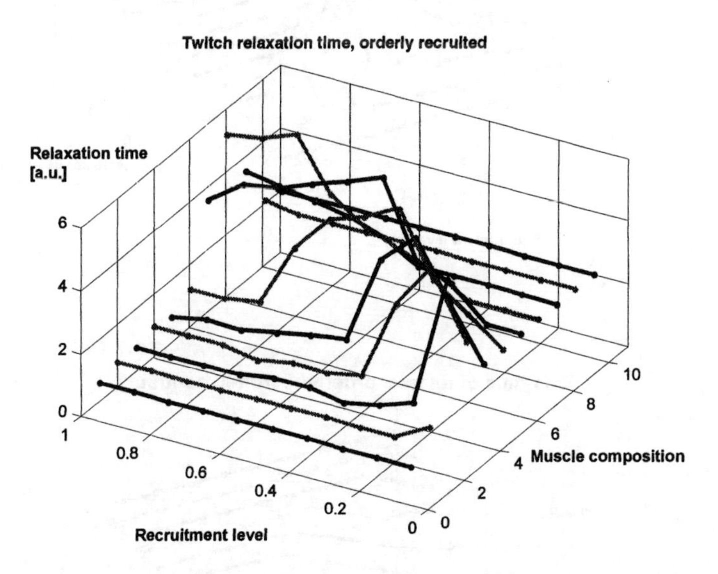

Figure 13.4. The relationship between relaxation time and recruitment level for different fibre-type compositions using FF and S fibres that are quite different in maximal force amplitude (Figure 13.1B). Contrast this to the relationship presented in Figure 13.2A, in which FF and S fibres were used that produce similar maximal force amplitudes (Figure 13.1D).

The global phenomenon—relaxation times depending on both muscle composition and recruitment level—was not affected when FR fibres were also used to compose the simulated muscles. The temporal characteristics of the intermediate FR fibres, whether they look more like those of the fast or like those of the slow fibres, determined whether the relaxation times for a given recruitment level look more like that of fast or slow muscles. The pattern of relaxation times of the twitches and tetanic contractions differed only minimally and only quantitatively, relaxation times of tetanic contractions being longer (respectively, Figure 13.2A and Figure 13.5A). The pattern of the rise times of tetanic contractions showed smoother changes with muscle composition and recruitment level than that of the twitches (Figure 13.2B and Figure 13.5B).

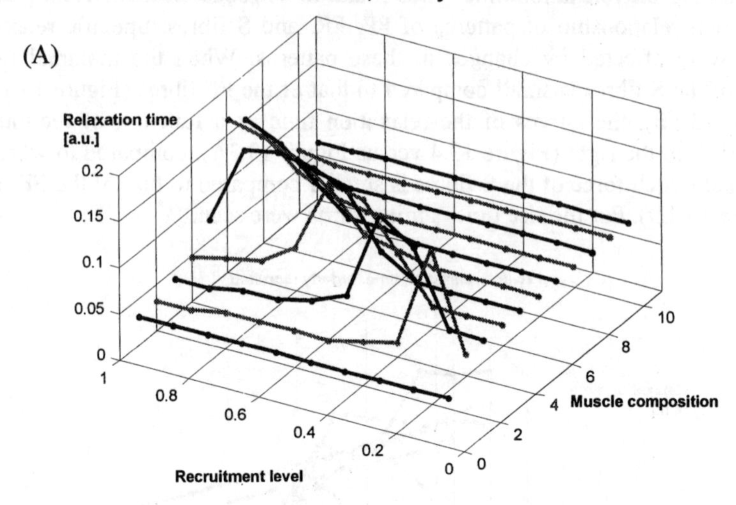

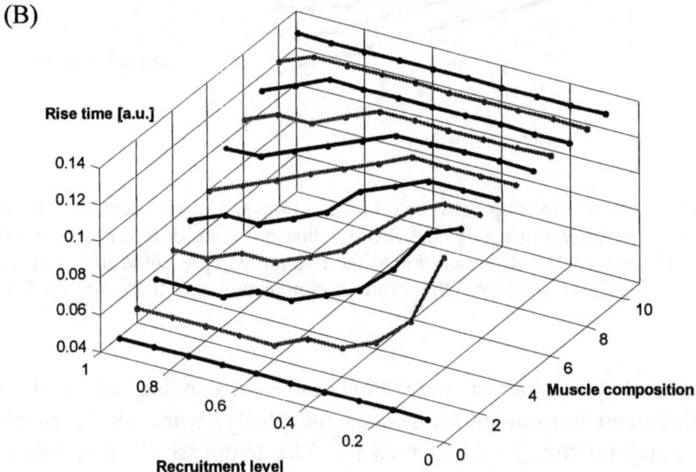

Figure 13.5. (A) The relaxation times, and (B) rise times for tetanic contractions at different recruitment levels and for different muscle compositions of orderly recruited muscles. To calculate these data, the same parameter set was used (Maenhout et al., 2000; Figure 13.1E) as was applied for the twitches. Grey shading of the lines is only meant to improve legibility of the figure. Times are given in arbitrary units (a.u.) to allow comparison among Figure 13.2, Figure 13.5, and Figure 13.6.

For mixed muscles containing both fast and slow muscles, the simulation of the reversely recruited motor units showed clear differences in relaxation and rise times with that of recruitment according to the size principle. For reversely recruited muscles, the previously mentioned interaction between muscle composition and recruitment level was not found. Rather, the relaxation time was found to increase with recruitment level and muscle containing more slow fibres. At low recruitment levels, relaxation times were shorter in reversely recruited muscles than in orderly recruited muscles; the changes in relaxation times with recruitment level was more abrupt in the reversely recruited muscles (Figure 13.2A and Figure 13.6A). When the recruitment level increases, fast fibres are recruited first. Increasing recruitment first leads to a condition where the concerted action of fast and slow fibres causes an elongated relaxation time, like in orderly recruited muscle. Subsequently, a proceeding increase of the recruitment level brings in the activity of even more slow fibres, which at the already high force level leads to a stabilization of the relaxation time (Figure 13.7). For the rise time, it was seen that in the muscle compositions where the rise time changed with changing recruitment, the changes in reversely and orderly recruited muscle were opposite. In reversely recruited muscles, rise time increased with increasing recruitment levels in muscles containing a high amount of slow fibres (Figure 13.2B and Figure 13.6B). These differences between reversely and orderly recruited muscles were independent of the mutual relationships between twitch patterns of different fibre types. As Figure 13.3B and Figure 13.7 illustrate, the force-time profiles of reversely and orderly recruited muscle differ enormously.

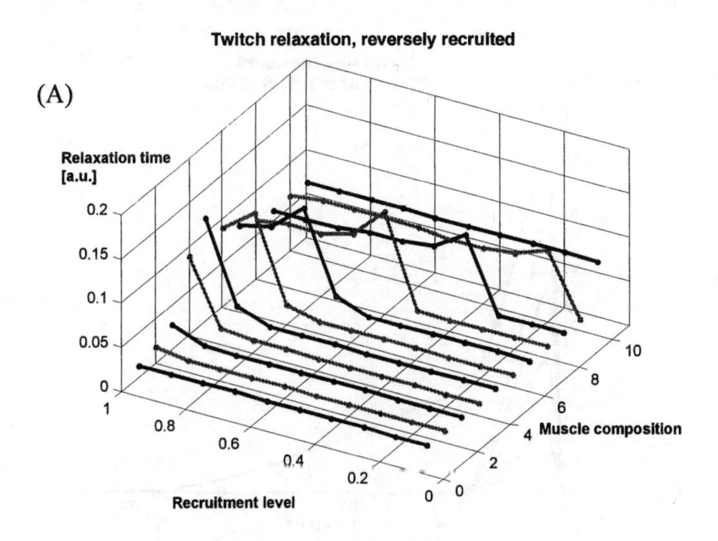

Figure 13.6. The relaxation (A) times of twitches as a function of recruitment level and muscle composition of reversely recruited muscle. See also Figure 13.6 (B), next page.

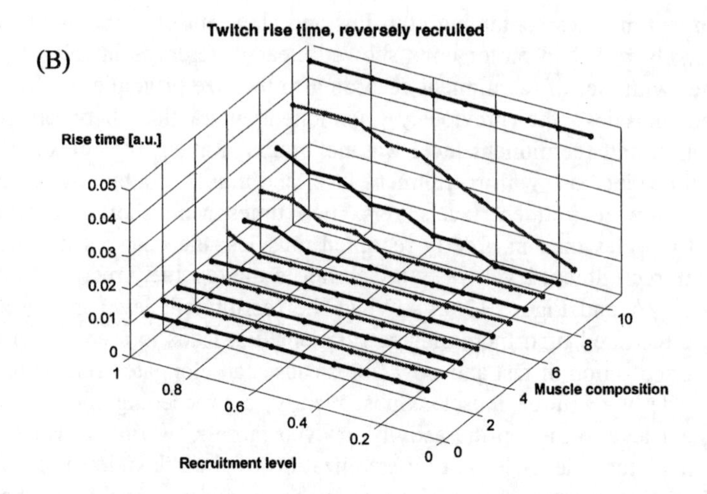

Figure 13.6. The relaxation (A, previous page) and rise (B) times of twitches as a function of recruitment level and muscle composition of reversely recruited muscle. The twitch pro-files for the fibres were obtained from a one-dimensional, Huxley-based simulation model (Maenhout et al., 2000; Figure 13.1D). Different lines represent different muscle compositions. Grey shading of the lines is only meant to improve legibility of the figure. The muscle compositions are ordered from muscles containing 100% FF fibres (0) to muscles containing 100% S fibres (10). A recruitment level of 1 denotes full recruit-ment, 0.1 denotes the recruitment of the number of motor units to generate 10% of the maximal twitch force. Times are given in arbitrary units (a.u.) to allow comparison among Figure 13.2, Figure 13.6, and Figure 13.5.

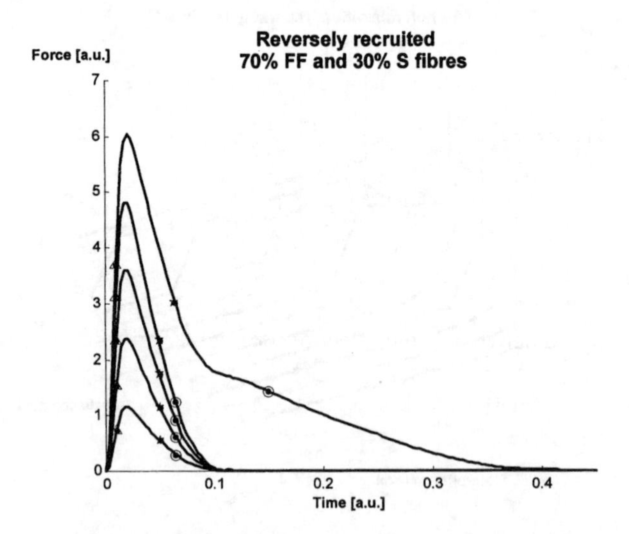

Figure 13.7. Twitch profile showing five curves representing recruitment levels of 20, 40, 60, 80 and 100%. Triangles indicate the point from which the rise time was determined (80% of the maximal force). Pentagrams and circles point to 50% and 25% of

the maximal force, respectively. The time from a pentagram to the corresponding circle was used as a measure of relaxation time. Twitches of the mixed muscle for five recruitment levels when the muscle was recruited inversely to the size principle are shown.

DISCUSSION

This simulation study showed that the commonly proposed idea that longer rise and relaxation times indicate lower recruitment levels (Fang and Mortimer, 1991b) is not necessarily correct. The composition of fibre types in a muscle also affects rise and relaxation times. Moreover, it was shown that changes in the fibre-type composition of a muscle may offset the effect of recruitment level on relaxation times, thereby leading to increased relaxation times with increasing recruitment levels—a finding that appears counter-intuitive, is not in accordance with much of the published literature, but could be explained readily using the presented simulation approach. The results for the rise times agreed with ideas proposed in the literature: decreasing recruitment levels and increasing the percentage of slow fibres in a muscle led to increased rise times.

The patterns in rise and relaxation times were found to be robust for changing relationships between slow, intermediate, and fast fibres. However, the exact properties given to the different fibre types, of course, influenced the prediction of rise and relaxation times. Therefore, knowing the fibre-type composition of a muscle and the recruitment level will not allow for a good prediction of rise and relaxation times. Also, rise and relaxation times cannot be used to assess muscle composition from the known recruitment level, nor to determine recruitment level from the known fibre-type distribution. In order to determine fibre-type distribution from rise and relaxation times, the twitch characteristics of individual fibres must be known.

In this study, muscle was considered as consisting of a number of parallel, mutually independent fibres, that differ only in the twitch characteristics. Lateral connections between fibres that may facilitate force transmission and modulate the forces generated by muscle fibres (Patel and Lieber, 1997; Huijing, 1999; Monti et al., 1999) were ignored. These simplifications in the model are expected to affect the absolute values of the predictions but not the global phenomena found here. Also, the effect of series elasticity was not considered in the simulations. However, series elasticity would only influence the global results if fibres with different twitch characteristics were arranged in series with elastic elements of different elastic properties. Given regionalization of muscle fibres (Kernell, 1998), but spatial homogeneity in mechanical properties of tendon fibres (Parry et al., 1978), it is assumed that ignoring series elasticity will not have a major influence on the results predicted by the model. Furthermore, inhomogeneity of the sarcomere length distribution in muscle (Willems and Huijing, 1994; Ettema and Huijing, 1994) was not taken into account. Incorporating sarcomere-length inhomogeneity in the model would be similar to considering twitches with different force amplitudes. In this study, the influence of two or three different twitch profiles on rise and relaxation times was investigated. As it turned out, the number of fibre types considered and the mutual relationships between these fibre types did not affect the global results, therefore it might be assumed that ignoring sarcomere length-inhomogeneity was justified for the purposes of this study.

In an experimental study, Gonyea et al. (1981) found a correlation between fibre composition (ranging from 1 to 49% slow fibres in a muscle) and contractile characteristics for isometric contractions: the greater the percentage of slow fibres, the longer the relaxation times. The present model gave similar results when the percentage of slow fibres was increased (Figure 13.2A, Figure 13.6A, and Figure 13.4). Depending on the recruitment level and the exact twitch properties of the individual fibres, a decrease in relaxation time was predicted above a certain percentage of slow fibres. At high recruitment levels, the percentage of slow fibres causing a reduction in relaxation times was found to be above 50%. Fang and Mortimer (1991b) used orderly recruitment in the cat medial gastrocnemius muscle, and found decreasing twitch times with increasing recruitment levels. Knowing that the cat medial gastrocnemius is a predominantly fast twitch-fibred muscle (61% FF, 14% FR, and 25% S fibres, according to Ariano et al., 1973), the model would predict relaxation times to increase with decreasing recruitment levels (Figure 13.4). In conclusion, the model, although ignoring a number of factors, agrees conceptually with experimental results.

It was mentioned previously that given the interaction between muscle composition and recruitment level on relaxation times, relaxation times cannot be used directly as a measure of the recruitment level. Rise times appear better suited than relaxation times for estimating recruitment levels. For a given muscle composition, rise times are always longer at lower recruitment levels. However, experimentally, the rise time may not be an appropriate indicator of recruitment level, since stiffness of the measuring equipment and compliance of the muscle series elasticity affects the ascending part of the force-time curve. However, the pronounced differences in relaxation times of orderly and reversely recruited muscles (Figure 13.2A and Figure 13.5A) may be used to judge whether muscles are orderly recruited. Using this approach, an estimate of fibre-type composition might also be obtained.

In summary, it was shown that relaxation times of twitches and tetanic contractions were affected by fibre-type composition and recruitment level. To evaluate the effectiveness of approaches to orderly recruit motor units in muscle, differences between orderly and reversely recruited muscle in relaxation times as a function of recruitment level appear to be helpful. The differences predicted to exist between reversely and orderly recruited muscle indicate that measurements of orderly recruited muscles are necessary to understand the behaviour of muscle under submaximal conditions. Conversely, the technique for submaximal recruitment and corresponding measurements will allow validation of the presented model and assessment of the effect of ignored factors on the model predictions. Moreover, and of greater importance for understanding the behaviour of a muscle, the results suggest that metabolic, mechanical, and control aspects of a muscle are not independent magnitudes, but are highly related. Changes in fibre composition of a muscle will affect its mechanical characteristics at a certain level of activation, and thus the control demands required for a particular mechanical effect. Fibre composition should no longer be considered as a factor affecting only endurance and maximal force of a muscle.

REFERENCES

Ariano, M.A., Armstrong, R.B., Edgerton, V.R. (1973). Hindlimb muscle fibre populations of five mammals. *Journal of Histochemistry and Cytochemistry* **21**(1), 51–55.

Baratta, R., Ichie, M., Hwang, S.K., Solomonow, M. (1989). Orderly stimulation of skeletal muscle motor units with tripolar nerve cuff electrode. *IEEE Transactions on Biomedical Engineering* **36**(8), 836–843.

Burke, R.E., Levine, D.N., Tsairis, P., Zajac, F.E. (1973). Physiological types and histochemical profiles in motor units of the cat gastrocnemius. *Journal of Physiology (London)* **234**(3), 723–748.

Burke, R.E., Levine, D.N., Zajac, F.E. (1971). Mammalian motor units: Physiological-histochemical correlation in three types in cat gastrocnemius. *Science* **174**(10), 709–712.

De Ruiter, C.J., De Haan, A., Sargeant, A.J. (1995). Physiological characteristics of two extreme muscle compartments in gastrocnemius medialis of the anaesthetized rat. *Acta Physiologica Scandinavica* **153**(4), 313–324.

Ettema, G.J., Huijing, P.A. (1994). Effects of distribution of muscle fiber length on active length-force characteristics of rat gastrocnemius medialis. *Anatomical Record* **239**(4), 414–420.

Fang, Z.P., Mortimer, J.T. (1991a). Selective activation of small motor axons by quasi-trapezoidal current pulses. *IEEE Transactions on Biomedical Engineering* **38**(2), 168–174.

Fang, Z.P., Mortimer, J.T. (1991b). A method to effect physiological recruitment order in electrically activated muscle. *IEEE Transactions on Biomedical Engineering* **38**(2), 175–179.

Gonyea, W.J., Marushia, S.A., Dixon, J.A. (1981). Morphological organization and contractile properties of the wrist flexor muscles in the cat. *Anatomical Record* **199**(3), 321–339.

Gordon, D.A., Enoka, R.M., Stuart, D.G. (1990). Motor-unit force potentiation in adult cats during a standard fatigue test. *Journal of Physiology (London)* **421**, 569–582.

Henneman, E., Somjen, G., Carpenter, D.O. (1965). Functional significance of cell size in spinal motoneurons. *Journal of Neurophysiology* **28**, 560–580.

Huijing, P.A. (1999). Muscle as a collagen fiber reinforced composite: A review of force transmission in muscle and whole limb. *Journal of Biomechanics* **32**(4), 329–345.

Huijing, P.A., Baan, G.C. (1992). Stimulation level-dependent length-force and architectural characteristics of rat gastrocnemius muscle. *Journal of Electromyography and Kinesiology* **2**(2), 112–120.

Kernell, D. (1998). Muscle regionalization. *Canadian Journal of Applied Physiology* **23**(1), 1–22.

Latash, M.L. (1998). *Neurophysiological Basis of Movement.* Human Kinetics, Champaign, Illinois.

Ma, S.P., Zahalak, G.I. (1991). A distribution-moment model of energetics in skeletal muscle. *Journal of Biomechanics* **24**(1), 21–35.

Maenhout, M., Hesselink, M.K.C., Oomens, C.W.J., Drost, M.R. (2000). Parameter identification of a distribution-moment approximated two-state Huxley model of the rat tibialis anterior muscle. In: *Skeletal Muscle Mechanics: From Mechanisms to Function.* Herzog, W. (ed.), John Wiley & Sons, Ltd., Chichester, England.

McComas, A.J. (1996). *Skeletal Muscle: Form and Function.* Human Kinetics, Champaign, Illinois.

McComas, A.J., Thomas, H.C. (1968). Fast and slow twitch muscles in man. *Journal of Neurological Science* **7**(2), 301–307.

Monti, R.J., Roy, R.R., Hodgson, J.A., Edgerton, V.R. (1999). Transmission of forces within mammalian skeletal muscles. *Journal of Biomechanics* **32**(4), 371–380.

Mortimer, J.T. (1981). Motor prostheses. In: *Handbook of Physiology: The Nervous System.* Brooks, V.B. (ed.), American Physiological Society, Bethesda, Maryland, pp. 155–187.

Parry, D.A., Barnes, G.R., Craig, A.S.A. (1978). Comparison of the size distribution of collagen fibrils in connective tissues as a function of age and a possible relation between fibril size distribution and mechanical properties. *Royal Society, London B: Biological Sciences* **203**, 305–321.

Patel, T.J., Lieber, R.L. (1997). Force transmission in skeletal muscle: From actomyosin to external tendons. *Exercise and Sport Science Reviews* **25**, 321–363.

Petrofsky, J.S. (1978). Control of the recruitment and firing frequencies of motor units in electrically stimulated muscles in the cat. *Medical and Biological Engineering and Computing* **16**, 302–308.

Roszek, B. (1996). Effects of submaximal activation on skeletal muscle properties. Ph.D. thesis, Free University, Amsterdam.

Savelberg, H.H.C.M., Van Bolhuis, A.I. (1999). A direct current block to recruit motor units in a physiological order in the rat gastrocnemius-plantaris complex. In *Proceedings of the XVII International Society of Biomechanics Congress.* University of Calgary, Canada, p. 621.

Willems, M.E., Huijing, P.A. (1994). Heterogeneity of mean sarcomere length in different fibres: Effects on length range of active force production in rat muscle. *European Journal of Applied Physiology* **68**(6), 489–496.

Zajac, F.E., Faden, J.S. (1985). Relationship among recruitment order, axonal conduction velocity, and muscle-unit properties of type-identified motor units in cat plantaris muscle. *Journal of Neurophysiology* **53**(5), 1303–1322.

Zhou, B.-H., Baratta, R., Solomonow, M. (1987). Manipulation of muscle force with various firing rate and recruitment control strategies. *IEEE Transactions on Biomedical Engineering* **34**, 128–139.

14 Linear and Non-linear Analysis of Lower-limb Behaviour With Heat and Stretching Routines Using a Free Oscillation Method

J. SPRIGGS AND G. D. HUNTER
Faculty of Health and Social Care, University of the West of England, Bristol, England

INTRODUCTION

The muscle-tendon unit (MTU) contributes to efficient co-ordinated movement of the human body by its ability to generate, dissipate, and recuperate mechanical energy. An essential feature of the MTU's ability to perform these tasks is its ability to regulate mechanical stiffness. In addition, the ability of the MTU to elongate is one of the most important functional parameters characterizing the muscle-tendon unit's responsibility for producing complicated, co-ordinated movement patterns. The ability of the MTU to elongate has been given the synonymous terms of 'flexibility' or 'extensibility' (Gleim and Malachy, 1997).

The flexibility of the muscle-tendon unit can be both static or dynamic, with static flexibility relating to the range of motion available at a joint or combination of joints, and dynamic flexibility being the ease of movement within an obtainable range of joint motion (Gleim and Malachy, 1997). Measurements of flexibility are used by clinicians to estimate the ability of the MTU to lengthen. Etiologically, these measurements have been related to the risk of MTU injury with the ubiquitous notion that 'tight' muscles are more likely to be strained (Worrell and Perrin, 1992). The assumption is that a more compliant muscle can be stretched further and to a higher ultimate strain, and is therefore less susceptible to injury (Safran et al., 1988). This has led to the opinion that stretching is important to reduce the risk of MTU injury (Anderson and Burke, 1991; Hubley-Kozey and Stanish, 1990; Smith, 1994; Sports Medicine Australia, 1997), although currently, no substantive evidence exists to support this claim.

Studies investigating muscle-tendon unit flexibility have predominantly used static measurement techniques, which, arguably, are not functional, and have

Skeletal Muscle Mechanics: From Mechanisms to Function. Edited by W. Herzog.

yielded inconclusive results regarding the relationship between flexibility, injury, and performance (Gleim and Malachy, 1997).

In contrast, dynamic flexibility more closely resembles functional activity and, with regard to injury, the measurement of active stiffness may be a more valid measurement than static flexibility, since the MTU is active during the measurement, and the active stiffness determines the effectiveness of force transmission through the MTU. The limitations of this method are that the measurement procedure is more demanding than measuring static flexibility and, although plausible, currently no published study has investigated the relationship between active stiffness and injury.

Passive stiffness is measured by recording joint angle at the same time as passive torque generation (Gadjosik,1991; Gadjosik et al., 1990; McHugh et al, 1996). The measurement, however, involves passive elongation of the MTU, which fails to represent the state of the MTU during dynamic activity. The measurement of active muscle stiffness and its interpretation typically involves the use of phenomenological approaches (Frizen et al., 1969). A related approach has been used to measure active MTU stiffness directly from the biological tissues (Hein and Vain, 1998; Wilson et al., 1991a), in relation to knee-joint stability and active hamstring stiffness (Jennings and Seedhom, 1998; McNair et al., 1992), and for the measurement of joint stiffness (Oatis, 1993).

Attempts have been made to quantify MTU stiffness in relation to its importance in regulating 'normal' movement (Toft et al., 1991; Sinkjaer et al., 1988; Hunter and Kearney, 1982; Kearney and Hunter, 1982), alteration with regards to pathology (Toft, 1995; Sinkjaer, 1997; McNair et al., 1992; Jennings and Seedhom, 1998), and as a predictive factor with regard to injury (Wilson et al., 1991a) involving approaches that attempt to measure the components of MTU stiffness using load displacement analysis (Sale et al., 1982; Blanpied and Smidt, 1993). An alternative approach is to model the MTU as a component of a single degree of freedom system, which, when perturbed from a position of equilibrium by a transient force, will oscillate at its natural frequency. The effective overall stiffness of the system is determined by examining the frequency at which it oscillates (Frizen et al., 1969).

Viscous damping in the system produces an exponential rate of decay in oscillation, the rate being determined by the amount of damping present. Since the MTU has been shown to exhibit stress, relaxation, and creep, the use of a single degree of freedom system with mass (m), spring (k), and viscous damper (c) to model the leg (including the MTU; Figure 14.1) appears to have some validity. The technique, used by a number of authors (McHugh et al., 1992; Taylor et al., 1990; Aruin and Zatsiorsky, 1984; Bach et al., 1983; Oatis, 1993; Jennings and Seedhom, 1998; McNair et al., 1992; McNair and Stanley, 1996; Wilson et al., 1991b, 1992; Wilson et al., 1994), has been shown to achieve stiffness load characteristics similar to those observed in isolated muscle preparations (Cavagna, 1970), and to be valid and reliable (Walshe et al., 1996).

However, protagonists of this approach have assumed that the behaviour of the system is linear.

Using a model relating to the MTU of the calf complex, the purpose of our studies was to investigate the assumption of linearity via a simple non-linear model. The study, based upon pilot results, focuses on dynamic behaviour and effective stiffness of the lower limb using a single degree of freedom model with (a) linear and (b) non-linear approaches on a wide range of subjects, and addresses the effects of heat and stretching routines on experimental data obtained. In the dynamic studies, use is generally made of the equation,

$$k = 4\pi^2 f_d^2 m + \frac{c^2}{4m} \tag{14.1}$$

where *m*, *k*, and *c* are as indicated in Figure 14.1, and f_d is the damped natural free oscillation frequency of the system. A value for damping (*c*) is usually assumed.

LINEAR THEORY AND MODEL

The equation of motion for a linear, single degree of freedom, damped system (Figure 14.1) in free oscillation is:

$$m\ddot{x} + c\dot{x} + kx = 0 \tag{14.2a}$$

or

$$\ddot{x} + 2\zeta\omega_o\dot{x} + \omega_o^2 x = 0 \tag{14.2b}$$

where *x* is the displacement of the mass, ω_o is the undamped natural frequency, and ζ the damping ratio.

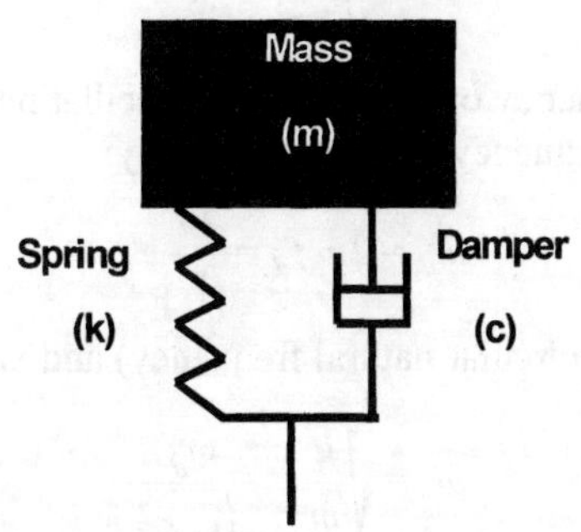

Figure 14.1. Mass, spring, and damper model.

Because of linearity, the differential equation for the force (*F*) transmitted to the 'ground' (e.g., a force plate) through the spring and dashpot elements is evidently:

$$\ddot{F} + 2\zeta\omega_o\dot{F} + \omega_o^2 F = 0 \tag{14.3}$$

If the system is set into free oscillation, the displacement amplitude at the beginning of the n_{th} cycle (X_n) and at the beginning of the $(n+1)^{th}$ cycle (X_{n+1}) can be related by (Rao, 1995):

$$\frac{X_{n+1}}{X_n} = \frac{X_{n+2}}{X_{n+1}} = e^{-\lambda} \tag{14.4}$$

where λ is the logarithmic decrement of the system.

In addition, from Equation 14.3, and because the solution to Equation 14.2a can be expressed in terms of an exponent of a system constant multiplied by time since the initial disturbance,

$$\frac{F_{n+1}}{F_n} = \frac{A_{n+1}}{A_n} = \frac{X_{n+1}}{X_n} = e^{-\lambda} \tag{14.5}$$

therefore

$$\lambda = \ln F_n - \ln F_{n+1} \tag{14.6}$$

where A_n and F_n are the amplitudes of acceleration and transmitted force, respectively. The damping ratio of the system, ζ, is given by:

$$\zeta = \frac{c}{2\sqrt{m\,k}} \tag{14.7}$$

with λ and ζ related by:

$$\lambda = \frac{2\pi\,\zeta}{\sqrt{1-\zeta^2}} \tag{14.8}$$

or

$$\zeta = \frac{\lambda}{\sqrt{4\pi^2 + \lambda^2}} \tag{14.9}$$

In addition, the frequency of a free damped oscillation (f_d), the period (T), and the damped circular frequency (ω_d) are related by:

$$\omega_d = 2\pi\, f_d = \frac{2\pi}{T} \tag{14.10}$$

and ω_o (the undamped circular natural frequency) and ω_d by:

$$\omega_o = \sqrt{\frac{k}{m}} = \frac{\omega_d}{\sqrt{1-\zeta^2}} \tag{14.11}$$

(Note that substitution of Equation 14.7 and Equation 14.10 into Equation 14.11 yields Equation 14.1.)

A typical linear oscillatory trace for the single degree of freedom system shown in Figure 14.1 is shown in Figure 14.2, illustrating time periods (T), force amplitudes (F_1, F_2, and F_3), and exponential decay.

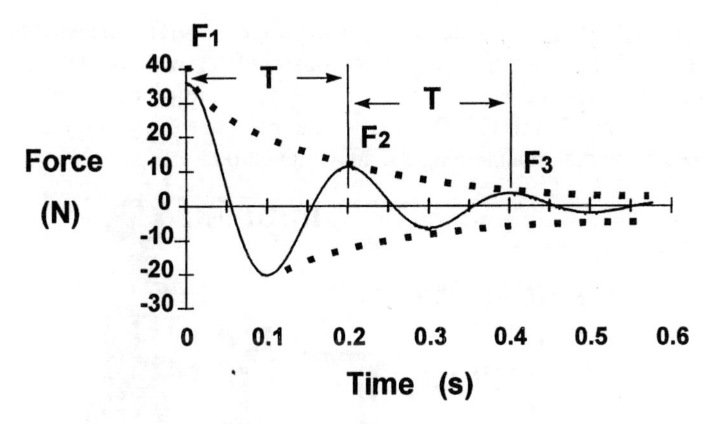

Figure 14.2. Linear oscillation trace.

EXPERIMENTAL PROCEDURE

Free vibration tests were performed on the right legs of a total of 45 consenting subjects (21 females and 24 males) with an age range of 19-48. Each subject was randomly allocated into one of three groups: control, stretch, or heat.

The control group subjects were tested, and then re-tested after 10 minutes.

The stretch group subjects were tested, after which stretching protocol based on the work of Taylor et al. (1990) was undertaken. Ten 30-second sustained stretches were performed by each subject, held at the point of mild discomfort, with 30-second rest periods between each stretch. The first set of five stretches involved ankle dorsiflexion with the knee flexed to bias the stretch towards the soleus. The second set of five stretches involved ankle dorsiflexion with the knee in extension to bias the stretch towards the gastrocnemius. The subjects were retested after the 10-minute stretching protocols.

The heat group subjects were tested, after which the subjects lay prone with the lower leg exposed and a thermister attached over the gastrocnemius muscle. Hydrocollatot steam packs (Chattanooga Group Inc.) were applied to cover the lower leg from the Achilles tendon to the heads of the gastrocnemius, the packs in turn being covered by 2 cm of towelling. After a period of 10 minutes, the packs were removed and the subjects retested. Skin temperature varied between 37°C (initial contact) to 43°C at the end of the time period.

METHOD

The experimental model (Figure 14.1) and set-up (Figure 14.3) was adopted from Shorten (1987), with the calf complex modelled as a single degree of freedom

(SDOF) mass, spring, and damper system (Figure 14.1). Each subject was placed in a sitting position (knee at approximately 90°), with the ball of the foot placed on a wooden block, the block itself being centrally located on the force plate (Kistler, type 9281B12). The foot was kept as near horizontal as possible, with the calf complex maintaining the ankle in neutral.

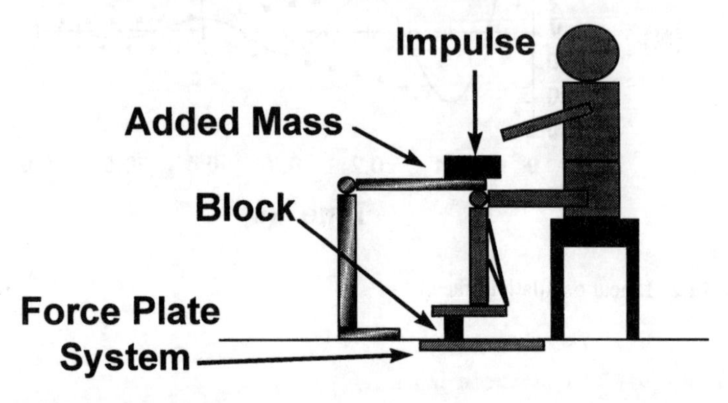

Figure 14.3. Experimental set-up.

A platform to carry the applied masses was subsequently placed on top of the knee. The subjects were asked to maintain a steady-state muscle contraction and not to react to any stimulus, then a small downward impulse was applied manually to the top of the knee to set the leg into oscillation, with the resulting signal captured by a Bioware data-acquisition system at a sampling rate of 500 Hz. A total of seven masses were used on each subject in increments of 2.5 kg, starting with 1.3 kg (the mass of the supporting platform) up to a maximum of 16.3 kg. The tests were performed four times for each subject (twice before and twice after the heat and stretching routines), resulting in a total of 1260 tests. The sequence of the tests was randomized in an attempt to alleviate the possibility of continuous incremental muscle loading that could influence the results. Since the ISB co-ordinate system configuration had been used, the F_y (vertical) force component was required and extracted from the resulting raw data for each test.

DATA ANALYSIS

The signals of vertical force (F_y) against time recorded from the force plate were imported into a spreadsheet in reduced format, and Fourier analysis was performed using a custom-written macro program. Frequencies above 25 Hz were rejected in the reconstructed signal to filter out the noise (predominantly at 100 Hz and above). Examples of signals before and after filtering are shown in Figure 14.4.

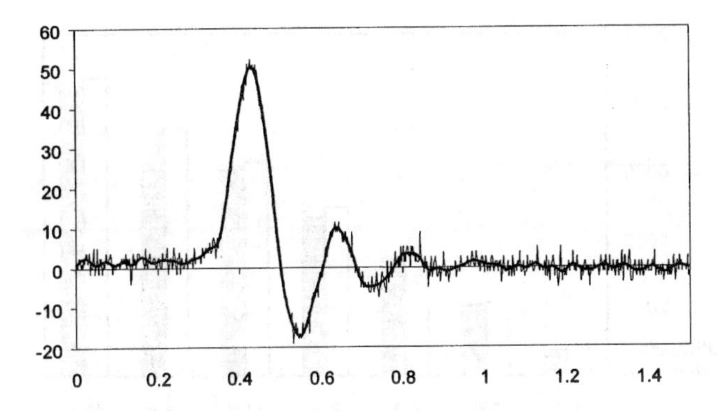

Figure 14.4. Signals before and after filtering and Fourier analysis.

After calculating the mean level of the datum signal before oscillation and subtracting the values of F_1 and F_2, the time period (T) between the first two peaks was determined. Use of Equation 14.6 and Equation 14.9 for one full cycle enabled the logarithmic decrement and damping ratio to be found, with the damped and undamped circular natural frequencies determined using Equation 14.10 and Equation 14.11, respectively.

RESULTS

Stiffness values (mean) against added mass for the three groups based on the first oscillation cycle are shown in Figure 14.5, Figure 14.6, and Figure 14.7 for pre- and post-intervention data. Table 14.1 and Table 14.2 give pre- and post-values for damping ratios and natural circular frequencies for each group and added mass.

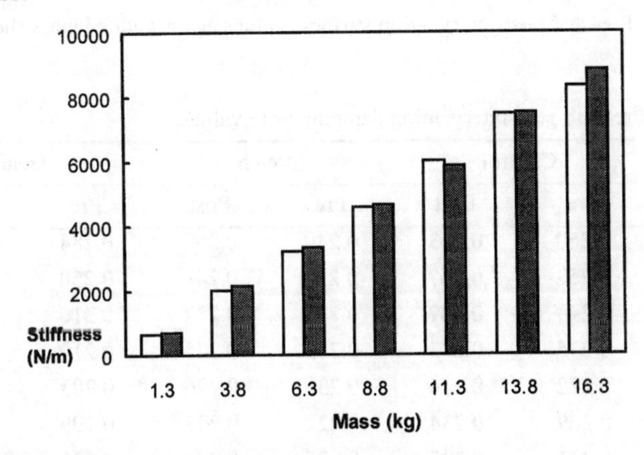

Figure 14.5. Pre- and post-intervention stiffness values against added mass (control group).

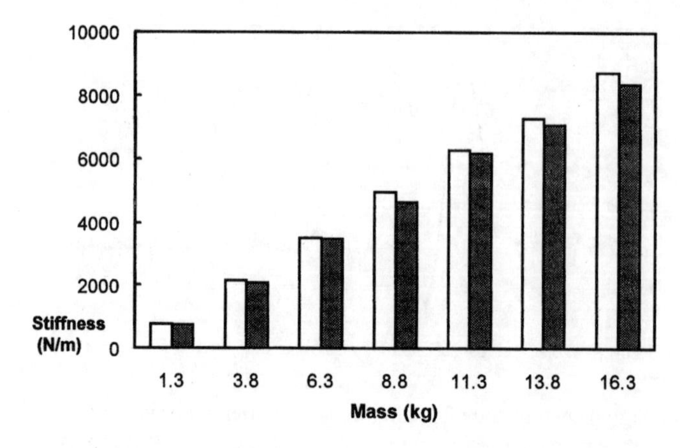

Figure 14.6. Pre- and post-intervention stiffness values against added mass (stretch group).

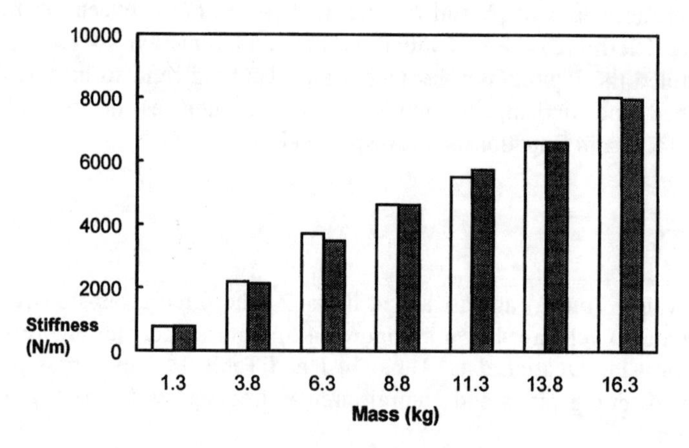

Figure 14.7. Pre- and post-intervention stiffness values against added mass (heat group).

Table 14.1. Pre- and post-intervention damping ratio values.

ζ	Control		Stretch		Heat	
Mass	Pre	Post	Pre	Post	Pre	Post
1.3	0.260	0.263	0.240	0.267	0.284	0.294
3.8	0.252	0.267	0.235	0.248	0.250	0.254
6.3	0.244	0.247	0.234	0.232	0.316	0.230
8.8	0.235	0.243	0.233	0.225	0.219	0.240
11.3	0.222	0.246	0.236	0.226	0.203	0.201
13.8	0.209	0.234	0.222	0.222	0.199	0.209
16.3	0.211	0.225	0.205	0.223	0.196	0.185

Table 14.2. Pre- and post-intervention natural frequencies.

ω_0	Control		Stretch		Heat	
Mass	Pre	Post	Pre	Post	Pre	Post
1.3	22.157	23.105	24.324	24.074	24.542	24.570
3.8	23.062	23.821	23.798	23.358	24.108	23.743
6.3	22.611	23.048	23.618	23.557	23.769	23.541
8.8	22.828	23.012	23.812	23.012	22.952	23.004
11.3	23.101	22.805	23.612	23.392	22.113	22.570
13.8	23.284	23.352	23.002	22.702	21.897	21.925
16.3	22.637	23.279	23.182	22.700	22.221	22.121

The results show an increase in stiffness with increased added mass, however the results showed no obvious difference between pre- and post-intervention values, or for that matter among groups. This is also true for the natural undamped circular frequency data given in Figure 14.9. The values for the damping indicate a decreasing trend with mass increment, but no significant differences among groups or pre- and post- values were found.

Further inspection of the experimental data based on successive oscillation cycles indicated a decrease in the time period (and hence increase in undamped natural circular frequency) per cycle. This trend was evident for all sets of experimental data. Figure 14.8, Figure 14.9, and Figure 14.10 show the mean pre-intervention stiffness values plotted against added mass over two successive cycles (K1-2 and K2-3) for the three groups. Figure 14.11 shows a typical experimental trace.

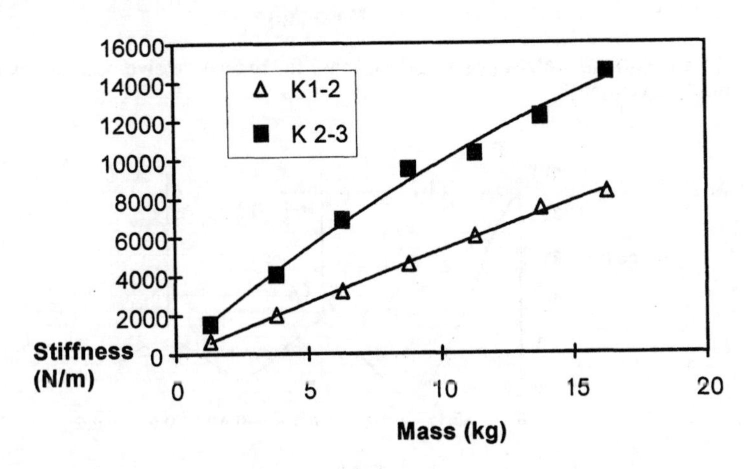

Figure 14.8. Stiffness values against added mass for two successive oscillations (control group, pre-intervention).

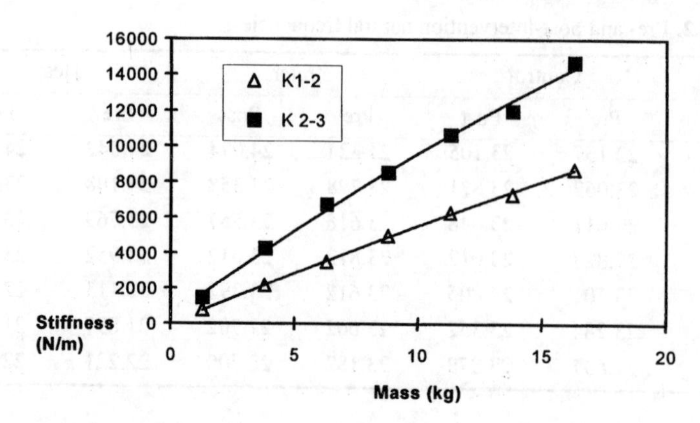

Figure 14.9. Stiffness values against added mass for two successive oscillations (stretch group, pre-intervention).

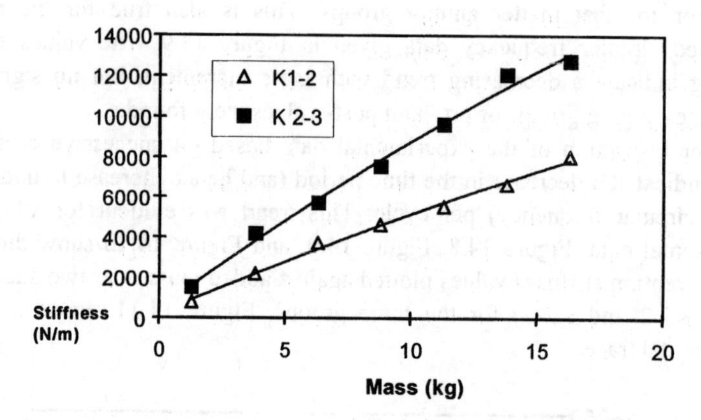

Figure 14.10. Stiffness values against added mass for two successive oscillations (heat group, pre-intervention).

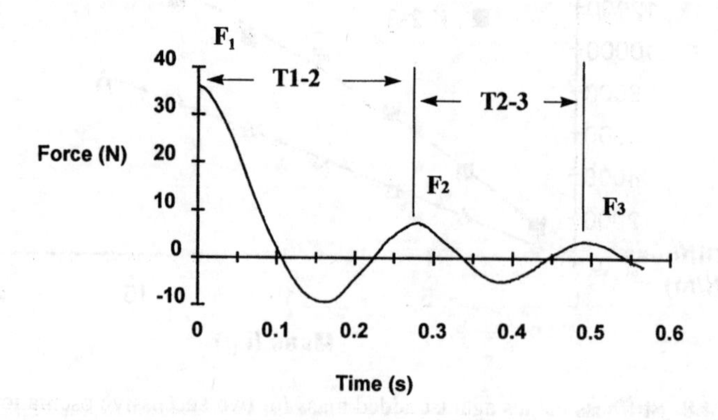

Figure 14.11. A typical experimental trace.

NON-LINEAR MODEL

The original linear model was amended in an attempt to simulate the non-linear behaviour observed in the experimental data. Addition of a term (βF^3) to Equation 14.3 results in an equation which serves as a model for systems with cubic non-linearities:

$$\ddot{F} + 2\zeta\omega_o \dot{F} + \omega_o^2 F + \beta F^3 = 0 \qquad (14.12)$$

Following the approach of Rao (1995), an approximate solution for Equation 14.12 (a form of Duffing's equation), and associated expressions can be obtained. The solution is as that for the linear case (Equation 14.3), with the exception that ω is now the circular damped *amplitude-dependent* natural frequency given by:

$$\omega^2 = \omega_0^2(1-\zeta^2) + \frac{\beta}{2}F^2 \qquad (14.13)$$

The values of ω_o^2 and β may be estimated by solving a pair of simultaneous equations for the i_{th} and $(i+1)_{th}$ oscillation cycles once ζ has been estimated as in the linear case. Hence:

$$\omega_i^2 = \omega_0^2(1-\zeta^2) + \frac{\beta}{2}F_i^2 \qquad (14.14a)$$

$$\omega_{i+1}^2 = \omega_0^2(1-\zeta^2) + \frac{\beta}{2}F_{i+1}^2 \qquad (14.14b)$$

SIMULATION

Equation 14.3 and Equation 14.12 were used to simulate linear and non-linear free oscillation of the lower leg, respectively, for comparison with experimental data. Time domain solutions for the differential equations were obtained by use of a custom-written routine incorporating a fourth-order Runge-Kutta algorithm. Measured experimental parameters (F_1, F_2, F_3 and $T_{1\text{-}2}$, $T_{2\text{-}3}$) were used to calculate model parameters ζ, ω_o, and β (Table 14.3). These model parameters were subsequently substituted into Equation 14.3 and Equation 14.12 (linear and non-linear models, respectively). The resulting plots and model data are shown in Figure 14.12 and Table 14.3, respectively.

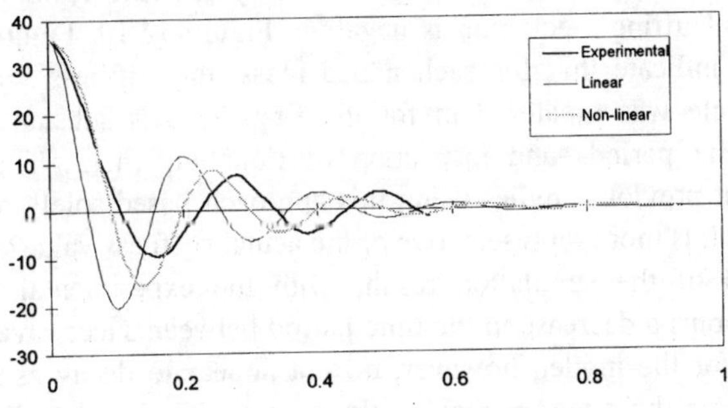

Figure 14.12. Comparison of experimental with linear and non-linear results.

Table 14.3. Model data and experimental results.

Parameter (Units)	Experimental Data	Linear Model	Non-linear Model
F_1 (N)	35.938	35.938	35.938
F_2 (N)	7.443	11.365	8.394
F_3 (N)	3.533	3.601	2.622
$T_{1\text{-}2}$ (s)	0.280	0.204	0.242
$T_{2\text{-}3}$ (s)	0.212	0.204	0.200
ζ		0.18	0.18
ω_0 (rad/s)		31.890	31.890
β ($N^{-2}s^{-2}$)	/	/	−0.606

DISCUSSION

The results for the linear model based on Equation 14.2a and Equation 14.2b show identical time periods, and hence undamped natural circular frequencies for successive cycles.

If the results for the non-linear simulation are compared with the results for the linear model, it is evident from Figure 14.12 and from Table 14.3 that the time periods vary from cycle to cycle. This is in contrast to the linear model. Moreover, the non-linear simulation illustrates the effect of the sign of the non-linearity parameter, β. With β<0 ('softening' spring), the time period between successive oscillations decreases with cycle number, hence natural frequency increases. (A 'softening' spring is one that has decreasing stiffness with increasing deformation amplitude. Conversely, a positive value of β denotes a 'stiffening' spring, which exhibits the opposite behaviour.)

In order to maintain the leg in the test position as the added mass increased, values for stiffness increased with increments in added mass. The experimental trace (Figure 14.12) exhibits trends typical of all 1260 experimental data sets. Table 14.1 shows T decreasing with successive cycle number, implying that the value of β in Duffing's equation is negative. Figure 14.10, Figure 14.11, and Figure 14.12 indicate that for each added mass, the stiffness for the second oscillation cycle was greater than for the first. This is substantiated by the decreasing time periods and inspection of Equation 14.12. This raises the question about previous studies using this approach based solely on first cycle analysis. Which is more representative of the actual stiffness value?

Comparison of the simulation results with the experimental data reveals similar behaviour; a decrease in the time period between successive cycles. The time periods for the model, however, do not appear to decay as rapidly with cycle number as the experimental results. Comparison of data in Table 14.3

shows that the force amplitudes measured experimentally and estimated mathematically are very similar. The experimental data and non-linear simulations also exhibit similar asymmetry within single cycles *uncharacteristic* of viscous-damped oscillation of a linear system.

Comparison of the three plots in Figure 14.12 indicates that the non-linear model is indeed an improvement on the linear one. Thus, it appears that the use of a non-linear modelling approach outlined here may be a useful method for assessing the mechanical stiffness of the MTU, and that the oscillations produced are representative of the contractile, non-contractile, and reflex components of the MTU. Using this non-linear approach, further studies are required to determine which of these components may contribute to the non-linearity and to the 'softening spring' effect that has been demonstrated in this study.

The failure to reveal significant stiffness value differences between the control, stretch, and heat groups (pre- and post-intervention), was something of a mystery, since previous authors have found such differences. This may be due to the oscillation method inducing protective servo-motor components into the system, which are unaffected by stretching and heat. These reflex components require further investigation to validate the model. Another possibility is that the added masses were not sufficiently large for detection of parameter changes, even though the largest of these was approximately 30% MVC. With respect to the argument of how functionally relevant the adopted test procedure was, the landing of a subject on the ball of the foot when running would produce a similar oscillatory motion through the calf complex.

CONCLUSIONS

The free oscillation data exhibit non-linear characteristics, implying that the current approach is an improvement on the previous use of a linear model for estimation of effective muscle stiffness. The approach provides an additional (non-linearity) parameter for use in studying lower-leg behaviour and the condition of muscle-tendon unit groups. However it is not known what this parameter represents in terms of the MTU. Further studies are required to investigate the physiological origins of the phenomenon observed.

The free oscillation data obtained experimentally exhibit non-linear characteristics, implying that previous use of a linear expression for estimation of effective muscle stiffness and lower-leg behaviour is unrealistic. The use of Duffing's equation appears to replicate more closely the behaviour of the lower leg under free oscillation conditions, albeit some differences in the reduction of time periods for successive cycles persist. Although the approach adopted could be considered a 'hybrid' one and needs refining, the method shows promise in the monitoring and modelling of lower-leg behaviour.

FUTURE WORK

Further studies should be undertaken to confirm how well Duffing's equation (with displacement and force as the dependent variables) describes lower-leg oscillation. Additional studies are being undertaken to investigate the sensitivity of the method with respect to detection of changes in parameters such as β and ζ. A difficulty with using this approach is that the experimental data obtained represent the behaviour of the whole lower-leg system, including contractile components, non-contractile components, stretch reflex, and numerous control systems, in addition to different forms of damping (viscous, coulombic, and hysteretic). The authors are currently refining the experimental method in terms of accuracy of data collection, isolation of individual muscle groups, and capture of both displacement and force data.

A further study has just been completed aimed at investigating the relationship between static flexibility and active stiffness in the lower limb. The aim of this study was to investigate this relationship in the lower limb, using primarily the soleus, deep posterior tibial, and peroneal muscles, with the purpose of identifying whether measures of static flexibility can be used to yield useful information about the active stiffness of these lower-limb muscles.

REFERENCES

Anderson, B., and Burke, E.R. (1991). Scientific, medical and practical aspects of stretching. *Clinics in Sports Medicine* **10**, 63–86.

Aruin, A.S., and Zatsiorsky, V.M. (1984). Biomechanical characteristics of human ankle joint muscles. *European Journal of Applied Physiology* **52**, 400–406.

Bach, T.M., Chapman, A.E., and Calvert, T.W. (1983). Mechanical resonance of the human body during voluntary oscillations about the ankle joint. *Journal of Biomechanics* **16**, 85–90.

Blanpied, P., and Smidt, G.L. (1993). The difference in stiffness of the active plantarflexors between young and elderly human females. *Journal of Gerontology* **48**(2), M58–M63.

Cavagna, G.A. (1970). Elastic bounce of the body. *Journal of Applied Physiology* **29**, 279–282.

Frizen, M., Magi, M., Sonnerup, L., and Viidik, A. (1969). Rheological analysis of soft collagenous tissue. Part 1, Theoretical considerations. *Journal of Biomechanics* **12**, 13–20.

Gadjosik, R.L. (1991). Passive compliance and length of clinically short hamstring muscles of healthy men. *Clinical Biomechanics* **6**, 239–244.

Gadjosik, R.L., Giuliani, C.A., and Bohanon, R.W. (1990). Passive compliance and length of the hamstring muscles of healthy men and women. *Clinical Biomechanics* **5**, 23–29.

Gleim, G.W., and Malachy, P.M. (1997). Flexibility and its effects on sports injury and performance. *Sports Medicine* **24**(5), 289–299.

Hein, V., and Vain, A. (1998). Joint mobility and the oscillation characteristics of muscle. *Scandinavian Journal of Medical Science in Sport* **8**, 7–13.

Hubley-Kozey, C.L., and Stanish, W.D. (1990). Can stretching prevent athletic injuries? *Journal of Musculo-skeletal Medicine* **7**, 21–31.

Hunter, I.W., and Kearney, R.E. (1982). Dynamics of human ankle stiffness: Variation with mean ankle torque. *Journal of Biomechanics* **15**(10), 747–752.

Jennings, A.G., and Seedhom, B.B. (1998). The measurement of muscle stiffness in anterior cruciate injuries: An experiment revisited. *Clinical Biomechanics* **13**(2), 138–140.

Kearney, R.E., and Hunter, I.W. (1982). Dynamics of human ankle stiffness: Variation with displacement amplitude. *Journal of Biomechanics* **15**(10), 753–756.

McHugh, M.P., Kremenic, I.J., Fox, M.B., et al. (1996). The relationship of linear stiffness of human muscle to maximum joint range of motion. *Medicine and Science in Sports and Exercise* **28**(5) Supplement, S77.

McHugh, M.T., Magnusson, S.P., Gleim, G.W., and Nicholas, J.A. (1992). Viscoelastic stress relaxation in human skeletal muscle. *Medicine and Science in Sports and Exercise* **24**, 1375–1382.

McNair, P.J., and Stanley, S.N. (1996). Effect of passive stretching and jogging on the series elastic muscle stiffness and range of motion of the ankle joint. *British Journal of Sports Medicine* **30**, 313–318.

McNair, P.J., Wood, G.A., and Marshall, R.N. (1992). Stiffness of the hamstring muscles and its relationship to function in anterior cruciate ligament deficient individuals. *Clinical Biomechanics* **7**, 131–137.

Oatis, C.A. (1993). The use of a mechanical model to describe the stiffness and damping characteristics of the knee joint in healthy adults. *Physical Therapy* **73**(11), 740–749.

Rao, S.S. (1995). *Mechanical Vibrations*. Addison-Wesley Publishing Company, ISBN 0-201-59289-4, pp 770–775.

Safran, M.R., Garrett, W.E., and Seaber, A.V. (1988). The role of warm-up in muscular injury prevention. *American Journal of Sports Medicine* **16**(2), 123–129.

Sale, D., Quinlan, J., Marsh, E., McComas, A.J., Belanger, A.Y. (1982). Influence of joint position on ankle plantarflexion in humans. *Journal of Applied Physiology* **52**, 1636–1642.

Shorten, M.R. (1987). Muscle elasticity and human performance. *Medicine and Sports Science* **25**, 1–18.

Sinkjaer, T. (1997). Muscle, reflex and central components in the control of the ankle joint in healthy and spastic man. *Acta Neurologica Scandinavica, Supplementum* **170**(96), 1–21.

Sinkjaer, T., Toft, E., Andreassen, S., and Hornemann, B.C. (1988). Muscle stiffness in human ankle dorsiflexors: Intrinsic and reflex components. *Journal of Neurophysiology* **60**(3), 1110–1121.

Smith, C.A. (1994). The warm-up procedure: To stretch or not to stretch. A brief review. *Journal of Orthopaedic and Sports Physical Therapy* **19**, 12–17.

Sports Medicine Australia (1997). Safety guidelines for children in sport and recreation. Sports Medicine Australia, Canberra.

Taylor, D.C., Dalton, J.D., Seaber, A.V., and Garrett, W.E. (1990). Viscoelastic properties of muscle tendon units. *American Journal of Sports Medicine* **18**, 300–309.

Toft, E. (1995). Mechanical and electromyographic stretch responses in spastic and healthy subjects. *Acta Neurologica Scandinavica, Supplementum* **163**(92), 1–24.

Toft, E., Sinkjaer, T., Andreassen, S., Larsen, K. (1991). Mechanical and electromyographic responses to stretch of the human ankle extensors. *Journal of Neurophysiology* **65**(6), 1402–1410.

Walshe, A.D., Wilson, G.J., and Murphy, A.J. (1996). The validity and reliability of a test of lower body musculotendinous stiffness. *European Journal of Applied Physiology* **73**, 332–339.

Wilson, G.J., Elliot, B.C., and Wood G.A. (1992). Stretch shorten cycle performance enhancement through flexibility training. *Medicine and Science in Sports and Exercise* **24,** 116–123.

Wilson, G.J., Murphy, A.J., and Pryor, J.F. (1994). Musculotendinous stiffness: Its relationship to eccentric isometric and concentric performance. *Journal of Applied Physiology* **76**, 2714–2719.

Wilson, G.J., Wood, G.A., and Elliott, B.C. (1991a). The relationship between stiffness of the musculature and static flexibility: An alternative explanation for the occurrence of muscular injury. *International Journal of Sports Medicine* **12**, 403–407.

Wilson, G.J., Wood, G.A., and Elliott, B.C. (1991b). Optimal stiffness of the series elastic component in a stretch shorten activity. *Journal of Applied Physiology* **70**, 825–833.

Worrell, T.W., and Perrin, D.H. (1992). Hamstring muscle injury: The influence of strength, flexibility, warm-up, and fatigue. *Journal of Orthopaedic and Sports Physical Therapy* **16**, 12–18.

III *In Vivo* Muscle Function (Human)

15 Considerations on *In Vivo* Muscle Function

W. HERZOG
Human Performance Laboratory, University of Calgary, Calgary, Alberta, Canada

The study of skeletal muscle is rewarding in many aspects. Muscle's active contractility has inspired biophysicists and physiologists to study aspects of molecular force production, to contemplate the transfer of chemical energy into work and heat, and to investigate how muscles function within the living system. Neurophysiologists have considered muscles as the actuators of the brain and have determined the activation patterns to gain insight into the control of voluntary movements. Biomechanists have studied the mechanical properties of muscles with the aim to optimize movement performance in sport and work situations, to simulate limb or whole body movement, and to understand the basic principles underlying force production.

Since the emergence of the cross-bridge theory (Huxley, 1957), research in muscle mechanics and muscle physiology has been focussed on frog single-fibre preparations. It is probably fair to state that more is known about the physiology and mechanics of single fibres from frogs than any other muscle preparation. By comparison, little research has been done on determining the mechanical properties of *in vivo* human or animal skeletal muscle. Many of the results obtained from frog single fibres have been transferred directly to *in vivo* muscle properties so, naturally, the question arises whether such extrapolations from single fibres to whole (*in vivo*) muscles are appropriate and justified.

TRANSFER OF PROPERTIES ACROSS STRUCTURAL LEVELS

It can be shown on an intuitive level that extrapolation across organizational levels in a system may give disastrous results. Imagine, for example, that you are asked to calculate the deflection of a uniform rectangular beam subjected to a constant force. Undergraduate beam theory (a 'systems' approach) gives a good solution quite readily. Imagine now that beam theory is not available, and the same problem has to be solved using quantum mechanics exclusively (a 'molecular' approach). Needless to say, the problem would become incomprehensibly complex, and a final answer would likely be inaccurate at best.

Similarly, in muscle mechanics, we must ask ourselves continuously whether it is valid to apply findings from single actin-myosin interactions or from single-fibre

Skeletal Muscle Mechanics: From Mechanisms to Function. Edited by W. Herzog.

experiments to whole muscle. Obviously, there are distinct differences in the experiments performed at different structural levels. For example, in single fibres, all sarcomeres are activated simultaneously by a pair of electrodes running along the fibre. During voluntary contractions of a whole muscle, fibres are innervated at a specific point—the neuromuscular junction—and it takes a discrete amount of time until all sarcomeres of the fibre are activated sequentially. Is this difference, and the many others that exist between artificial activation and voluntary contraction, important? I would like to consider this question using the force-length property of skeletal muscle.

Likely, the most often cited (and used) force-length property of skeletal muscle is that obtained by Gordon et al. (1966). In this experiment, single fibres from frog striated muscle were stimulated supramaximally and isometrically using a 'sarcomere clamp' (a way to keep sarcomeres within a specific region of the fibre at uniform and constant length).[1] The resulting force-length relationship is shown below in Figure 15.1.

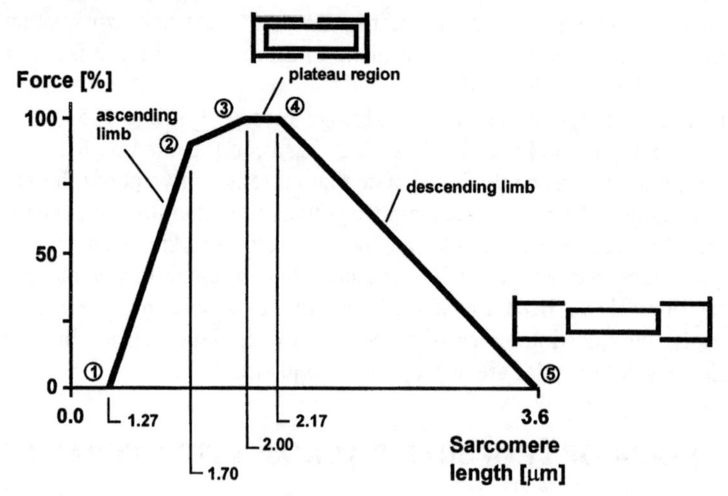

Figure 15.1. Sarcomere force-length relationship for frog striated muscle. Myofilament lengths in frog are approximately 1.6 μm (thick) and 0.95 μm (thin). The bare zone in the middle of the thick (myosin) filament is thought to be about 0.17 μm, and the Z-band width is approximately 0.1 μm. (Adapted from Gordon et al., 1966, and reproduced with the permission of The Physiological Society.)

Whole muscle force-length relationships, however, are distinctly different from that obtained by Gordon et al. (1966). In a classic study, Rack and Westbury (1969)

[1] However, since the 'sarcomere clamp' technique works on a feedback system, sarcomere lengths are nearly, but not absolutely, constant when using this technique, and the effect of the resulting small sarcomere length perturbations on force production are not known. It has always been assumed, possibly incorrectly, that the approach by Gordon et al. (1966) gave the exact isometric sarcomere force-length relationship.

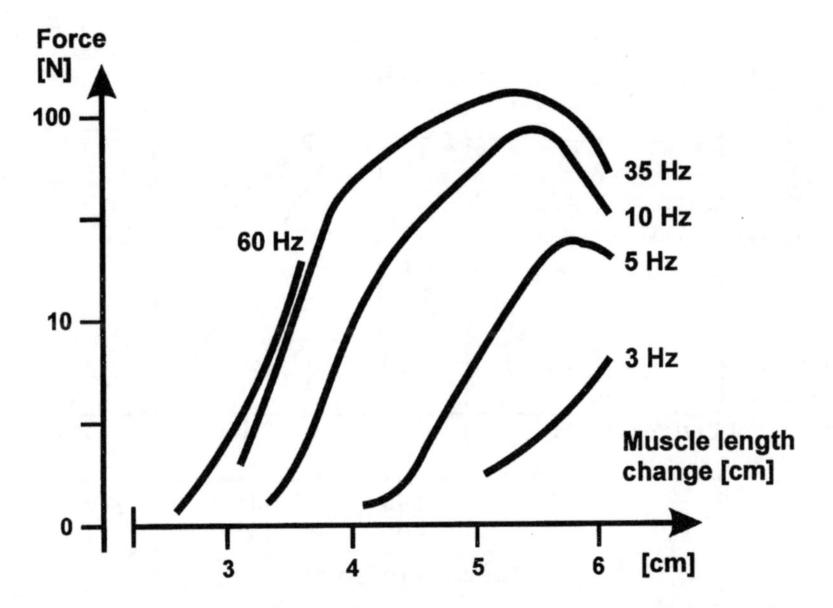

Figure 15.2. Force versus change in length relationship for whole cat soleus, stimulated at the frequencies shown in the figure. Note the non-linear change in shape at different stimulation frequencies, and the loss in force and increase in optimal length with decreasing stimulation frequencies. (Adapted from Rack and Westbury, 1969, and reproduced with the permission of The Physiological Society.)

determined the force-length property of whole cat soleus (Figure 15.2). There are many interesting differences between the single fibre (sarcomere) and the whole muscle force-length relationship, some that have been explicitly pointed out, and others that have not.

Compared to the frog sarcomere force-length relationship, the whole soleus force-length relationship is smoother; its optimal length does not occur at the expected plateau region (based on cat actin-myosin lengths, Herzog et al., 1992b), and optimal length varies as a function of activation: it is shifted towards increasing lengths for decreasing levels of activation. Also, much of the descending limb of the force-length relationship is missing. The smoothness of the whole-muscle force-length relationship is likely associated with structural non-uniformities, for example, non-uniform sarcomere lengths within a fibre and non-uniform average sarcomere lengths across fibres, thus removing the sharp corners that one would expect in a single-sarcomere force-length relationship. It has been known for a long time that sarcomere length non-uniformities exist in muscle fibres (Huxley and Peachey, 1961), and that these non-uniformities, if not controlled, as done by Gordon et al. (1966), produce a fibre force-length relationship (Ramsey and Street, 1940) that is quite different from the sarcomere force-length relationship (Figure 15.3).

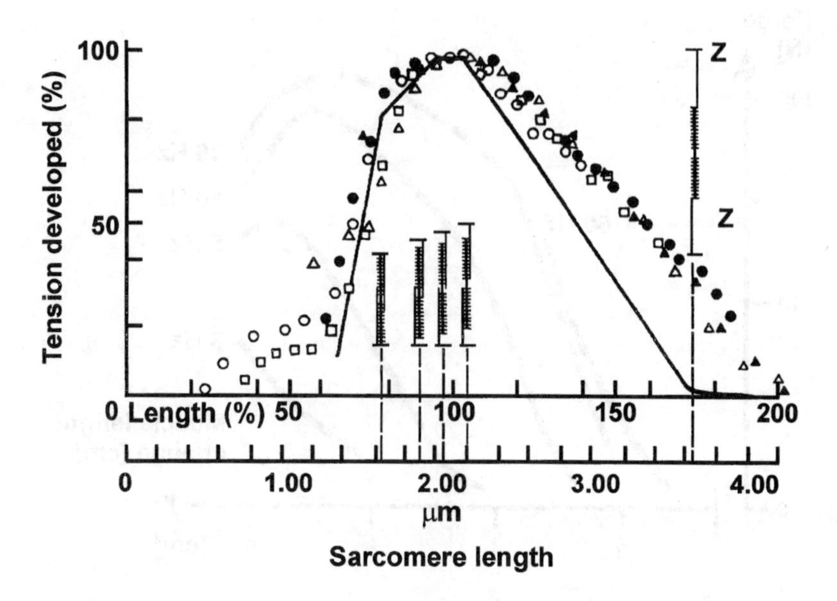

Figure 15.3. Sarcomere force-length relationship for frog striated muscle, as in Figure 15.1. However, in Figure 15.1, sarcomere lengths were controlled to remain isometric and at approximately uniform lengths within a mid-segment of the fibre, whereas in this figure, no attempt was made to correct for sarcomere length, i.e., the relationship was obtained by fixing the ends of the fibres. (Adapted from Ramsey and Street, 1940, and reproduced with the permission of Wiley-Liss, Inc.)

The whole-muscle force plateau of cat soleus occurs at a greater average sarcomere length than one would expect, based on the cross-bridge theory (Huxley, 1957). Actin and myosin filament lengths were shown to be 1.6 μm and 1.12 μm, respectively, in cat skeletal muscle (Herzog et al., 1992b), therefore the theoretical plateau should be at an average sarcomere length of about 2.34 μm to 2.51 μm, whereas, experimentally, it was found to occur at sarcomere lengths of around 3.0 μm. This result has been explained with the idea that sarcomere length non-uniformities might change the length at which the plateau occurs in whole muscle, as it had been shown in single fibres (see Pollack, 1990, for a review). Furthermore, the shift of the plateau region of the whole-muscle force-length relationship to unexpected lengths has been associated with the idea that individual fibres in a muscle might be at different average sarcomere lengths, thereby decreasing the absolute maximal force that can be reached by a muscle, but increasing its active range of force production and shifting (potentially) the optimal length to unexpectedly large values. Also, muscles deform considerably during contraction (as illustrated later in this book in chapters 16-19), therefore fibre lengths and the corresponding average sarcomere lengths may have been overestimated when muscle deformations were not at all, or not fully, accounted for.

The optimum length of cat soleus was found to increase for decreasing levels of activation, i.e., decreasing frequencies of stimulation. This shift in optimal length has been associated with a length dependence of Ca^{2+} sensitivity, and the idea that average cross-bridge forces and rate constants of cross-bridge attachment and detachment may vary as a function of thick and thin interfilament spacing (which in turn depends directly on fibre length). For a complete discussion of length-dependent activation and force production, please refer to chapter 5 of this book.

Single-fibre force-length relationships are typically given for the entire active range of force production (Gordon et al., 1966). This representation may give the impression that muscles operate over the entire range of their force-length relationship *in vivo*. However, this is not the case. For example, cat soleus, gastrocnemius, and plantaris appear to function primarily on the ascending limb and plateau of the force-length relationship (Herzog et al., 1992a). Frog semimembranosus was said to occupy mainly the plateau region (Lutz and Rome, 1993), and frog semitendinosus was found to work primarily on the descending limb of the force-length relationship during normal everyday activities (Mai and Lieber, 1990). Therefore, when considering the contractile properties of *in vivo* muscle, it is important to know on what part of the force-length relationship a muscle works when constrained within the anatomically possible configurations of the musculoskeletal system.

Summarizing, it is rather difficult to take properties of single fibres, such as the force-length relationship discussed above, and transfer them into an *in vivo* whole-muscle system.

IN VIVO WHOLE MUSCLE PROPERTIES

One area of research that has received great attention in recent years is the determination of mechanical properties of whole, *in vivo* muscles. These properties are the force-length relationship, the force-velocity relationship, and the history dependence of force production. When attempting to determine these properties, researchers have invariably encountered some basic problems: (a) muscles typically work in functional groups (e.g., the knee extensors), so it becomes difficult to isolate the function and force (moment) contribution of an individual muscle from its synergist group; (b) muscles are known to deform greatly during contraction, therefore the determination of the contractile conditions of the force-producing elements (the muscle fibres) is difficult. For example, it has been shown that during an 'isometric' contraction of a muscle (the whole muscle-tendon unit length remains constant), the corresponding muscle fibres may shorten up to 30% of their initial length (Kawakami et al., 1998; Narici et al., 1996) and the shortening speeds may be considerable; (c) fibres are typically not arranged parallel to the long axis of the muscle. It has been shown that the angle a fibre forms relative to the long axis of the muscle may change by 25-30° in some cases, reaching angles of pinnation >60° during contraction (Kawakami et al., 1998; Carvalho et al., 1999), and this realignment of the contractile material causes problems in the analysis of the contractile properties. Moreover, when calculating muscle forces from fibre forces, it has typically been assumed that the two are related by the cosine of the angle of pinnation in such a way that internal force constraints (for example, those arising due to isovolumetricity) are negligible (Otten, 1988). However, these constraint forces cannot be neglected, as has been shown theoretically (Epstein and Herzog, 1998) and experimentally (Herzog et al., submitted).

In order to overcome some of the difficulties associated with determining the properties and functions of *in vivo* muscles during voluntary movements, a series of experimental techniques has been developed and successfully applied. Some of these techniques are described in detail in the following chapters. For example, ultrasound imaging has become a standard tool for observing structural deformations of contracting muscles. Chapters 16-19 are devoted to research using ultrasound imaging. These studies show the power of this technique when determining the contractile conditions and functional properties during *in vivo* muscle contraction, but they also demonstrate the limitations: measurements during normal movements are hard to obtain, and the three-dimensional observation of muscle deformation remains a time-intensive endeavour. Although the measurement of individual tendon forces during unrestrained movements in animals has become a standard procedure (Herzog and Leonard, 1991; Walmsley et al., 1978; Abraham and Loeb, 1985; Hodgson, 1983), individual muscle-force measurements in human muscles are still rare. In chapter 25, such measurements obtained in human finger tendons are shown.

A powerful approach aimed at elucidating the function of individual muscles *in vivo* is the combination of direct and simultaneous muscle force measurement, EMG recording, and muscle-fibre length measurement from a series of muscles during unrestrained movements. We have started making such combined measurements in the cat soleus, medial gastrocnemius, and tibialis anterior. Although only in its preliminary stage (we have succeeded in making complete measurements in three animals for a variety of movement conditions), the results are promising. Below is one exemplar result from a cat trotting on a treadmill at a nominal speed of 1.4 m/s (Figure 15.4). Further results from these types of measurements for fish are presented in chapter 27 of this book.

Although much of the work in muscle mechanics in the past 50 years has been focussed on (frog) single-fibre preparations, we must be careful to not extrapolate the fibre data directly to the whole muscle. Since 1994, exciting techniques have been used to determine the mechanics and biochemistry of single cross-bridge attachments and ATP hydrolysis, respectively. Although this work is of utmost importance and, hopefully, will provide many answers to questions about the molecular interaction and force production between thick (myosin) and thin (actin) filaments and the corresponding breakdown of ATP that provides the energy for the work produced by these molecular motors, work on the single cross-bridge level will not provide many answers about the mechanics of whole-muscle force production, properties, and function. *In vivo* whole-muscle mechanics has been

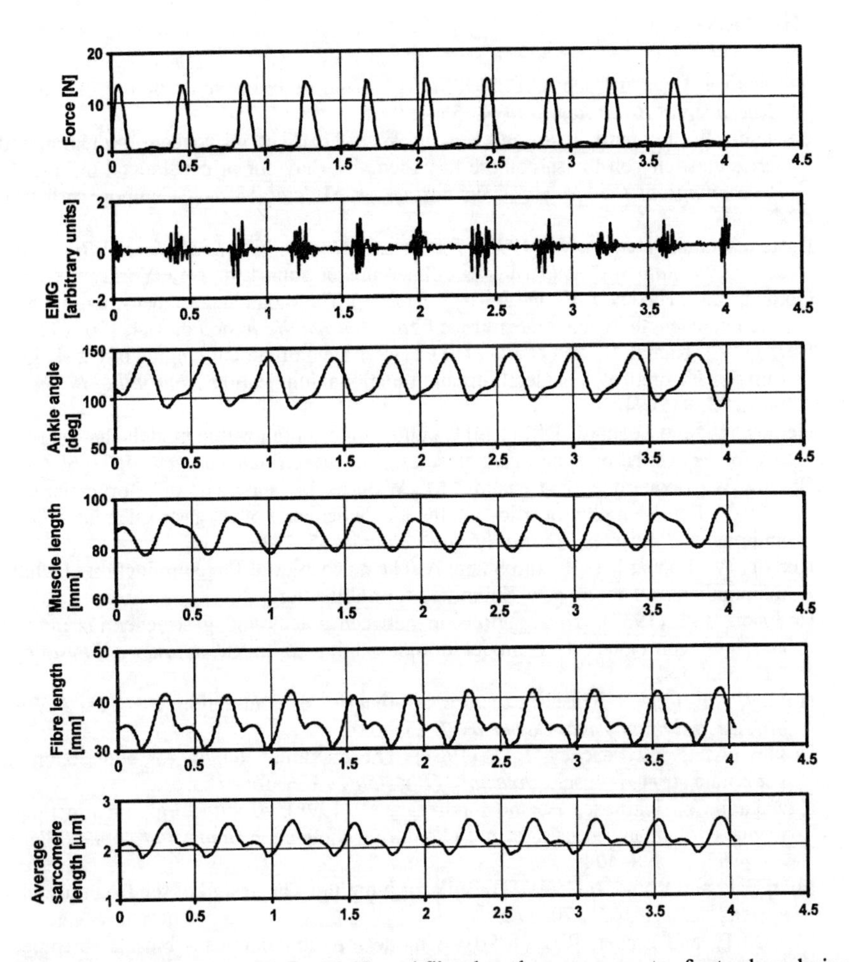

Figure 15.4. Direct *in vivo* force, EMG, and fibre-length measurements of cat soleus during unrestrained locomotion (exemplar results of 10 consecutive steps from one animal trotting at 1.4 m/s). The directly measured results are supplemented by measurements of the ankle angle (from high-speed video), muscle lengths (from post-sacrifice tendon-travel measurements), and average sarcomere lengths (from post-sacrifice sarcomere-number measurements using a laser diffraction technique).

neglected for many years, likely because of the difficulties associated with the determination of muscle forces and fibre contractile properties. However, these difficulties have been largely overcome in the past few years, so that now the time may have come where whole, human, *in vivo* muscle mechanics can make major contributions to our understanding of human (and animal) movement mechanics and control. The excellent contributions in parts III and IV of this book give testimony to the great and exciting possibilities that advances in technology have brought.

REFERENCES

Abraham, L.D., and Loeb, G.E. (1985). The distal hindlimb musculature of the cat. *Experimental Brain Research* **58**, 580–593.

Carvalho, W., Leonard, T. R., and Herzog, W. (1999). The influence of pinnation and series elasticity on the sarcomere force-length behaviour of cat skeletal muscle. In *Proceedings of the Canmore Symposium on Skeletal Muscle*. Canmore, Alberta, Canada.

Epstein, M., and Herzog, W. (1998). *Theoretical Models of Skeletal Muscle: Biological and Mathematical Considerations*. John Wiley & Sons, Ltd., New York.

Gordon, A.M., Huxley, A.F., and Julian, F.J. (1966). The variation in isometric tension with sarcomere length in vertebrate muscle fibres. *Journal of Physiology* **184**, 170–192.

Herzog, W., Kamal, S., and Clarke, H.D. (1992b). Myofilament lengths of cat skeletal muscle: Theoretical considerations and functional implications. *Journal of Biomechanics* **25**, 945–948.

Herzog, W., and Leonard, T.R. (1991). Validation of optimization models that estimate the forces exerted by synergistic muscles. *Journal of Biomechanics* **24**, 31–39.

Herzog, W., Leonard, T.R., Renaud, J.M., Wallace, J., Chaki, G., and Bornemisza, S. (1992a). Force-length properties and functional demands of cat gastrocnemius, soleus, and plantaris muscles. *Journal of Biomechanics* **25**, 1329–1335.

Herzog, W., Leonard, T. R., and Stano, A. The mechanics of force production in whole mammalian skeletal muscle. Submitted for publication.

Hodgson, J.A. (1983). The relationship between soleus and gastrocnemius muscle activity in conscious cats: A model for motor unit recruitment? *Journal of Physiology* **337**, 553–562.

Huxley, A.F. (1957). Muscle structure and theories of contraction. *Progress in Biophysics and Biophysical Chemistry* **7**, 255–318.

Huxley, A.F., and Peachey, L.D. (1961). The maximum length for contraction in vertebrate striated muscle. *Journal of Physiology* **156**, 150–165.

Kawakami, Y., Ichinose, Y., and Fukunaga, T. (1998). Architectural and functional features of human triceps surae muscles during contraction. *Journal of Applied Physiology* **85**(2), 398–404.

Lutz, G.J., and Rome, L.C. (1993). Built for jumping: The design of the frog muscular system. *Science* **263**, 370–372.

Mai, M.T., and Lieber, R.L. (1990). A model of semitendinosus muscle sarcomere length, knee, and hip joint interaction in the frog hindlimb. *Journal of Biomechanics* **23**, 271–279.

Narici, M.V., Binzoui, T., Hiltbrand, E., Fasel, J., Terrier, F., and Cerretelli, P. (1996). *In vivo* human gastrocnemius architecture with changing joint angle at rest and during graded isometric contraction. *Journal of Physiology* **496**(1), 287–297.

Otten, E. (1988). Concepts and models of functional architecture in skeletal muscle. *Exercise and Sport Science Reviews* **16**, 89–137.

Pollack, G.H. (1990). *Muscles and Molecules: Uncovering the Principles of Biological Motion*. Ebner and Sons, Seattle, Washington, USA.

Rack, P.M.H., and Westbury, D.R. (1969). The effects of length and stimulus rate on tension in the isometric cat soleus muscle. *Journal of Physiology* **204**, 443–460.

Ramsey, R.W., and Street, S.F. (1940). The isometric length-tension diagram of isolated skeletal muscle fibers of the frog. *Journal of Cellular Composition* **15**, 11–34.

Walmsley, B., Hodgson, J.A., and Burke, R.E. (1978). Forces produced by medial gastrocnemius and soleus muscles during locomotion in freely moving cats. *Journal of Neurophysiology* **41**, 1203–1215.

16 *In Vivo* Mechanics of Maximum Isometric Muscle Contraction in Man: Implications for Modelling-based Estimates of Muscle Specific Tension

CONSTANTINOS N. MAGANARIS
Scottish School of Sport Studies, University of Strathclyde, Glasgow, Scotland

VASILIOS BALTZOPOULOS
TEFAA, University of Thessaly, Karyes, Trikala, Greece
Department of Sport and Exercise Science, Manchester Metropolitan University, Alsager, England

SUMMARY

In the present study, we estimated the specific tension of the human soleus (SOL) and tibialis anterior (TA) muscles, taking into account muscle-specific joint moment and contraction-specific musculoskeletal geometry measurements, and we tested the hypothesis that conventional estimates of specific tension based on net joint moment and resting/cadaver-based musculoskeletal geometry are unrealistic. Measurements were taken in six healthy males at the neutral anatomical ankle position, with the knee flexed at 60°. Dynamometry-based measurements of maximum isometric plantarflexion and dorsiflexion joint moments were taken upon maximum voluntary contraction (MVC) and by means of maximal-voltage percutaneous tetanic electrical stimulation. Pennation angles were measured *in vivo* during MVC using ultrasonography. Moment arms were measured *in vivo* at rest and during MVC using sagittal-plane magnetic resonance imaging (MRI). Fibre lengths were measured *in vivo* at rest using ultrasonography. Muscle volume estimations were derived from axial-plane MRI. Muscle physiological cross-sectional area (PCSA) estimates were derived by dividing the estimated muscle volumes by the respective fibre lengths. Tendon forces were calculated from the moment equilibrium equation around the ankle joint. Muscle forces were calculated from tendon forces employing a planar geometric muscle model with no angulation between aponeuroses and tendons. Specific tensions were estimated by dividing the calculated muscle forces by the respective PCSAs. Two approaches were followed

Skeletal Muscle Mechanics: From Mechanisms to Function. Edited by W. Herzog.

to estimate muscle specific tension. First, joint moments during electrical stimulation, pennation angles and moment arms during MVC, and *in vivo* measured fibre lengths were taken into account. Using the traditional approach, net MVC joint moments, moment arms at rest, and cadaveric pennation angle and fibre length data were taken into account. The first approach yielded TA and SOL specific tensions of 23.3 and 17.8 N·cm^{-2} ($P<0.01$), respectively, which are in line with previous animal-based reports. The difference of 30% found between TA and SOL is in line with the reported difference of 50% in specific tension between type II and type I muscle fibres. In contrast, using the second, traditional approach, the TA specific tension was found to be 530% higher than that of SOL (65.8 versus 10.4 N·cm^{-2}, $P<0.01$), a difference which is unrealistic and cannot be explained by mechanical or physiological considerations. These results show the importance of accurate parameter input in musculoskeletal modelling applications in order to have realistic and valid simulations of human movement.

INTRODUCTION

It is generally accepted that a major determinant of the muscle force-generating capacity is muscle size. The maximum muscle force normalized to muscle size is an index of the intrinsic muscle strength, and it is known as maximal muscle stress or specific tension. Accurate information on muscle specific tension is essential for accurate predictions of the force and joint moment generating potential of human muscle in modelling applications (e.g., Zajac, 1989; Loren and Lieber, 1995).

HISTORICAL BACKGROUND AND CURRENT PROBLEMS

Reports on human muscle specific tension estimation appeared fairly early in the scientific literature. In 1846, Weber estimated the ankle plantarflexors' specific tension from the maximum weight that a man could lift with the calf muscles, gross anthropometric estimates of moment arms around the ankle joint, and cadaver-based estimates of muscle size in terms of cross-sectional area (Weber, 1846). Experiments employing similar methodologies followed by Henke (1865) on wrist flexors and ankle plantarflexors, Hermann (1898) and Reys (1915) on ankle plantarflexors, von Recklinghausen (1920) on finger flexors, Franke (1920) on elbow flexors and extensors, Haxton (1944) on ankle plantarflexors, Morris (1948) on knee flexors and extensors and elbow flexors and extensors, and Hettinger (1961) on elbow flexors. Ikai and Fukunaga (1968, 1970) were the first to measure muscle force by means of strain gauges, and muscle cross-sectional area *in vivo* by means of ultrasound to derive specific tension estimates for the elbow flexors. The above studies yielded values for specific tension between 37 and 107 N·cm^{-2}, indicating that the intrinsic muscle strength may be muscle-specific.

Surprisingly, however, there are large interstudy differences in specific tension estimates not only across muscles, but also within a given muscle (Table 16.1, for ankle plantarflexors). Differences in defining and measuring maximum force and expressing muscle size may be the reasons for this inconsistency. In some studies, maximum force has been considered as the recorded maximum isometric force exerted externally by the studied segment (e.g., Young et al., 1984; Davies et al., 1988), while in other studies, concentric joint moments were used to quantify muscle output (e.g., Kanehisa et al., 1996). With respect to muscle size, either muscle physiological cross-sectional area (PCSA) (e.g., Fukunaga et al., 1996; Narici et al., 1996) or anatomical cross-sectional area (ACSA) (e.g., Narici et al., 1989; Kanehisa et al., 1996) has been used when estimating muscle specific tension.

Table 16.1. Summarized reports in the literature on ankle plantarflexor specific tension.

Referencc	Force measurements	CSA estimations	Specific tension ($N \cdot cm^{-2}$)
Fukunaga et al., 1996	External moment → muscle force	PCSA from MRI and cadaveric values	10.8
Narici et al., 1996	External force → muscle force	PCSA from MRI and *in vivo* measures	9.7 *
Davies et al., 1986	External force	ASCA from anthropometry	32.9
Haxton, 1944	External force → muscle force	PCSA from cadavers	38.3
Reys, 1915	External force → muscle force	ACSA from cadavers	54.9
Hermann, 1898	External force → muscle force	ACSA from cadavers	62.8
Weber, 1846	External force → muscle force	ACSA from cadavers	41.2

* gastrocnemius medialis muscle
Notice the discrepancy in defining and measuring force and muscle cross-sectional area (CSA) between studies. ACSA = anatomical cross-sectional area; PCSA = physiological cross-sectional area.

A meaningful expression of the intrinsic force-generating capacity of muscle should be based on the maximum isometric muscle force, which can be estimated from the muscle joint moment, knowing the moment arm and pennation angle of the muscle, and the muscle PCSA, which can be estimated by dividing the muscle volume by the muscle fibre length. Although there are studies in which maximum isometric muscle force and PCSA measurements have been used to estimate specific tension, the validity of the results is questionable, since they have been derived from net resultant joint moments and musculoskeletal geometry measurements either taken at rest or based on cadaveric data. The following factors need to be considered when analyzing maximum isometric loads: a) Antagonist muscles may co-contract during an agonistic voluntary contraction (e.g., Maganaris et al., 1998c) which results in a negative moment relative to the moment generated by the agonists, and a decreased neural activation of the agonists (Tyler and Hutton, 1989). Therefore, muscle-specific moments, as opposed to net resultant joint moments, should be measured. b) Moment arms and pennation angles at rest or from cadaver-based studies do not represent actual values during contraction (Maganaris et al., 1998a, 1998b, 1999; Maganaris and Baltzopoulos, 1999). Therefore, contraction-specific musculoskeletal geometry, as opposed to resting state/cadaver-based data, should be used for calculating musculo-tendon forces. c) Cadaveric muscles may shrink during fixing (Friederich and Brand, 1990; Yamaguchi et al., 1990), and therefore, cadaver-based estimates of PCSA may not represent actual *in vivo* dimensions. *In vivo* PCSA estimates (e.g., Narici et al., 1996) should thus be used in the estimation of specific tension.

The aim of the present study was to examine the effects of the above factors on estimates of specific tension. The hypothesis was that the traditionally followed approach results in unrealistically high muscle specific tension values. In order to test this hypothesis, we estimated the specific tension of human soleus (SOL) and tibialis anterior (TA): a) using muscle-specific joint moment and *in vivo* contraction-specific musculoskeletal geometry, and b) following the traditional approach, using net resultant joint moments and resting/cadaver-based musculoskeletal geometry.

METHODS

Six healthy males, with (mean±SD) age, height, and body mass of 28±4 years, 175±8 cm, and 75±7 kg, respectively, from whom informed consent had previously been obtained, volunteered to participate in this study. All subjects were physically active and none had any history of musculoskeletal injury or any orthopaedic abnormality in the lower extremities. The study was approved by the local ethics committee.

The protocol involved dynamometry, muscle electrical stimulation, ultrasonography, and magnetic resonance imaging (MRI). All measurements were taken on the right leg, at the neutral anatomical ankle position with the tibia at right angles to the foot. In this position: a) no passive forces were detected in the plantarflexion and dorsiflexion directions (Siegler et al., 1984), and b) the ankle muscles are at about their optimal length (Cutts, 1988). It is important to realize that maximal contractile forces in intact muscles should be estimated at optimal fibre lengths, which may not correspond to joint angles where maximum joint moments are generated. At joint angles corresponding to longer or shorter than optimal fibre lengths, submaximal contractile forces (but not necessarily submaximal joint moments) are generated. Thus, comparative data for muscle specific tension estimates at optimal joint angles rather than fibre lengths (e.g., Kawakami et al., 1994; Buchanan, 1995; Fukunaga et al., 1996) should be considered carefully.

The following steps were followed in this study to estimate specific tension (Figure 16.1): 1) measurement of joint moment; 2) estimate of moment arm; 3) calculation of tendon force from *1)* and *2)*; 4) measurement of pennation angle and fibre length; 5) calculation of muscle force from *3)* and *4)*; 6) estimation of PCSA; and 7) calculation of specific tension from *5)* and *6)*.

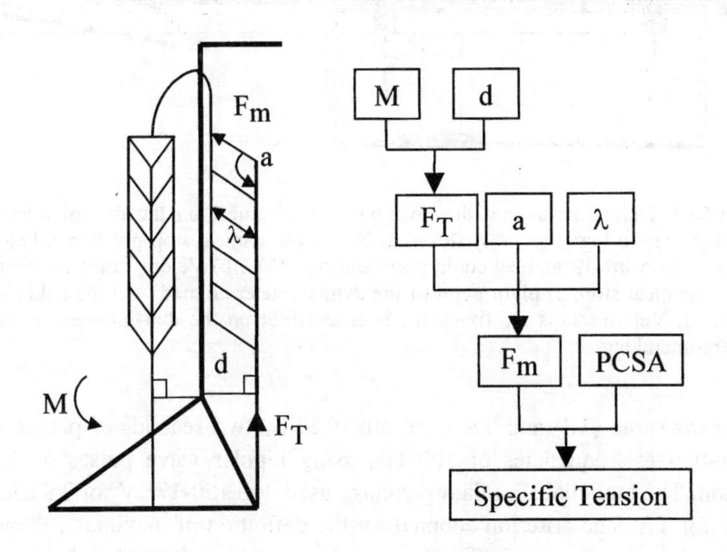

Figure 16.1. Estimate of muscle specific tension. Schematic representation of the two-dimensional muscle model and the steps and parameters used for estimating muscle specific tension.

JOINT MOMENT MEASUREMENTS

Ankle joint moments were measured with the knee of the tested leg flexed at 60° (Figure 16.2). Subjects were securely fixed in the prone position on an isokinetic dynamometer (Lido Active, Loredan Biomedical, Davis, CA). The pivot point of the lever arm of the dynamometer was aligned with the anatomical rotation axis of the ankle joint.

Isometric plantarflexion and dorsiflexion maximum voluntary contraction (MVC) joint moments (the highest out of three efforts) were measured, having compensated for the effect of gravity. Three minutes after the completion of all voluntary efforts, isometric joint moments were generated in the above position by means of percutaneous electrical stimulation. Maximal tetanic contractions of

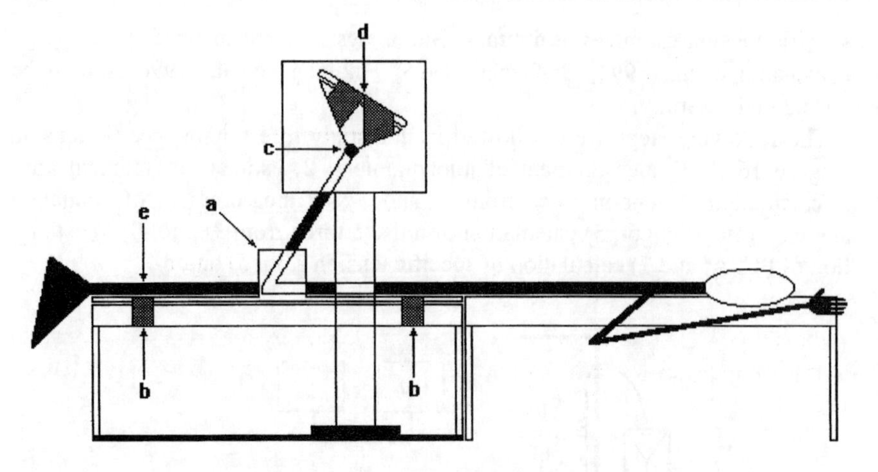

Figure 16.2. Diagram showing the position of the subject on the dynamometer. The knee of the tested leg is securely flexed at 60°. a) mechanical stop positioned below the knee so that no lift of the heel could occur during MVC; b) Velcro straps for fixing the knee mechanical stop; c) pivot point of the dynamometer aligned with the ankle axis of rotation; d) Velcro straps for fixing the heel and foot on the dynamometer foot-plate; e) contra-lateral leg.

the triceps surae (TS) and TA were produced by two seconds of percutaneous stimulation at frequencies of 100 Hz, using bipolar wave pulses of 100 μs duration. The maximal stimulating voltage used was 150-175 V for TS and 125-150 V for TA. The criterion adopted for the definition of maximal voltage was that no further increase in joint moment could be achieved with a voltage increase of 25 V, with plateau values to agree within ±5%. The electrodes consisted of two aluminium foil pads, 10×10 cm for TS and 6×4 cm for TA, covered in paper tissue which was soaked in water. The electrodes for TS were bandaged over the gastrocnemius muscle belly and distal part of SOL muscle belly. The electrodes for TA were bandaged over the proximal and distal halves of the TA muscle belly. Stimuli were delivered by a custom-built high-voltage stimulator controlled by a purpose-developed computer software. Surface EMG signals from the nearby SOL and peroneus tertius muscle during TA stimulation, and TA during TS stimulation indicated minimal or no current leakage. Furthermore, at the adopted knee position of 60°, the bi-articular gastrocnemius muscle is slack, providing negligible contribution to the generated plantarflexion moment (Hof and van den Berg, 1977; Gravel et al., 1987; Herzog et al., 1991). Thus, the plantarflexion and dorsiflexion joint moments recorded in the study were considered to have been generated exclusively by the SOL and TA muscles, respectively.

Joint moment data collection was repeated after three days. There was no difference ($P > 0.05$, student's t-test) in the measurements between tests.

MOMENT ARM ESTIMATES

In vivo moment arm estimates in the sagittal plane were derived from image analysis of MRI scans at rest and during MVC by means of digitization (TDS, Blackburn, UK; Maganaris et al., 1998a, 1999). Sagittal-plane MRI scans (G.E. Signa Advantage, Milwaukee 1.5 T/64 Hz; Fast-GRASS scanning with: 15 ms repetition time, 6.7 ms echo time, 24 cm field of view, 1.0 excitation, 256×128 matrix, 5 mm slice thickness, and 2 s scanning time) of the foot were taken at rest and during a series of plantarflexion and dorsiflexion MVCs at ankle angles corresponding to 15° of dorsiflexion, neutral ankle position, and 15° of plantarflexion. The ankle joint complex was represented by the tibio-talar joint and the orientation of the Achilles tendon was considered representative of that of the SOL tendon. In the MRI at the neutral ankle position, the Achilles and TA tendon action lines were identified, and from the MRIs at 15° of dorsiflexion and 15° of plantarflexion, the centre of rotation in the tibio-talar joint at the neutral ankle position was calculated using the Reuleaux method (Reuleaux, 1875). The perpendicular distance from the centre of rotation to the tendon action line was considered as the muscle-tendon unit moment arm (Figure 16.3 and Figure 16.4).

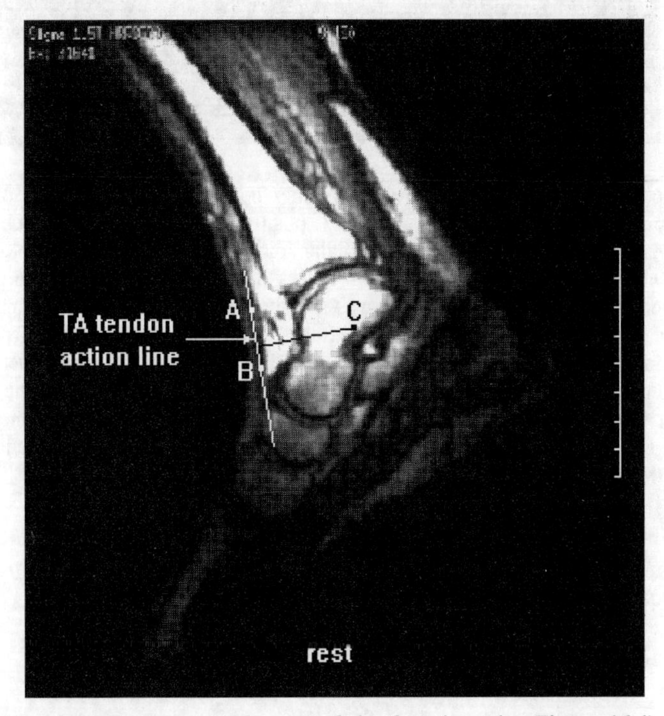

Figure 16.3. Magnetic resonance images of the foot (see also Figure 16.4). Sagittal-plane MRI of the foot at rest. (Reproduced from Maganaris et al., 1999, with the permission of Elsevier Science.)

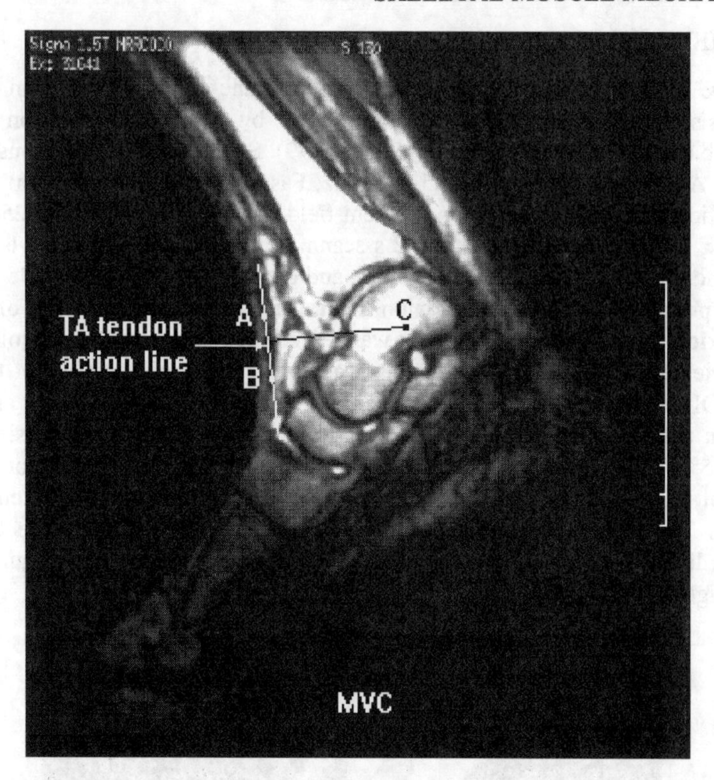

Figure 16.4. Magnetic resonance images of the foot. Sagittal-plane MRIs of the foot during dorsiflexion MVC at the neutral ankle position in one of the tested subjects. C is the instant centre of rotation in the tibio-talar joint, calculated from the ankle rotation $-15° \rightarrow +15°$ using the Reuleaux method. Points A and B on the TA tendon action line are markers whose horizontal position was measured in relation to a common reference point for quantifying the displacement of the tendon in the transition from rest to dorsiflexion MVC. The perpendicular black line from C to the TA tendon action line is the TA tendon moment arm. Notice the displacement of A, B, and C in the transition from rest to dorsiflexion MVC, which results in an increased TA tendon moment arm during dorsiflexion MVC as compared with rest. (Reproduced from Maganaris et al., 1999, with the permission of Elsevier Science.)

TENDON FORCE CALCULATIONS

Tendon forces were calculated from the moment equation:

$$F_T = M \cdot d^{-1} \tag{16.1}$$

where F_T is the tendon force, M is the joint moment measured, and d is the muscle-tendon unit moment arm (Figure 16.5).

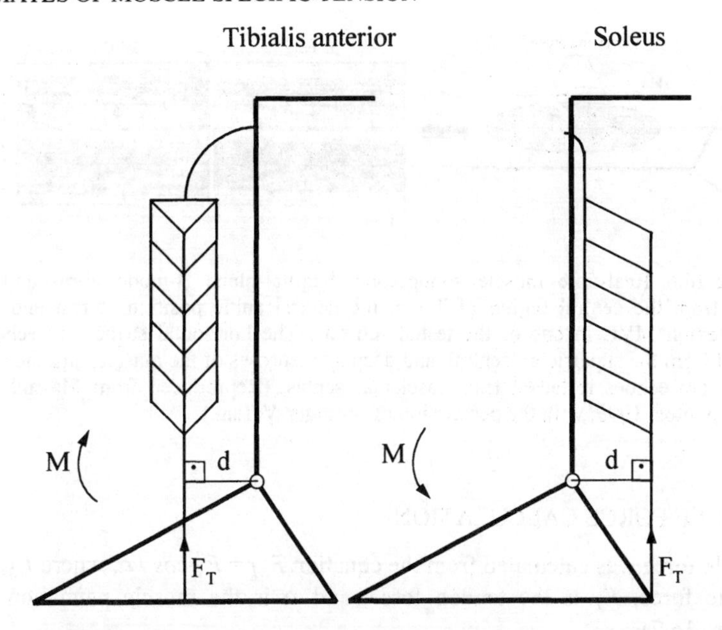

Figure 16.5. The lower extremity musculoskeletal model used for calculating tendon forces. The foot is considered a rigid body neglecting inter-tarsal, tarso-metatarsal, metatarso-phalangeal, and inter-phalangeal joints. According to this model, $M = F_T{\cdot}d$, where M is the moment generated around the ankle joint, F_T is the force acting on the tendon, and d is the moment arm.

FIBRE LENGTH AND PENNATION ANGLE MEASUREMENTS

Muscle geometry characteristics were estimated from image analysis of sagittal-plane sonographs taken with a 7.5 MHz linear-array B-mode probe (Esaote Biomedica AU3 Partner, Florence). Sonographs were taken from the central region along the muscle mid-sagittal axis where architectural characteristics were representative of those taken from different sites along and across the muscles (Maganaris et al., 1998b; Maganaris and Baltzopoulos, 1999). Scanning was performed at rest and during MVC at the neutral ankle position. Muscle fascicle length at rest was considered to represent fibre length, and it was measured as the length of the straight line drawn along the echoes parallel to the fascicles from the deep to the superficial aponeuroses. The muscle pennation angle was measured during MVC as the angle between the fascicles and the aponeuroses echoes. Pennation angle measurements were taken at the fascicle insertions into the superficial and deep aponeuroses of the muscle, and the two values were averaged. For TA, which is symmetrically bi-pennate (Figure 16.6), fibre length and pennation angle of each uni-pennate part were measured separately and average values across the two halves were used for further analyses.

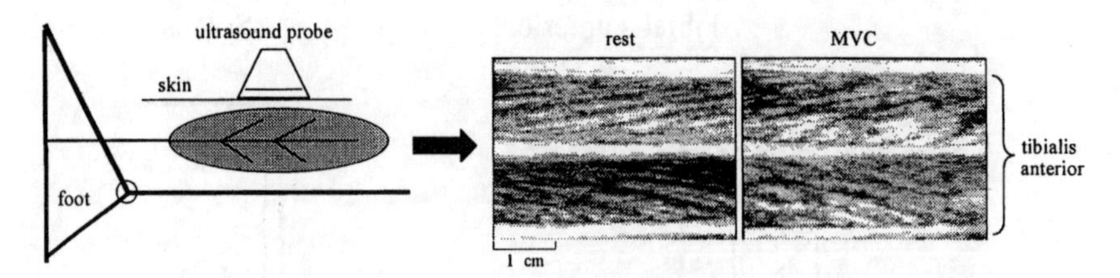

Figure 16.6. Real-time muscle sonographs. Sagittal-plane B-mode ultrasound scans taken from the central region of TA at the neutral ankle position at rest and during dorsiflexion MVC in one of the tested subjects. The horizontal stripes are echoes reflected from the superficial, central, and deep aponeuroses of the muscle, and the oblique stripes are echoes reflected from fascicular septas. (Reproduced from Maganaris and Baltzopoulos, 1999, with the permission of Springer-Verlag.)

MUSCLE FORCE CALCULATION

Muscle force was calculated from the equation $F_M = F_T \cdot cos^{-1} \alpha$, where F_M is the muscle force, F_T is the tendon force, and α is the muscle pennation angle (Figure 16.7).

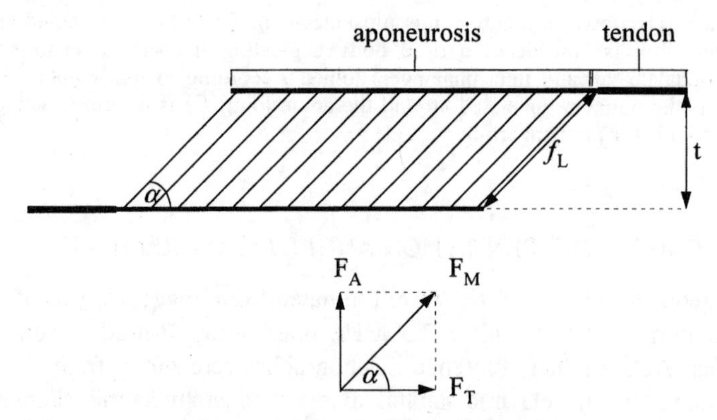

Figure 16.7. The planimetric muscle model used in the study for calculating muscle forces. The model represents SOL and one of the two uni-pennate halves of TA. Notice that aponeuroses and tendons act on the same line. According to the vectorial analysis of forces: $F_M = F_T \cdot cos^{-1} \alpha$, where F_M is the muscle force during contraction, F_T is the tendon force, and α is the pennation angle. F_A is the perpendicular to the aponeurosis muscle force component, f_L is the muscle fibre length, and t is the muscle thickness.

PCSA ESTIMATES

Estimates of PCSA were obtained from the equation $PCSA = m \cdot \rho^{-1} \cdot f_L^{-1}$, where m is the muscle mass, ρ is the muscle density, and f_L is the muscle fibre length (see Alexander and Vernon, 1975; Narici et al., 1992). As the product $m \cdot \rho^{-1}$ yields

the muscle volume, $PCSA = V \cdot f_L^{-1}$, where V is the muscle volume. Muscle volumes were estimated from a series of continuous axial-plane MRI scans (G.E. Signa Advantage, Milwaukee; 1.5 T/64 Hz, Fast Spin Echo scanning with: 500 ms repetition time, 18 ms effective echo time, 1 excitation, 256×128 matrix, field of view 48 cm, 10 mm slice thickness, 0 mm inter-slice gap, and 37 s scanning time) from the proximal to the distal end of the muscle (Fukunaga et al., 1992). All MRIs were taken at rest in the supine position with the knee of the scanned leg fully extended. In each slice, the ACSA of SOL and TA were digitized, and the respective muscle volumes were calculated as the product of the sum of all slices' ACSA, multiplied by the MRI slice thickness.
SOL was treated as a uni-pennate muscle, neglecting its relatively small portio-anterior bi-pennate part (see also Alexander and Vernon, 1975; Fukunaga et al., 1992). Tibialis anterior was treated as two separate equidimensional uni-pennate parts, each one occupying half the whole muscle volume. The whole TA PCSA was then obtained by adding the PCSAs of the two uni-pennate halves (Narici et al., 1992).

All digitizing measurements and image analyses were performed three times by the same investigator, and average values were used for further analysis.

SPECIFIC TENSION CALCULATION

Muscle specific tension was calculated as the ratio of muscle force to PCSA using two data sets. First, joint moments during electrical stimulation, *in vivo* pennation angles and moment arms during MVC, and *in vivo* fibre lengths were taken into account (data set A). Following the traditional approach, muscle stress was calculated using net MVC joint moments, cadaver-based pennation angles and fibre lengths, and resting moment arms (data set B). In this data set, average cadaveric pennation angles of 25° for SOL and 5° for TA reported by Wickiewicz et al. (1983) were used. Cadaver-based estimates of muscle fibre length were taken by assuming that the muscle fibre length to muscle length ratio is constant for a given muscle across different cadavers (Kawakami et al., 1994; Fukunaga et al., 1996). Average muscle fibre length to muscle length ratios of 0.06 and 0.25 reported by Wickiewicz et al. (1983) were used for SOL and TA, respectively. Muscle length was measured *in vivo* as the distance between the most proximal and the most distal axial-plane MRIs in which the muscle was visible (Kawakami et al., 1994; Fukunaga et al., 1996).

STATISTICS

Values are presented as means±SD. Differences in specific tension estimates between a) SOL and TA using a given data set, and b) different data sets for given muscle, were tested using two-way ANOVA. Statistical difference was set at a level of $P<0.05$.

RESULTS

The net isometric plantarflexion and dorsiflexion MVC joint moments were 121.5±5.8 and 44.7±3.6 Nm, respectively. The maximum isometric plantarflexion and dorsiflexion moments recorded during electrical stimulation were 107±7.3 (88% of net plantarflexion MVC) and 22.5±1.8 Nm (50% of net dorsiflexion MVC), respectively. SOL and TA pennation angles during MVC were 44.3±3.5 and 19.7±3.6°, respectively. SOL and TA fibre lengths were 3.5±0.3 and 7.3±0.4 cm, respectively. SOL and TA PCSA were 141.5±13.8 and 21.0±1.1 cm^2, and 264±14.9 and 20.3±1.2 cm^2 when using *in vivo* and cadaver-based fibre lengths, respectively. The Achilles and TA tendon moment arms increased from 4.9±0.4 to 6.0±0.3 cm and from 3.4±0.3 to 4.9±0.4 cm in the transition from rest to MVC. The estimated Achilles and TA tendon forces were 1776±118 and 458±22 N when using electrically elicited joint moments and moment arms during MVC, respectively, and 2470±176 and 1324±83 N when using net MVC joint moments and resting moment arms, respectively. The estimated SOL and TA muscle forces were 2500±260 and 489±28 N, respectively, when using *in vivo* pennation angles during MVC, and 2745±195 and 1337±83 N, respec-tively, when using cadaveric pennation angles. The estimated SOL and TA specific tensions were 17.8±3.0 and 23.3±1.8 $N \cdot cm^{-2}$, when using data set A, and 10.4±0.7 and 65.8±4.2 $N \cdot cm^{-2}$, respectively, when using data set B (Figure 16.8).

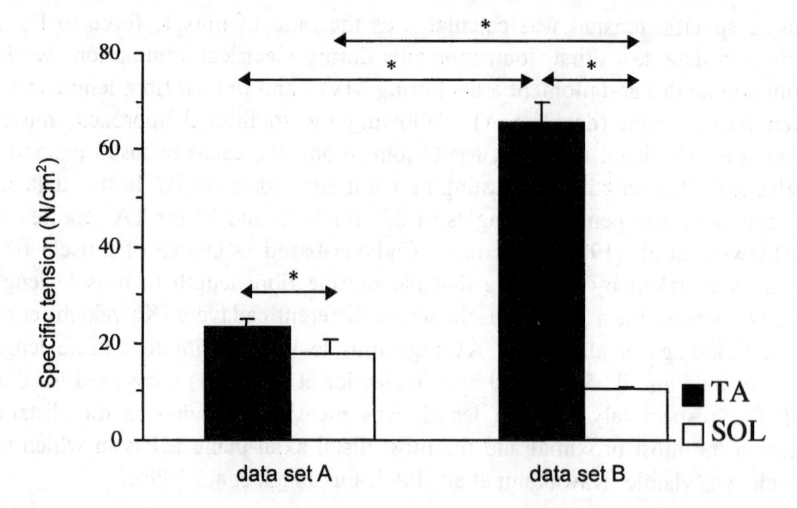

Figure 16.8. Muscle specific tension estimates. Estimates using data set A were derived using joint moments elicited from electrical stimulation of a single muscle, *in vivo* measured pennation angles and moment arms during MVC, and *in vivo* measured fibre lengths. Estimates using data set B were derived following the traditional approach, i.e., using net MVC joint moments, cadaver pennation angles and fibre lengths, and moment arm measurements at rest. Values are means±SD. (* denotes $P<0.01$ between the values indicated.)

Using data set A, the estimated specific tension in TA was higher by 30% ($P<0.01$) compared with SOL. Using data set B, the estimated specific tension in TA was higher by 530% ($P<0.01$) compared with SOL. Using data set B for SOL resulted in an underestimation of specific tension by 42% ($P<0.01$) as compared with data set A. Using data set B for TA resulted in overestimation of specific tension by 84% ($P<0.01$) as compared with data set A.

DISCUSSION

In the present study, three factors were considered for estimating muscle-specific tension: antagonistic co-activation during MVC; changes in musculo-skeletal geometry from rest to MVC; and *in vivo* estimates of PCSA. Each is discussed in the sections that follow.

ANTAGONISTIC CO-ACTIVATION DURING MVC

Any agonist muscle belongs to a group of synergists that is not the sole contributor to the mechanical output generated across the joint studied. In fact, the net joint moment measured during voluntary contraction represents less than the actual moment-generating capacity of the agonists tested, due to antagonistic co-activation that results in a negative moment relative to the moment generated by the agonists. Decreases of up to 35% in the agonistic joint moment have been reported due to the negative mechanical action of antagonists (Maganaris et al., 1998c). The effect of antagonistic co-activation on agonistic and net joint moments has been neglected when estimating tendon forces (e.g., Kawakami et al., 1994; Fukunaga et al., 1996; Narici et al., 1996; Ito et al., 1998), mainly due to the difficulties in quantifying the antagonistic co-activation joint moment. Optimization methodologies (e.g., Baltzopoulos and Kellis, 1999) or interpolation of EMG data from the antagonist muscle (Kellis and Baltzopoulos, 1997; Maganaris et al., 1998c) could give an estimate of the antagonistic joint moment. This joint moment would then have to be added to the net moment output to give an estimate of the joint moment generated by the agonists only. However, antagonistic co-activation may impair, through reciprocal inhibition, full neural agonistic activation (Tyler and Hutton, 1989), and this indicates that even corrected joint moment values for antagonistic co-activation could still be erroneous.

The percutaneous stimulation protocol followed in this study was assumed to activate, selectively and fully, the studied muscles, and the moments recorded were considered to represent the true muscle joint moment generating capacity. Some evidence of selective muscle activation was obtained in our experiments from EMG recordings of non-stimulated muscles. The possibility of partial muscle activation, however, should be carefully considered. For a given frequency, the muscle force generated during electrical stimulation is a function of the number of motoneuron branches activated (Hultman et al., 1983). A safe way to ensure activation of all motor units in SOL and TA would be to deliver electrical current to the tibial and peroneal nerves, respectively. This approach, however, would generate forces from all muscles supplied from the same nerve. Stimulation of the tibial nerve branch that supplies SOL would result in contraction of all ankle plantarflexors, while stimulation of the peroneal nerve branch that supplies TA would result in contraction of all ankle dorsiflexors and two plantarflexors, the peroneus longus and peroneus brevis muscles. When applying percutaneous stimulation, the number of motor units activated depends on the location of motoneuron branches in relation to the current field (Knaflitz et al., 1990; Lieber and Kelly, 1991). In muscles where motor points are located deeply, away from the skin, percutaneous stimulation would result in only partial activation, independent of the size of the stimulating electrodes and the voltage or current used. This is not the case for TA, a superficial muscle. SOL, however, is proximally covered by the gastrocnemius, obstructing a direct contact between the stimulating electrodes and a major part of the muscle. Thus, the SOL moments recorded in our experiments may represent less than the actual isometric ankle moment-generating capacity of this muscle.

CHANGES IN MUSCULOSKELETAL GEOMETRY FROM REST TO MVC

In this study, changes in two musculoskeletal parameters in the transition from rest to MVC were examined: the moment arm, and the pennation angle. Each is discussed in the sections that follow.

Changes in moment arm

Moment arms at rest have been used for calculating forces acting along the tendon from the moment equilibrium equation (e.g., Kawakami et al., 1994; Narici et al., 1996; Fukunaga et al., 1996; Ito et al., 1998). However, moment arms of muscle-tendon units at rest may be substantially smaller than those during MVC due to: a) increase in muscle thickness (e.g., gastrocnemius lateralis and SOL; Maganaris et al., 1998a), b) displacement of the tendon action line due to stretch of retinaculum systems surrounding the muscle-tendon unit (e.g., tibialis anterior; Maganaris et al., 1999), and c) displacement of the joint centre of rotation caused by joint forces and ligament stretch (e.g., tibio-talar joint; Maganaris et al., 1998a, 1999) during contraction. This contraction effect upon resting moment arms should be taken into account when analyzing maximal static musculoskeletal loads.

Changes in pennation angle

The effective component of the contractile muscle force transmitted to the tendon depends on the angulation of the fibres with respect to the tendon, i.e., the muscle pennation angle (Figure 16.7). The pennation angle is larger during contraction than at rest (Narici et al., 1996; Kawakami et al., 1998; Maganaris et al., 1998b; Maganaris and Baltzopoulos, 1999). This contraction effect should be taken into account in the vectorial analysis of muscle force but it has been neglected, and pennation angles from embalmed muscles that do not reflect changes due to contraction have often been used (e.g., Kawakami et al., 1994; Fukunaga et al., 1996). In the present study, the pennation angles measured were those of the fascicles with respect to the aponeuroses of the muscle. These angles result in a realistic vectorial analysis of muscle forces only if the aponeuroses of the muscle lie on the action line of the muscle-tendon complex. Using ultrasonography, we could not identify any angulation between the aponeuroses and tendons studied, and therefore, we represented the muscle by the geometric model shown in Figure 16.7, an approach introduced by Niels Stensen (Steno) 300 years ago (Stensen, 1667). This type of muscle model has been used repeatedly for estimating forces (Ker et al., 1988; Maganaris et al., 1998b; Maganaris and Baltzopoulos, 1999; Kawakami et al., 1998; Ito et al., 1998). However, some researchers have developed muscle models in which the aponeuroses lie at an angle with respect to the tendons (Huijing and Woittiez, 1984; Woittiez et al., 1984). For a given tendon force, an angulation between the aponeuroses and tendons would result in a smaller contractile force as compared with the force calculated from the muscle model shown in Figure 16.7. An angulation of the aponeuroses with respect to the muscle-tendon action line needs further investigation and appropriate modelling. It must be remembered, however, that oversimplified planar muscle models, with or without an angulation between aponeuroses and tendons, may be unrealistic and mechanically unstable during contraction (van Leeuwen and Spoor, 1992, 1993).

In the present study, pennation angles and moment arms were estimated from two-dimensional image analysis. Furthermore, musculoskeletal geometry during MVC was considered to be representative of that during maximal tetanic stimulation. We have observed no differences in pennation angle and fibre length between MVC and maximal muscle tetanic stimulation (C.N. Maganaris and J.P. Paul, unpublished observations). Any differences in moment arm between MVC and maximal muscle tetanic stimulation cannot be quantified with the current MRI technology.

In vivo estimates of PCSA

Cadaver-based PCSA estimates have been used in the past when calculating muscle specific tension (e.g., Kawakami et al., 1994; Fukunaga et al., 1996). The assumptions made are that: a) the muscle fibre length to muscle length ratio of a given cadaveric muscle is constant across subjects, and b) muscle fibres do not undergo shrinkage during fixation (Kawakami et al., 1994; Fukunaga et al., 1996). Indeed, there is evidence to suggest that the first assumption may be valid; Wickiewicz et al. (1983) and Friederich and Brand (1990) reported a consistency in the muscle fibre length to muscle length ratio in several muscles of the human lower extremity across different cadavers. The second assumption is unrealistic. It is well known that preserved muscles undergo remarkable shrinkage during the fixing process (Yamaguchi et al., 1990). Friederich and Brand (1990) reported a maceration-induced muscle bundle shrinkage of up to 20% in human cadaveric material. Such sizeable shrinkage levels indicate that the muscle fibre length to muscle length ratio may differ between living and cadaveric muscles. From the *in vivo* muscle length and muscle fibre length data in our study, it was calculated that the muscle fibre length to muscle length ratio was 0.11 for SOL, as opposed to 0.06 reported by Wickiewicz et al. (1983), and 0.24 for TA, very close to the reported value of 0.25. These comparisons indicate that the use of a single cadaver-based muscle fibre length to muscle length ratio for a given muscle may be inappropriate for estimating the muscle PCSA. In our study, use of cadaver-based muscle fibre length to muscle length ratios resulted in a realistic estimation of PCSA in TA, but an overestimated PCSA by 87% in SOL.

Neglecting the above three factors resulted in an underestimation by 42% in SOL specific tension and overestimation by 84% in TA specific tension. Such sizeable differences indicate that large errors are made when calculating muscle specific tension using net MVC joint moments and resting state/cadaver-based musculoskeletal geometry. A comparison between the estimates derived using data set B showed that the TA specific tension was higher by 530% compared with SOL, a difference that is unrealistic. The data set A estimates of 17.8 $N{\cdot}cm^{-2}$ for SOL and 23.3 $N{\cdot}cm^{-2}$ for TA are in line with the respective values of ~16 and 23 $N{\cdot}cm^{-2}$ obtained from animal model-based experiments (Rowe and Goldspink, 1969; Witzmann et al., 1983; Powell et al., 1984). In contrast to the values obtained using data set B, the TA specific tension was higher by only 30% compared with SOL. This difference is in line with the reported difference of ~50% between the specific tensions of type II and type I human skinned fibres (Bottinelli et al., 1996). Given that TA has more type II fibres than SOL (Johnson et al., 1973; Henriksson-Larsen et al., 1985), the difference found in our study is reasonable. Differences in muscle fibre composition between TA and SOL might be associated not only with differences in the intrinsic contractile force, but also with differences in the efficiency of force transmission. A more effective force transmission to the aponeuroses through the cytoskeletal matrix, e.g., due to intermediate connective tissue attachments (Jones et al., 1989) would result in a higher total force for a given amount of contractile material. However, methodological limitations, such as assumptions that SOL was fully activated during electrical stimulation, and that TA and SOL operate on the plateau of the force-length relationship at the neutral ankle position, may have accounted for the difference between TA and SOL specific tensions.

FUTURE RESEARCH

The assumption that the TA and SOL operate on the plateau of the force-length relationship at the neutral ankle position requires further investigation. We have relied on previous cadaver-based studies for optimal fibre length in lower limb muscles and derived estimates of specific tension at the neutral anatomical position. *In vivo* studies should be carried out to confirm the validity of cadaver-based studies. Lieber and Baskin (1985) and Fleeter et al. (1985) developed a laser diffraction-based method for *in vivo* human sarcomere length measurement at rest. Using this methodology during maximal isometric contraction, the joint angle corresponding to the optimal sarcomere length of ~2.8 μm (Walker and Schrodt, 1973) could be identified for any given human muscle and considered as the optimal joint angle for muscle force generation. Future research should also focus on: a) quantification of the error made using conventional dynamometry and assuming that muscle-tendon units generate forces in the measurement plane only, b) three-dimensional moment arm, pennation angle, and fibre length estimates during contraction (Scott et al., 1993), and c) development of realistic, yet computationally efficient, three-dimensional models of muscle geometry.

In the present study, we estimated *in vivo* the specific tension of two human muscles, taking into account muscle-specific joint moments and contraction-specific musculoskeletal geometry measurements. We showed that by following the traditional approach of using net resultant joint moments and resting or cadaver-based musculoskeletal geometry, unrealistic conclusions would have been reached. Despite the good agreement of our specific tension estimates with previous reports, major assumptions had to be made and their validity requires further investigation.

REFERENCES

Alexander, R. McN., and Vernon, A. (1975). The dimensions of knee and ankle muscles and the forces they exert. *Journal of Human Movement Studies* **1**, 115–123.

Daltzopoulos, V., and Kellis, E. (1999). Prediction of antagonist muscle moment during dynamic isokinetic knee extension using non-linear optimization. In *XVIIth Congress of the International Society of Biomechanics*, University of Calgary, Calgary, Canada. p. 392.

Bottinelli, R., Canepari, M., Pellefrino, M.A., and Reggiani, C. (1996). Force-velocity properties of human skeletal muscle fibres: Myosin heavy chain isoform and temperature dependence. *Journal of Physiology* **495**, 573–586.

Buchanan, T.S. (1995). Evidence that maximum muscle stress is not a constant: Differences in specific tension in elbow flexors and extensors. *Medical Engineering and Physics* **17**, 529–536.

Cutts, A. (1988). The range of sarcomere lengths in the muscles of the human lower limb. *Journal of Anatomy* **160**, 79–88.

Davies, C.T.M., Thomas, D.O., and White, M.J. (1986). Mechanical properties of young and elderly human muscle. *Acta Medica Scandinavica* **711**, 219–26.

Davies, J., Parker, D.F., Rutherford, O.M., and Jones, D.A. (1988). Changes in strength and cross-sectional area of the elbow flexors as a result of isometric strength training. *European Journal of Applied Physiology* **57**, 667–670.

Fleeter, T.B., Adams, J.P., Brenner, B., and Podolsky, R.J. (1985). A laser diffraction method for measuring muscle sarcomere length *in vivo* for application to tendon transfers. *Journal of Hand Surgery* **10A**, 542–546.

Franke, F. (1920). Die Kraftkurve menschlicher Muskeln bei willkuerlicher Innervation und die Frage der absoluten Muskelkraft. *Pflügers Archiv* **184**, 300–322.

Friederich, J.A., and Brand, R.A. (1990). Muscle fibre architecture in the human lower limb. *Journal of Biomechanics* **23**, 91–95.

Fukunaga, T., Roy, R.R., Shellock, F.G., Hodgson, J.A., Day, M.K., Lee, P.L., Kwong-Fu, H., and Edgerton, V.R. (1992). Physiological cross-sectional area of human leg muscles based on magnetic resonance imaging. *Journal of Orthopaedic Research* **10**, 926–934.

Fukunaga, T., Roy, R.R., Shellock, F.G., Hodgson, J.A., and Edgerton, V.R. (1996). Specific tension of human plantarflexors and dorsiflexors. *Journal of Applied Physiology* **80**, 158–165.

Gravel, D., Arsenault, A.B., and Lambert, J. (1987). Soleus-gastrocnemius synergies in controlled contractions produced around the ankle and knee joints: An EMG study. *Electromyography and Clinical Neurophysiology* **27**, 405–413.

Haxton, H.A. (1944). Absolute muscle force in the ankle flexors of man. *Journal of Physiology* **103**, 267–273.

Henke, W. (1865). Die Groesse der absoluten Muskelkraft aus Versuchen neu berechnet. *Z. f. rat. Med.* **3**, 247–260.

Henriksson-Larsen, K., Friden, J., and Wretling, M-L. (1985). Distribution of fibre sizes in human skeletal muscle. An enzyme histochemical study in m. tibialis anterior. *Acta Physiologica Scandinavica* **123**, 171–177.

Hermann, L. (1898). Zur Messung der Muskelkraft am Menschen. *Pflügers Archiv* **73**, 429–437.

Herzog, W., Read, L.J., and ter Keurs, H.E.D. (1991). Experimental determination of force-length relations of intact human gastrocnemius muscles. *Clinical Biomechanics* **6**, 230–238.

Hettinger, T. (1961). *Physiology of Strength.* Published by Charles C. Thomas.

Hof, A.L., and Van Der Berg, J.W. (1977). Linearity between the weighted sum of the EMGs of the human triceps surae and the total torque. *Journal of Biomechanics* **10**, 529–539.

Huijing, P.A., and Woittiez, R.D. (1984). The effect of muscle architecture on skeletal muscle performance: A simple planimetric model. *Netherlands Journal of Zoology* **35**, 521–525.

Hultman, E., Sjoholm, H., Jadeholm-Ek, I., and Krynicki, J. (1983). Evaluation of methods for electrical stimulation of human skeletal muscle *in situ. Pflügers Archiv European Journal of Physiology* **398**, 139–141.

Ikai, M., and Fukunaga, T. (1968). Calculation of muscle strength per unit cross-sectional area of human muscle by means of ultrasonic measurement. *Internationale Zeitschrift fuer angewandte Physiologie* **26**, 26–32.

Ikai, M., and Fukunaga, T. (1970). A study of training effect on strength per unit cross-sectional area of muscle by means of ultrasonic measurement. *Internationale Zeitschrift fuer angewandte Physiologie* **28**, 173–180.

Ito, M., Kawakami, Y., Ichinose, Y., Fukashiro, S., and Fukunaga, T. (1998). Nonisometric behavior of fascicles during isometric contractions of a human muscle. *Journal of Applied Physiology* **85**, 1230–1235.

Johnson, M.A., Polgar, J., Weightman, D., and Appleton, D. (1973). Data on the distribution of fibre types in thirty-six human muscles. An autopsy study. *Journal of the Neurological Sciences* **18**, 111–129.

Jones, D.A., Rutherford, O.M., and Parker, D.F. (1989). Physiological changes in skeletal muscle as a result of strength training. *Quarterly Journal of Experimental Physiology* **74**, 233–256.

Kanehisa, H., Nemoto, I., Okuyama, H., Ikegawa, S., and Fukunaga, T. (1996). Force generation capacity of knee extensor muscles in speed skaters. *European Journal of Applied Physiology* **73**, 544–551.

Kawakami, Y., Ichinose, Y., and Fukunaga, T. (1998). Architectural and functional features of human triceps surae muscles during contraction. *Journal of Applied Physiology* **85**, 398–404.

Kawakami, Y., Nakazawa, K., Fujimoto, T., Nozaki, D., Miyashita, M., and Fukunaga, T. (1994). Specific tension of elbow flexor and extensor muscles based on magnetic resonance imaging. *European Journal of Applied Physiology* **68**, 139–147.

Kellis, E., and Baltzopoulos, V. (1997). The effects of antagonistic moment on the resultant knee joint moment during isokinetic testing of the knee extensors. *European Journal of Applied Physiology* **76**, 253–259.

Ker, R.F., Alexander, R.Mcn., and Bennett, M.B. (1988). Why are mammalian tendons so thick? *Journal of Zoology, London* **216**, 309–324.

Knaflitz, M., Merletti, R., and Deluca, C.J. (1990). Inference of motor unit recruitment order in voluntary and electrically elicited contractions. *Journal of Applied Physiology* **68**, 1657–1667.

Lieber, R.L., and Baskin, R.J. (1985). Surgical myometer method. U.S. Patent Number: 4-570-641.

Lieber, R.L., and Kelly, M.J. (1991). Factors influencing quadriceps femoris muscle torque using transcutaneous neuromuscular electrical stimulation. *Physical Therapy* **71**, 715–723.

Loren, G.J., and Lieber, R.L. (1995). Tendon biomechanical properties enhance human wrist muscle specialization. *Journal of Biomechanics* **28**, 791–799.

Maganaris, C.N., Baltzopoulos, V., and Sargeant, A.J. (1998a). Changes in Achilles tendon moment arm from rest to maximum isometric plantarflexion: *In vivo* observations in man. *Journal of Physiology* **510**, 977–985.

Maganaris, C.N., Baltzopoulos, V., and Sargeant, A.J. (1998b). *In vivo* measurements of the triceps surae complex architecture in man: Implications for muscle function. *Journal of Physiology* **512**, 603–614.

Maganaris, C.N., Baltzopoulos, V., and Sargeant, A.J. (1998c). Differences in antagonistic ankle dorsiflexor co-activation between legs: Can they explain the moment deficit in the weaker plantarflexor leg? *Experimental Physiology* **83**, 843–855.

Maganaris, C.N., Baltzopoulos, V., and Sargeant, A.J. (1999). Changes in the tibialis anterior tendon moment arm from rest to maximum isometric dorsiflexion: *In vivo* observations in man. *Clinical Biomechanics* **14**, 661–666.

Maganaris, C.N., and Baltzopoulos, V. (1999). Predictability of *in vivo* changes in pennation angle of human tibialis anterior human muscle from rest to maximum isometric dorsiflexion. *European Journal of Applied Physiology* **79**, 294–297.

Morris, C.B. (1948). The measurement of the strength of muscle relative to the cross-section. *Research Quarterly* **19**, 295–303.

Narici, M.V., Binzoni, T., Hiltbrand, E., Fasel, J., Terrier, F., and Cerretelli, P. (1996). *In vivo* human gastrocnemius architecture with changing joint angle at rest and during graded isometric contraction. *Journal of Physiology* **496**, 287–297.

Narici, M.V., Landoni, L., and Minneti, A.E. (1992). Assessment of human knee extensor muscle stress from *in vivo* physiological cross-sectional area and strength measurements. *European Journal of Applied Physiology* **65**, 438–444.

Narici, M.V., Roi, G.S., Landoni, L., Minetti, A.E., and Cerretelli, P. (1989). Changes in force, cross-sectional area and neural activation during strength training and detraining of the human quadriceps. *European Journal of Applied Physiology* **59**, 310–319.

Powell, P.L., Roy, R.R., Kanim, P., Beloo, M.A., and Edgerton, V.R. (1984). Predictability of skeletal muscle tension from architectural determinations in guinea pig hindlimbs. *Journal of Applied Physiology* **57**, 1715–1721.

Reuleaux, F. (1875). Theoretische Kinematik: Grundzuege einer Theorie des Maschinenwesens. F. Vieweg; Braunschweig, MacMillan, London (1963). *The Kinematics of Machinery: Outline of a Theory of Machines.* (Translated by Kennedy, A.B.W.). Dover, New York, pp. 56–70.

Reys, J.H.O. (1915). Ueber die absolute Muskelkraft im menschlichen Koerper. *Pflügers Archiv* **160**, 133–204.

Rowe, W.D., and Goldspink, G. (1969). Surgically induced hypertrophy in skeletal muscles of the laboratory mouse. *Anatomical Record* **161**, 69–75.

Scott, S.H., Engstrom, C.M., Loeb, G.E. (1993). Morphometry of human thigh muscles. Determination of fascicle architecture by magnetic resonance imaging. *Journal of Anatomy* **182**, 249–257.

Siegler, S., Moskowitz, G.D., and Freedman, W. (1984). Passive and active components of the internal moment developed about the ankle joint during ambulation. *Journal of Biomechanics* **19**, 647–652.

Stensen, N. (1667). Elementorum Myologiae Specimen, sue Musculi Descriptio Geometrica. In: *Opera Philosophico*, Volume II. Stellae, Florence, pp. 61–111. Quoted by Kardel, T. (1990). *Journal of Biomechanics* **23**, 953–965.

Tyler, A.E., and Hutton, R.S. (1989). Was Sherrington right about co-contractions? *Brain Research* **370**, 171–175.

van Leeuwen, J.L., and Spoor, C.W. (1992). Modelling mechanically stable muscle architecture. *Philosophical Transactions of the Royal Society of London B* **336**, 275–292.

van Leeuwen, J.L., and Spoor, C.W. (1993). Modelling the pressure and force equilibrium in unipennate muscles with in-line tendons. *Philosophical Transactions of the Royal Society of London B* **342**, 321–333.

Von Recklinghausen, H. (1920). *Gliedermechanik und Laehmungsprothesen.* Verlag J. Springer, Berlin.

Walker, S.M., and Schrodt, G.R.I. (1973). Segment lengths and thin filament periods in skeletal muscle fibers of the rhesus monkey and humans. *Anatomical Record* **178**, 63–82.

Weber, W., and Weber, E. (1836). *Mechanik der menschlichen Gehwerkzeuge.* W. Fischer Verlag, Goettingen.

Wickiewicz, T.L., Roy, R.R., Powel, P.L., and Edgerton, V.R. (1983). Muscle architecture of the human lower limb. *Clinical Orthopaedics and Related Research* **179**, 275–283.

Witzmann, F.A., Kim, D.A., and Fitts, R.H. (1983). Effect of hindlimb immobilization on the fatigability of skeletal muscle. *Journal of Applied Physiology* **54**, 1242–1248.

Woiitiez, R.D., Huijing, P.A., Boom, H.B.K., and Rozendal, R.H. (1984). A three-dimensional muscle model: A quantified relation between form and function of skeletal muscles. *Journal of Morphology* **182**, 95–113.

Yamaguchi, G.T., Sawa, A.G.U., Moran, D.W., Fessler, M.J., and Winters, J.M. (1990). A survey of human musculotendon actuator parameters. In: *Multiple Muscle Systems: Biomechanics and Movement Organization.* Winters, J.M., and Woo, S.L-Y. (eds.), Springer-Verlag, New York, pp. 713–773.

Young, A., Stokes, M., and Crowe, M. (1984). Size and strength of the quadriceps muscles of old and young women. *European Journal of Clinical Investigations* **14**, 282–287.

Zajac, E.F. (1989). Muscle and tendon: Properties, models, scaling, and application to biomechanics and motor control. *CRC Critical Reviews in Biomedical Engineering* **17**, 359–411.

17 Effect of Elastic Tendon Properties on the Performance of Stretch-shortening Cycles

TETSUO FUKUNAGA, KEITARO KUBO, YASUO KAWAKAMI, AND HIROAKI KANEHISA

Department of Life Science (Sports Sciences), University of Tokyo, Komaba, Tokyo, Japan

ABSTRACT

In order to investigate the influence of tendon properties on running and jumping performance, elongation of the tendon and aponeurosis (L) of the human vastus lateralis muscle (VL) was directly measured by ultrasonography during isometric knee extension. The relationship between knee extension force and L was fitted to a linear regression above 50% MVC, the slope of which was defined as the compliance of the tendon structures. Although compliance was not significantly related to absolute jump height, it was significantly correlated to the pre-stretch augmentation, i.e., the difference in height when jumping with and without counter-movement. Maximum elongation of tendon structures during MVC were significantly higher in sprinters and lower in long-distance runners compared to control subjects. For sprinters, the relationship between compliance and 100 m performance was positive, i.e., the better the performance, the more compliant the tendon structure. These results suggest that compliance of tendon structures has a favourable effect on stretch-shortening cycle exercise, possibly because of improved storage and release of elastic energy.

Key words: vastus lateralis muscle, compliance, ultrasonography, *in vivo* measurement, stretch-shortening cycle exercise.

INTRODUCTION

The stretch-shortening cycle (SSC) can be defined as a sequence of an eccentric muscle action immediately followed by a concentric action (Komi and Bosco, 1978). The SSC is a natural component of muscle function in many daily activities, such as running and jumping. It is well known that if an activated muscle is stretched prior to shortening, its performance is enhanced during the concentric

Skeletal Muscle Mechanics: From Mechanisms to Function. Edited by W. Herzog.

phase. Many previous studies have indicated that this phenomenon is caused by strain energy stored in tendons. This notion implies that SSC performance is influenced not only by the force and power provided by the muscles, but also by the elastic elements of tendon structures (Cavagna, 1977). However, there is no study regarding the influence of tendon structures on SSC performance in humans. Also, the precise mechanism of SSC exercise remains to be determined.

The elastic properties of the muscle-tendon complex (MTC) can be associated primarily with the series elastic component (SEC). The importance of tendon structures and their influence on the mechanical performance of muscle have been recognized. Although the elastic properties of tendon structures have been determined from cadaver and animal experiments so far (Benedict et al., 1968; Woo et al., 1981), it is likely that tendon structures in living humans differ substantially from those of cadavers and animals, both in dimensions and mechanical properties because of differences in species and age. Information on elastic properties of tendon structures *in vivo* is essential for the understanding of muscle contractile properties and function.

Recent progress in technology has made it possible to study the dynamics of MTC *in vivo* with the use of ultrasonography (Fukunaga et al., 1997; Ichinose et al., 1997; Kawakami et al., 1998). Fukashiro et al. (1995) and Ito et al. (1998) showed that one can determine the *in vivo* elastic properties of human tendon structures from observations of lengthening of the tendon and aponeurosis during isometric ramp contractions. However, the effect of tendon lengthening on exercise performance, as well as the variations of tendon and aponeurosis properties across people, have not been studied. Recently, Kawakami et al. (1998) showed that tendon elasticity plays an important role during isometric contractions. Exercise performance during SSC may be greatly influenced by the tendon structures. The purpose of this research was to describe the methodology for evaluating the elastic properties of tendon structures in human knee extensor muscles *in vivo*, and to investigate the influence of these elastic tendon structures on running and jumping performance.

IN VIVO ESTIMATION OF ELASTIC PROPERTIES OF TENDON STRUCTURES

Recently developed B-mode ultrasonic imaging allows for observation of the dynamics of tendon structures during contraction. In order to evaluate the elastic properties of tendon structures, elongation of the tendon and aponeurosis of vastus lateralis (VL) was measured by ultrasonography during isometric knee extension.

MEASUREMENT OF ELONGATION OF TENDON STRUCTURES

A real-time ultrasonic apparatus was used to obtain a longitudinal image of VL at the mid-thigh level. The experimenter visually confirmed the echoes from the aponeurosis and VL fascicles. The point at which one fascicle was attached to the aponeurosis (P) was marked. Subjects were requested to perform isometric knee extensor contractions up to maximum level effort. During contraction, P moved proximally (Figure 17.1). Ultrasonic images recorded on the videotape were printed, frame by frame, onto calibrated recording films (SSZ-305, Aloka, Japan). A marker was placed between the skin and the ultrasonic probe to confirm that the probe did not move during measurements. The cross-point between superficial aponeurosis and fascicles did not move. Therefore, the displacement of P was considered to indicate the lengthening of the deep aponeurosis and the distal tendon (L) (Ito et al., 1998).

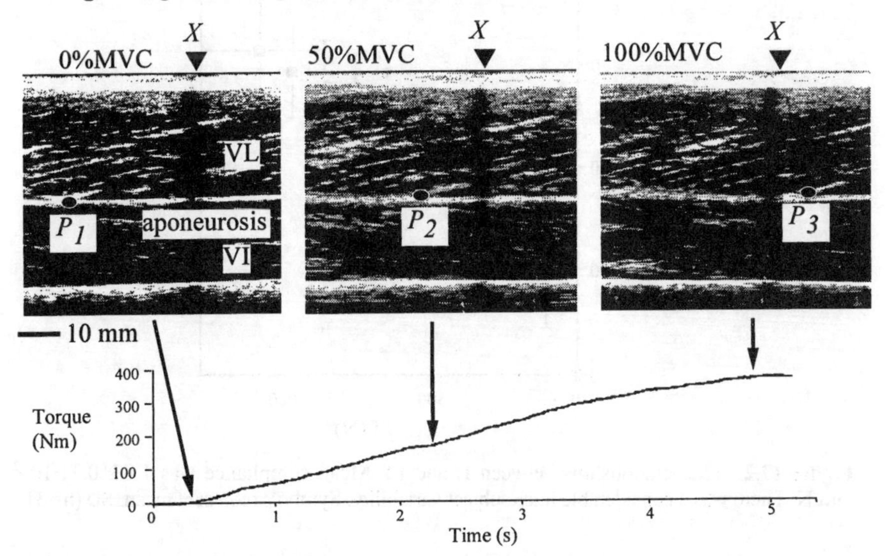

Figure 17.1. Ultrasonic images of longitudinal sections of vastus lateralis muscle during isometric contraction. The ultrasonic transducer was placed on the skin over the muscle at a 50% distance from the greater trochanter to the lateral epicondyle of the femur. The cross-point (p) was determined from the echoes of the deep aponeurosis and fascicles. The distance travelled by p (L) was defined as the length change of the tendon and aponeurosis during contraction.

CALCULATION OF THE ELASTIC PROPERTIES

The knee joint torque (TQ) measured by the dynamometer was converted to muscle force (Ff) by the following equation:

$$Ff = k \cdot TQ \cdot MA^{-1} \tag{17.1}$$

where k is the relative contribution of VL to force production during knee extension, expressed as the percentage of its physiological cross-sectional area relative to that of the quadriceps femoris muscle, 22% (Narici et al., 1992); and MA is the moment arm length of the quadriceps femoris muscle at the knee when flexed at an angle of 80°. MA was estimated from the thigh length of each subject as described by Visser et al. (1990).

The resulting L-Ff relationship (Figure 17.2) was non-linear as previously reported for animal and human tendons *in vitro* (Abrahams, 1967; Benedict et al., 1968; Woo et al., 1981). The initial region of the relationship (the so-called toe region) was characterized by a large increase in L for small increases in Ff. The ratio was calculated at every 10% MVC (Figure 17.3). The linear region between 50-100% MVC has an approximately constant modulus of elasticity that was used for the determination of MTC compliance (mm/N).

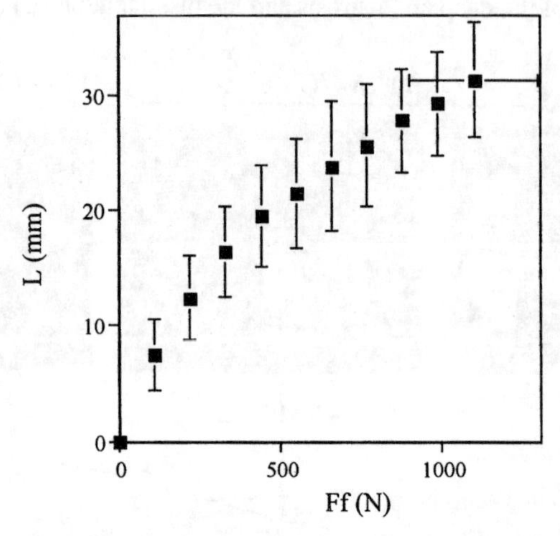

Figure 17.2. The relationships between L and Ff. Mean compliance was $1.83 \pm 0.31 \cdot 10^{-2}$ mm/N. There was a considerable inter-subject variability. Symbols represent mean±SD (n=31).

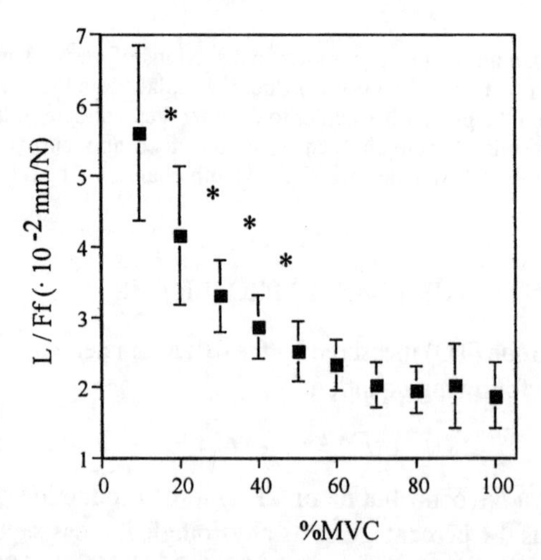

Figure 17.3. The relationships between % MVC and L/Ff at every 10% MVC. While the L/Ff tended to decrease curvilinearly with increasing force, the changes of L/Ff above 50% MVC did not significantly differ from each other. Symbols represent mean±SD (n=31).
* significant decrease in L/Ff with increasing force.

MEASUREMENT OF PATELLAR EXCURSION

In order to confirm whether the joint angle remained unchanged during isometric contractions, patellar excursion during the isometric contraction was examined using radiography. Furthermore, to test the possibility of movement of the probe relative to the femur during measurements, an anti-radioactive marker was attached to the skin at the mid-thigh level. Based on results by Marshall et al. (1990), the displacement of the superior margin of the patella was used to approximate length changes of quadriceps femoris muscle. Figure 17.4 shows the patellar excursion during isometric contractions. The position of the reference marker did not change relative to the femur during contractions. The patella and tibia moved during contraction, but only at force levels below 50% MVC. Furthermore, the distance between patella and tibia was constant during contraction. This indicated that there was a slight knee extension at contraction levels <50% MVC, but not at levels above 50% MVC. Therefore, the L values in the range from 50-100% MVC were considered elongations of the tendon structures, and compliance was determined over this range.

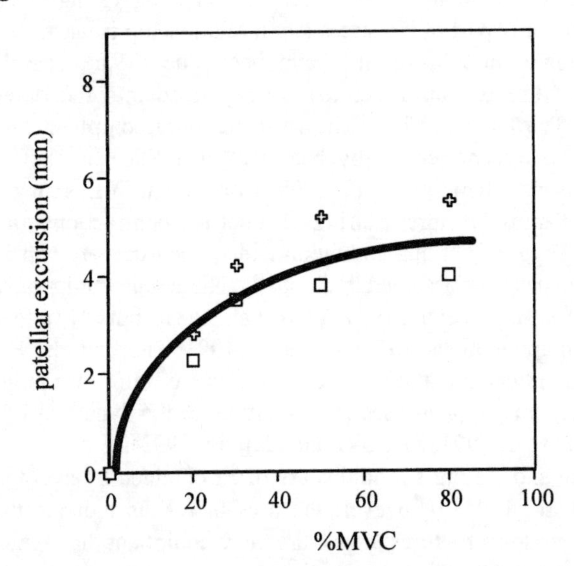

Figure 17.4. Patellar excursions during isometric contraction. There were slight patellar excursions at levels <50% MVC, but not at levels above 50% MVC. (Kubo et al., 1999.)

COMPLIANCE OF TENDON STRUCTURES IN HUMAN VASTUS LATERALIS

Elasticity is a function of force generation and length changes and is usually measured as stiffness (force generation/length change) or as its reciprocal, compliance. In the present study, compliance was calculated to be $1.83 \pm 0.31 \cdot 10^{-2}$ mm/N on average, with significant inter-individual differences. These values lie among those reported for *in vitro* studies. However, the variability in previously reported tendon stiffness was larger than found here, ranging from 7.5 to 2400 N/mm (Aruin et al., 1979; Green and McMahon, 1979; Luhtanen and Komi, 1980). In order to make appropriate comparisons, the present L-Ff relationship was converted to a stress-strain relationship. Cross-sectional areas of tendon were obtained from a previous report (200 mm^2 for the quadriceps tendon; Yamaguchi et al., 1990), and the length of the tendon was defined as the distance between the measurement site and the estimated insertion of the muscle over the skin. The Young's modulus, i.e., the slope of the stress-strain curve, was 0.25 GPa, which was lower than previously reported values for human tendons (0.6 and 1.8 GPa) (Abrahams, 1967; Benedict et al., 1968; Vigt et al., 1995). There are two possible reasons for this discrepancy: first, previous reports were based on tendon failure tests, therefore loads far exceeded the physiological range (2%; Lieber, 1991; Zajac, 1989). Lieber (1991) reported that the Young's modulus for tendon was 0.188 GPa, which is less than the modulus reported for most mammalian and human cadaver tendons. The present value was comparable to Lieber's (1991), suggesting that in the physiological range, only part of the stress-strain relationship is used. Second, the low *in vivo* Young's modulus might have been due to the compliance of the 'aponeurosis'. The tendon structures are separated into the outer-tendon and aponeurosis. The latter has been shown to be more compliant than the former (Lieber, 1991). Recent reports by Narici et al. (1996) on the human medial gastrocnemius and Ichinose et al. (1997) on human VL showed about 30% shortening of muscle fibres during 'isometric' contractions *in vivo*. These observations suggest that human tendon and aponeurosis are considerably more compliant than has been assumed traditionally. Numerous studies have investigated the outer-tendon (Aruin et al., 1979; Woo et al., 1981), but only a few studies have directly tested aponeurosis stiffness (Lieber, 1991; Zuurbier and Huijing, 1993). These studies indicated that aponeuroses strains are four times larger than outer-tendon strains, thus, aponeuroses greatly influence the dynamics of fibres during contraction (Lieber, 1991; Zuurbier and Huijing, 1993).

Compliance and Young's modulus are often obtained by averaging cadaveric data (Vigt et al., 1995). However, the present results indicate that the elastic properties of tendon structures *in vivo* are more compliant than typically assumed and they vary significantly among individuals.

THE EFFECT OF ELASTIC TENDON PROPERTIES ON JUMPING PERFORMANCE

The height achieved in a vertical jump depends on the power output of the appropriate muscles and the capacity of muscles to store elastic energy. It is speculated that the properties of tendon structures affect performance of counter-movement jumps. In order to investigate the influence of tendon properties on jumping performance, subjects performed maximal vertical jumps on a force plate (Kistler 9281B) with and without counter-movement, i.e., counter-movement jumps (CMJ) and squatting jumps (SJ), respectively.

MEASUREMENT OF JUMP PERFORMANCE

For the SJ, subjects were positioned on the force plate with the knee angle at 90°. Knee angle was accurately controlled by an electrogoniometer (Penny and Giles). Subjects were instructed to jump for maximum height. By measuring the flight time (T_{air}) from the force record, the vertical take-off velocity (Vv) of the centre of gravity was calculated as follows (Komi and Bosco, 1978):

$$Vv = 1/2 \cdot T_{air} \cdot g \quad (17.2)$$

Where g is the acceleration of gravity, jump height (H) can then be computed as:

$$H = Vv^2 \cdot (2\,g)^{-1} \quad (17.3)$$

For the CMJ, the knee angle at the lowest position was measured as 91±2°. Thus, the range of motion was considered identical between CMJ and SJ. The SSC performance was evaluated from the jump heights of CMJ and SJ as an augmentation by prior stretch (Walshe et al., 1996), that is:

$$\textit{Pre-stretch augmentation}\ (\%) = (CMJ - SJ) \cdot SJ^{-1} \cdot 100 \quad (17.4)$$

MORE COMPLIANT TENDONS GIVE HIGHER PRE-STRETCH AUGMENTATION

The jump height in both SJ and CMJ was significantly correlated to the body weight, muscle thickness of quadriceps femoris, and the peak reaction force at the push-off phase divided by body weight (Table 17.1). These results indicate that the jump height in both SJ and CMJ depended on the muscle volume and/or the force output during ballistic movement.

Table 17.1. Correlation coefficients between jump height and selected take-off variables. (Kubo et al., 1999.)

Variables	versus SJ Height	versus CMJ Height
Body height	−0.07	−0.08
Body mass	0.42 *	0.45 *
Muscle thickness	0.43 *	0.47 *
MVC	0.34	0.33
Compliance	−0.18	−0.07
Peak force at push-off phase	0.75 *	0.74 *

* P<0.05

An interesting finding of this study was that the pre-stretch augmentation was significantly correlated to the MTC compliance (Figure 17.5), even though the compliance was not significantly related to jump height in vertical jumps with or without prior stretch. In a previous study in which a damped oscillation technique was applied to human MTC, a significant negative correlation was observed between MTC stiffness and pre-stretch augmentation during SSC exercise (Walshe et al., 1996; Wilson et al., 1991). Cavagna (1977) suggested that a compliant MTC is better suited to utilize elastic energy, allowing for a better performance during SSC.

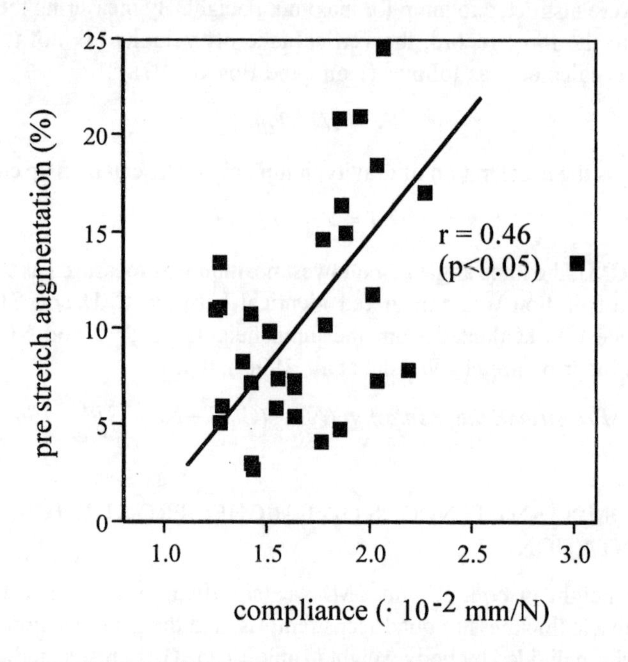

Figure 17.5. The relationship between compliance and pre-stretch augmentation. Whilst compliance was not significantly related to jump height in the vertical jumps with and without prior stretch, it was significantly correlated to the percentage difference between the two jumps (r=0.46, $P<0.05$). (Data from Kubo et al., 1999.)

Other factors must be considered when evaluating the pre-stretch augmentation. The correlation coefficient, r=0.46, between compliance and pre-stretch augmentation (Figure 17.5) suggests that the elastic properties of tendon structures could account for 21% of the variance in pre-stretch augmentation. The remaining 79% might be accounted for by differences of hip and ankle joint movements, contribution of other muscle groups to knee extension, and jumping technique. Kilani et al. (1989) stated that the myoelectric response accounted for up to 85% of the increase in jump height following a counter-movement.

These results indicate that the pre-stretch effect was more pronounced in subjects with highly compliant tendon structures.

ELASTICITY OF TENDON STRUCTURES IN LONG-DISTANCE RUNNERS AND SPRINTERS

Previous studies have provided evidence that the morphological and mechanical properties of connective tissues change with training and immobilization (Booth and Gould, 1975; Butler et al., 1978; Stone, 1988; Viidik, 1986). In these studies, the mechanical properties of connective tissues were evaluated on the basis of stress-strain curves. It was found that the ultimate tensile strength of bone-ligament or bone-tendon preparations increased as a result of endurance training (Stone, 1988). Woo et al. (1981) observed that the stiffness of swine tendon was significantly increased by endurance training. However, Viidik (1986) failed to find a significant change in the stiffness of the anterior crurate ligament of rabbits following endurance training. Tipton et al. (1975) did not find significant changes in stiffness of ligament-bone junctions after endurance or sprint training. These discrepancies in findings can be explained partially by differences in species and/or specimen type (Viidik, 1986).

Sprint running induces SSC in lower limb MTCs. Previous studies showed that the compliance of MTC has important effects on the mechanics of eccentric muscle actions (Cook and McDonagh, 1995), and contributes to the performance of SSC activities (Ettema et al., 1992).

These findings lead us to hypothesize that, as a result of long-term participation in sprint training or endurance training, the tendon structures of the lower limbs will become more compliant and stiffer, respectively.

LONG-DISTANCE RUNNERS HAVE STIFFER TENDON STRUCTURES THAN SPRINTERS

Thirteen male long-distance runners (LDR, age=19.8±0.4 years, height=171.1±4.7 cm, body mass=57.2±3.8 kg, mean±SD) and 10 male sprint runners (SPR, age=21.4±1.8 years, height=171.2±5.6 cm, body mass=68.6±7.2 kg) volunteered to participate in this study. Subjects had participated in their sports for 3 to 10 years. The best official times of the LDRs in a 5000 m race was 14.29±0.27 minutes, and for the SPRs was 11.0±0.2 s in a 100 m race within one year prior to these tests.

Comparisons between the L-Ff relations of LDR, SPR, and control subjects (CON) are presented in Figure 17.6. The L values at force levels beyond 880 N were significantly larger in SPR and smaller in LDR than in CON, respectively.

The present results indicated that the MTC of VL in LDR was less compliant than that in CON. This agrees with findings obtained in animal experiments, which indicate that endurance training increases the failure load and stiffness of tendons (Goubel and Marinni, 1987; Woo et al., 1981). Moreover, comparison of the elastic energy, which is given by the area below the L-Ff curve, showed that long-distance runners store less energy than control subjects during MTC lengthening. These results rule out the possibility that the MTCs of long-distance runners become more compliant with training, and so enhance the energy-storage capacity of MTCs.

However, in sprinters, the tendon structures elongated more than in the control subjects. The difference between the two groups was statistically significant at force levels above 500 N (about 50% of MVC). The maximum elongation (Lmax) at MVC for VL was significantly greater in SPR (41 mm) than in CON (33 mm). The mechanisms that produced these differences are unknown, but it is speculated that the increased elongation may play a role in reducing the risk of knee injury.

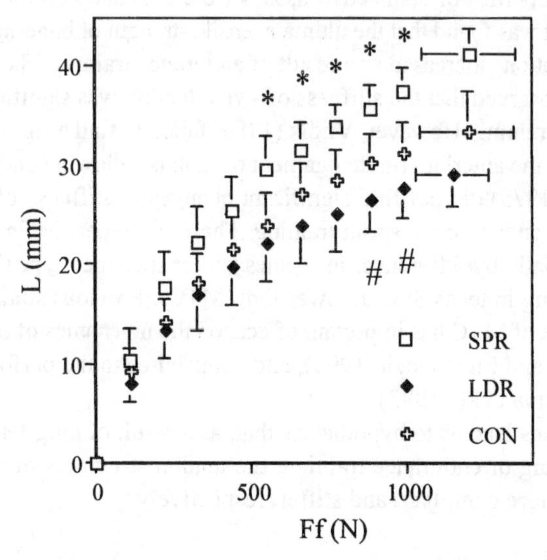

Figure 17.6. Comparisons of L-Ff relation among sprinters (SPR), long-distance runners (LDR), and untrained subjects (CON). Symbols represent mean±SD. * denotes that the value of SPR is significantly higher than that of CON at $P<0.05$. # denotes that the value of LDR is significantly shorter than that of CON at $P<0.05$.

The relation between % MVC and L/Ff decreased with increasing force from 10-50% of MVC, and became almost constant in the range 50-100% MVC (Figure 17.7). The L/Ff at 10% and 20% MVC were significantly higher in SPR than in LDR and CON. The elongation of tendon structures at low levels of contraction is caused by crimping of the collagen fibres, and tendon compliance is high in this region (Rigby et al., 1959). Hence, the high L/Ff at force levels below 20% MVC in the SPRs suggests that the arrangements of collagen fibres in the tendon component for SPRs may differ from those for CON and LDR.

However, we cannot distinguish whether the low (LDR) and high (SPR) MTC compliance are due to training or genetic reasons. It is well known that elite long-distance runners and sprinters show a higher distribution of slow-twitch (ST) and fast-twitch (FT) fibres than untrained subjects, respectively (Costill et al., 1976; Thorstensson et al., 1977). The elastic properties of ST fibres differ from those of FT fibres (Galler et al., 1996; Petti et al., 1990). It was found that slow motor units had a greater dynamic stiffness than fast motor units in peroneus longus muscle of

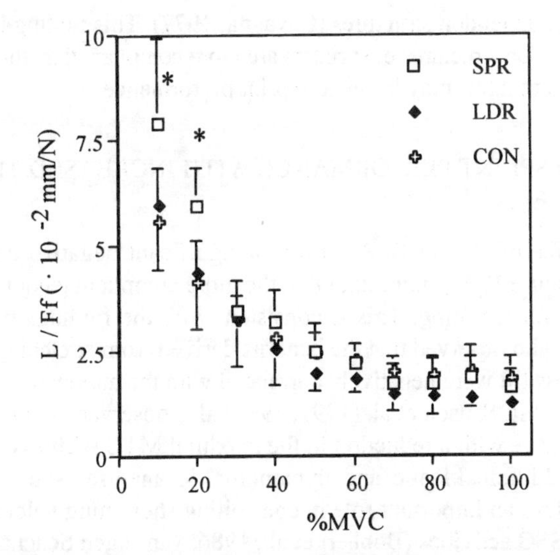

Figure 17.7. Comparisons of % MVC versus L/Ff relations among sprinters (SPR), long-distance runners (LDR), and untrained subjects (CON). Symbols represent mean±SD. * denotes that the value of SPR is significantly higher than that of CON at $P<0.05$.

the cat (Petti et al., 1990). Furthermore, the transition from FT to ST fibres was associated with an increase in stiffness in the rat soleus (Goubel and Marini, 1987). Considering these findings, it seems that the difference in MTC compliance between LDR and SPR might be attributed to the difference in muscle fibre composition, which has been considered to be influenced largely by genetic factors (Komi et al., 1977).

Some studies have been aimed at assessing the effects of training on elasticity of tendons. Physical loading of tendons increased the number, diameter, and degree of alignment of the constituent collagen fibres (Michna, 1984; Vilarta and Vidal, 1989). Nakagawa et al. (1989) observed that the cross-sectional area of the Achilles tendon was significantly greater in long-distance runners than in untrained individuals. Hence, it is likely that the mechanical stress, rather than genetic factors, is responsible for the difference in MTC compliance between LDR and CON.

The present results indicate that the MTC of VL is less compliant and more compliant in LDR and SPR than in untrained individuals, respectively. These elastic properties of VL may be associated with lower/higher performance during SSC exercises.

EFFECT OF TENDON ELASTICITY ON SPRINT PERFORMANCE

The training of sprinters consists primarily of sprint running and other high intensity exercise. Sprint running induces SSC in lower limb MTC. In SSC exercises, elastic energy is stored in tendon structures (Cavagna, 1977). This finding leads to the hypothesis that tendon structures of sprinters are more compliant than those of untrained individuals and that this may influence sprint performance.

INCREASED SPRINT PERFORMANCE WITH INCREASED TENDON COMPLIANCE

The compliance of VL for SPR showed a significant negative correlation with sprint time (Figure 17.8), suggesting that the more compliant tendon of VL may be advantageous for sprinting. This is consistent with the findings of Wilson et al. (1991, 1992), who observed that the benefits derived from stretching muscles prior to a bench-press lift were negatively correlated with the maximal stiffness of MTC involved in the lift. Wilson et al. (1991, 1992) also observed an increase in bench-press performance with a reduction in the maximal MTC stiffness after flexibility training. In addition, kinetic and dynamometric analyses showed that tendon compliance plays an important role in controlling shortening velocities of muscle fibres during SSC activities (Bobbert et al., 1986; van Ingen Schenau et al., 1985). They suggested that a significant amount of shortening can be taken up by recoil of a previously stretched tendon when MTCs are shortening and tension is falling. The shortening of the tendon reduces the speed of shortening of the fibres in the MTC, and so provides better contractile conditions for force production in the fibres compared to a situation in which the entire MTC shortening is accommodated by fibre shortening. Given these points, it is reasonable to assume that a more compliant tendon in the knee extensor muscles may help sprint runners to develop higher forces during the stance phase and so help reduce sprinting time.

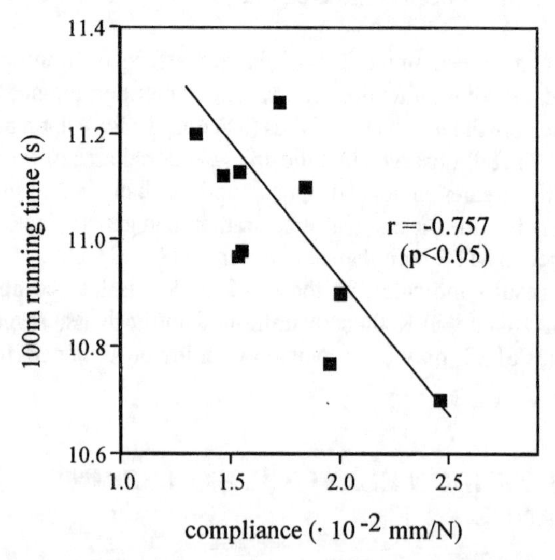

Figure 17.8. Relationship between the compliance of vastus lateralis muscle and 100 m running time for sprint runners. (Kubo et al., 2000.)

The present results demonstrated that the elasticity of tendon structures of VL is higher in sprinters than in untrained individuals. For sprint runners, the tendon structures of VL were more compliant at low force levels, and their elasticity at high force levels may enhance sprint performance.

SUMMARY

Using ultrasonography, it was possible to quantify the elastic properties of *in vivo* human tendon structures. The results underlined the possibility that SSC performance might be affected by the compliance of tendon structures. The elastic properties of tendon structures could be an important index of physical performance.

REFERENCES

Abrahams, M. (1967). Mechanical behavior of tendon *in vitro*: A preliminary report. *Medical and Biological Engineering* **5**, 433–443.

Aruin, A.S., Prilutski, B.I., Raitsun, L.M., and Savelev, I.A. (1979). Biomechanical properties of muscles and efficiency of movement. *Human Physiology* **5**, 426–434.

Benedict, J.V., Walker, L.B., and Harris, E.H. (1968). Stress-strain characteristics and tensile strength of unembalmed human tendon. *Journal of Biomechanics* **1**, 53–63.

Bobbert, M.F., Huijing, P.A., and van Ingen Schenau, G.J. (1986). A model of the human triceps surae muscle-tendon complex applied to jumping. *Journal of Biomechanics* **19**, 887–898.

Booth, F.W., and Gould, E.W. (1975). Effects of training and disuse on connective tissue. In: *Exercise and Sport Sciences Reviews, Volume 3*. Wilmore, J.H., and Keogh, J.F. (eds.), Academic Press, New York, pp. 83–112.

Butler, D.L., Grood, E.S., Noyes, F.K., and Zernicke, R.F. (1978). Biomechanics of ligaments and tendons. In: *Exercise and Sport Sciences Reviews, Volume 6*. Hutton, R.S. (ed.), Academic Press, New York, pp. 83–112.

Cavagna, G.A. (1977). Storage and utilization of elastic energy in skeletal muscle. *Exercise and Sport Sciences Reviews* **5**, 89–129.

Cook, C.S., and McDonagh, M.J.N. (1995). Force responses to controlled stretches of electrically stimulated human muscle-tendon complex. *Quarterly Journal of Experimental Physiology* **80**, 477–490.

Costill, D.J., Daniels, J., Evans, W., Fink, W., Krahenbuhl, G., and Saltin, B. (1976). Skeletal muscle enzymes and fiber composition in male and female track athletes. *Journal of Applied Physiology* **40**, 149–154.

Ettema, G.J.C., Huijing, P.A., and Haan, A. (1992). The potentiating effect of prestretch on the contractile performance of rat gastrocnemius medialis muscle during subsequent shortening and isometric contractions. *Journal of Experimental Biology* **165**, 121–136.

Fukashiro, S., Itoh, M., Ichinose, Y., Kawakami, Y., and Fukunaga, T. (1995). Ultrasonography gives directly but non-invasively elastic characteristic of human tendon *in vivo*. *European Journal of Applied Physiology* **71**, 555–557.

Fukunaga, T., Kawakami, Y., Kuno, S., Funato, K., and Fukashiro, S. (1997). Muscle architecture and function in humans. *Journal of Biomechanics* **30**, 457–463.

Galler, S., Hilber, K., and Pette, D. (1996). Force responses following stepwise length changes of rat skeletal muscle fibre types. *Journal of Physiology* **493**, 219–227.
Goubel, F., and Marinni, J.F. (1987). Fibre type transition and stiffness modification of soleus muscle of trained rats. *Pflügers Archiv* **410**, 321–325.
Green, P.R., and McMahon, T.A. (1979). Reflex stiffness of man's antigravity muscles during knee bends while carrying extra weights. *Journal of Biomechanics* **12**, 881–891.
Ichinose, Y., Kawakami, Y., Ito, M., and Fukunaga, T. (1997). Estimation of active force-length characteristics of human vastus lateralis muscle. *Acta Anatomica* **159**, 78–83.
Ito, M., Kawakami, Y., Ichinose, Y., Fukashiro, S., and Fukunaga, T. (1998). Non-isometric behavior of fascicles during isometric contractions of a human muscle. *Journal of Applied Physiology* **85**, 1230–1235.
Kawakami, Y., Ichinose, Y., and Fukunaga, T. (1998). Architectural and functional features of human triceps surae muscles during contraction. *Journal of Applied Physiology* **85**, 398–404.
Kilani, H.A., Palmer, S.S., Adrian, M.J., and Gapsis, J. (1989). Block of the stretch reflex of vastus lateralis during vertical jumps. *Human Movement Science* **8**, 247–269.
Komi, P.V., and Bosco, C. (1978). Utilization of stored elastic energy in leg extensor muscle by men and women. *Medicine and Science in Sports* **10**, 261–265.
Komi, P.V., Viitasalo, J.H.T., Havu, M., Thorstensson, A., Sjodin, B., and Karsson, J. (1977). Skeletal muscle fibres and muscle enzyme activities in monozygous and dizygous twins of both sexes. *Acta Physiologica Scandinavia* **100**, 385–392.
Kubo, K., Kanehisa, H., Kawakami, Y., and Fukunaga, T. (2000). Elastic properties of tendon structures in sprinters. *Acta Physiologica Scandinavia* (in press).
Kubo, K., Kawakami, Y., and Fukunaga, T. (1999). The influence of elastic properties of tendon structures on jump performance in humans. *Journal of Applied Physiology* **87**, 2090–2096.
Lieber, R.I. (1991). Frog semitendinosis tendon load-strain and stress-strain properties during passive loading. *American Journal of Physiology* **261**, C86–C92.
Luhtanen, P., and Komi, P.V. (1980). Force-, power-, and elasticity-velocity relationships in walking, running, and jumping. *European Journal of Applied Physiology* **44**, 279–289.
Marshall, R.N., Mazur, S.M., and Taylor, N.A.S. (1990). Three-dimensional surfaces for human muscle kinetics. *European Journal of Applied Physiology* **61**, 263–270.
Michna, H. (1984). Morphometric analysis of loading-induced changes in collagen-fibril populations in young tendons. *Cell and Tissue Research* **236**, 456–470.
Nakagawa,Y., Ikegawa, S., Abe, T., Fukunaga, T., Inoue, T., Totsuka, M., Sekiguchi, O., and Hirota, K. (1989). The morphology and SEM in Achilles tendon. *Japanese Journal of Physical Fitness and Sports Medicine* **38**, p. 424 (abstract).
Narici, M.V., Binzoni, T., Hiltbrand, E., Fasel, J., Terrie, F., and Cerretelli, P. (1996). *In vivo* human gastrocnemius architecture with changing joint angle at rest and during graded isometric contraction. *Journal of Physiology* **496**, 287–297.
Narici, M.V., Roi, C.C., Landoni, L., Minetti, A.E., and Cerretelli, P. (1992). Assessment of human knee extensor muscles stress from *in vivo* physiological cross-sectional area and strength measurements. *European Journal of Applied Physiology* **65**, 438–444.
Petti, J., Filippi, G.M., Emonet-Denand, F., Hunt, C.C., and Laporte, Y. (1990). Changes in muscle stiffness produced by motor units of different types in peroneus longus muscle of cat. *Journal of Neurophysiology* **63**, 190–197.
Rigby, B.J., Hirai, N., Spikes, J.D., and Eyring, H. (1959). The mechanical properties of rat tail tendon. *Journal of General Physiology* **43**, 265–283.
Stone, M.L. (1988). Implications for connective tissue and bone alterations resulting from resistance exercise training. *Medicine and Science in Sports and Exercise* **20**, S162–S168.

Thorstensson, A., Larsson, L., Tesch, P., and Carlson, J. (1977). Muscle strength and fibre composition in athletes and sedentary men. *Medicine and Science in Sports and Exercise* **9**, 26–30.

Tipton, C.M., Mathes, R.D., Maynard, J.A., and Carey, R.A. (1975). The influence of physical activity on ligaments and tendons. *Medicine and Science in Sports* **7**, 165–175.

van Ingen Schenau, G.J., Bobbert, M.F., Woittiez, R.D., and Huijing, P.A. (1985). The instantaneous torque-angular velocity relation in plantar flexion during jumping. *Medicine and Science in Sports and Exercise* **17**, 422–426.

Vigt, M., Moller, F.B., Simonsen, E.B., and Poulsen, P.D (1995). The influence of tendon Young's modulus, dimensions and instantaneous moment arms on the efficiency of human movement. *Journal of Biomechanics* **28**, 281–291.

Viidik, A. (1986). Adaptability of connective tissue. In: *Biochemistry of Exercise VI.* Saltin, B. (ed.), Human Kinetic Publishers, Champaign, Illinois, pp. 545–562.

Vilarta, R., and Vidal, B.C. (1989). Anisotropic and biomechanical properties of tendons modified by exercise and denervation: Aggregation and macromolecular order in collagen bundles. *Matrix* **9**, 55–61.

Visser, J.J., Hoogkamer, J.E., Bobbert, M.F., and Huijing, P.A. (1990). Length and moment arm of human leg muscles as a function of knee and hip-joint angles. *European Journal of Applied Physiology* **61**, 453–460.

Walshe, A.D., Wilson, G.J., and Murphy, A.J. (1996). The validity and reliability of a test of lower body musculotendinous stiffness. *European Journal of Applied Physiology* **73**, 332–339.

Wilson, G.J., Elliott, B.C., and Wood, G.A. (1992). Stretch shorten cycle performance enhancement through flexibility training. *Medicine and Science in Sports and Exercise* **24**, 116–123.

Wilson, G.J., Wood, G.A., and Elliot, B.C. (1991). Optimal stiffness of the series elastic component in a stretch shorten cycle activity. *Journal of Applied Physiology* **70**, 825–833.

Woo, S.L., Gomez, M.A., Amiel, D., Ritter, M.A., Gelberman, R.H., and Akeson, W.H. (1981). The effects of exercise on the biomechanical and biochemical properties of swine digital flexor tendons. *Journal of Biomechanical Engineering* **103**, 51–56.

Yamaguchi, G.T., Sawa, A.G.U., Moran, D.W., Fessler, M.J., and Winters, J.M. (1990). A survey of human musculotendon actuator parameters. In: *Multiple Muscle Systems: Biomechanics and Movement Organization.* Winters, J.M., and Woo, S.L-Y. (eds.), Springer-Verlag, New York, pp. 717–773.

Zajac, F.E. (1989). Muscle and tendon: Properties, models, scaling and application to biomechanics and motor control. *CRC Critical Reviews in Biomedical Engineering* **17**, 359–411.

Zuurbier, C.J., and Huijing, P.A. (1993). Changes in geometry of actively shortening unipennate rat gastrocnemius muscle. *Journal of Morphology* **218**, 167–180.

18 A Non-invasive Approach for Studying Human Muscle-tendon Units *In Vivo*

DAVID HAWKINS
Human Performance Laboratory, University of California (Davis), Davis, California, USA

INTRODUCTION

Muscle-tendon units are complex biological actuators able to impart considerable force to bones to stabilize and/or move segments of the body and to absorb energy imparted to the body. Muscles are controlled through neural inputs and generate force by converting chemical energy into mechanical energy. A muscle's mechanical behaviour is directly linked to its macroscopic, cellular, and molecular structure; the properties of its specific proteins; and the structural properties of its associated tendons. Muscle-tendon units are highly adaptable, modifying their structure and protein forms in response to changes in environmental stimuli. Due to the integral role muscle-tendon (MT) units play in human function and the quality of life, there is considerable interest in understanding (1) the basic structural properties of MT units, (2) how MT properties change throughout life and in response to altered environmental stimuli (e.g., disuse, exercise), (3) risk factors associated with MT injury, (4) how best to maintain or restore normal function throughout life, and (5) how to compensate for a loss in MT function.

The purpose of this chapter is to describe the merit and limitations of using ultrasound and joint testing procedures to non-invasively quantify the structure and mechanical behaviour of human muscle and tendon *in vivo*. A brief review is given of ultrasound and its ability to quantify muscle and tendon architecture and deformation during muscle contraction. Results from preliminary studies conducted to determine the effects that ageing, temperature variations, and exercise have on human muscle-tendon performance are presented to illustrate applications of ultrasound combined with joint testing.

Skeletal Muscle Mechanics: From Mechanisms to Function. Edited by W. Herzog.

HISTORICAL PERSPECTIVE AND STATEMENT OF THE PROBLEM

The force that skeletal muscle can generate depends non-linearly on such factors as the muscle's length, velocity, level of neural excitation, and past loading and activation history. The structural properties of tendons and, to a lesser extent bones, influence the length and velocity of the muscle and thus the force developed by the muscle. The deformations that occur within specific regions of a muscle-tendon-bone (MTB) unit such as the muscle, the muscle-tendon junction, the tendon, and the tendon-bone junction depend on the force developed and the stiffness or force-deformation behaviour of each particular region.

Testing of muscle, tendon, and bone indicate that the force-deformation characteristics of these tissues are affected by ageing, physical activity, temperature, strain-rate, and loading history (Walker et al., 1964; Elliott, 1965; Abrahams, 1967; Benedict et al., 1968; Tipton et al., 1970; Butler et al., 1978; Watt et al., 1982; Woo et al., 1982; Saltin and Gollnick, 1983; Woo et al., 1986; Woo et al., 1987a, 1987b; Li et al., 1990; Wakata et al., 1990; Staron et al., 1991; Carter and Orr, 1992; O'Brien, 1992). The effects of ageing and altered physical activity occur over time as the architecture, organization, and/or composition of the various regions of the MTB unit change. The effects of temperature, strain-rate, and loading history are acute and result from the visco-elastic nature of MTB tissues, as well as transient architectural changes (e.g., a reduction in collagen cross-linking within connective tissue).

Much is known about the force-deformation behaviour of muscle, tendon, and bone acting as individual tissues, and the mechanisms responsible for this behaviour. However, very little is known about the relative force-deformation characteristics of these tissues acting as a functional unit, the chronic and acute changes experienced by these tissues, and how relative changes in the properties of one tissue affect muscle-tendon performance and joint mechanics *in vivo*.

The primary reason for the paucity of data in these areas has been the inability to simultaneously quantify tissue deformation and loading *in vivo*. Understanding the interaction between these tissues requires simultaneous quantification of the architecture, force, and deformation of each region of the MTB complex during muscle action and/or limb movement. Tendon force and tissue deformation have been quantified using buckle transducers (Gregor et al., 1988; Herzog et al., 1992; Komi, 1992) and piezoelectric crystals implanted in tendon and/or muscle (Askew and Marsh, 1998). However, these approaches are invasive and of limited utility for studies involving humans. One approach that appears promising for studying intact human muscle-tendon units combines ultrasonography with joint-testing techniques.

CAPABILITIES AND LIMITATIONS OF ULTRASONOGRAPHY FOR STUDYING MUSCLE AND TENDON *IN VIVO*

Current state-of-the-art methods used to non-invasively study muscle-tendon units rely on ultrasound and force or joint torque measuring devices. Ultrasound refers to any sound frequency beyond our auditory sensing range (i.e., $>$ 20,000 Hz). Most clinical ultrasound systems use frequencies between 1 and 10 MHz. Typically, clinical ultrasound signals are generated by exciting disc-shaped piezoelectric crystals with voltage impulses. The crystals respond to an impulse by either expanding or constricting in thickness in a very short time. Following a voltage impulse, the crystal attempts to return to its original shape and oscillates in an underdamped response. This mechanical oscillation is transmitted from the crystal to the tissue against which the transducer is directed. Underlying tissues respond by oscillating and transmitting their own characteristic pressure waves. Because piezoelectric crystals resonate and generate an electric signal in response to pressure waves, they may be used as receivers as well as transmitters. Thus, ultrasound transducers induce and record ultrasonic radiation. Complex algorithms convert the electric signals generated by the receiver into an image of the underlying tissues. The depth of tissue that can be imaged and the signal resolution depend on the ultrasound frequency; higher frequencies provide greater resolution but lower depth of penetration.

During the past 10 years, advances in ultrasonography and joint-testing procedures have provided the tools necessary to quantify muscle and tendon architecture, deformation, and force *in vivo*. Ultrasound has been used to quantify muscle fascicle pennation angles (Rutherford and Jones, 1992; Kawakami et al., 1993; Fukunaga et al., 1997a; Ito et al., 1998), muscle fascicle length (Fukashiro et al., 1995; Narici et al., 1996; Fukunaga et al., 1997a; Ito et al., 1998), tendon cross-sectional area (Gillis et al., 1995; Ito et al., 1998), tendon excursion (Fukashiro et al., 1995; Fukunaga et al., 1996; Ito et al., 1998), and tendon stiffness (Fukashiro et al., 1995).

Results from studies of the gastrocnemius muscle/Achilles tendon complex are presented to illustrate the capabilities, limitations, and applications of ultrasound for studying muscle-tendon behaviour. Results from experiments designed to test the resolution and sensitivity of ultrasound for quantifying muscle-tendon architecture and deformation during muscle contraction are presented first, followed by results from three specific applications.

The set-up that was used to test the human gastrocnemius muscle/Achilles tendon complex consisted of a force transducer (Omega Model LCCA 50, 200 N capacity, or LCC 500, 2000 N capacity) attached to a foot strap, to quantify the effort level; a Hitachi EUB 405 Ultrasound System, with a 64-mm, 7.5 MHz linear probe, to record and analyze images of the gastrocnemius muscle/Achilles tendon complex; and a custom bench designed to prevent movement of the lower limb (Figure 18.1). Subjects knelt on the custom bench with their lower leg parallel to the floor and their knee bent 90°. Their right foot was placed into a stirrup, attached via a steel cable to a force transducer mounted to the bench. The knee was positioned against a rigid support to prevent movement of the lower leg horizontally. The length of the steel cable was adjusted to position the foot at 90° relative to the lower leg. A 2-mm diameter plastic tube was taped to the skin distal to the junction of the lateral head of the gastrocnemius and the Achilles tendon. The tube provided a shadow in the ultrasound images that was used as a reference for quantifying tendon excursion.

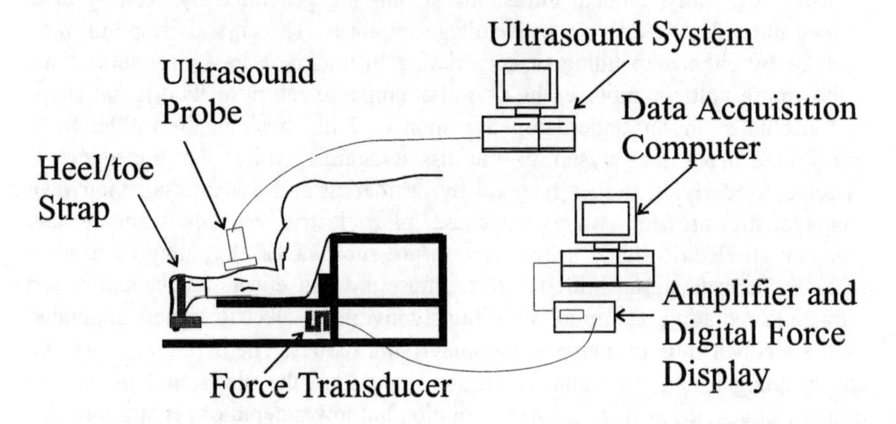

Figure 18.1. The basic set-up used to quantify muscle-tendon architecture and tendon excursion during various isometric muscle efforts. The force transducer was used to quantify the ankle plantarflexion torque developed by the triceps surae muscles. The ultrasound system was used to quantify muscle pennation angle, tendon cross-sectional area, and tendon excursion during muscle contraction.

CHARACTERIZATION OF THE ULTRASOUND SYSTEM ACCURACY AND RESOLUTION

A commercial calibration phantom (Computerized Imaging Reference Systems Inc., Norfolk, VA) was used to test the accuracy and resolution of the Hitachi Ultrasound System. The distance between preset 1-mm diameter nylon rods was determined by digitizing the centroid of each rod as it appeared in the ultrasound images. Both near (within 1 cm of the surface) and far (3 cm from the surface) field depths were tested (Figure 18.2). The Hitachi system calculated the distance between reference rods within ±0.2 mm. It was able to resolve two rods separated by 0.5 mm, the smallest separation distance provided in the phantom.

In addition to understanding the system's capabilities relative to an inert system, it is important to know the accuracy, repeatability, and sensitivity of the system relative to the biological system and the measurements of interest. Results from tests designed to characterize the system capabilities for quantifying pennation angle, fascicle length, tendon cross-sectional area, and tendon excursion are provided in the following sections.

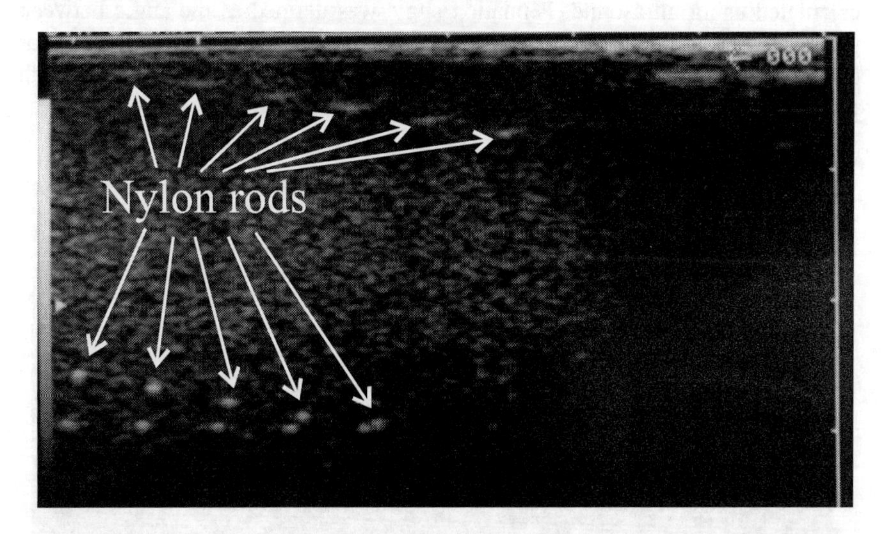

Figure 18.2. Ultrasound image of a calibration phantom. The ultrasound transducer was located along the top surface. 2-mm diameter nylon rods were located within 1 cm of the surface, oriented along an oblique line relative to the surface, and approximately 3 cm below the surface oriented along lines both parallel and oblique to the surface. The rods located at the bottom of the image along a line parallel with the surface are separated by 6 mm (centre-to-centre). The two closest rods are separated by 0.5 mm (edge-to-edge). The ultrasound system determined the distance between rods within 0.2 mm and was able to resolve the rods separated by 0.5 mm.

PENNATION ANGLE

Several experiments were performed to evaluate various methods of quantifying gastrocnemius pennation angle and the repeatability and reliability of these methods. The methods employed were similar to those used and demonstrated to be reliable by others (Rutherford and Jones, 1992; Narici et al., 1996; Fukunaga et al., 1997a; Fukunaga et al., 1997b; Kawakami et al., 1998). The ultrasound probe was placed along the posterior aspect of the lower leg near the gastrocnemius muscle/Achilles tendon junction (MTJ) (Figure 18.1). Ultrasound images were obtained with the probe in this location (Figure 18.3). The skin is located to the top of each image in Figure 18.3, the foot is to the right of the image, and the knee is to the left. The connective tissue surrounding the muscles and fascicles appears white. The lines superimposed on the image represent the line of action of the muscle fascicles and the muscle aponeurosis. The lower line is located over the aponeurosis and the upper line is located over a fascicle. It is evident from Figure 18.3 that fascicles can be observed and a pennation angle calculated using ultrasound. Pennation angle was defined as the angle between the fascicles and the aponeurosis. The pennation angle was found to be 15° for the relaxed muscle (top image in Figure 18.3). These results are consistent with those reported by Wickiewicz et al. (1983) and Kawakami et al. (1998).

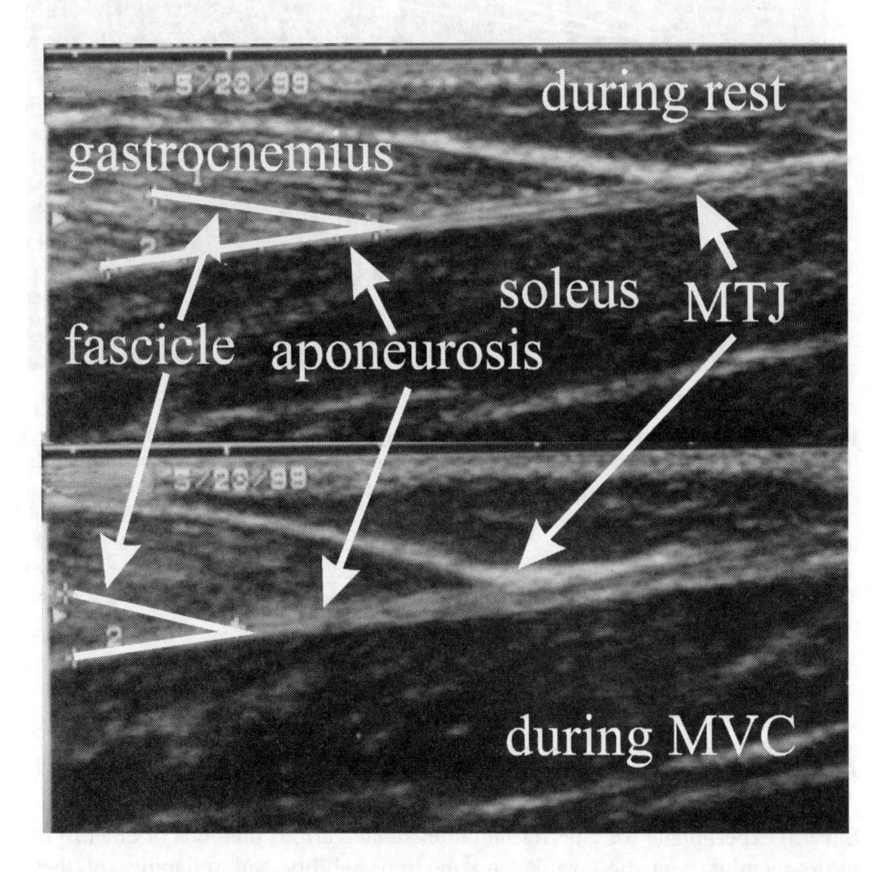

Figure 18.3. Longitudinal images of the gastrocnemius muscle of a five-year-old girl, just proximal to the muscle-tendon junction (MTJ) for a relaxed state (top) and 100% of maximum voluntary contraction (MVC, bottom). The skin is located to the top of the image, the foot to the right, and the knee to the left. The white lines are superimposed on the image to illustrate the line of action of the gastrocnemius muscle fascicles. The pennation angle was calculated to be 15° for the relaxed state and 21° for the activated state.

One subject was tested on three separate occasions over a one-year period to determine the repeatability of pennation angle measurements. Multiple images were analyzed from each test session. Repeat analyses of the same image produced angles within ±2°. Average results from analysis of multiple images taken at different times yielded the same pennation angle within ±2°. Though repeatable results were obtained, it was observed that pennation angle depended on the orientation of the probe and the location of the probe along the muscle belly. Rotation of the probe by 5° or 10° resulted in a 5° change in the pennation angle. Further, the pennation angle tended to be slightly larger for the midbelly of the muscle compared to the distal end, which is consistent with findings by Narici et al. (1996). It appears that, with proper transducer placement and alignment, ultrasonography can be used to quantify pennation angles within 2°. This resolution is sufficient for detecting angle changes that may result during muscle contraction and in response to ageing (illustrated later in this chapter).

Understanding the magnitude of pennation angle changes that occur during dynamic muscle actions, and determining if these angle changes can be detected using ultrasound was also of interest. Illustrated in the lower image in Figure 18.3 is the same muscle as in the top image, but during a maximum isometric plantarflexion effort. The pennation angle increased from 15° in the relaxed state to 21° in the activated state. The pennation angle during muscle activation for this subject is slightly less than the 29° average angle reported by Kawakami et al. (1998).

FASCICLE LENGTH

Quantifying muscle fascicle length and length changes is necessary for understanding muscle performance. Several investigators have used ultrasonography to quantify muscle fascicle length in a variety of muscles (Kawakami et al., 1995; Fukunaga et al., 1997a; Fukunaga et al., 1997b; Ito et al., 1998; Kawakami et al., 1998). Ultrasound images were taken of the midbelly of the right lateral gastrocnemius muscle of a 24-year-old male (Figure 18.4). Fascicle insertion points were digitized (represented by the ends of the superimposed oblique curves in Figure 18.4) and the distance between these points calculated. Fascicle lengths were found to be 6.7 cm in the relaxed state and 5.2 cm in the activated state. These values are higher than the average values reported by Kawakami et al. (1998) for relaxed and maximally activated gastrocnemius muscles, 4.6 cm and 3.1 cm respectively. The fascicle lengths at rest are within the range of values reported by Wickiewicz et al. (1983), however, in this study the knee was extended and the ankle dorsiflexed, which would cause the fascicle lengths to be longer. As with pennation angle, fascicle length can be distorted by the orientation of the ultrasound probe, so care must be taken to align the probe with the orientation of the fascicles. It was also found that fascicle length varied along the length of the muscle, thus it is important to quantify fascicle length at the same location if comparisons are to be meaningful.

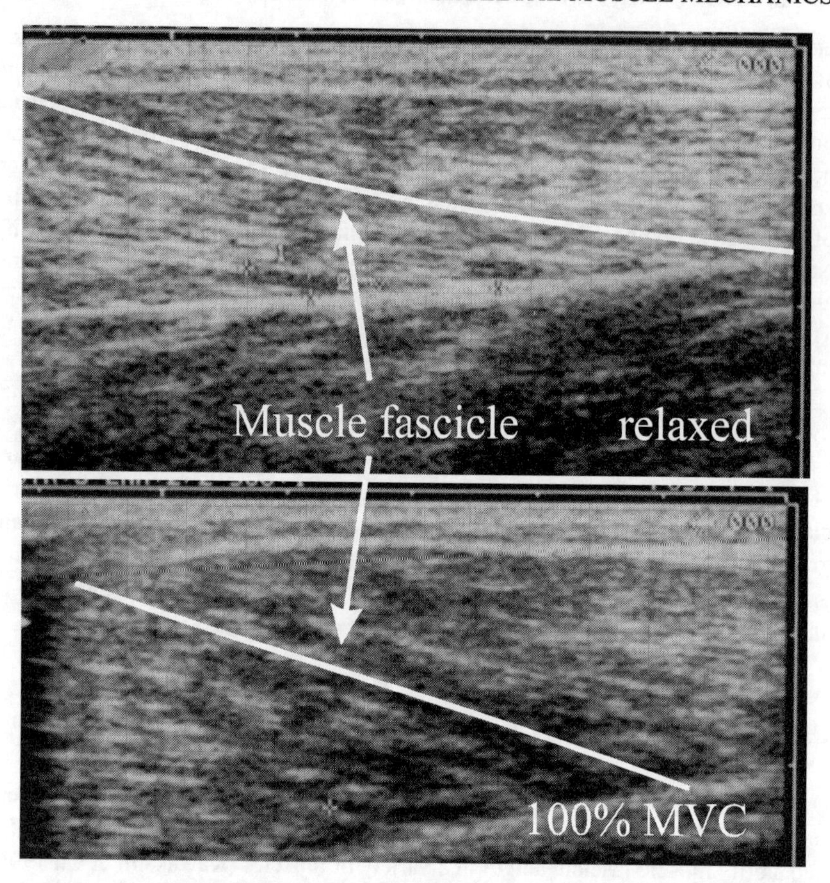

Figure 18.4. A longitudinal image of the midbelly of the lateral gastrocnemius muscle from a 24-year-old male in the rest state (top) and maximally activated state (bottom). The skin is located to the top of the image, the foot to the right, and the knee to the left. The total image width is slightly greater than 6 cm. The white curves are superimposed on the image to highlight muscle fascicles. Fascicle length was 6.7 cm in the relaxed state and 5.2 cm in the activated state. This corresponds to a 22% shortening of the muscle fascicle during contraction.

TENDON CROSS-SECTIONAL AREA

Quantifying tendon cross-sectional area is important if tendon material properties are to be determined. Gillis et al. (1995) and Ito et al. (1998) utilized ultrasound to quantify tendon cross-sectional area in horse and human tendons, respectively. The cross-sectional area of the Achilles tendon in young adults (Figure 18.5) was determined using ultrasound and found to vary between 0.5

and 0.8 cm^2. These values compare very well to values reported by Komi (1992), ranging from 0.44 cm^2 to 0.81 cm^2. Though we were able to estimate tendon cross-sectional area, we believe that further refinement in these procedures is necessary. Our cross-sectional area results varied by as much as 20% depending on the grey-scale contrast used.

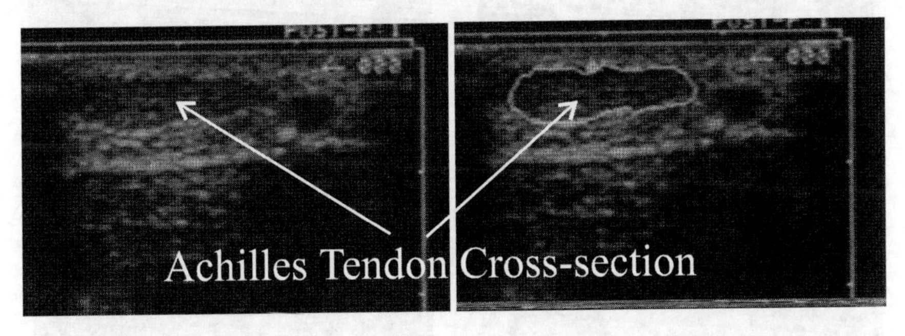

Figure 18.5. An axial image of the right lower leg approximately 2 cm proximal to the calcaneus bone. The image on the left shows the raw image. The image on the right shows the Achilles tendon highlighted with a superimposed white trace that was used to calculate tendon cross-sectional area. The cross-sectional area was calculated to be 0.5 cm^2. Skin is to the top of the image, lateral is to the right.

TENDON EXCURSION

The ability to track tendon excursion is critical for quantifying tendon structural properties. Tendon excursion was quantified as a young adult performed six ankle plantarflexion efforts (Figure 18.6). Skin appears at the top of each image in Figure 18.6, the knee to the left, and the foot to the right. The distance between the MTJ and the midline of the shadow, created by the tube taped to the skin, was used to determine tendon excursion during various muscle efforts. The MTJ translated nearly 3 cm as the effort level increased from rest to a maximum voluntary contraction (MVC). These values are of similar magnitude as those observed by Fukunaga et al. (1996) and Ito et al. (1998), 1.5 cm to 2.0 cm, for translation of the tibialis anterior muscle-tendon junction during maximum voluntary contractions.

Results from the fundamental studies described above demonstrate the feasibility of using ultrasonography to quantify muscle and tendon architecture and deformation during loading. It should be noted that quality ultrasound images were not obtainable from all people. The data presented above were obtained from relatively fit people with average lean body mass. However, even within this population, it was occasionally difficult to obtain clear and consistent ultrasound images. It is expected that image quality would degrade when testing

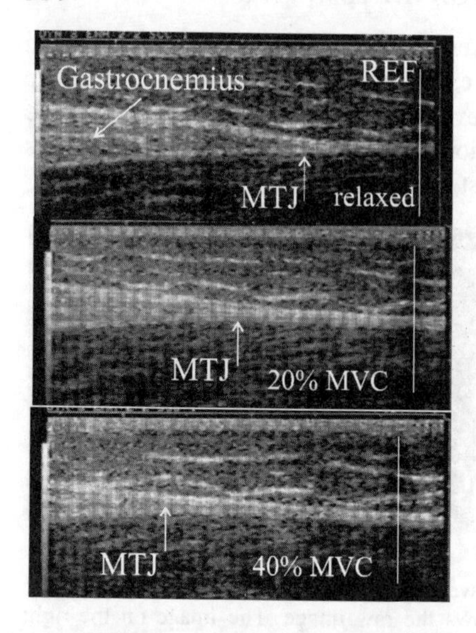

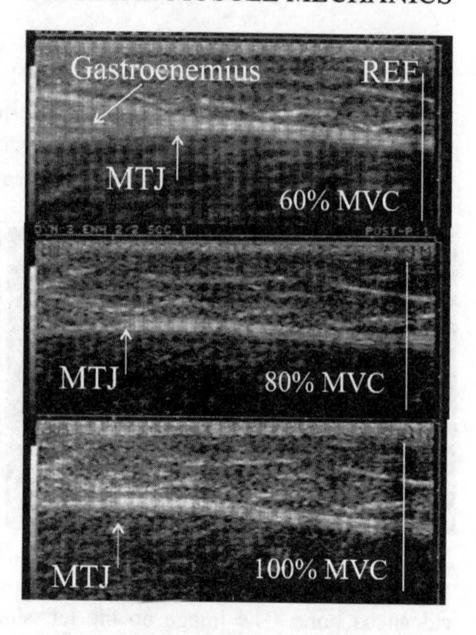

Figure 18.6. Longitudinal images of the lower leg obtained near the gastrocnemius muscle/Achilles tendon junction (MTJ) during six isometric plantarflexion effort levels, expressed as a percentage of the subject's maximum voluntary contraction (MVC). The skin is located to the top of the image, the foot to the right, and the knee to the left. The total image width is slightly greater than 6 cm. The black shadow approximately 1-2 cm from the right border was created by a 2-mm diameter tube taped to the skin. The left arrow on each image identifies the MTJ. The tendon excursion resulting from each effort level was determined by taking the difference between the distance from the MTJ to the reference line for each effort level, and the same distance measured from the rest condition.

people with lower lean body mass. We have also noted that muscle architecture can be easily discerned in certain muscles, but not in others. This is probably due to differences in the connective tissue composition and organization within different muscles. Provided an appropriate muscle is selected, ultrasonography and joint-testing devices can be used to obtain unique and fundamental information about muscle-tendon structure and performance; how structure and performance change with respect to maturation, ageing, and altered physical activity; and how these changes alter joint mechanics and the risk of tissue damage. Examples of three such applications are presented next.

APPLICATIONS

Experiments designed to test three hypotheses pertaining to force-deformation behaviour of the gastrocnemius/soleus-Achilles tendon complex are presented to illustrate the value of the methods described above. The first hypothesis involves chronic changes that occur in the MT unit as a function of the maturation and ageing processes. The second and third hypotheses consider acute changes that occur in the MT unit due to temperature changes and exercise. The three hypotheses are:

1. The Achilles tendon deforms less with ageing relative to the force-generating capacity of the gastrocnemius and soleus muscles.
2. The Achilles tendon deforms more following exercise relative to the force-generating capacity of the gastrocnemius and soleus muscles.
3. The Achilles tendon deforms less following icing relative to the force-generating capacity of the gastrocnemius and soleus muscles.

The bases for these hypotheses stem from isolated mechanical and biochemical testing of ligament and tendon. The mechanical properties of tendon have been found to be associated with both the amount and orientation of collagen, and the interstitial matrix which binds fasciculi together (Elliot, 1965; O'Brien, 1992). Results from mechanical testing of mature and immature ligaments showed that ligaments from older animals tend to be stiffer than ligaments from younger animals (Woo et al., 1986). These data suggest that tendon becomes stiffer with age and thus may deform less during muscle contractions in older individuals. The second hypothesis was based on results from cyclic loading of tendon and ligament. Ligaments and tendons tend to soften during cyclic loading, developing less force per cycle when stretched to the same length, or stretching more per cycle when loaded with the same force (Woo et al., 1982; Woo et al., 1990). The third hypothesis was based on temperature studies of isolated ligaments. Ligaments become stiffer as the temperature decreases (Woo et al., 1987a).

TESTING

Twenty subjects (six children aged 2-8 years, 10 adults aged 18-40 years, and four seniors over 70 years of age) participated in the study to test the first hypothesis. Ten adults participated in the studies to test the second and third hypotheses. The University of California Human Subjects Review Committee approved all protocols, and all subjects gave written consent to participate in these studies.

The testing set-up was the same as that described previously (Figure 18.1). Subjects performed isometric plantarflexion efforts of 20%, 40%, 60%, 80% and 100% of their maximum voluntary contraction (MVC). Effort level was determined from the force output of the force transducer. Three trials were performed at each effort level and the results averaged. The maximum efforts were performed first and used to establish the target force values for the other effort levels. All subsequent effort levels were performed in a random order. During each effort level, the ultrasound system was used to record images of the gastrocnemius muscle/Achilles tendon complex. Still images were digitized to determine muscle fascicle pennation angle and the amount of gastrocnemius muscle/Achilles tendon junction (MTJ) excursion during each effort level. Data were compared between age groups. The effects of icing and exercise were determined by comparing results from the tests described above prior to exercise, following 15-20 minutes of jogging and stretching, and following 10 minutes of ice applied to the skin covering the Achilles tendon.

Pennation angles appear to increase during adulthood and during muscle activation (Figure 18.7). Pennation angle changes very little during childhood, increasing during adulthood by 5° to 10°, and increasing during muscle activation by 5° to 15°.

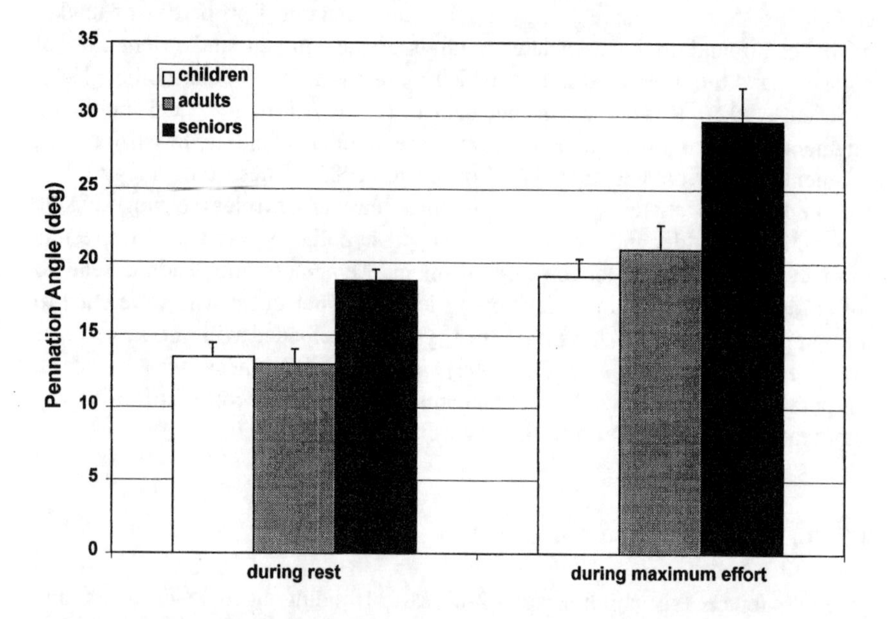

Figure 18.7. Histogram indicating the average pennation angles (±1 Standard Error) for children, adults, and seniors during rest and during a maximum effort. Pennation angle does not change during maturation but increases during adulthood. Pennation angle increases during muscle activation.

Normalized excursion of the MTJ appears to decrease with age (Figure 18.8). MTJ excursion was normalized with respect to tendon rest length to make comparisons between age groups more meaningful. Tendon rest length was defined as the distance from the calcaneus bone to the MTJ for the ankle at 90° and the muscles relaxed. Children were separated into two age groups because it was noticed that young children (<5 years of age) showed different responses than the older children (5-8 years of age). The young children had the greatest normalized MTJ excursion relative to their normalized force-producing capability.

There was very little difference between the five-to-eight-year-old children and the adults. The senior adults showed slightly less normalized MTJ excursion than the younger adults did.

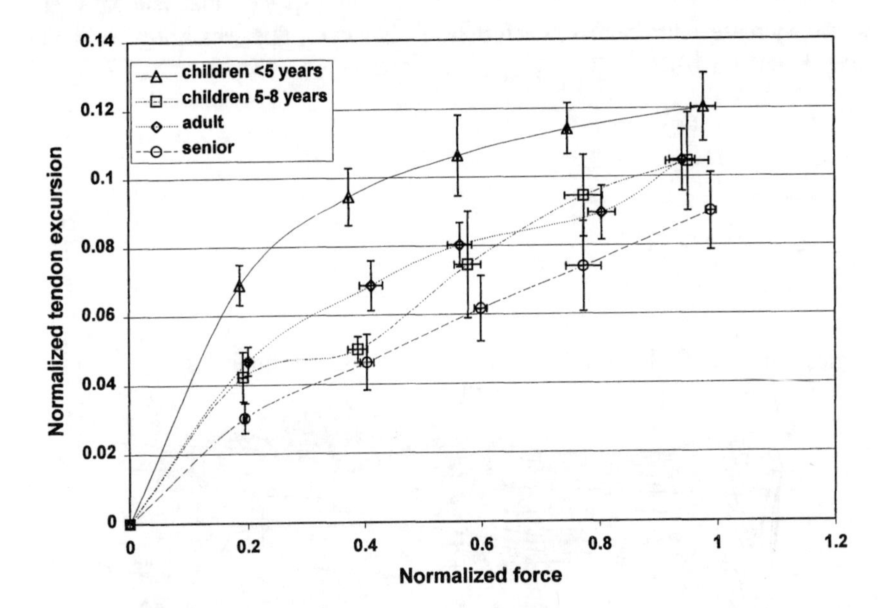

Figure 18.8. Average normalized muscle-tendon junction (MTJ) excursion (±1 Standard Error) as a function of normalized muscle force for different age groups. The seniors had the least normalized MTJ excursion relative to their maximum muscle force, whereas the young children had the greatest normalized MTJ excursion. There was very little difference between the five-to-eight-year-old children and the young adults.

It should be recognized that normalized MTJ excursion does not represent tendon strain. Even though the portion of the foot attached to the foot-plate did not move, the calcaneus bone moved proximally by up to 2 cm during an MVC. This movement occurred in part because of deformation of soft tissue both at the knee support and the foot-plate. However, the primary cause of this movement was translation and rotation of the calcaneus bone. The magnitude of this movement reflects the structural properties of the ligaments and connective tissue around the ankle and the combined force of the soleus and gastrocnemius muscles. The soleus attaches to the Achilles tendon over a large portion of the total Achilles tendon length (Figure 18.9). Thus, the soleus muscle generates a portion of its force distal to the MTJ. A non-uniform force profile is created along the length of the Achilles tendon, with the greatest force being applied at the distal end. Translation of the Achilles tendon/calcaneus bone junction (TBJ) was quantified in an effort to estimate an average Achilles tendon strain.

Translation of the TBJ during various 'isometric' plantarflexion efforts was quantified using a tape measure placed along the back of the leg. The difference between MTJ and TBJ translation, normalized with respect to tendon rest length, was defined as the average tendon strain. It was expected that tendon strain would increase with increased effort level, however, this was not always the case (Figure 18.10).

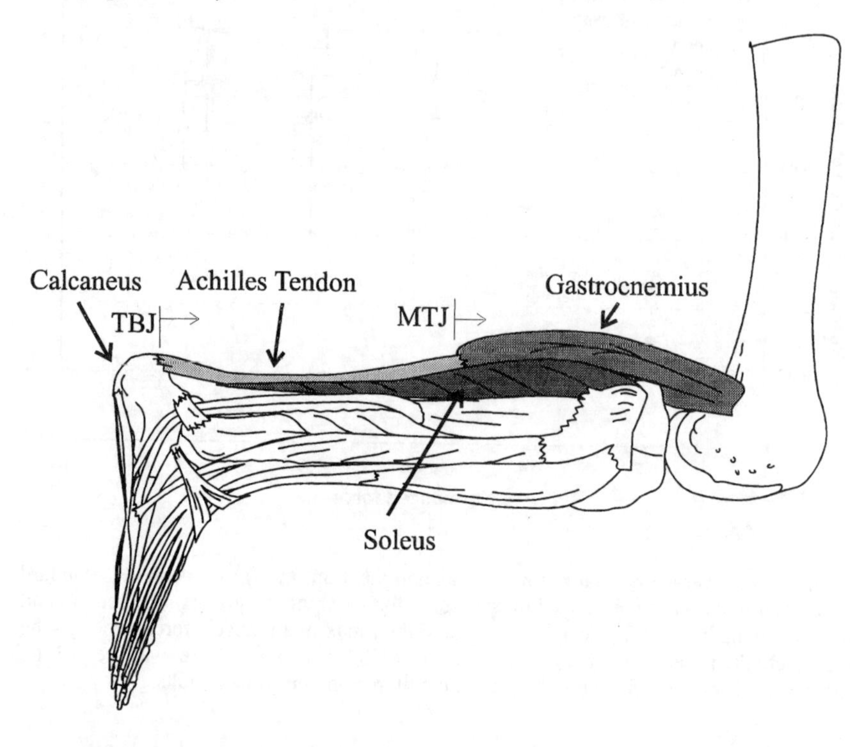

Figure 18.9. Illustration of the lower leg. Translation of the gastrocnemius/Achilles tendon muscle-tendon junction (MTJ) during various 'isometric' plantarflexion efforts was quantified from the ultrasound images. Translation of the Achilles tendon/calcaneus bone junction (TBJ) during various 'isometric' plantarflexion efforts was quantified using a tape measure placed along the back of the leg.

The relationship between tendon strain and normalized force differed among subjects and among age groups. Several subjects showed a peak strain between 40% and 60% of MVC, with a decrease in strain thereafter. Other subjects showed an increase in strain to 40% and very little change in strain thereafter. Several of the children, as well as a few adults, actually showed negative strain at the higher effort levels. The average tendon strains that were observed for the seniors (0 to 3.2%) were greater than those of the adults (0 to 2.7%), which were greater than those of the children (–1% to 1%). These results disprove the first hypothesis and are opposite to what was expected based on the MTJ excursion data. There are at least two possible explanations for these results. First, there may be differences in the stiffness of connective tissue within the ankle joint, along with differences in the Achilles tendon preload between age groups. The Achilles tendon of growing children may experience a greater preload (due to the stretch imposed on the tendon during bone growth) than the Achilles tendon in mature individuals. This would cause the tendon to act along a steeper portion of its force-length curve (i.e., it would be stiffer) and therefore not deform as much during muscle activation. The connective tissue within the ankle joint may be more compliant in the children, allowing the calcaneus bone to translate more during muscle activation, causing the muscle to act along a less optimal portion of the force-length relationship. If the ankle becomes stiffer with ageing, then Achilles tendon strain should decrease for the seniors, provided the relative strength of muscles and tendon remained the same. However, the seniors showed the greatest Achilles tendon strain for the muscle force they could develop. Connective tissue has been shown to become stiffer with age, thus, these results suggest that muscle strength may be greater relative to tendon strength in the seniors.

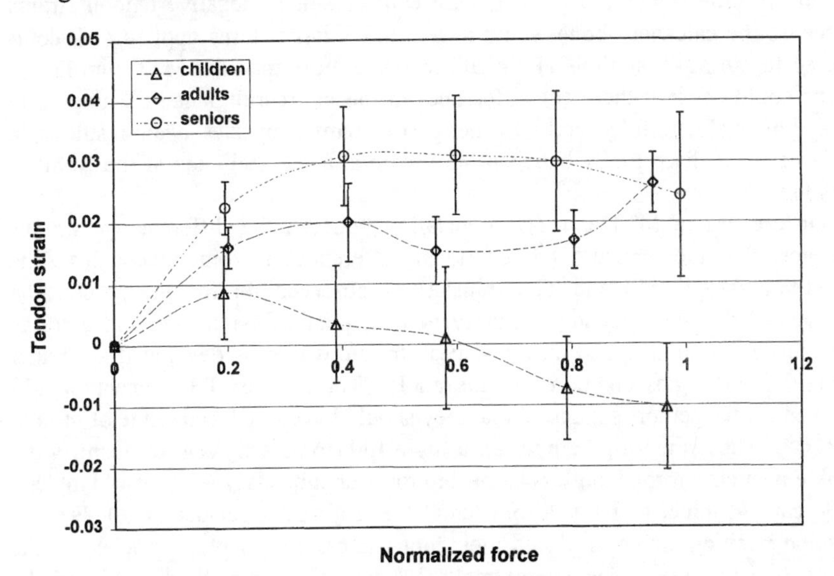

Figure 18.10. Average normalized Achilles tendon strain (±1 Standard Error) for different age groups. The seniors had the greatest average Achilles tendon strains, approximately 3%. Young adults had average strains of about 2%. Children had average strains of 1% during low effort levels and –1% for high effort levels. This suggests that the children had a large force gradient along the length of the Achilles tendon, created by the force developed by the soleus muscle.

Differences in muscle recruitment strategies and muscle morphology may also explain the different tendon strain profiles. The gross appearance of the gastrocnemius and soleus muscles of our subjects varied considerably. Some subjects had very small gastrocnemius muscles, while others had very large muscles. Some subjects had a large portion of their soleus muscle distal to the MTJ, while others appeared to have very little soleus muscle volume distal to the MTJ. Some subjects had a large soleus relative to the gastrocnemius, while others had a large gastrocnemius relative to the soleus. If the strength of the gastrocnemius and soleus muscles varies greatly between subjects, or if these muscles are recruited differently, then the average tendon strain should also vary between subjects. For example, it would be expected that two people able to produce the same force at the distal end of the Achilles tendon, having the same Achilles tendon structural properties, and having different ratios of gastrocnemius to soleus strength, would have different Achilles tendon strain during an MVC. The person with the large gastrocnemius muscles would develop a large force at the proximal end of the Achilles tendon near the MTJ. The person with the large soleus muscle would generate a large portion of the total muscle force distal to the MTJ and thus generate a large force gradient along the length of the Achilles tendon. The latter case could result in negative tendon strains because the calcaneus bone would have a much larger force applied to it compared to the MTJ, and the TBJ could translate more than the MTJ. Similar results could occur if there are differences in muscle recruitment. A young child may not be able to fully recruit his/her gastrocnemius muscles, again resulting in a large force discrepancy between the proximal and distal ends of the Achilles tendon.

Differences in MTJ and TBJ translation between the different age groups suggest that there should be functional differences in joint mechanics. The relatively large MTJ and TBJ translations observed for the young children suggested that they would generate very little plantarflexion torque at extreme plantarflexion angles, and that their peak torque would be developed at greater dorsiflexion angles compared to the adults. The smaller TBJ translation observed for the seniors suggested that they should have a reduced range of motion (ROM). These theories were tested using a Lido Active System to quantify the ankle isometric torque-angle relationship for four subjects (a 72-year-old male, a 38-year-old male, a five-year-old female, and a three-year-old male). From a supine position, subjects placed their foot against a steel plate attached to the Lido dynamometer. The lateral malleolus was aligned with the axis of the dynamometer. Their normal range of ankle motion was determined. Subjects performed maximum isometric plantarflexion efforts in 10° increments over their ROM. The senior adult had the smallest ROM (Figure 18.11). The youngest child was unable to produce significant torque at the plantar-flexed positions as was proposed. However, the five-year-old child had a normalized torque-angle relationship very similar to the adults. These data suggest that there may be functional differences in joint mechanics during early maturation, but that relatively early in life, and throughout life, muscles and tendons adapt to maintain relatively constant normalized joint torque-angle behaviour.

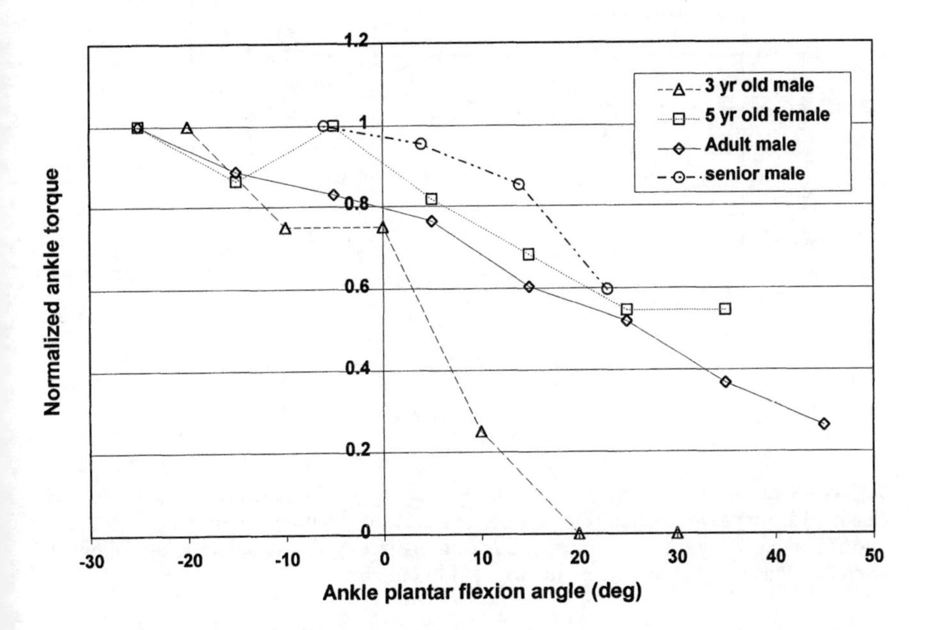

Figure 18.11. Isometric ankle torque-angle data for four subjects of different age. The senior adult had the least range of motion (5° dorsiflexion to 25° plantarflexion). All subjects generated their greatest torque in the greatest dorsiflexed position. The youngest child was unable to generate torque in the fully plantar-flexed position and thus had a steeper torque-angle relationship compared to the other subjects.

The Achilles tendon became more compliant following exercise and icing (Figure 18.12 and Figure 18.13). MTJ excursion increased following exercise and also following icing (Figure 18.12). There was approximately a 0.5% increase in tendon strain following exercise, and an additional 0.5% increase in strain following icing (Figure 18.13). These results support the second hypothesis (i.e., relative tendon deformation would increase following exercise) and disprove the third hypothesis (i.e., relative tendon deformation would decrease following icing). The temperature effect may be due to increased blood flow to the Achilles tendon resulting from a shunting of blood flow from the skin during icing. It may also be the result of a decrease in the tendon rest length. If the tendon constricted slightly in response to the icing, then loading of the tendon was initiated at a shorter tendon rest length, and thus, the tendon acted along a more compliant region of its force-length curve.

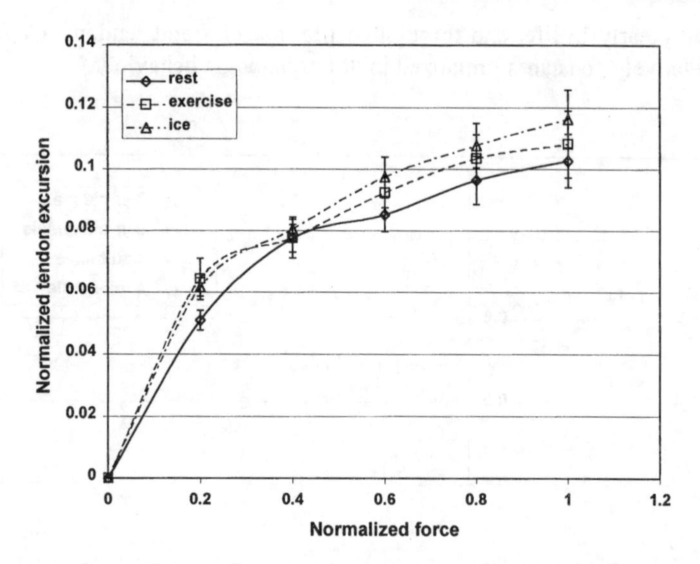

Figure 18.12. Average normalized muscle-tendon junction (MTJ) excursion (±1 Standard Error) as a function of normalized muscle force before exercise, after exercise, and following icing. Exercise increased the normalized MTJ excursion. Icing following exercise tended to increase the normalized MTJ excursion.

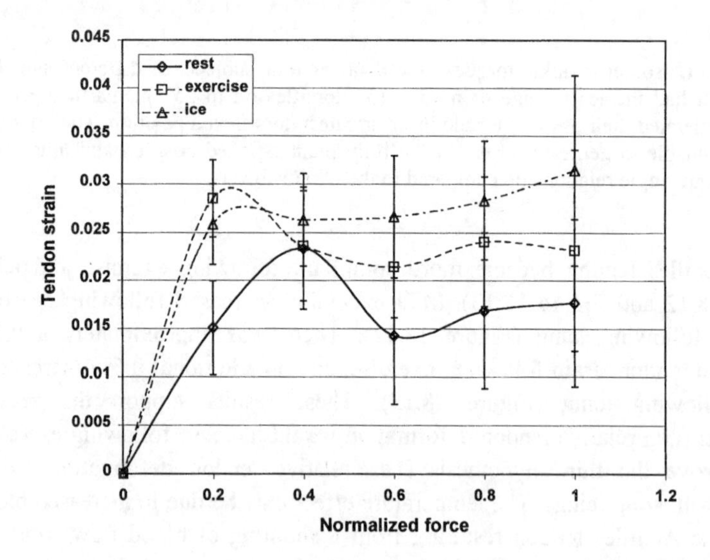

Figure 18.13. Average Achilles tendon strain (±1 Standard Error) as a function of normalized muscle force before exercise, after exercise, and following icing. The average strain induced in the Achilles tendon increased following exercise and following icing.

DISCUSSION

The fundamental studies described in this chapter demonstrate the feasibility of using ultrasonography to quantify muscle and tendon architecture and deformation during loading. The Hitachi Ultrasound System can accurately determine the distance between reference rods within ±0.2 mm and resolve two rods separated by less than 0.5 mm. Ultrasonography can be used to quantify pennation angles, fascicle lengths, tendon cross-sectional area, and translation of specific sites within a muscle-tendon unit (e.g., the muscle-tendon junction). However, these measurements are sensitive to transducer orientation and thus care must be taken to ensure proper transducer alignment and placement. Further, cross-sectional area measurements are sensitive to image contrast, which must be controlled if repeatable results are to be obtained. Provided an appropriate muscle-tendon unit is selected, ultrasonography and joint testing devices can be used to non-invasively study muscle-tendon structure and performance *in vivo*. This approach has several advantages, but also some disadvantages compared to other approaches that are used to study muscle-tendon behaviour.

Muscle-tendon properties have been studied using a variety of invasive and non-invasive techniques applied to animal and human models. There are pros and cons to all approaches. Animal models allow the ultimate force or stress of various MTB structures to be determined. Studies of this nature can be performed to quantify muscle-tendon structural properties as a function of ageing and altered physical activity. This type of information could be highly relevant for understanding injury. However, with animal models there is always concern about extrapolating the findings to humans, and it is not always easy to regulate the physical activity of the animals. Human studies have the advantage in that the data collected pertain directly to the population of interest, the activity level of the subjects can be regulated, and during testing the subjects are relatively co-operative. However, human subjects are not very receptive to having their tissues removed for research purposes, thus requiring the use of non-invasive and non-destructive techniques. The advantage of the procedures described in this chapter are that human tissues are investigated non-invasively *in vivo*. The disadvantages are that the ultimate strengths of specific zones within an MT unit cannot be determined, tracking the displacement of the same site within an MT unit during dynamic activities is challenging, and fractionating joint torques among individual tissue forces can be difficult.

Ultrasonography, combined with joint testing, allows the measurement of important quantities needed to characterize muscle-tendon structure and function *in vivo*. Results from the ageing, exercise, and temperature studies demonstrate the value of this approach, and illustrate that tendon properties are not static and thus muscle force production capabilities and joint mechanics may change with maturation, ageing, exercise, and tissue temperature. These properties must be understood and represented in muscle-tendon models if these models are to provide reasonable estimates of muscle force and joint mechanics *in vivo*.

FUTURE DIRECTIONS

It is anticipated that variations of the procedures described above will be employed in future studies to address a variety of issues related to muscle-tendon function, adaptation, and injury. Specific studies may include quantifying (1) muscle-tendon interactions during stretch-shortening cycles, (2) structural property changes of muscle and tendon in response to strength training, endurance training, and de-training, (3) the effects various diseases have on muscle and tendon structural properties, (4) changes in tissue structural properties leading to overuse injuries, and (5) muscle-tendon architectural and structural changes following muscle-tendon transfers. Knowledge from such studies is fundamental for prescribing exercise programs that promote health without inducing injury, for identifying training programs that will facilitate recovery following an injury, for designing equipment or training programs to prevent such injuries, and for developing surgical interventions that restore joint movement.

Future studies should attempt to combine ultrasound and joint-testing devices with other technologies to study a variety of muscle-tendon units. Many of the hypotheses that were proposed to explain the results presented in this chapter are easily testable with additional information. For example, it was proposed that differences in muscle morphology and muscle recruitment might explain differences in average Achilles tendon strain observed for different age groups. Magnetic resonance imaging and electromyography could be used to characterize differences in muscle morphology and differences in muscle recruitment strategies, respectively, in subjects eliciting different Achilles tendon strain. The triceps surae-Achilles tendon unit, having a relatively complex MT structure, was considered in this chapter. It would be beneficial to study other muscle-tendon units that are structurally less complicated. This would simplify the task of quantifying and interpreting tissue force and deformation.

Future research should focus on developing new techniques and devices. It would be beneficial to have lightweight ultrasound probes that can be adhered to the skin to facilitate MT imaging during dynamic activities. It may be possible to devise biodegradable markers that provide good ultrasound image contrast within biological tissues, and that can be injected into a tissue using a needle. Such markers would eliminate the difficult task of tracking a fuzzy and often arbitrary site within a tissue.

The foundation exists for testing a variety of hypotheses related to muscle-tendon function *in vivo*. Future advances in this area will involve technological innovations, methodological improvements, and insightful data interpretation.

REFERENCES

Abrahams, M. (1967). Mechanical behavior of tendon *in vitro*: A preliminary report. *Medical and Biological Engineering* **5**, 433–443.

Askew G.N., and Marsh, R.L. (1998). The mechanical power output of quail pectoralis muscle *in vitro* during simulated *in vivo* flight strain trajectories. Presented at the 1998 Society for Experimental Biology Meeting in York, England.

Benedict, J., Walker, L., and Harris, E. (1968). Stress-strain characteristics and tensile strength of unembalmed human tendon. *Journal of Biomechanics* **1**, 53–63.

Butler, D., Grood, E., Noyes, F., and Zernicke, R. (1978). Biomechanics of ligaments and tendons. *Exercise and Sports Science Reviews* **6**, 125–181.

Carter, D., and Orr, T. (1992). Skeletal development and bone functional adaptation. *Journal of Bone and Mineral Research* **7**(S2), S389–S395.

Elliott, D. (1965). Structure and function of mammalian tendon. *Biological Review* **40**, 392–421.

Fukashiro, S., Itoh, M., Ichinose, Y., Kawakami, Y., and Fukunaga, T. (1995). Ultrasonography gives directly but non-invasively elastic characteristics of human tendon *in vivo*. *European Journal of Applied Physiology* **71**(6), 555–557.

Fukunaga, T., Ichinose, Y., Ito, M., Kawakami, Y., and Fukashiro, S. (1997a). Determination of fascicle length and pennation in a contracting human muscle *in vivo*. *Journal of Applied Physiology* **82**(1), 354–358.

Fukunaga, T., Ito, M., Ichinose, Y., Kuno, S., Kawakami, Y., and Fukashiro, S. (1996). Tendinous movement of a human muscle during voluntary contractions determined by real-time ultrasonography. *Journal of Applied Physiology* **81**(3), 1430–1433.

Fukunaga, T., Kawakami, Y., Kuno, S., Funato, K., and Fukashiro, S. (1997b). Muscle architecture and function in humans. *Journal of Biomechanics* **30**(5), 457–463.

Gillis, C., Sharkey, N., Stover, S., Pool, R., Meagher, D., and Willits, N. (1995). Ultrasonography as a method to determine tendon cross-sectional area. *American Journal of Veterinary Research* **56**(10), 1270–1274.

Gregor, R.J., Roy, R.R., Whiting, W.C., Lovely, R.G., Hodgson, J.A., and Edgerton, V.R. (1988). Mechanical output of the cat soleus during treadmill locomotion: *In vivo* vs. *in-situ* characteristics. *Journal of Biomechanics* **21**, 721–732.

Herzog, W., Leonard, T.R., Renaud, J.M., Wallace, J., Chaki, G., and Bornemisza, S. (1992). Force-length properties and functional demands of cat gastrocnemius, soleus, and plantaris muscles. *Journal of Biomechanics* **25**(11), 1329–1335.

Ito, M., Kawakami, Y., Ichinose, Y., Fukashiro, S., and Fukunaga, T. (1998). Non-isometric behavior of fascicles during isometric contractions of a human muscle. *Journal of Applied Physiology* **85**(4), 1230–1235.

Kawakami, Y., Abe, T., and Fukunaga, T. (1993). Muscle-fiber pennation angles are greater in hypertrophied than in normal muscles. *Journal of Applied Physiology* **74**, 2740–2744.

Kawakami, Y., Abe, T., Kuno, S-Y., and Fukunaga, T. (1995). Training induced changes in muscle architecture and specific tension. *European Journal of Applied Physiology* **72**(1), 37–43.

Kawakami, Y., Ichinose, Y., and Fukunaga, T. (1998). Architectural and functional features of human triceps surae muscles during contraction. *Journal of Applied Physiology* **85**(2), 398–404.

Komi, P.V. (1992). Biomechanical loading of Achilles tendon during normal locomotion. *Clinics in Sports Medicine* **11**(3), 521–531.

Li, JX., Jee, W.S., Chow, S.Y., and Woodbury, D.M. (1990). Adaptation of cancellous bone to aging and immobilization in the rat: A single photon absorptiometry and histomorphometry study. *Anatomical Record* **227**(1), 12–24.

Narici, M., Binzoni, T., Hiltbrand, E., Fasel, J., Terrier, F., and Cerretelli, P. (1996). *In vivo* human gastrocnemius architecture with changing joint angle at rest and during graded isometric contraction. *Journal of Applied Physiology* **496**(1), 287–297.

O'Brien, M. (1992). Functional anatomy and physiology of tendons. *Clinics in Sports Medicine* **11**(3), 505–520.

Rutherford, O.M., and Jones, D.A. (1992). Measurement of fibre pennation using ultrasound in the human quadriceps *in vivo*. *European Journal of Applied Physiology* **65**, 433–437.

Saltin B., and Gollnick, P. (1983). Skeletal muscle adaptability: Significance for metabolism and performance. Chapter 19 in *Handbook of Physiology: Section 10, Skeletal Muscle*. American Physiological Society, Bethesda, Maryland.

Staron, R., Leonardi, M., Karapondo, D., Malicky, E., Fakel, J., Hagerman, F., and Kikida, R. (1991). Strength and skeletal muscle adaptations in heavy-resistance-trained women after detraining and retraining. *American Journal of Physiology* **70**, 631–640.

Tipton, C., James, S., Mergner, W., and Tcheng, T-K. (1970). Influence of exercise and strength of medial collateral knee ligaments of dogs. *American Journal of Physiology* **218**(3), 894–902.

Wakata, N., Kawamura, Y., Kobayashi, M., and Kinoshita, M. (1990). Biochemical and histochemical studies on skeletal muscle in rat during the course of development. The biochemical properties of type 2c muscle fiber. *Comparative Biochemistry and Physiology* **97B**(1), 201–204.

Walker, L., Harris, E., and Benedict, J. (1964). Stress-strain relationship in human cadaveric plantaris tendon: A preliminary study. *Medical Electronics and Biological Engineering* **2**, 31–38.

Watt, P., Kelly, F., Goldspink, D., and Goldpsink, G. (1982). Exercise-induced morphological and biochemical changes in skeletal muscles of rat. *Journal of Applied Physiology* **53**(5), 1144–1151.

Wickiewicz, T.L., Roy, R.R., Powell, P.L., and Edgerton, V.R. (1983). Muscle architecture of the human lower limb. *Clinical Orthopaedics and Related Research* **179**, 275–283.

Woo, S.L-Y., Gomez, M., Sites, T., Newton, P., Orlando, C., and Akeson, W. (1987b). The biomechanical and morphological changes in the medial collateral ligament of the rabbit after immobilization and remobilization. *Journal of Bone and Joint Surgery* **69A**(8), 1200–1201.

Woo, S.L-Y., Gomez, M., Woo, Y-K., and Akeson, W. (1982). Mechanical properties of tendons and ligaments. II. The relationship of immobilization and exercise on tissue remodeling. *Biorheology* **19**, 397–408.

Woo, S.L-Y., Lee, T.Q., Gomez, M.A., Sato, S., and Field, F.P. (1987a). Temperature dependent behavior of the canine medial collateral ligament. *Journal of Biomechanical Engineering* **109**, 68–71.

Woo, S.L-Y., Orlando, C., Gomez, M., Frank, C., and Akeson, W. (1986). Tensile properties of the medial collateral ligament as a function of age. *Journal of Orthopaedic Research* **4**, 133–141.

Woo, S.L-Y., Weiss, J.A., Gomez, M.A., and Hawkins, D.A. (1990). Measurement of changes in ligament tension with knee motion and skeletal maturation. *Journal of Biomedical Engineering* **112**, 46–51.

19 The Length-Force Characteristics of Human Gastrocnemius and Soleus Muscles *In Vivo*

YASUO KAWAKAMI
Department of Life Sciences (Sports Sciences), University of Tokyo, Komaba, Tokyo, Japan

KENYA KUMAGAI
Department of Sports Sciences, Tokyo Metropolitan University, Tokyo, Japan

PETER A. HUIJING
Faculty of Kinesiology and Institute for Basic and Clinical Kinesiology, Free University of Amsterdam; and Integrated Biomedical Engineering for Restoration of Human Function, Biomedical-Technical Institute, University of Twente, Enschede, The Netherlands

TAKAO HIJIKATA
Faculty of Medicine, Gunma University School of Medicine, Gunma, Japan

TETSUO FUKUNAGA
Department of Life Sciences (Sports Sciences), University of Tokyo, Komaba, Tokyo, Japan

INTRODUCTION

The maximal isometric force exerted by a skeletal muscle is a function of its length. This length-dependence of muscle force production reflects length-force characteristics of muscle fibres and individual sarcomeres, and is converted at the joint into the relationship between joint position and torque. In humans, because of the difficulty in observing actual muscle behaviour, the relationship between joint position and torque has been frequently studied to represent the length-force characteristics of muscle (Ismail and Ranatunga, 1978; Sale et al., 1982; Marsh et al., 1981). However, the force exerted by muscle fibres is modified by their geometric arrangement, the structure of the joint, and the location of the tendon with respect to the bone, before it appears at the joint as torque. Thus, observations of total joint performance give little information about individual muscle behaviour.

Skeletal Muscle Mechanics: From Mechanisms to Function. Edited by W. Herzog.

Among the few studies on human muscle mechanics that accounted for the above considerations, Cutts (1988), Herzog et al. (1990), and Lieber et al. (1994) reported that the operating range of sarcomere lengths differs between muscles, and in some muscles, only parts of the length-tension curve of sarcomeres were used for the whole range of joint motion. Furthermore, in many cases, joint angles at which peak muscle force and peak torque are observed differ (Marshall et al., 1990; Leedham and Dowling, 1995), suggesting that the joint position-torque relationship is the result of different length-force curves of the agonist muscles. These studies demonstrated the necessity for investigating the behaviour of individual muscles during joint actions.

However, previous studies only estimated the behaviour of contracting muscles from theoretical models or anatomical measurements performed on human cadaver specimens; detailed study of actively contracting muscle fibres has been neglected. Recently, it has been shown (Fukunaga et al., 1997; Kawakami et al., 1998) that there is considerable change in human muscle architecture (geometrical arrangement of muscle fibres) during contraction. Therefore, data on human muscle architecture derived from fixed tissues of cadavers might not accurately represent the architecture of actively contracting muscles. Consequently, there are advantages in using non-invasive techniques to determine the architecture of human muscles *in vivo*.

In this chapter, recent data obtained by the authors on architecture of human triceps surae muscles are presented. Based on the results, length-force relationships of these muscles were determined. It was found that internal shortening of muscle fibres during isometric contractions resulted in a large shift of the operating range of sarcomeres to shorter lengths, and that changes in pennation angle further resulted in modification of the length-force curve.

ARCHITECTURE OF THE HUMAN TRICEPS SURAE MUSCLES *IN VIVO*

The triceps surae muscles are the main synergists for plantarflexion (Fukunaga et al., 1992; Murray et al., 1976), but they have different architectural properties, such as muscle length, fascicle length, and pennation angles (Cutts, 1988; Friedrich and Brand; 1990, Wickiewicz et al., 1983). In addition, the gastrocnemii are two-joint muscles crossing the knee and ankle joints, while the soleus is a single-joint plantarflexor. Consequently, the relationships between joint angles (knee and ankle), muscle (fascicle) lengths, and pennation angles are specific to the individual muscles. We studied the relationships between joint angles and muscle architecture (lengths and angles of fascicles) of human triceps surae muscles *in vivo* in passive (relaxed) and active (contracting) conditions to explore the functional implications.

ULTRASONIC MEASUREMENT

We have developed a technique to visualize human skeletal muscles based on ultrasound. With the use of B-mode ultrasonography, sectional images of muscles and tendinous tissues (tendons and aponeuroses) are visualized in a two-dimensional plane (Figure 19.1). The precision of ultrasonic image reconstruction has been confirmed (Kawakami et al., 1993). When we place the ultrasonic transducer in the longitudinal direction of the muscle, echoes from inter-fascicle connective tissues are visualized, and fascicle length can be measured. The angles at which fascicles arise from the aponeuroses can also be measured as fascicle angles. The measurements are fairly reproducible (Fukunaga et al., 1997; Kawakami et al., 1993, 1995). The reliability of fascicle length and angle measurement have been confirmed from a comparison with manual measurements on human cadavers (Kawakami et al., 1993; Narici et al., 1996).

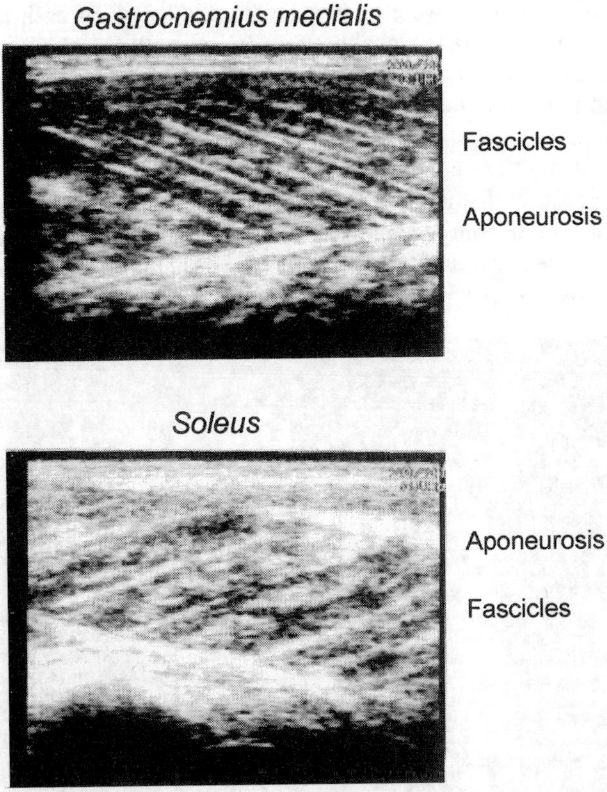

Gastrocnemius medialis

Fascicles

Aponeurosis

Soleus

Aponeurosis

Fascicles

Figure 19.1. Ultrasonic images of the human gastrocnemius and soleus muscles. The ultrasonic probe was aligned on the longitudinal axis of the fascicles. Inter-fascicle connective tissues are visualized, from which fascicle length can be measured. Also, angulation of fascicles with respect to the aponeuroses can be measured as fascicle angles.

MUSCLE FIBRE AND FASCICLE

Before studying fascicle architecture in detail, it is important to determine the differences between fascicle and muscle fibre. Muscle fibres are packed in bundles (fascicles) which extend from the proximal to the distal tendon. In many cases, when investigators refer to muscle-fibre length, they actually refer to fascicle length (Friedrich and Brand, 1990; Scott et al., 1993, 1996; Spoor et al., 1991). However, in some muscles, fibres have been shown to terminate mid-fascicularly and intrafascicularly; thus terminating fibres overlap with each other (Loeb et al.; 1987, Trotter, 1993). In this case, fascicle length and fibre length do not correspond, and the changes in fascicle length might not be the same as those in muscle fibres.

We investigated muscle-fibre architecture within fascicles by looking at the geometry of motor end-plates of the fibres in a human cadaver. The triceps surae muscles (medial and lateral gastrocnemius and soleus muscles) were dissected and blocks of the muscle were removed along the length of each muscle. The blocks were stained for acetylcholinestrase (Cewis, 1961), which was concentrated at the motor end-plate of each muscle fibre. It was found that end-plates were located in the middle of the fascicles, and formed a band in the middle of the muscle in the longitudinal direction (Figure 19.2). Further microscopic observation of the fibre bundles revealed that there was no further motor end-plate zone except for that in the middle of the fascicle. This finding suggests that individual fibres run from tendon to tendon in human triceps surae, and that the lengths of fascicle and muscle fibres are the same; therefore ultrasonography can be used to study changes in muscle-fibre length.

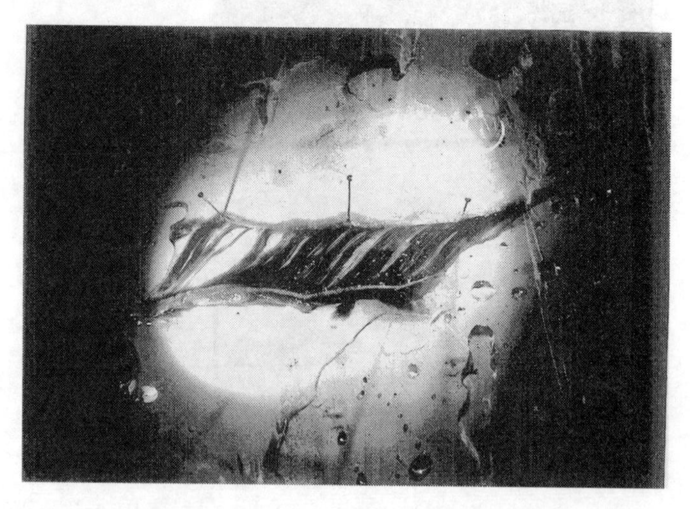

Figure 19.2. A block of the soleus muscle of a human cadaver. The block was cut from the midbelly of the muscle in the longitudinal direction, and stained for acetylcholinestrase, shown as black shades in the middle of the fascicle in this picture.

VARIATION OF FASCICLE ARCHITECTURE WITHIN MUSCLE

Muscle architecture is typically studied at a certain position within a muscle (Fukunaga et al., 1997; Kawakami et al., 1993; Lieber et al., 1994). However, there might be variability in architecture within a muscle, as suggested previously (Willems and Huijing, 1994; Scott et al., 1993). Therefore, behaviour of muscle fibres during contraction might vary within a muscle. In addition, although muscle fibres are arranged in three-dimensional space (Lam et al., 1991; Otten, 1988; Scott et al., 1993), not much is known about muscle architecture in three dimensions.

We studied intramuscular variation in fascicle arrangement of the medial gastrocnemius muscle (MG) in humans (Kawakami et al., 2000). We measured fascicle lengths and fascicle angles of MG with the knee extended and the ankle at 90°. Measurements were carried out with the plantarflexor muscles relaxed, and activated to produce isometric plantarflexion at a level of 50% of maximum voluntary force. Ultrasonic images were obtained from eight positions of the midbelly of MG, separated by 2-3 cm. Fascicle lengths were virtually uniform in the relaxed and contracted conditions (54-57 mm [relaxed] and 36-39 mm [contracted], average of six subjects). Fascicle lengths shortened substantially because of contraction (Figure 19.3), and the amount of shortening (30-34%) did not differ between the eight positions. The thickness of the muscle varied among positions but did not change during contraction.

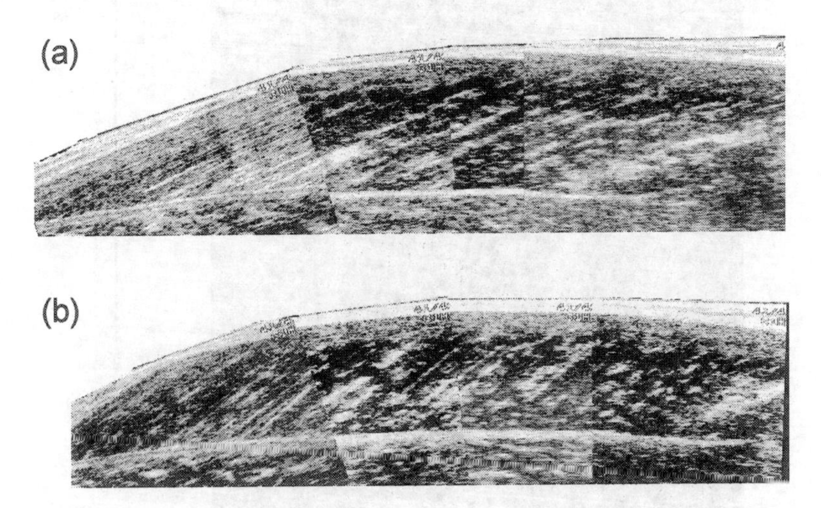

Figure 19.3. Longitudinal ultrasonic images of the medial gastrocnemius muscle. The images were obtained from the midbelly. The distal end of the muscle is on the left. (a) Relaxed condition, (b) during contraction. Fascicle length decreased during contraction, which was accompanied by an increase in fascicle angle.

This study showed that muscle fibres shorten during fixed-end (isometric) contraction, which is in agreement with previous reports from animal muscles (Hoffer et al., 1989; Griffiths, 1987; Trestik and Lieber, 1993). The shortening of the muscle fibres might occur at the expense and stretching of elastic tendinous structures (tendons and aponeuroses) that anchor muscle fibres to bones. To further study the three-dimensional changes in muscle architecture caused by contraction, attempts are presently being undertaken to reconstruct ultrasonic images three-dimensionally (Figure 19.4).

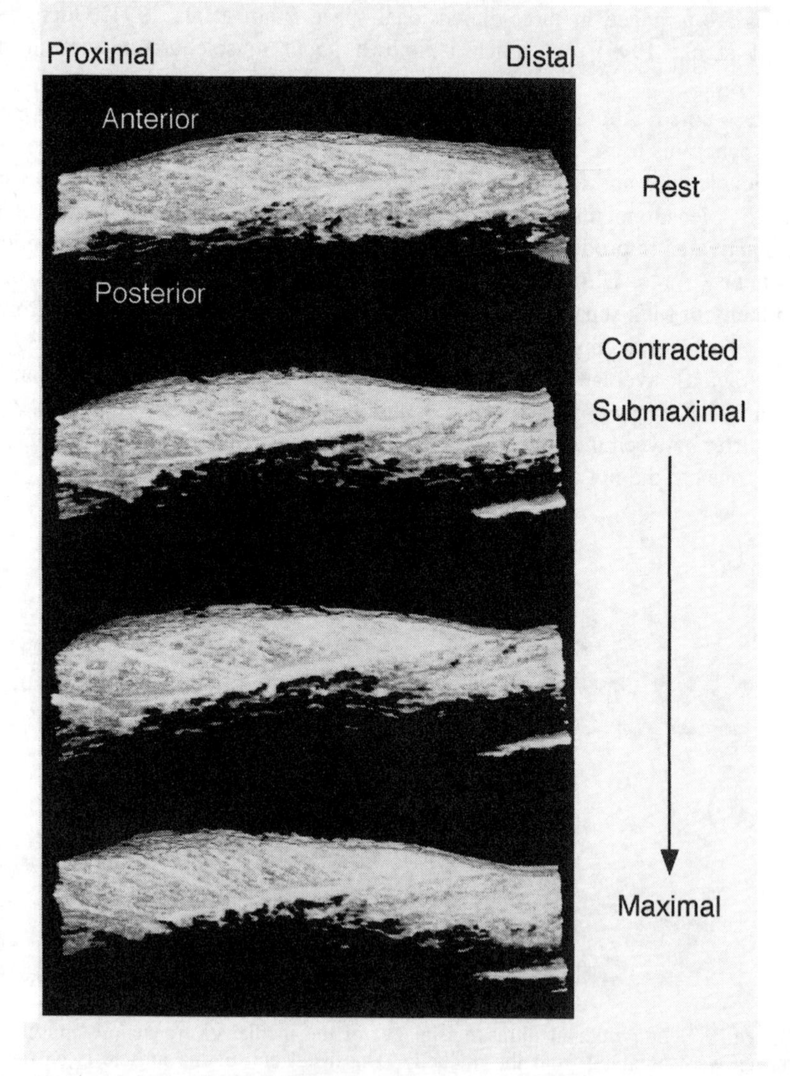

Figure 19.4. Three-dimensional reconstruction of ultrasonic images of the medial gastrocnemius muscle. From top to bottom, relaxed to maximal contraction.

LENGTH-FORCE CHARACTERISTICS OF THE GASTROCNEMIUS AND SOLEUS MUSCLES *IN VIVO*

We studied the relationship between fascicle architecture and joint position for the midbellies of the medial gastrocnemius (MG) and soleus (Sol) muscles while the subject was relaxed (passive condition), and performed maximal isometric plantarflexor contractions (active condition) (Kawakami et al., 1998). The ankle was kept at 15° dorsiflexion, and 0°, 15°, and 30° plantarflexion, with the knee at 0°, 45°, and 90°. From the ultrasonic images, fascicle lengths and fascicle angles were measured. Fascicle lengths differed at the different joint positions in MG and Sol, and decreased when the muscle was contracted (Figure 19.5). Fascicle angles increased during contraction, with MG angles showing greater variability than Sol angles, ranging from 22° to 67° (Figure 19.6).

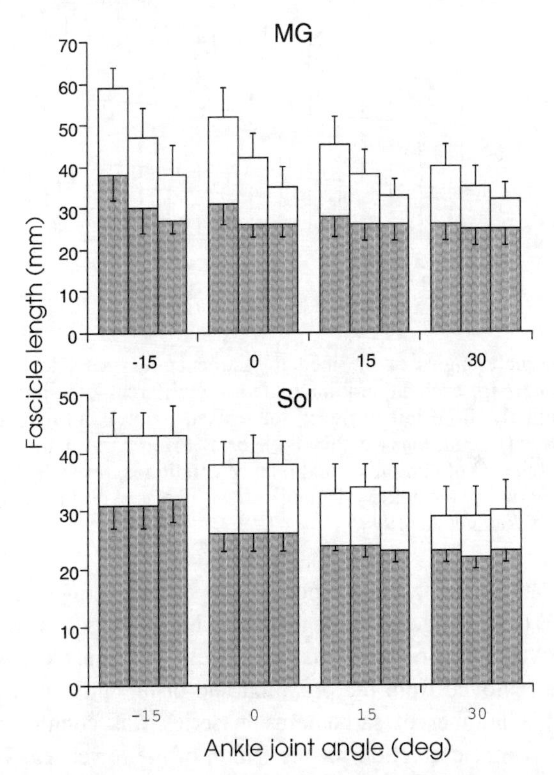

Figure 19.5. Fascicle lengths of the medial gastrocnemius (MG, top) and soleus (Sol, bottom). Three bars for each ankle-joint angle show different knee-joint angles (0°, 45°, and 90°, respectively, from left to right), for relaxed (white plus shaded) and maximally contracted (shaded) conditions (within each bar). Error bars in the upper and lower directions represent SD of relaxed and contracted conditions, respectively. Fascicle length changed as a function of the ankle-joint angles, and shortened during contraction. (Drawn from data of Kawakami et al., 1998.)

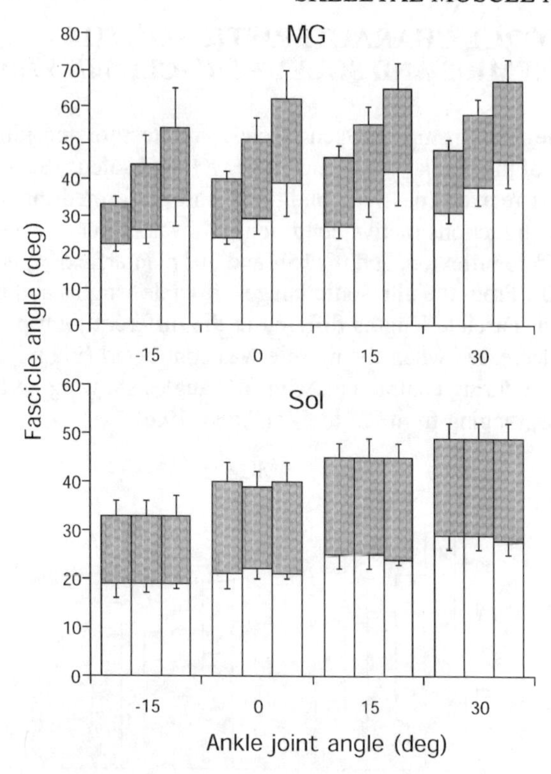

Figure 19.6. Fascicle angles of the medial gastrocnemius (MG, top) and soleus (Sol, bottom). Three bars for each ankle-joint angle show different knee-joint angles (0°, 45°, and 90°, respectively, from left to right), for relaxed (white) and maximally contracted (white plus shaded) conditions (within each bar). Error bars in the upper and lower directions represent SD of contracted and relaxed conditions, respectively. Fascicle angle changed as a function of the ankle-joint angles, and increased during contraction. (Drawn from data of Kawakami et al., 1998.)

From the changes in fascicle length, changes in average sarcomere length were estimated to investigate the length-force characteristics of MG and Sol. For this purpose, MG and Sol were dissected from one male cadaver, blocks of fascicles were removed from the proximal and distal ends and midbelly of the muscle, and the number of sarcomeres in series was counted and averaged. These numbers, after correction for the length of the lower leg, were combined with the directly measured fascicle lengths to estimate average sarcomere length changes of MG and Sol *in vivo*. Average sarcomere length ranged from 1.86 to 3.40 μm for MG, and from 2.23 to 3.19 μm for Sol in the passive condition. These lengths are on the ascending, plateau, and descending portions of the length-tension curve of sarcomeres for humans (Walker and Schrodt, 1974). In the active condition, because of internal shortening of fascicles, sarcomere lengths were shifted to shorter lengths, and they ranged from 1.48-2.24 μm and

1.69-2.30 μm for MG and Sol, respectively. Thus, in the active condition, only the ascending limb of the length-force curve was used in both muscles (Figure 19.7).

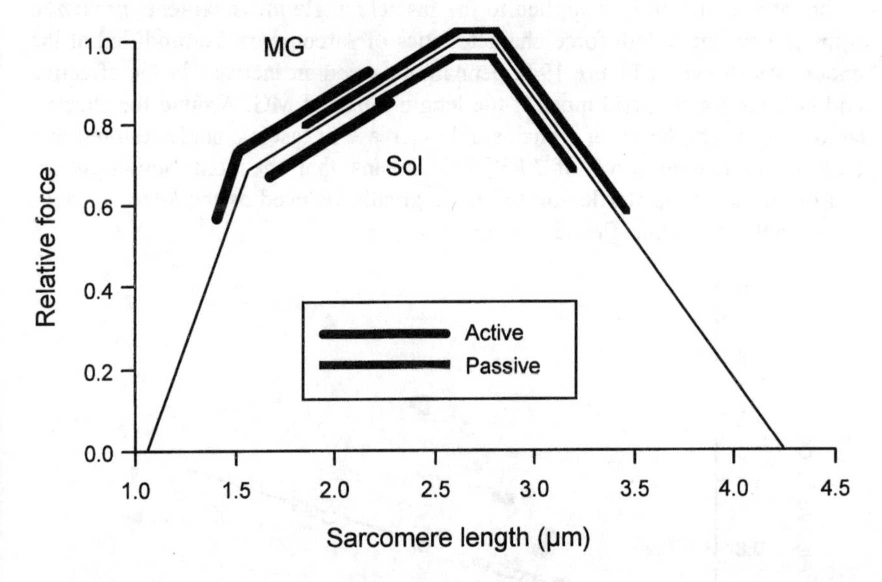

Figure 19.7. The length-force curve of human sarcomeres and operating ranges of the medial gastrocnemius (MG) and soleus (Sol) muscles. For the passive condition, the sarcomere length range of both muscles was on the ascending, plateau, and descending portions of the length-tension curve. For the active condition, because of fascicle shortening, the sarcomere length range was shifted to the ascending limb of the length-force curve for both muscles.

PENNATION EFFECT

There are many models proposed to explain the effect of pennation on force-producing characteristics of skeletal muscle (e.g., Gans and de Vree, 1987; Huijing, 1992; Otten, 1988; van Leeuwen and Spoor, 1993). Based on muscle geometry and volume constraints during contraction, Epstein and Herzog (1998) mathematically deduced the following equation,

$$F_f = F_t \cos(\beta) / 2 \cos(\alpha + \beta) \quad (19.1)$$

where F_f and F_t are the force component in the direction of muscle fibre and tendon, and α and β are angles of muscle fibres and aponeurosis with respect to the line of action of the muscle, respectively. If we assume that $\beta=0$ for the gastrocnemius and soleus muscles (Kawakami et al., 1998), the above equation may be rewritten as:

$$F_t = 2 \cdot F_f \cos(\alpha) \quad (19.2)$$

This means that the force exerted by muscle fibres is reduced by a factor of the cosine of the pennation angle and enhanced by a factor of two when it is transmitted to tendon.

The latter equation was applied to the fascicle angle measurements *in vivo* to estimate how the length-force characteristics of sarcomeres are modified at the tendon. As shown in Figure 19.8, pennation caused an increase in the effective tendon force for Sol, and most of the length range of MG. Around the shortest sarcomere length, however, considerable increase in fascicle angle resulted in a decrease in tendon force of MG. This means that the contribution of the gastrocnemius to plantarflexion torque is greatly reduced as the knee is flexed and the ankle is plantar-flexed.

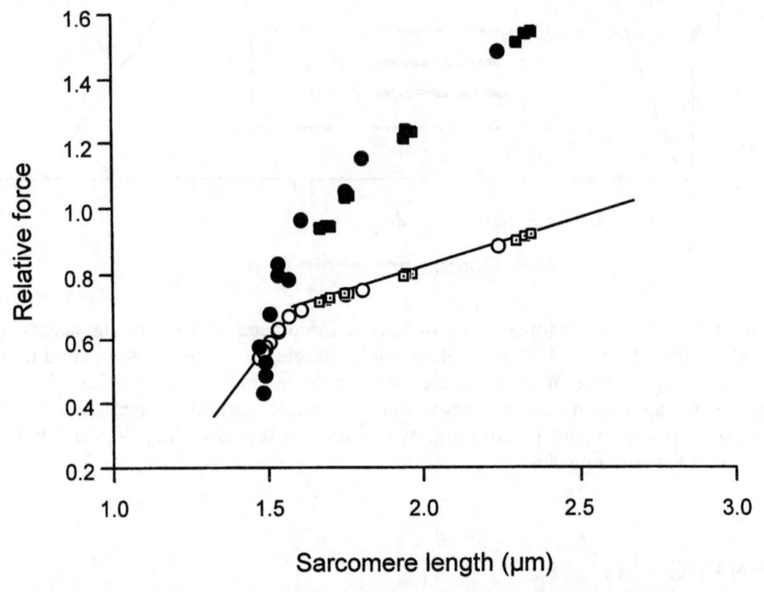

Figure 19.8. The length-force curve of the medial gastrocnemius (open circles) and soleus (open squares). The length range of the two muscles in the active condition (Figure 19.7) was enlarged. Due to the pennation effect (see text), the force exerted by sarcomeres was enhanced at the tendon (closed symbols) for the soleus. For the gastrocnemius, there is a loss of force at the tendon when the muscle is at its shortest length.

Based on these results and the physiological cross-sectional area (CSA) of the triceps muscles (Fukunaga et al., 1992), muscle forces of MG and Sol were estimated. We assumed that the force of each muscle is proportional to its physiological CSA (Close, 1972). Figure 19.9 shows the estimated relative tendon force of MG and Sol at different joint angles. Also plotted is the Achilles tendon force estimated from the plantarflexion torque and previously reported values of moment arm length of the Achilles tendon (Rugg et al., 1990). The combined tendon force of MG and Sol agreed well with the estimated Achilles tendon force; both were increasing with increasing muscle length.

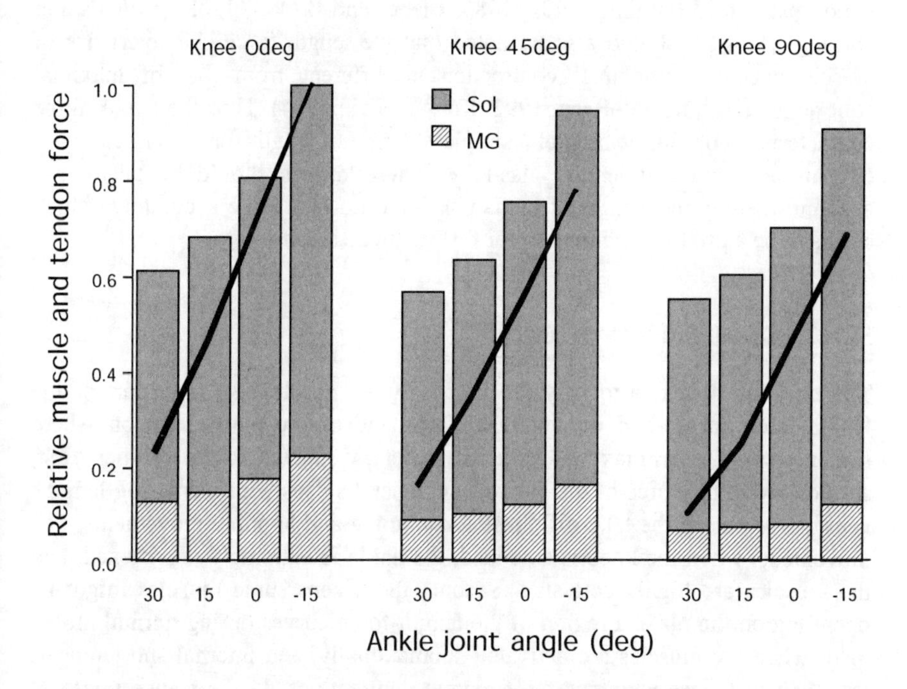

Figure 19.9. Estimated relative tendon force of the medial gastrocnemius (MG) and soleus (Sol) (bars). Data are normalized to the maximal value when the knee is extended (0°) and the ankle is dorsiflexed at −15°. Lines in the bars are estimated Achilles tendon forces (also normalized to the maximal value).

Cutts (1988) reported that the operating range of sarcomere length of the triceps surae covers the entire range of the length-force curve. On the contrary, Herzog et al. (1990) suggested that there are knee/ankle joint configurations where the gastrocnemius muscle is not able to produce active force. These studies were based on anatomical measurements of human cadavers. The present study showed that both the gastrocnemius and soleus muscles use only the ascending limb of the length-force curve. This is because of the tendon elasticity, which causes substantial internal fibre shortening during isometric contractions. Thus, there is a considerable difference between passive and active muscles, and care must be taken when incorporating data from cadaver muscles into the analysis and interpretation of *in vivo* muscle function.

The above calculations have many assumptions, some of which might be inaccurate or too simplified. For example, in many muscles, the aponeurosis angle β is not zero (Zuurbier and Huijing, 1993); thus the effect of this angle should also be considered in muscle models (Huijing, 1992). In addition, the homogeneity of sarcomere length and series sarcomere numbers within a muscle is not guaranteed (Huijing, 1981, 1985; Lieber and Baskin, 1983; Willems and Huijing, 1994). It should also be noted that the length-force characteristics of muscle during submaximal contraction is different from that of maximal contraction (Huijing and Baan, 1992; Roszek et al., 1994). Thus the length-force characteristics of muscle are not a simple function of length-force characteristics of muscle fibres and tendon elasticity. These factors should be taken into account to study mechanics of human skeletal muscles more accurately, which seems to be a promising direction for further investigation.

FUNCTIONAL IMPLICATIONS

The question arises as to why the triceps surae muscles use only part of the length-force curve, and do not take advantage of the plateau region where sarcomeres can exert maximal force. At present, we have two possibilities: first, an increase in the force of the gastrocnemius and soleus with increasing lengths is effective during the take-off phase for jumping and hopping movements, i.e., movements when the knee is extended, the ankle is extremely dorsiflexed, and the muscles are highly activated. Second, the triceps surae muscles might be operating on the plateau region of the length-force curves during normal movement when the muscles are activated submaximally, and internal shortening is less than that observed here for maximal contractions. In the stance phase of walking, the knee-joint angle is between 0° and 30°, and the ankle-joint angle ranges between 15° dorsiflexion and 0° plantarflexion (Winter, 1984). It appears that the gastrocnemius and soleus operate in the high-force region of the length-force relationship during locomotion. Studies are currently under way to test these two speculations. Future studies should focus on *in vivo* behaviour of muscles during various human movements. These studies might greatly contribute to our understanding of the link between muscle behaviour and joint performance.

ACKNOWLEDGEMENTS

The authors would like to acknowledge the excellent technical support provided by Guus C. Baan (Institute for Basic and Clinical Kinesiology, Free University of Amsterdam) in the determination of the number of sarcomeres in series within fibres. We also express our gratitude to Dr. Harunori Ishikawa (Department of Anatomy, Gunma University School of Medicine) for his assistance with cadaveric allocations and assistance.

REFERENCES

Cewis, P.R. (1961). Histochemistry of cholinestrase. *Bibliotheca Anatomica* **2**, 11–20.

Close, R.I. (1972). Dynamic properties of mammalian skeletal muscle. *Physiological Reviews* **52**, 129–197.

Cutts, A. (1988). The range of sarcomere lengths in the muscles of the human lower limb. *Journal of Anatomy* **160**, 79–88.

Epstein, M., and Herzog, W. (1998). *Theoretical Models of Skeletal Muscle: Biological and Mathematical Considerations.* John Wiley & Sons Ltd., Chichester, England.

Friedrich, J.A., and Brand, R.A. (1990). Muscle fiber architecture in the human lower limb. *Journal of Biomechanics* **23**, 91–95.

Fukunaga, T., Ichinose, Y., Ito, M., Kawakami, Y., Fukashiro, S. (1997). Determination of fascicle length and pennation in a contracting human muscle *in vivo*. *Journal of Applied Physiology* **82**, 354–358.

Fukunaga, T., Roy, R.R., Shellock, F.G., Hodgson, J.A., Day, M.K., Lee, P.L., Kwong-Fu, H., and Edgerton, V.R. (1992). Physiological cross-sectional area of human leg muscles based on magnetic resonance imaging. *Journal of Orthopaedic Research* **10**, 926–934.

Gans, C., and de Vree, F. (1987). Functional bases of fiber length and angulation in muscle. *Journal of Morphology* **192**, 63–85.

Griffiths, R.I. (1987). Ultrasound transit time gives direct measurement of muscle fiber length *in vivo*. *Journal of Neuroscience Methods* **21**, 159–165.

Herzog, W., Abrahamse, S.K., and ter Keurs, H.D.J. (1990). Theoretical determination of force-length relations of intact human skeletal muscles using the cross-bridge model. *Pflügers Archiv* **416**, 113–119.

Hoffer, J.A., Caputi, A.A., Pose, I.E., and Griffiths, R.I. (1989). Roles of muscle activity and load on the relationship between muscle spindle length and whole muscle length in the freely walking cat. *Progress in Brain Research* **80**, 75–85.

Huijing, P.A. (1981). Bundle length, fibre length and sarcomere number in human gastrocnemius. *Journal of Anatomy* **133**, p. 132.

Huijing, P.A. (1985). Architecture of the human gastrocnemius muscle and some functional consequences. *Acta Anatomica* **123**, 101–107.

Huijing, P.A. (1992). Mechanical muscle models. In: *Strength and Power in Sport.* Komi, P.V. (ed.), Blackwell Scientific Publications, London, pp. 130–150.

Huijing, P.A., and Baan, G.C. (1992). Stimulation level-dependent length-force and architectural characteristics of rat gastrocnemius muscle. *Journal of Electromyography and Kinesiology* **2**, 112–120.

Ismail, H.M., and Ranatunga, K.W. (1978). Isometric tension development in a human skeletal muscle in relation to its working range of movement: the length-tension relation of biceps brachii muscle. *Experimental Neurology* **62**, 595–604.

Kawakami, Y., Abe, T., and Fukunaga, T. (1993). Muscle-fiber pennation angles are greater in hypertrophied than in normal muscles. *Journal of Applied Physiology* **74**, 2740–2744.

Kawakami, Y., Abe, T., Kuno, S., and Fukunaga, T. (1995). Training-induced changes in muscle architecture and specific tension. *European Journal of Applied Physiology* **72**, 37–43.

Kawakami, Y., Ichinose, Y., and Fukunaga, T. (1998). Architectural and functional features of human triceps surae muscles during contraction. *Journal of Applied Physiology* **85**, 398–404.

Kawakami, Y., Ichinose, Y., Kubo, K., Ito, M., and Fukunaga, T. (2000). Architecture of contracting human muscles and its functional significance. *Journal of Applied Biomechanics* **16**, 88–97.

Lam, E.W.N., Hannam, A.G., and Christiansen, E.L. (1991). Estimation of tendon-plane orientation within human masseter muscle from reconstructed magnetic resonance images. *Archives of Oral Biology* **36**, 845–853.

Leedham, J.S., and Dowling, J.J. (1995). Force-length, torque-angle and EMG-joint angle relationships of the human *in vivo* biceps brachii. *European Journal of Applied Physiology* **70**, 421–426.

Lieber, R.L., and Baskin, R.J. (1983). Intersarcomere dynamics of single muscle fibers during fixed-end tetani. *Journal of General Physiology* **82**, 347–364.

Lieber, R.L., Loren, G.J., and Friden, J. (1994). *In vivo* measurement of human wrist extensor muscle sarcomere length changes. *Journal of Neurophysiology* **71**, 874–881.

Loeb, G.E., Pratt, C.A., Chanaud, C.M., and Richmond, F.J.R. (1987). Distribution and innervation of short, interdigitated muscle fibers in parallel-fibered muscles of the cat hindlimb. *Journal of Morphology* **191**, 1–15.

Marsh, E., Sale, D., McComas, A.J., Quinlan, J. (1981). Influence of joint position on ankle dorsiflexion in humans. *Journal of Applied Physiology: Respiratory, Environmental, and Exercise Physiology* **51**, 160–167.

Marshall, R.N., Mazur, S.M., and Taylor, N.A.S. (1990). Three-dimensional surfaces for human muscle kinetics. *European Journal of Applied Physiology* **61**, 263–270.

Murray, M.P., Guten, G.N., Baldwin, J.M., and Gardner, G.M. (1976). A comparison of plantar flexion torque with and without the triceps surae. *Acta Orthopaedica Scandinavica* **47**, 122–124.

Narici, M.V., Binzoni, T., Hiltbrand, E., Fasel, J., Terrier, F., and Cerretelli, P. (1996). *In vivo* human gastrocnemius architecture with changing joint angle at rest and during graded isometric contraction. *Journal of Physiology* **496**, 287–297.

Otten, E. (1988). Concepts and models of functional architecture in skeletal muscle. In: *Exercise and Sport Sciences Reviews*. Pandolf, K.B. (ed.), MacMillan Publishing, New York, pp. 89–137.

Roszek, B., Baan, G.C., and Huijing, P.A. (1994). Decreasing stimulation frequency-dependent length-force characteristics of rat muscle. *Journal of Applied Physiology* **77**, 2115–2124.

Rugg, S.G., Gregor, R.J., Mandelbaum, B.R., and Chiu, L. (1990). *In vivo* moment arm calculation at the ankle using magnetic resonance imaging (MRI). *Journal of Biomechanics* **23**, 495–501.

Sale, D., Quinlan, J., Marsh, E., McComas, A.J., and Belangar, A.Y. (1982). Influence of joint position on ankle plantarflexion in humans. *Journal of Applied Physiology: Respiratory, Environmental, and Exercise Physiology* **52**, 1636–1642.

Scott, S.H., Brown, I.E., Loeb, G.E. (1996). Mechanics of feline soleus: I. Effect of fascicle length and velocity on force output. *Journal of Muscle Research and Cell Motility* **17**, 207–219.

Scott, S.H., Engstrom, C.M., and Loeb, G.E. (1993). Morphometry of human thigh muscles. Determination of fascicle architecture by magnetic resonance imaging. *Journal of Anatomy* **182**, 249–257.

Spoor, C.W., van Leeuwen, J.L., van der Meulen, W.J.T.M., and Huson, A. (1991). Active force-length relationship of human lower-leg muscles estimated from morphological data: A comparison of geometric muscle models. *European Journal of Morphology* **29**, 137–160.

Trestik, C.L., and Lieber, R.L. (1993). Relationship between Achilles tendon mechanical properties and gastrocnemius muscle function. *Journal of Biomedical Engineering* **115**, 225–230.

Trotter, J.A. (1993). Functional morphology of force transmission in skeletal muscle. A brief review. *Acta Anatomica* **146**, 205–222.

van Leeuwen, J.L., and Spoor, C.W. (1993). Modelling the pressure and force equilibrium in unipennate muscles with in-line tendons. *Philosophical Transactions of the Royal Society of London* **342**, 321–333.

Walker, S.M., and Schrodt, G.R. (1974). I segment lengths and thin filament periods in skeletal muscle fibers of the Rhesus monkey and the human. *Anatomical Record* **178**, 63–81.

Wickiewicz, T.L., Roy, R.R., Powell, P.L., and Edgerton, V.R. (1983). Muscle architecture of the human lower limb. *Clinical Orthopaedics and Related Research* **179**, 275–283.

Willems, M.E.T., and Huijing, P.A. (1994). Heterogeneity of mean sarcomere length in different fibres: Effects on length range of active force production in rat muscle. *European Journal of Applied Physiology* **68**, 489–496.

Winter, D.A. (1984). Kinematic and kinetic patterns in human gait: Variability and compensating effects. *Human Movement Science* **3**, 51–76.

Zuurbier, C.J., and Huijing, P.A. (1993). Changes in geometry of actively shortening unipennate rat gastrocnemius muscle. *Journal of Morphology* **218**, 167–180.

20 *In Vivo* Measures of Musculoskeletal Dynamics Using Cine Phase Contrast Magnetic Resonance Imaging

F. T. SHEEHAN
Mechanical Engineering Department, The Catholic University of America, Washington, D.C., USA

G. PAPPAS
Diagnostic Radiology Center and Rehabilitation R&D Center, Veterans Affairs, Palo Alto Health Care System, Palo Alto, California, USA

J. E. DRACE
Diagnostic Radiology Center, Veterans Affairs, Palo Alto Health Care System, Palo Alto, California, USA

INTRODUCTION

An understanding of how the neurological, muscular, and skeletal systems work together to complete a specific task has long been a major goal of musculoskeletal experimental and modelling studies. In part, this goal has been fuelled by the need for accurate characterization of normal and pathological musculoskeletal function in order to provide improved diagnosis and treatment for persons with movement deficits. Unfortunately, most kinematic and kinetic properties of the human musculoskeletal system could not be measured directly without the use of invasive techniques. Thus, musculoskeletal dynamics typically have been derived from models that transform external measurements into estimates of the internal state of muscle, bone, and connective tissue. Explicitly or implicitly, these models assumed that architectural parameters and dynamic properties obtained from cadaver and animal studies accurately represented the *in vivo* function of the human musculoskeletal system. Yet, the validity and accuracy of these and other modelling assumptions are still in question.

Until recent advances in medical imaging, *in vivo* musculoskeletal kinematic and kinetic data has been acquired in a highly invasive manner. For example, sarcomere length has been measured intra-operatively using laser diffraction (Ljung et al., 1999), myotendinous tissue strain has been measured using injected

Skeletal Muscle Mechanics: From Mechanisms to Function. Edited by W. Herzog.

radio-opaque markers (Amis et al., 1987), and tendon force has been tracked using buckle transducers (Fukashiro et al., 1995) and fibre optic cables (Finni et al., 1998). Thus, the bulk of our understanding of the force-length (Gordon et al., 1966) and force-velocity (Hill, 1938) properties of muscle has been derived through extracted animal fibres. Similarly, the stress-strain (Zajac, 1989) relationship of human tendon has been primarily quantified through testing of extracted cadaver tendons.

Ultrasound imaging techniques currently allow *in vivo* musculo-tendinous architecture and displacement data to be acquired non-invasively (Fukunaga et al., 1997; Narici et al., 1996). However, studies using ultrasound were limited by a short depth of view and by the need to manually track anatomic landmarks such as the intersection of an aponeurosis with a fascicle (Kuno and Fukunaga, 1995). Thus, *in vivo* tendon strain could only be approximated from the displacement of a single point on the aponeurosis, and strain distribution over large areas in muscle and tendon during dynamic motion could not be acquired using ultrasound alone.

Traditionally, biomechanical models have assumed that whole muscle contraction dynamics were simply scaled versions of sarcomere dynamics (Huijing, 1998). However, in reality, muscle that contracts in a perfectly uniform manner is not likely to exist (Huijing, 1995). Complex musculo-tendon architecture may lead to inhomogeneous shortening (Lieber et al., 1994; van Bavel et al., 1996; Zuurbier et al., 1995), resulting in functional consequences such as alteration of the force-length properties of a muscle. Thus, if not properly modelled, heterogeneity could have introduced inaccuracy in the prediction of muscle function.

Nearly all tendons have been modelled as springs operating in two regions: a compliant toe region (0-2% strain), and stiff linear region (2-6% strain). The validity of this description for *in vivo* function could greatly affect our understanding of energy storage during rhythmic movements (such as walking), and thus the neurological control of such movements. Further, tendon elasticity has often been completely ignored and the tendon modelled as a rigid link, a common modelling assumption used for the patellar tendon; surprising, since this tendon has shown high maximum strain (14-30%) in *ex vivo* testing. Neglecting tendon compliance did simplify modelling, however it added a degree of uncertainty. For example, based on studies that refuted the validity of patellar olecranization (Richter et al., 1996; Rungee et al., 1995), a model using a rigid patellar tendon should have produced erroneously large values for posterior cruciate ligament strain, increased patello-femoral contact force, and altered femorotibial kinematics.

In the clinical realm, the study of pathologies has typically been confined to an investigation of either skeletal or muscle dynamics. For example, patello-femoral pain, associated with abnormal tracking of the patella within the femoral groove (Grabiner et al., 1994; Nisell and Ekholm, 1985), has been attributed to an imbalance in vasti strength (Boucher, 1992). Clinical studies typically focussed on strength testing and muscle activation patterns to assess function, whereas imaging studies concentrated on quantifying the 2D orientation of the patella within the femoral groove during static positioning or active knee extensions. The latter seemed to offer the most promise for truly defining pathologies, since abnormalities in tracking may be most evident during active extension (Brossmann et al., 1993). Such studies cannot shed light on the interplay between muscle and patellar dynamics. Such a combined study may be key to understanding the true nature of patello-femoral and other musculoskeletal disorders.

The studies reviewed in this chapter set out to acquire accurate, non-invasive, 3D *in vivo* musculoskeletal data during dynamic movements using Cine-phase contrast (Cine-PC) magnetic resonance imaging (MRI). Previous studies (Drace and Pelc, 1994a, 1994c) have shown the accuracy and feasibility of using Cine-PC MRI to track deformable muscle and rigid bone motion. The original publication of the biceps brachii and patellar tendon studies can be found in Pappas, et al. (in review), and Sheehan and Drace (2000), respectively.

CINE-PHASE CONTRAST MRI (CINE-PC MRI)

Cine-PC MRI (Pelc et al., 1991), originally developed to measure blood flow and heart motion, shows promise as a non-invasive technique to directly measure human bone motion and muscle fibre velocity *in vivo,* during dynamic tasks. For each time frame, Cine-PC MRI provides an anatomic image and velocity images in three orthogonal directions ($\mathbf{m}_x$, $\mathbf{m}_y$, $\mathbf{m}_z$). The anatomic images show the familiar gradient echo tissue contrasts (Figure 20.1), while the 3D velocity vector, for any point in the imaging plane, is represented by the pixel values in the 3D velocity images. Through integration, estimates of tendon length (Sheehan et al., 1998), muscle fibre length, and moment arms can be derived.

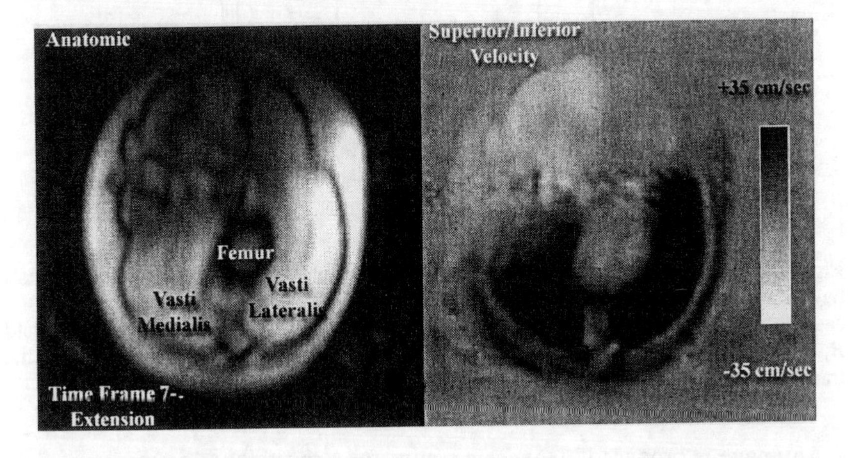

Figure 20.1. Magnitude and superior/inferior velocity image. The extensor (flexor) compartment is at the bottom (top) of the image. Black (white) indicates maximum superior/inferior motion. Thus, in frame 7, the leg is extending and we can see the superior motion of the knee extensors with a reciprocal inferior motion of the knee flexors.

CONVENTIONAL MAGNETIC RESONANCE IMAGING

MRI is based on the interaction of atomic nuclei with three magnetic fields (Allen, 1993; Keller, 1988): the main field (B_o), the radio-frequency field (B_1), and linear gradient fields (Gz and Gy). Atomic nuclei, such as hydrogen, have a magnetic moment m (Figure 20.2A) and angular momentum. With the application of an external magnetic field B_o, individual protons (spins) align at an angle $\pm\Theta$ (low/high energy state) with B_o (Figure 20.2B), causing them to precess about B_o (Figure 20.2C), at a frequency proportional to B_o, described by the Larmor equation (Bracewell, 1993):

$$\boldsymbol{\omega}_o = \gamma \boldsymbol{B}_o \tag{20.1}$$

where:

$\boldsymbol{\omega}_o = \omega_o \cdot \boldsymbol{m}_y$ ($|\omega_o|$ is the magnitude of precessional [Larmor] frequency)

$\boldsymbol{m}_i$ ($i=x,y,z$) is three unitary orthogonal basis vectors fixed in the magnet (Figure 20.3)

γ = the magnetogyric ratio (related to the nuclei's magnetic moment)

$\boldsymbol{B}_o = B_o \cdot \boldsymbol{m}_y$ ($|B_o|$ is the strength of the main magnetic field)

When numerous spins are exposed to B_o, they precess out of phase with each other and the total magnetic moment (M) is parallel to B_o (Figure 20.2C) in the longitudinal direction. The net magnetization in the direction of B_o of all spins is referred to as the longitudinal magnetization (M_l).

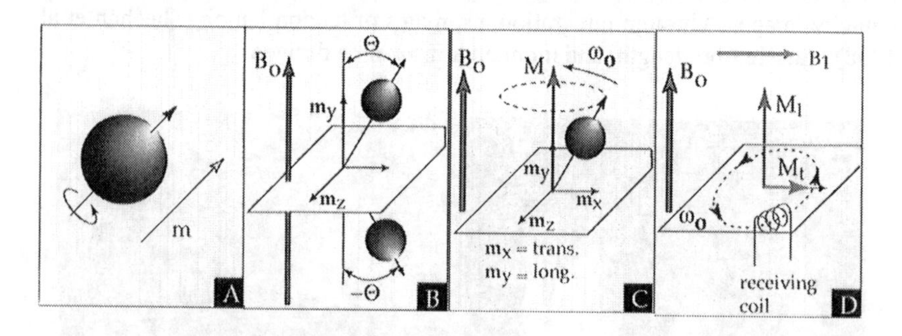

Figure 20.2. The interaction of spins with magnetic fields. A) Hydrogen atoms ('spins') have a magnetic moment m, B) which tends to align with the main magnetic field B_o, C) causing the spins to precess at a frequency ω_o. D) With the application of a rotating field B_1, longitudinal magnetization (M_l) is converted to transverse (M_t), which rotates in the transverse plane, creating a signal in the receiving coil.

An image is produced by exposing spins for a short period to magnetic field B_1, which is rotating in the transverse ($\boldsymbol{m}_x$-$\boldsymbol{m}_z$) plane at the Larmor (ω_o) frequency. This results in some of the longitudinal magnetization being 'tipped' into the transverse plane. This new component M_t (Figure 20.2D), the transverse magnetization, is perpendicular to B_o and rotates in the transverse plane at ω_o. As M_t rotates (Figure 20.2D), it can generate a signal—the MR signal—in a receiving coil as explained by Faraday's law. The transverse (longitudinal) magnetization decays (recovers) exponentially with a time constant *T2* (*T1*) to its equilibrium state. The variation in *T1* and *T2* among different tissues is responsible for much of the contrast in MR images.

Since the receiving coil is sensitive over a large volume, the measured signal is a sum of all MR signals from the volume. Gradient waveforms are used to separate the signal for individual spins or groups of spins according to their spatial localization (Peters, 1993). Any arbitrary plane can be imaged but, for simplicity, the discussion is formulated to match the 2D imaging plane $\boldsymbol{m}_y$-$\boldsymbol{m}_z$ (sagittal), used in the knee joint study (Figure 20.3). A 3D volume can be imaged as a series of 2D images.

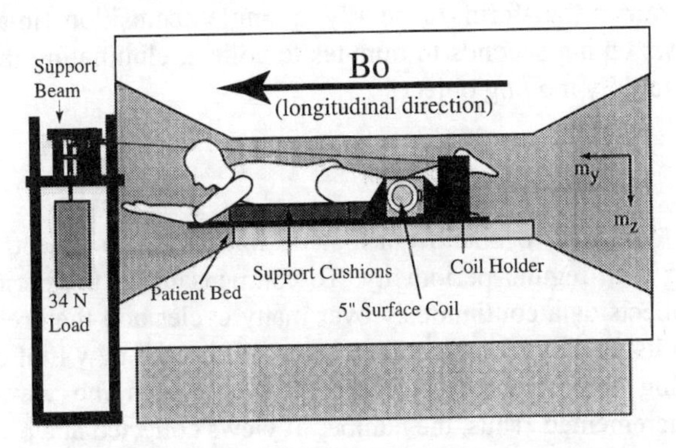

Figure 20.3. Cross-sectional view of the magnet resonance imager. Reproduced from Sheehan et al. (1999) by permission of the American Society of Mechanical Engineers.

Let's begin with a simple case: a series of spins distributed in the $\boldsymbol{m}_z$ direction, each at a location $(z_i \boldsymbol{m}_z)$ with a signal equal to $s(z_i)$ $(i = 1 \ldots n)$. Thus, the measured signal equals $\Sigma s(z_i)$. To spatially isolate individual signals in the $\boldsymbol{m}_z$ (read) direction, a magnetic field gradient, $G_z(t)$, is applied while the signal is being received. G_z is in the same direction as B_o, but its magnitude varies linearly in the readout ($\boldsymbol{m}_z$) direction.

$$\boldsymbol{G}_z = G_z\, z\, \boldsymbol{m}_y \tag{20.2}$$

Thus, the total magnetic field (Equation 20.2) becomes:

$$\boldsymbol{B}_o'(z) = (B_o + G_z\; z)\, \boldsymbol{m}_y \tag{20.3}$$

and

$$\omega(z) = \gamma(B_o + G_z\; z)\, \boldsymbol{m}_y \tag{20.4}$$

creating a dependency between the spin's resonant frequency and its relative position (z) in the $\boldsymbol{m}_z$ direction. The MR signal as a function of position can be derived by applying an inverse spatial Fourier transform (Pelc and Glover, 1993) to the signal as a function of time measured in the presence of G_z.

Next, let's look at a 2D image. The received signal is now $S = \Sigma s(z_i, y_i)$. The magnetic field gradient (G_z) is used to localize spins in the m_z direction, but the signal must be localized in the $\boldsymbol{m_y}$ direction as well. This is accomplished by applying a magnetic field gradient ($G_y(t) = G_y\, y\, m_y$) for a time t_y before the signal is read, and then turning it off. Similar to G_z, G_y is in the same direction as B_o, but its magnitude varies linearly in the $\boldsymbol{m_y}$ (phase encoding) direction. While G_y is non-zero, a spin's frequency will vary according to its position (y) in the $\boldsymbol{m_y}$ direction. When G_y is turned off, after a time t_y, the spins return to their original frequency, but their phase has been altered by ($\gamma G_y t_y y$). Each signal received using a different G_y will provide information about the image along a separate horizontal line in Fourier space, $Ky = \gamma G_y t_y$ (Figure 20.3). By collecting a series of signals (views), each at a different G_y, all other parameters being equal, the entire data set in Fourier space is acquired. The spatial image is derived using a 2D inverse Fourier transform. Generally, a single acquisition (image) requires 128-256 views, taking seconds to minutes to collect, eliminating the possibility of imaging a rapidly moving object.

CINE MRI

Cine MRI is designed to obtain quasi-static images of an object, such as the heart, moving with regular periodicity. To compensate for the periodic motion, Cine MRI collects data continuously over many cycles and then retrospectively sorts the data using a synchronization trigger. With each new motion cycle, the phase encoding amplitude (Ky_i, Figure 20.4) at which the views are being collected is incremented. Thus, the number of views collected at a predefined Ky_i

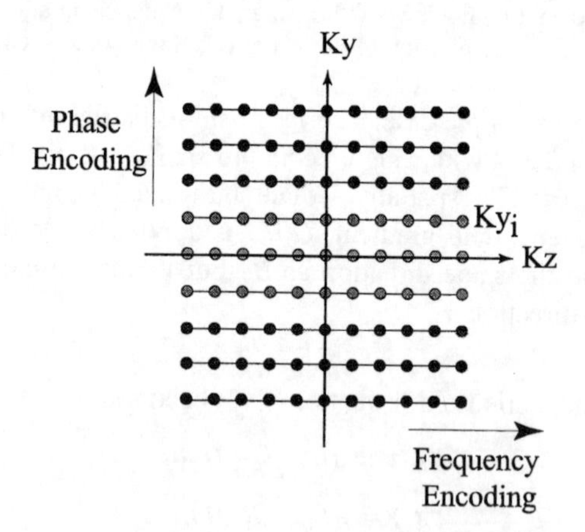

Figure 20.4. K-space map. Data are collected in the Fourier domain and the final image is derived using the inverse Fourier transform. Each view is a single line in the Fourier domain (K-space). If data are collected along lines of constant *Ky*, then *Ky* is the phase encoding direction and *Kz* is the frequency encoding (read) direction.

is the cycle period (e.g., 1/heart rate) divided by the repetition time (*TR*). *TR* is the time between each signal acquisition (i.e., the temporal resolution). The data are then retrospectively interpolated based on individual cycle lengths, which compensates for small variations in cycle period. Thus, a series of quasi-static anatomic images, portraying various phases of the motion cycle, are collected during a single acquisition lasting on the order of 128-256 motion cycles.

PHASE CONTRAST (PC) MRI

The MRI signal is a complex quantity, having both magnitude and phase information. In conventional MR imaging, only the magnitude of the transverse magnetization contributes to the image intensity, but in PC MRI, velocity-sensitive pulse sequences are used and the velocity is extracted from the phase of the signal.

As stated earlier (Equation 20.4), in the presence of a G_z gradient there is a direct relationship between the spin's resonant frequency, ω_o, and the spin's location in the $\boldsymbol{m_z}$ direction. We now examine what happens if the 1D position of spins is dependent on time (i.e., $z(t)$, Equation 20.5). By taking the Taylor series expansion of $z(t)$, we can derive the following relationships:

$$z(t) = (z_o + vt + at^2/2 + ...) \tag{20.5}$$

$$\omega_o(t) = \gamma\,[B_o + G_z(t)\,(z_o + vt + at^2/2 + ...)] \tag{20.6}$$

where z_o, v, and a are the spin's position, velocity, and acceleration, respectively, in the m_z direction, at $t = 0$. The phase of the signal from a group of spins is the integral of their instantaneous frequency (Nayler et al., 1986; Pelc et al., 1994):

$$\phi(t') = \int_0^{t'} \omega_o(t)dt \tag{20.7}$$

Assuming B_o (the main magnetic field) is homogeneous and that the phase dependency on B_o is removed through demodulation of the MRI signal, the spin's phase at time t' is:

$$\phi(t') = \gamma M_o z_o + \gamma M_1 v + \gamma M_2 a/2 + ... \tag{20.8}$$

M_i ($i = 0,1,2$) are the zeroeth, first, and second moments of the gradient waveform, respectively:

$$M_0 = \int_0^{t'} G_z\,(t)\,dt \qquad M_1 = \int_0^{t'} G_z\,(t)\,t\,dt \qquad M_2 = \int_0^{t'} G_z\,(t)\,t^2\,dt \tag{20.9}$$

The M_0 term is responsible for spatial localization. Ignoring acceleration and higher order terms, which is commonly done in phase contrast MRI, the velocity

(v) can be calculated using two gradient waveform sequences with identical spatial localization properties (M_0 constant), but which differ in their sensitivity to motion in the m_z direction (M_1 varies).

$$\Delta\phi = \gamma \Delta M_1 v \tag{20.10}$$

$$v = \frac{\Delta\phi}{\gamma \Delta M_1} \tag{20.11}$$

The maximum velocity, v_{enc} (determined by ΔM_1), that can be measured unambiguously produces a phase shift of π radians:

$$v_{enc} = \frac{\pi}{\gamma \Delta M_1} \tag{20.12}$$

The phase is converted into velocity by:

$$v = \frac{v_{enc}}{\pi} \Delta\phi \tag{20.13}$$

The 3D velocity vector is measured using four gradient waveform sequences, G_i ($i = 1$-4), with differing velocity sensitivities. One possible sequence collects the first data set with M_1 equal to zero (i.e., velocity insensitive). The next three have a value of M_1 that is non-zero in one of the three velocity directions of interest. The phase of the first data set is then subtracted from that of the last three to obtain three single-direction velocity images.

CINE-PC MRI

Cine-PC MRI is the marriage of Cine and PC MRI, resulting in a reliable method (Drace and Pelc, 1994b) of measuring the velocity of an object experiencing periodic motion. During each cardiac movement cycle, views are still acquired at a single phase encoding amplitude (K_{yi}), but the four gradient waveform sequences of varying velocity sensitivities are interleaved (i.e., $G_1, G_2, G_3, G_4, G_1, G_2$...). This results in a diminished temporal resolution, by a factor of four with no increase in scan time, as compared to Cine MRI. As with Cine, the phase-encoding amplitude is updated with the beginning of each new cycle. These data are then sorted and interpolated to produce a complete MR raw data set for each velocity-sensitivity at each time frame. From these data, separate MR magnitude and velocity images that uniformly span the motion cycle are produced (Figure 20.1). The magnitude images are essentially equivalent to those produced by the conventional Cine MRI. The three velocity images, within which pixel intensity is directly proportional to velocity, depict the 3D velocity and how it changes throughout the motion cycle.

INTEGRATION OF THE CINE-PC MRI DATA

The 3D position trajectory of any material point can be estimated by integrating its velocity trajectory as a function of time. Velocity along the trajectory is unknown, *a priori*, but can be calculated along with displacement. In the simplest method, the location of point $P(t_k)$ and its velocity $v(P(t_k),t_k)$ are used to estimate the location at a future time $(t = t_{k+1})$:

$$P(t_{k+1}) = P(t_k) + v(P(t_k),t_k) \cdot \Delta t \tag{20.14}$$

$v(P(t_k),t_k)$ is obtained from the 3D velocity images. Once $P(t_{k+1})$ is estimated, $v(P(t_{k+1}),t_{k+1})$ can be derived from the 3D velocity images. Carrying this process through the entire cycle provides a displacement and velocity trajectory of the point. Enhancements to this method, such as using the velocity for an entire region (a contiguous collection of pixels) and the fact that the motion is cyclical (i.e., $\oint v(t) \cdot dt = 0$), have been developed to reduce noise and eddy-current errors, thus improving position and velocity estimates (Pelc et al., 1995; Zhu et al., 1996).

While ideally we should have velocity data spanning an entire volume, it is feasible to estimate the 3D position and velocity trajectories using data from a single 2D imaging plane. If a point being tracked moves out of the imaging plane ($\boldsymbol{m}_x$ direction), its velocity $v(x,y,z,t)$ is estimated using the velocity of its nearest neighbour in the imaging plane $v(x_o,y,z,t)$, where x_o is the slice location. This produces a reasonable estimate if either of two assumptions holds true:

1) the out-of-plane motion is small in comparison to slice thickness, or
2) the slice direction ($\boldsymbol{m}_x$) motion is primarily rigid translation, ensuring that $v(x,y,z,t) \approx v(x_o,y,z,t)$.

Unlike other imaging techniques, the tracking accuracy can be well below one pixel. Since the signal-to-noise ratio (*SNR*) improves with larger pixel sizes (lower spatial resolution), the variance in velocity decreases when spatial resolution is lowered. To be exact, the variance in the measured velocity (σ_v) is inversely proportional to the *SNR* of the data, and the square root of the number of independent pixels in the region being analyzed (N_p), and directly proportional to the v_{enc} (maximum measurable velocity) (Pelc et al., 1995):

$$\sigma_V = \left(\sqrt{\frac{2}{N_P}}\right)\left(\frac{v_{enc}}{\pi SNR}\right) \tag{20.15}$$

METHODS

All studies followed the same general skeleton protocol with specializations added for each specific study. Upon entering the study, subjects were informed of the protocol and an informed consent was obtained from each subject in accordance with institutional policy. All studies required repetitive limb motion using a cycle rate of 35 cycles/minute. An auditory metronome (two beats/cycle) was provided in order to help subjects maintain a consistent motion. All Cine-PC data acquisitions used a flip angle of 30°, a maximum encoding velocity of 35 cm/s, a 10-mm slice thickness, and *TR* equal to 21 ms (minimum available). The temporal resolution was 84 ms and the data were divided into 24 time frames. Data collection was synchronized to the motion cycle using an optical transducer, which was triggered at full (elbow or knee) extension. All protocols were approved by the Institutional Review Board, the Administrative Panel on Human Subjects in Medical Research at Stanford University, and informed consent was obtained from each subject in accordance with institutional policy.

BICEPS BRACHII STUDY

To briefly summarize the methodology described by Pappas et al. (in review), six unimpaired volunteers (five males and one female, 30-47 years, 5'9" to 6'2") performed elbow flexion (against a two-pound resistance) and extension within a 1.5 Tesla GE magnetic resonance imager. Approximately 60° of supination was maintained to ensure activation of the biceps brachii (Buchanan et al., 1989). The maximum flexion angle achieved by the subjects ranged from 65° to 90°.

Cine-PC MR images were acquired as each subject performed the elbow extension/flexion task. The arm was oriented parallel to the centre line of the MR imager. Prior to the movement study, an oblique-sagittal imaging plane bisecting the biceps along its length was selected using static axial images (Figure 20.5). The imaging plane was chosen to be approximately parallel to the muscle fascicle direction in order to minimize out-of-plane motion. Cine-PC images were acquired using 256 × 128 (or 256 × 192) pixels matrix with a typical imaging time of 2 minutes, 45 seconds. In order to maximize the SNR, the MR signal was acquired using a flexible extremity coil placed around the upper arm, using a custom-built cylindrical coil holder.

A closed-form Fourier integration method contained within the PCMotion software (Zhu et al., 1996) was used to calculate the 3D displacement trajectories of regions of interest (ROIs) within the muscle belly. ROIs (1 cm × 1 cm) were graphically prescribed in a distal-to-proximal direction along the centre line of the biceps brachii in the first anatomic image (time frame 1). The humerus and the anterior and posterior surfaces of the muscle belly helped guide the placement of the ROIs. The most distal ROI was approximately 2 cm proximal to the distal biceps tendon.

Longitudinal strain was estimated from the position trajectories of the ROIs prescribed along the centre line of the biceps brachii muscle. Strain was calculated between every second ROI using the following definition:

$$\varepsilon = \Delta l / l_o \qquad (20.16)$$

where:
ε = longitudinal strain
Δl = the change in distance between two ROIs
l_o = the initial distance between the ROIs in time frame 1 (full extension)

Given the contiguous arrangement of the ROIs along the muscle centre line, l_o was always ~2 cm. A negative (positive) strain indicated local muscle shortening (lengthening) in the longitudinal direction, as compared to the state of the muscle during full elbow extension. For each subject, the strains at the central (10 cm from the distal end) and distal biceps (4 cm from the distal end) were compared.

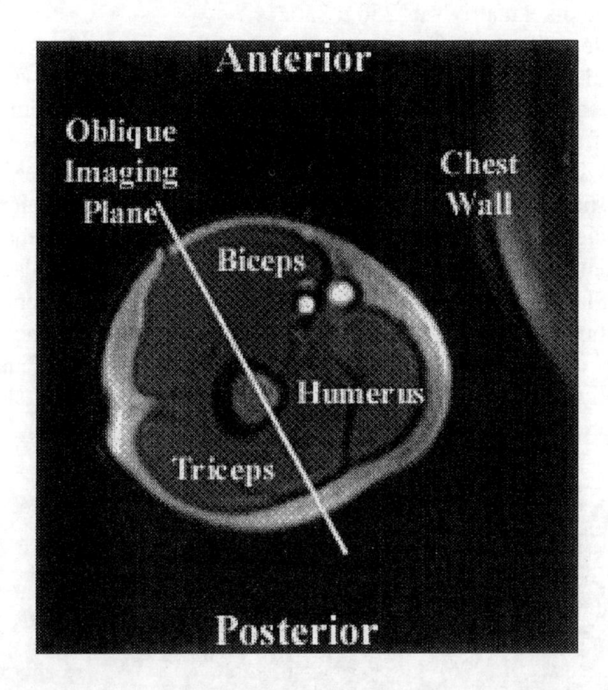

Figure 20.5. Static axial localizer for the biceps brachii study. Static axial MR image of right upper arm of a representative subject. The Cine-PC MR imaging plane is prescribed using a static axial image. This oblique-sagittal imaging plane bisects the biceps into essentially medial and lateral sections. The arrowheads indicate the viewing direction of the Cine-PC imaging plane.

PATELLAR KINEMATICS AND TENDON STRAIN

To briefly summarize the methodology described by Sheehan et al. (1999), subjects cyclically extended from maximum knee flexion angle to full extension while in a prone position within the imager (Figure 20.3). Eighteen knees (eight right and 10 left) were studied. All subjects (six females and seven males, height $=1.7\pm0.1$ m, weight$=67\pm13$ kg, and age$=24\pm2.4$ years) had no history of knee pain or pathologies. Both knees were studied if the subject's time permitted. To allow the subjects to reach full extension, cushioning was placed under the torso and thighs. This cushioning eliminated external contact forces on the patella. A 34 N weight resisting extension required subjects to produce a maximum knee moment of ~10 Nm at the beginning of extension (~35° to 45° of knee flexion).

After subjects were aligned within the imager using static images, Cine-PC MR images were taken in a sagittal plane ($\boldsymbol{m}_y$-$\boldsymbol{m}_z$, Figure 20.3) at the approximate centre line of the femur and patella. This imaging plane allowed all three bones (patella, femur, and tibia) to be imaged and minimized through-plane ($\boldsymbol{m}_x$) motion. The pixel matrix was 256×256.

ROIs selected on the femur, tibia, and patella were tracked throughout the motion cycle. In order to minimize errors due to noise, regions were selected to be as large as possible (minimum size = 300 pixels) while remaining within the confines of a single bone and within the receiving coil field. Since optimization algorithms insured consistency amongst all pixels within a region, only the region vertices' displacements were used for analysis. For reference frames fixed in the patella, femur, and tibia, the direction cosine matrix and 3D displacement of the origin were calculated based on the 3D displacement of the vertices (Sheehan et al., 1999). From these data, the 3D orientation angles, specific clinical angles, and tendon length were calculated. The tendon slack length was estimated by viewing the 24 anatomic images in a Cine movie loop and visually determining when the tendon transitioned from a slack to a taut state (Figure 20.6).

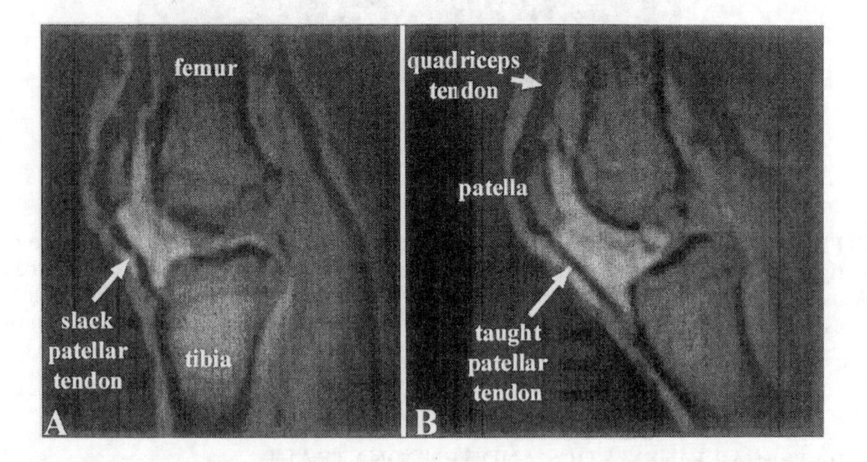

Figure 20.6. Tendon slack length (anatomical images). Sagittal anatomical images from frame 1 (A) when the tendon is slack, and frame 15 (B) when the tendon is taut. Reproduced from Sheehan and Drace (2000) with the permission of Lippincott Williams & Wilkins.

From these data the tendon strain was calculated:

$$Strain = \frac{1_t - 1_{to}}{1_{to}};\ 1_t \geq 1_{to} \qquad (20.17)$$
$$= 0;\ 1_t < 1_{to}$$

where:
l_t = tendon length, the distance from the insertion of the patellar tendon into the patella and tibial tuberosity, and l_{to} = tendon slack length (tendon length when there is minimal force on the tendon).

VASTI KINEMATICS

The vasti study was nearly identical to the knee joint study, except subjects (five unimpaired and three patients diagnosed with patellar tracking problems) raised three different weights (5 N, 34 N, and 64 N) during the extension phase of the knee exercise. The order of experiments was randomly assigned. An axial ($\boldsymbol{m_x m_z}$) plane just superior to the myotendinous junction of the vastus lateralis was used. Phase encoding was in the medial/lateral ($\boldsymbol{m_x}$) direction in order to limit blood vessel artefacts from affecting the data.

The 3D displacement and velocity of ROIs (typically 17 × 17 mm), graphically prescribed in the centre of the vasti lateralis and medialis, were tracked using the Fourier integration algorithms. Anatomical landmarks such as the femur and blood vessels were used to maintain consistent region placement between exams. The speed, the 3D velocity, and displacement vectors as functions of time were compared across the three loading conditions.

RESULTS

The results of the biceps brachii (for the complete results, see Pappas et al., in review) and patellar tendon studies (for the complete results, see Sheehan and Drace, 2000) called into question the long-standing assumptions about muscle and tendon strain, whereas the vasti study demonstrated the potential of Cine-PC MRI as a clinical diagnosis tool.

BICEPS BRACHII STUDY

The mechanical strain distribution (Figure 20.7) along the centre line of the contracting biceps brachii at maximum elbow flexion (65° to 90°) was highly non-uniform for all subjects (Pappas et al., 2000). Maximum strain ranged from –21.7% to –47.2% and was lower in those subjects who were unable to reach full 90° elbow flexion.

One of the key findings in this study was the substantially greater strain magnitude at the central and proximal regions of the biceps brachii as compared to the strain in the distal muscle. The average strain was less than 2% at the distal end and just under 30% in the mid-section. In all subjects, strain in the central and proximal portion of the muscle remained negative over the entire flexion motion, consistent with concentric contraction.

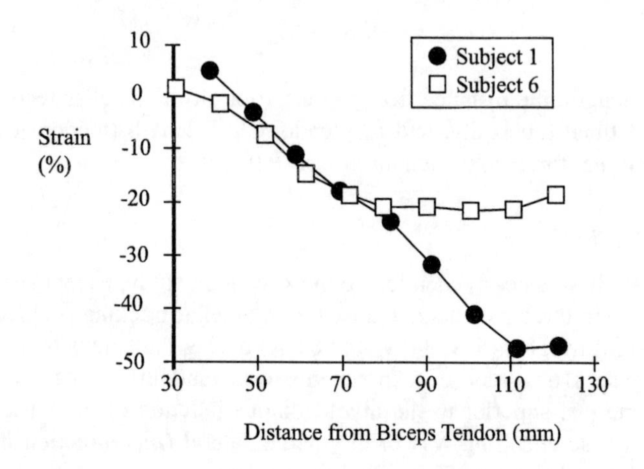

Figure 20.7. Biceps strain (Subjects 1 and 6). Longitudinal strain (i.e., strain in the direction of the longitudinal axis, or centre line of the biceps) is plotted as a function of distance from the distal biceps tendon; negative (positive) strain indicates shortening (lengthening). Strain is computed at maximum elbow flexion. Subject 1 (6) achieved the largest (smallest) maximum strain of all six subjects. Subject 6 achieved the smallest maximum flexion angle of 65°. This is the most likely reason for the variation in maximum strain between these two subjects.

PATELLAR KINEMATICS AND TENDON STRAIN

The average maximum tendon strain (Figure 20.8) was 6.6%, with some subjects reaching 11% strain (Sheehan and Drace, 2000). Not all subjects could reach the larger knee angles (>37°) due to the limited bore size. Thus, the sudden change in strain at ~37° is an artefact of the drop in the number of exams included in the average. No individual subject had such a sudden change. The average tendon slack length for the taller male subjects (height=1.78±0.07 m, n=7) was 48.8 mm, and for the rest of the subjects (height=1.66±0.07, n=11) it was 42.8 mm.

In comparing the patellar tendon strain with the patellar kinematics, a correlation was found between the knee angle when the patellar tendon became slack (17°, Figure 20.8) and the knee angle when the slope of the superior displacement of the patellar origin with respect to the femoral origin sharply changed (Sheehan and Drace, 2000). For each examination (n = 18), the knee angle at which the slope of the superior displacement of point *Pf* versus knee angle changed to a negative slope, and the knee angle at which the tendon transitioned from taut to slack, was correlated (Sheehan and Drace, 2000). The correlation coefficient (r) between these two quantities was 0.82 (p<0.05).

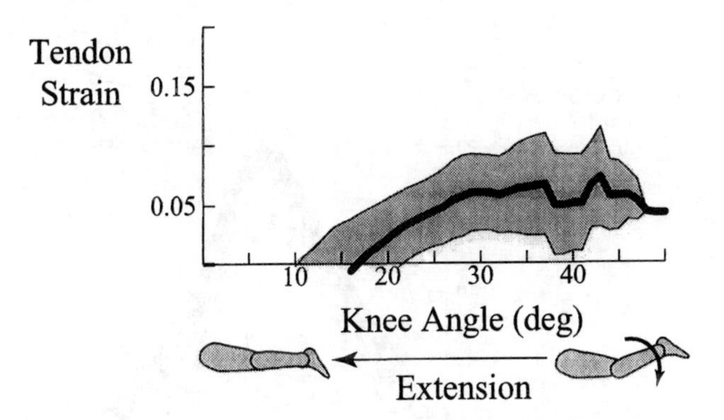

Figure 20.8. Tendon strain. Average strain ± one standard deviation (in grey). The large dip in the graph at ~37° is an artefact of the drop in the number of exams included in the average (all examinations are represented at knee angles from 14° to 26°). No single subject demonstrated such a dip. Reproduced from Sheehan and Drace (2000) with the permission of Lippincott Williams & Wilkins.

VASTI KINEMATICS

All subjects demonstrated similar superior/inferior displacement profiles for the vastus medialis and lateralis, but the variation in maximum values was different between the unimpaired and patient populations (Figure 20.9). The maximum value of displacement increased with increasing workload in the unimpaired population, whereas this pattern was not seen in those patients diagnosed with patellar tracking problems.

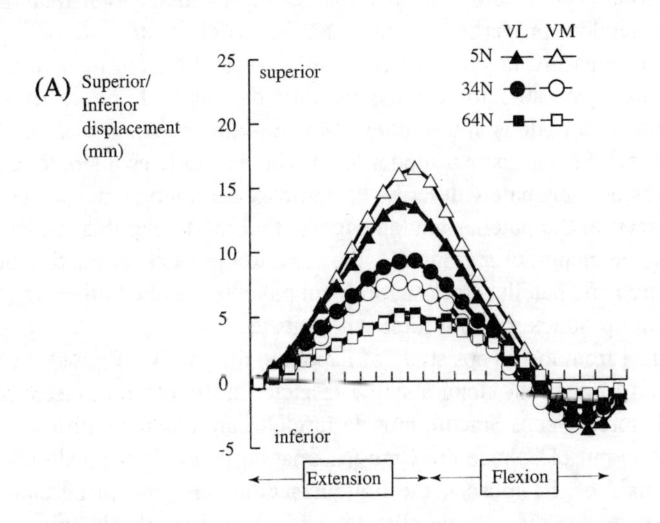

Figure 20.9: Vasti superior/inferior displacement. (A) Unimpaired subject (Subject J1). See also (B), next page.

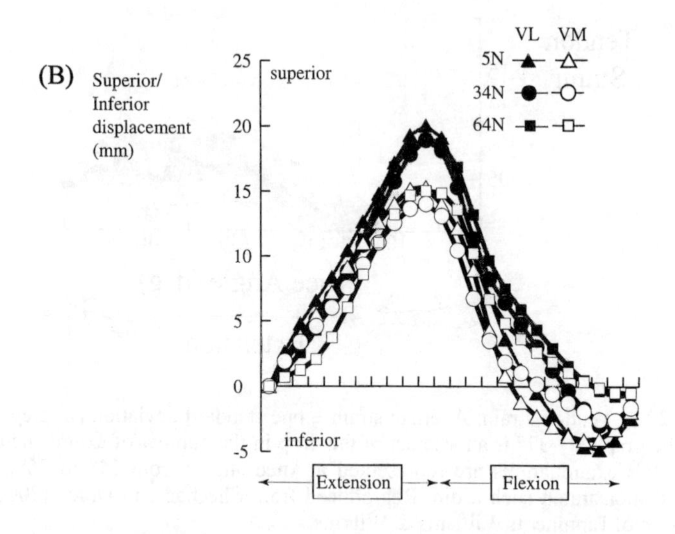

Figure 20.9. Vasti superior/inferior displacement. (A, previous page) Unimpaired subject (Subject J1), and (B) patient diagnosed with patellar tracking problems (Subject M1). The same shape profile can be observed for both subjects, but the maximum values do not vary with load cases for the patient Subject M1.

DISCUSSION

The experiments reviewed in this chapter represent the first quantification of musculo-tendon properties using Cine-PC MRI. With the ability to non-invasively measure *in vivo* musculoskeletal velocities during volitional movements, we were able to test the validity of some of the experimental and modelling assumptions in use today. The non-uniform strain measured along the biceps brachii suggests that modelling a whole muscle as a single scaled sarcomere may not adequately describe its complex contraction mechanics. The large strains seen in the patellar tendon suggest that modelling this tendon as a rigid link may be inappropriate. Finally, the variation in vasti contraction between the unimpaired and patellar-tracking problem patients may be indicative of activated muscle not producing useful force within its tendon.

The data from the biceps study of Pappas et al. (in review) raise the possibility of non-uniform strain along a single fascicle. In the central muscle belly of the parallel-fibred biceps brachii, muscle fascicles are aligned with the longitudinal axis of the muscle. Since the Cine-PC imaging plane is aligned with the longitudinal axis of the muscle, the oblique-sagittal imaging plane should contain muscle fascicles that are parallel to, and lie within, the imaging plane. This assertion is supported by the fact that maximum in-plane displacement is an order of magnitude greater than maximum out-of-plane displacement (Pappas et al., in review). Consequently, the longitudinal strain measured in the central region, where the fascicles are parallel to the longitudinal axis, should be a good approximation of the actual fascicle strain in that region.

Much of the observed strain heterogeneity along the biceps brachii is likely due to the architecture of the muscle, specifically the prominent distal aponeurosis. The stiffness of distal tendon and internal aponeurosis could produce lower levels of distal strain. Furthermore, even though the biceps is categorized as parallel-fibred muscle, fascicles insert at an angle into the distal aponeurosis. Not accounting for the effects of this insertion angle could result in a slight under-prediction of distal fascicle strain.

Although the patellar tendon strains reported here may appear high in comparison to other human tendons, previously reported maximum strains in the patellar tendon (14% to 30%) do support the findings of Sheehan and Drace (2000). Since Cine-PC has already been shown to be accurate in the tracking of trabecular bone motion (Sheehan et al., 1998), the largest possible source of error in estimating tendon strain is in determining the tendon slack length. Although this step is subjective, we believe it is accurate based on previous work. Van Eijden et al. (1987) reported an average patellar tendon slack length of 48 mm (five male cadavers, height not given), which agrees with the estimate found in this study for the taller male subjects (48.8 mm).

The high correlation between tendon slack length and knee kinematics is further evidence that tendon slack length estimates are reliable. A sudden change in the superior displacement of the patellar origin most likely indicates cessation in quadriceps force. First, with the knee extending, patellar tendon force is directly related to quadriceps force. The extensor force is low in comparison with maximum isometric force, so once this force is removed, it is likely that the patella would quickly decelerate due to residual strain energy in the patellar tendon and patellar contact forces, and contact with the synovial bursa, and infrapatellar fat pad.

Previous *ex vivo,* 3D knee kinematic studies have reported quite small estimates of patellar tendon strain (Heegaard, 1993; van Kampen and Huiskes, 1990). The differences between our strain values and those from the previous cadaver studies are attributable to numerous factors. First, we measured the *in vivo* tendon strain during active knee extensions of healthy volunteers. Thus, the loading conditions were dynamic, not static, and the load applied to the tendon was actively controlled by the quadriceps muscle. Also, the cadaver limbs were most likely left in a test rig outside any solution during imaging, which could have resulted in a drying and stiffening of the patellar tendon as well as other structures (Bechtold et al., 1994).

The patellar tendon clearly is not an inextensible link, and adding its spring-like properties could improve models. To begin, the ratio of patellar force to quadriceps force (*Fp/Fq*) would be more accurate if the actual patellar tendon length were considered. Typically *Fp/Fq* is reported maximum at 0° knee flexion angle (van Eijden et al., 1986; Yamaguchi and Zajac, 1989), but this would be offset somewhat if patellar tendon length changed with tendon force. Second, if the patellar tendon is slack, it cannot sustain force. Thus, allowing the modelled patellar tendon to transition to a slackened state may provide insights into the muscle activation patterns needed to compensate for delays (beyond muscle excitation delays) in extensor force.

The findings in the vasti study are still preliminary, but do show the enormous potential that Cine-PC MRI has as a research and diagnostic tool. The vasti displacement profiles in the unimpaired population agree with the notion that as larger extensor forces are needed, larger tendon strains will develop, and larger muscle displacements will result. One explanation for the lack of change in displacement extremes for the patient population is that the vasti lateralis and medialis may not generate useful force within the quadriceps tendon, even though they are contracting. Thus, the useful extensor force would have to come form the laterally directed vasti intermedialis or rectus femoris. Further study is warranted to see if these profiles are seen over a large population.

CONCLUSIONS AND FUTURE DIRECTIONS

The cumulative results of all three studies call into question numerous basic experimental and modelling assumptions. These studies represent the beginnings of an entire field of *in vivo*, non-invasive, human musculoskeletal function. Further studies are needed to characterize the strain throughout the entire muscle belly and to characterize not just tendon strain, but the tendon stress-strain relationship. The vasti profiles need to be collected over large patient and unimpaired populations, and the results should be correlated with *in vivo* patellar kinematics.

One major improvement in recent history has been the availability of new sequences (Fastcard and Fast-PC [Foo et al., 1995]) that allow for decreases in acquisition time along with a decrease in the number of required motion cycles. These sequences allow for numerous phase encodes (Kyi) to be taken during a single motion cycle. With the improvement in gradients, the minimum achievable TR has been drastically reduced. Thus, studies can be performed with reduced imaging times and no loss of temporal resolution. For example, the Cine-PC image acquisition for patellar kinematic study can now be done in less then two minutes, a huge saving in comparison to the original 7.5 minute acquisition time.

ACKNOWLEDGEMENTS

This work was supported by NIH grants 1 R29 HD31493 and P41 RR09784, the Rehabilitation R&D Center, and the Diagnostic Radiology Center of the

Department of Veterans Affairs (VA). We are grateful to Norbert Pelc, Yudong Zhu, Patty Spezia, Thomas Kane, Paul Mitiguy, and Felix Zajac for their help with this work and manuscript; and to Jim Anderson and Doug Schwandt for their help in the design and building of experimental equipment.

REFERENCES

Allen, P.S. (1993). Some fundamental principles of nuclear magnetic resonance. In: *The Physics of MRI 1992 AAPM Summer School Proceedings*. Bronskill and Sprawls (eds.), American Institute of Physics, Woodbury, NY, pp. 15–31.

Amis, A., Prochazka, A., Short, D., Trend, P.S., and Ward, A. (1987). Relative displacements in muscle and tendon during human arm movements. *Journal of Physiology (London)* **389**, 37–44.

Bechtold, J.E., Eastlund, D.T., Butts, M.K., Lagerborg, D.F., and Kyle, R.F. (1994). The effects of freeze-drying and ethylene oxide sterilization on the mechanical properties of human patellar tendon. *American Journal of Sports Medicine* **22**, 562–566.

Boucher, J. (1992). Quadricep femoris muscle activity in patellofemoral pain syndrome. *American Journal of Sports Medicine* **20**, 527–532.

Bracewell, R.N. (1993). *The Fourier Transform and its Application*. McGraw Hill, New York, New York.

Brossmann, J., Muhle, C., Schroder, C., Melchert, U.H., Bull, C.C., Spielmann, R.P., and Heller, M. (1993). Patellar tracking patterns during active and passive knee extension: Evaluation with motion-triggered cine MR imaging. *Radiology*. **187**, 205–212.

Buchanan, T.S., Rovai, G.P., and Rymer, W.Z. (1989). Strategies for muscle activation during isometric torque generation at the human elbow. *Journal of Neurophysiology* **62**, 1201–1212.

Drace, J.E., and Pelc, N.J. (1994a). Measurement of skeletal muscle motion *in vivo* with phase-contrast MR imaging. *Journal of Magnetic Resonance Imaging* **4**, 157–163.

Drace, J.E., and Pelc, N.J. (1994b). Skeletal muscle contraction: Analysis with use of velocity distributions from phase-contrast MR imaging. *Radiology* **193**, 423–429.

Drace, J.E., and Pelc, N.J. (1994c). Tracking the motion of skeletal muscle with velocity-encoded MR imaging. *Journal of Magnetic Resonance Imaging* **4**, 773–778.

Finni, T., Komi, P.V., and Lukkariniemi, J. (1998). Achilles tendon loading during walking: Application of a novel optic fiber technique. *European Journal of Applied Physiology* **77**, 289–291.

Foo, T.K., Bernstein, M.A., Aisen, A.M., Hernandez, R.J., Collick, B.D., and Bernstein, T. (1995). Improved ejection fraction and flow velocity estimates with use of view sharing and uniform repetition time excitation with fast cardiac techniques. *Radiology* **195**, 471–478.

Fukashiro, S., Komi, P.V., Jarvinen, M., and Miyashita, M. (1995). *In vivo* Achilles tendon loading during jumping in humans. *European Journal of Applied Physiology* **71**, 453–458.

Fukunaga, T., Kawakami, Y., Kuno, S., Funato, K., and Fukashiro, S. (1997). Muscle architecture and function in humans. *Journal of Biomechanics* **30**, 457–463.

Gordon, A.M., Huxley, A.F., and Julian, F.J. (1966). Tension development in highly stretched vertebrate muscle fibres. *Journal of Physiology (London)* **184**, 143–169.

Grabiner, M.D., Koh, T.J., and Draganich, L.F. (1994). Neuromechanics of the patellofemoral joint. *Medicine and Science in Sports and Exercise* **26**, 10–21.

Heegaard, J. (1993). *Large Slip Contact in Biomechanics*. Ph.D. thesis, Ecole Polytechnique Federale de Lausanne, Lausanne, Switzerland.

Hill, A. V. (1938). The heat of shortening and the dynamic constants of muscle. *Proceedings of the Royal Society of London B* **126**, 136–195.

Huijing, P.A. (1995). Parameter interdependence and success of skeletal muscle modelling. *Human Movement Science* **14**, 443–486.

Huijing, P.A. (1998). Muscle, the motor of movement: Properties in function, experiment and modelling. *Journal of Electromyography and Kinesiology* **8**, 61–77.

Keller, P.J. (1988). *Basic Principles of Magnetic Resonance Imaging*. GE Medical Systems, Milwaukee, WI.

Kuno, S., and Fukunaga, T. (1995). Measurement of muscle fibre displacement during contraction by real-time ultrasonography in humans. *European Journal of Applied Physiology* **70**, 45–48.

Lieber, R.L., Loren, G.J., and Friden, J. (1994). *In vivo* measurement of human wrist extensor muscle sarcomere length changes. *Journal of Neurophysiology* **71**, 874–881.

Ljung, B.O., Friden, J., and Lieber, R.L. (1999). Sarcomere length varies with wrist ulnar deviation but not forearm pronation in the extensor carpi radialis brevis muscle. *Journal of Biomechanics* **32**, 199–202.

Narici, M.V., Binzoni, T., Hiltbrand, E., Fasel, J., Terrier, F., and Cerretelli, P. (1996). *In vivo* human gastrocnemius architecture with changing joint angle at rest and during graded isometric contraction. *Journal of Physiology (London)* **496**, 287–297.

Nayler, G.L., Firmin, D.N., and Longmore, D.B. (1986). Blood flow imaging by cine magnetic resonance. *Journal of Computer Assisted Tomography* **10**, 715–722.

Nisell, R., and Ekholm, J. (1985). Patellar forces during knee extension. *Scandinavian Journal of Rehabilitation Medicine* **17**, 63–74.

Pappas, G., Sheehan, F.T., Zajac, F.E., and Drace, J.E. (2000). *In vivo* longitudinal strain measured along the biceps brachii during elbow flexion using Cine phase contrast MRI. *Journal of Biomechanics* (in press).

Pelc, N.J., Drangova, M., Pelc, L.R., Zhu, Y., Noll, D.C., Bowman, B.S., and Herfkens, R.J. (1995). Tracking of cyclic motion with phase-contrast cine MR velocity data. *Journal of Magnetic Resonance Imaging* **5**, 339–345.

Pelc, N.J., and Glover, G.H. (1993). A stroll through k-space. In: *The Physics of MRI 1992 AAPM Summer School Proceedings*. Bronskill and Sprawls (eds.), American Institute of Physics, Woodbury, NY, pp. 771–784.

Pelc, N.J., Herfkens, R.J., Shimakawa, A., and Enzmann, D.R. (1991). Phase contrast cine magnetic resonance imaging. *Magnetic Resonance Quarterly* **7**, 229–254.

Pelc, N.J., Sommer, F.G., Li, K.C., Brosnan, T.J., Herfkens, R.J., and Enzmann, D.R. (1994). Quantitative magnetic resonance flow imaging. *Magnetic Resonance Quarterly* **10**, 125–147.

Peters, T.M. (1993). An introduction to k-space. In: *The Physics of MRI 1992 AAPM Summer School Proceedings*. Bronskill and Sprawls (eds.), American Institute of Physics, Woodbury, NY, pp. 754–770.

Richter, M., Kiefer, H., Hehl, G., and Kinzl, L. (1996). Primary repair for posterior cruciate ligament injuries. An eight-year follow-up of 53 patients. *American Journal of Sports Medicine* **24**, 298–305.

Rungee, J.L., Fay, M.J., and Deberardino, T.M. (1995). Olecranization of the patella. *Orthopedics* **18**, 27–34.

Sheehan, F.T., and Drace, J.E. (2000). Human patellar tendon strain (a non-invasive, *in vivo* study). *Clinical Orthopaedics and Related Research* **370**, 201–207.

Sheehan, F.T., Zajac, F.E., and Drace, J.E. (1998). Using cine phase contrast magnetic resonance imaging to non-invasively study *in vivo* knee dynamics. *Journal of Biomechanics* **31**, 21–26.

Sheehan, F.T., Zajac, F.E., and Drace, J.E. (1999). *In vivo* tracking of the human patella using Cine phase contrast magnetic resonance imaging. *ASME Journal of Biomedical Engineering* **121:6**, 650–656.

van Bavel, H., Drost, M.R., Wielders, J.D., Huyghe, J.M., Huson, A., and Janssen, J.D. (1996). Strain distribution on rat medial gastrocnemius (MG) during passive stretch. *Journal of Biomechanics* **29**, 1069–1074.

van Eijden, T.M., Kouwenhoven, E., Verburg, J., and Weijs, W.A. (1986). A mathematical model of the patellofemoral joint. *Journal of Biomechanics* **19**, 219–229.

van Eijden, T.M., Kouwenhoven, E., and Weijs, W.A. (1987). Mechanics of the patellar articulation. Effects of patellar ligament length studied with a mathematical model. *Acta Orthopaedica Scandinavica* **58**, 560–566.

van Kampen, A., and Huiskes, R. (1990). The three-dimensional tracking pattern of the human patella. *Journal of Orthopaedic Research* **8**, 372–382.

Yamaguchi, G.T., and Zajac, F.E. (1989). A planar model of the knee joint to characterize the knee extensor mechanism. *Journal of Biomechanics* **22**, 1–10.

Zajac, F.E. (1989). Muscle and tendon: Properties, models, scaling, and application to biomechanics and motor control. *CRC Critical Reviews in Biomedical Engineering* **17**, 359–411.

Zhu, Y., Drangova, M., and Pelc, N.J. (1996). Fourier tracking of myocardial motion using cine-PC data. *Magnetic Resonance in Medicine* **35**, 471–480.

Zuurbier, C.J., Heslinga, J.W., Lee-de Groot, M.B., and van der Laarse, W.J. (1995). Mean sarcomere length-force relationship of rat muscle fibre bundles. *Journal of Biomechanics* **28**, 83–87.

21 Muscle Inhibition and Functional Deficiencies Associated with Knee Pathologies

ESTHER SUTER AND WALTER HERZOG
Human Performance Laboratory, University of Calgary, Calgary, Alberta, Canada

HISTORICAL BACKGROUND

Functional assessments of the knee extensors have typically focused on strength measurements using a dynamometer. Tests include isometric contractions at various joint angles, and isokinetic contractions of eccentric or concentric nature at different angular velocities. Strength measurements in combination with electromyographic (EMG) recordings have been used to assess motor drive during voluntary effort. This approach is well suited to study muscle function and neuromuscular activation in healthy subjects, since it is assumed that a healthy subject is able to drive a muscle to its full extent, and that the force produced during a voluntary effort represents a true maximum. In patients with neuromuscular deficiencies, estimates of the maximal strength obtained from force and EMG measurements do not always accurately reflect the maximal force potential. Typically, knee pathologies are associated with weakness of the quadriceps muscles. The reduced muscle strength has traditionally been attributed to disuse atrophy of the painful limb. A significant proportion of the strength deficit, however, may be caused by muscle inhibition, which constitutes an inability to activate all motor units of a muscle, or a group of muscles, to the full extent during voluntary contractions (Young, 1993). By only measuring knee extensor strength during maximal voluntary efforts, no information is gained on the extent of muscle activation, i.e., it is not known if a strength deficit is the result of muscle atrophy, or if the muscle has been incompletely activated. EMG is of limited use to evaluate muscle weakness because absolute values are difficult to interpret, and no meaningful comparisons between muscles from the injured and the contralateral leg can be made. Long-lasting functional deficiencies following knee pathologies are a major concern in rehabilitation research, and reduced muscle activation in patient populations may contribute to these deficiencies. In order to study activation patterns in patients, electrical muscle stimulation during maximal voluntary efforts has been used.

Skeletal Muscle Mechanics: From Mechanisms to Function. Edited by W. Herzog.

ASSESSING MUSCLE ACTIVATION

The degree of muscle activation can be assessed, at least qualitatively, using the interpolated twitch technique. This technique requires that the nerve innervating the target muscle is electrically stimulated while the muscle is maximally contracted (Belanger and McComas, 1981; Bülow et al., 1993; Gandevia et al., 1998; Kent-Braun and LeBlanc, 1996; Rutherford et al., 1986). A supramaximal stimulation activates all motor units, resulting in an increase in muscle force and corresponding joint torque from motor units that were not activated maximally during the voluntary contraction. This increase in torque can be measured using a strength-testing machine, and it has been referred to as the 'interpolated twitch torque' (ITT). Its magnitude is related non-linearly to the amount of inhibition in the target muscles. Using this technique, motor unit activation during voluntary contraction has been studied in healthy subjects and patient populations. Muscle groups investigated include the quadriceps group (Bülow et al., 1993; Rutherford et al., 1986; Suter et al., 1996), biceps brachii (Gandevia et al., 1998; Rutherford et al., 1986), dorsiflexors (Belanger and McComas, 1981; Kent-Braun and LeBlanc, 1996), and plantarflexors (Belanger and McComas, 1981). A curvilinear relationship between the size of the ITT and the level of voluntary effort has been found (Bülow et al., 1993; Belanger and McComas, 1981; Rutherford et al., 1986; Suter et al., 1996). A typical torque-time curve from a maximal voluntary isometric contraction of the quadriceps muscle with superimposed electrical stimulation is shown in Figure 21.1.

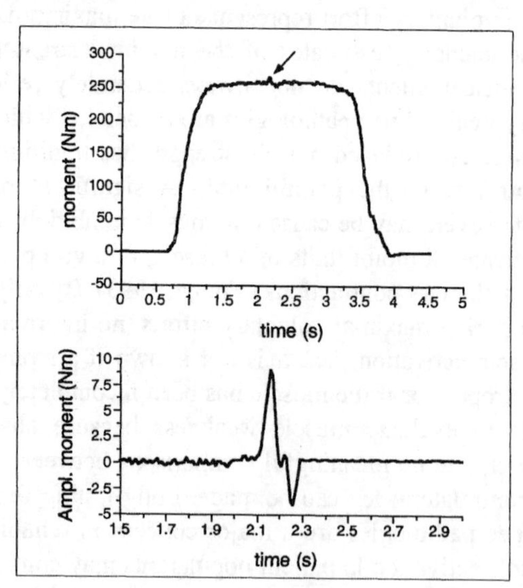

Figure 21.1. Maximal voluntary contraction with an electrically stimulated twitch (arrow) during the force plateau in a healthy subject. The raw signal is shown in the top graph, and the amplified twitch torque trace is shown in the bottom graph.

In Figure 21.1 the ITT is small, which indicates that motor unit activation was nearly maximal. In patients who are unable to maximally activate their knee extensors, the electrical stimulation will result in a substantial increase in the measured knee extensor torque.

An example of a torque-time curve from an isometric contraction with a superimposed electrical twitch from a patient with anterior cruciate ligament (ACL) deficiency is shown in Figure 21.2. The large ITT suggests that substantial muscle inhibition is present in the knee extensors of this particular patient.

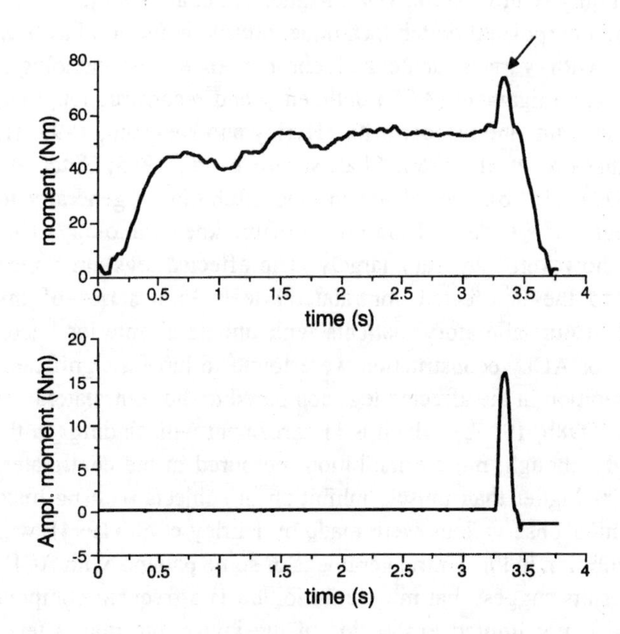

Figure 21.2. Maximal voluntary contraction with an electrically stimulated twitch (arrow) during the force plateau in a patient with knee pathology. The raw signal is shown in the top graph, and the amplified twitch torque trace is shown in the bottom graph. Note the large interpolated twitch torque indicating substantial muscle inhibition.

Although the interpolated twitch technique has been the preferred method of assessing motor unit activation, there is little agreement as to the optimal testing procedure. Differences in the measuring protocols influence muscle inhibition. For example, post-tetanic potentiation is known to occur and may influence the amount of muscle inhibition (Bülow et al., 1993, Suter and Herzog, 1997). Also, muscle inhibition has been found to be larger at a knee angle of 60° of flexion compared to 30° of flexion (Suter and Herzog, 1997). Because of the apparent dependency of muscle inhibition on joint angle and possible confounding factors such as potentiation, it is important to use a standardized approach when making comparisons between different populations.

MUSCLE INHIBITION FOLLOWING KNEE JOINT PATHOLOGIES

Weakness of the knee extensor muscles, loss of range of motion, and impaired joint proprioception are common functional deficiencies resulting from traumatic knee injuries. The rehabilitation process after an injury is often long and does not always lead to full functional recovery. In particular, strength deficits and muscle inhibition have been shown to persist for extended time periods following injury (Hurley et al., 1994; Rutherford et al., 1986).

Using the interpolated twitch technique, muscle inhibition has been evaluated in patients with various acute and chronic knee-joint pathologies, such as anterior cruciate ligament (ACL) deficiency and reconstruction, meniscectomy, anterior knee pain, and osteoarthritis (Hurley and Newham, 1993; Hurley et al., 1994; Rutherford et al., 1986; Shakespeare et al., 1985; Suter et al., 1998a, 1998b, 1999a, 1999b). Significant muscle inhibition is generally found in the affected legs, independent of the nature of the knee pathology. The amount of inhibition, however, may vary largely. The affected legs are typically weaker compared to the unaffected contralateral legs. In a series of investigations performed in our laboratory, patients with unilateral anterior knee pain, ACL deficiency, or ACL reconstruction were found to have a significant amount of muscle inhibition in the affected legs compared to the contralateral legs (Suter et al., 1998a, 1998b, 1999b), which is in agreement with findings in the literature. Interestingly, though, muscle inhibition measured in the contralateral legs was also 10-15% higher than muscle inhibition in subjects with no history of knee injury. Similar observations were made by Hurley et al. (1994), who measured muscle inhibition in the contralateral legs of some patients with ACL deficiency.

These results suggest that muscle inhibition is a frequent component of knee injury, and is not limited to the side of the injury, but may affect the contralateral side. Consequently, the contralateral leg in a unilateral knee injury does not necessarily represent a normal control. By using the contralateral leg as a normal reference (a practice often followed in clinical situations), the actual strength deficit and muscle inhibition of the injured legs may be underestimated. There are several mechanisms that could explain how unilateral joint injuries affect the functional integrity of the contralateral leg. Joint injuries often require a reduction in physical activity level, and gait patterns may change in an attempt to avoid pain and instability. Berchuk et al. (1990) observed that patients with ACL deficiency tend to reduce contraction of the quadriceps in order to avoid large flexor moments and associated anterior displacement of the tibia. Such adaptive behaviour may affect the neuromuscular control of the injured and the contralateral lower limb. Also, it has been shown that a unilateral inflammation can be transferred to the contralateral side through internuncial neural connections in the spinal cord (Levine et al., 1985).

MECHANISMS OF MUSCLE INHIBITION

The mechanisms responsible for muscle inhibition are not well understood. It has been suggested that pain receptors in the knee provide inhibitory input to the knee extensor muscles. This hypothesis was based on the observation that muscle inhibition decreases after local injection of anaesthetics into the knee (Shakespeare et al., 1985). However, reducing or eliminating pain through surgical or medical interventions was not always effective in improving muscle inhibition (Shakespeare et al., 1985; Suter et al., 1998a, 1998b). Also, muscle inhibition has been observed in complete absence of pain, indicating that there must be other contributing factors. Mechanical distension of the knee joint capsule may be another promoter of muscle inhibition. Evidence in support of this idea comes from studies in which muscle inhibition was experimentally induced by injecting saline into the knee (Iles et al., 1990). Knee distension consistently produced a dose-dependent decrease in the Hoffmann reflex (H-reflex). The H-reflex is a monosynaptic reflex and its amplitude is thought to reflect excitability of the α-motoneurons. A decrease in the H-reflex associated with increasing joint distension suggests a connection between the inhibitory input of joint afferents and muscle inhibition. In a recent study, a quadriceps avoidance gait pattern with typical inhibition of knee extensor activity was produced by artificial knee effusion (Torry et al., 2000), further supporting the hypothesis that inhibitory stimuli may mediate muscle inhibition.

LONG-TERM OUTCOME OF MUSCLE INHIBITION AND FUNCTIONAL DEFICIENCIES FOLLOWING KNEE PATHOLOGIES

In most studies aimed at investigating quadriceps inhibition in patients with knee pathologies, muscle function was evaluated within the first few weeks or months following injury and subsequent rehabilitation. Typically, rehabilitation programs following joint injury focus on muscle strengthening, proprioceptive training, and range-of-motion training, with the goal of achieving pre-injury strength and physical activity levels. Despite these efforts, recovery following injury is often incomplete and joint function at the end of the rehabilitation process is unsatisfactory. Persistent deficits in joint proprioception (Gillquist and Messner, 1999), neuromuscular performance (Pfeifer and Banzer, 1999), and quadriceps strength (Hurley et al., 1994) during the first year after injury or reconstruction of the ACL have been observed. Deficits at the end of the physical rehabilitation program appear to be particularly prominent in patients who show a large amount of initial muscle inhibition (Hurley et al., 1994). Patients with knee extensor inhibition of 30-45% at the onset of the physical rehabilitation program showed neither increases in knee extensor strength, nor improvements in muscle inhibition after one month of therapy (Hurley et al., 1994). If the initial muscle inhibition was small, significant improvements in strength were found, although some deficits always remained. It was concluded that minor muscle inhibition does not prevent strength increases, although full restoration of muscle strength may not be achieved. Severe knee extensor inhibition interferes with the rehabilitation process and compromises the functional recovery of the affected structures.

Long-term deficiencies following knee injury are not well described. In a number of studies, long-term recovery of knee extensor strength following ACL reconstruction was investigated. It was found that four to nine years following reconstructive surgery, the knee extensor strength of the injured legs remained lower compared to the contralateral legs or age-matched healthy controls (Arangio et al., 1997; Natri et al., 1996; Seto et al., 1988). Strength deficits exceeding 20% were observed. Shelbourne and Gray (1997) found long-term quadriceps strength deficits in the affected legs of approximately 6-9% compared to the contralateral legs in patients with ACL reconstruction. Factors that contribute to some of the differences in the results reported in long-term outcome studies of ACL injuries are severity of initial injury, time elapsed between ligament rupture and repair, the type of graft used for reconstruction, and/or intensity of the post-surgical rehabilitation. These variables are generally not controlled.

Muscle atrophy of the knee extensors partially explains the deficits in knee extensor strength of the affected compared to the contralateral legs. Differences in muscle mass between affected and contralateral sides in the order of 8-10% have been found in chronic ACL-deficient (Gerber et al., 1985) and ACL-reconstructed patients (Arangio et al., 1997). However, the strength deficits observed in patient populations are typically higher than one would expect from the loss in muscle mass, suggesting that incomplete muscle activation may be evident years after injury and may contribute to the strength deficit.

Little is known about the natural time course of muscle inhibition following knee injury. Long-term follow-up studies aimed at investigating how muscle inhibition may contribute to long-term strength deficits following knee injury are missing. The few cross-sectional studies assessing muscle inhibition after knee injury give conflicting results. A group investigating patients with acute ACL deficiency (less than three months following ACL rupture), chronic ACL deficiency (more than two years following ACL rupture), and ACL reconstruction only found knee extensor inhibition in the group with acute ACL injury (Snyder-Mackler et al., 1994). Little or no muscle inhibition was found in the ACL-reconstructed patients and the patients with a chronic ACL tear. These results suggest that muscle inhibition is transient, possibly related to changes in afferent input caused by the sudden instability of the knee after the ACL loss. Over time, adaptations occur, mechanisms to compensate for the instability come into effect, and normal activation patterns are restored. However, our own studies indicate that knee extensor weakness and inhibition may exist long after the rehabilitation program is completed (Suter et al., 1998a, 1999b). Anterior knee pain patients who underwent arthroscopic surgery and subsequent physical rehabilitation showed a large amount of knee extensor inhibition following surgery, and muscle inhibition did not decrease significantly during a six-week rehabilitation program. At the six-month follow-up, quadriceps weakness associated with knee extensor inhibition persisted, despite a significant improvement in subjective symptoms and pain (Suter et al. 1998a). Also, patients with ACL-reconstructed knees showed 20% muscle inhibition of the knee extensors at, on average, 22 months following reconstructive surgery (Suter et al., 1999b). The strength deficit between injured and contralateral legs was approximately 10%. Knee extensor inhibition in the ACL-deficient group measured at, on average, 44 months after ACL rupture was higher than that observed in the ACL-reconstructed patients.

Based on the findings of these studies, we propose that severe muscle inhibition of the knee extensors caused by traumatic knee injury may persist over years and may contribute to long-lasting strength deficits. The consequences of persisting muscle weakness and functional deficits are not clear. Decreased knee extensor force and reduced muscle activation may lead to changes in knee loading, which is a proposed cause for osteoarthritis in animal models (Brandt, 1997). Changes in joint mechanics have been associated with degeneration of musculoskeletal tissues in and around the joint in ACL-transected animals (Brandt, 1997). Such degeneration was also observed in long-term follow-up studies of ACL-injured humans. For example, there is a higher incidence of radiographically detected osteoarthritis in ACL-deficient knees than in the unaffected contralateral knees. According to a recent review (Gillquist and Messner, 1999), 50-70% of all patients with a complete ACL rupture and associated structural damage show radiographic signs of osteoarthritis 15 to 20 years after the injury.

There is increasing evidence that muscle weakness may be an important factor in human osteoarthritis (Brandt, 1997; Felson and Zhang, 1998). In a study involving more than 300 subjects (Slemenda et al., 1998), knee extensor weakness relative to body weight was a strong predictor for the occurrence of osteoarthritis in a follow-up examination. Women who developed osteoarthritis over a 30-month testing period had about 18% lower knee extensor strength at baseline compared to women who did not develop osteoarthritis. In cross-sectional investigations in senior populations, knee extensor strength was a significant predictor of radiographic and symptomatic osteoarthritis of the knee and of the extent of disability, even after adjusting for age, sex, and body weight (McAlidon et al., 1993; O'Reilly et al., 1998).

Although a direct link between muscle weakness/muscle inhibition and the occurrence of osteoarthritis has not been established, there is some indication that knee extensor weakness may constitute an independent risk factor for osteoarthritis and predispose a joint to degenerative disease.

IMPLICATIONS FOR REHABILITATION IN PATIENTS WITH KNEE PATHOLOGIES

The interpolated twitch technique is a useful tool for assessing functional deficiencies associated with joint pathology. Determining the amount of muscle inhibition provides valuable information beyond that obtained with strength and EMG measurements alone. Strength measurements allow for comparisons between left and right legs without giving any indication of the absolute force potential in each leg, or the level of muscle activation achieved. Also, by comparing the strength of the involved leg relative to the contralateral leg, it is often assumed that the contralateral leg is a normal, healthy control. According to results of recent studies (Hurley et al., 1994; Suter et al., 1998a, 1998b, 1999b), this assumption is not correct. The effects of unilateral joint injury may extend to the opposite side, and using the contralateral leg as a normal control may lead to an underestimation of functional deficiency in the involved leg.

Using the interpolated twitch technique, muscle weakness caused by muscular atrophy and weakness caused by incomplete activation of the muscle can be differentiated. From a rehabilitation point of view this difference is of importance, because if muscle atrophy is the main reason for the observed muscle weakness and muscle inhibition is absent or small, increases in muscle mass and associated strength can be achieved through systematic strength training. However, if the strength deficit is caused by muscle inhibition, treatment is more complicated. Strength training is likely to be ineffective as long as severe muscle inhibition is present (Hurley et al, 1994). It is not known through which mechanisms muscle inhibition may prevent strength increases of the affected muscle group.

FUTURE RESEARCH

Given the limited knowledge of the mechanisms involved in muscle inhibition and the lack of data on long-term consequences of persistent muscle inhibition and muscle weakness, several issues should be addressed in future research. First, identifying procedures that eliminate muscle inhibition in the early stages of rehabilitation should be a main focus. Results from studies on elite athletes with knee joint pathologies suggest that muscle inhibition can be overcome with a systematic and rigorous rehabilitation program in subjects who are highly motivated (Huber et al., 1998). However, the optimal exercise modalities to achieve this goal have yet to be determined. Maximal effort contractions appear to be ineffective in overcoming muscle inhibition. There are several treatment approaches that are potentially successful in eliminating muscle inhibition in patients. For example, knee extensor inhibition may be overcome by facilitating knee extensor pathways. Preliminary results suggest that there is a selective increase in activation of the rectus femoris or vastus medialis when knee extension is combined with hip flexion or hip extension, respectively, compared to an isolated knee extension (Suter et al., 1999b). Muscle inhibition may possibly be reduced using non-conventional interventions. We found that muscle inhibition in anterior knee pain patients was decreased following spinal manipulation of the sacroiliac joint (Suter et al., 1999a). Spinal manipulation has been proposed to activate receptors from structures in and around the manipulated joint (Murphy et al., 1995). The altered afferent input is thought to cause changes in motoneuron excitability with subsequent interruption of a proposed pain-spasm-pain cycle. Such changes in motoneuron excitability, measured as changes in H-reflex amplitude, have been observed in the soleus after sacroiliac joint manipulation (Murphy et al., 1995). The results of these studies suggest that pain management in combination with specific exercise protocols may be effective in reducing muscle inhibition, and thus increase the efficiency of post-surgical treatment and rehabilitation in patients with joint pathology.

Second, reflex loops and possible neural pathways involved in muscle inhibition require further investigation. Injection of saline into the joint induced a gait pattern known as 'quadriceps avoidance gait' (Torry et al., 2000). This gait pattern is frequently observed in patients with ACL deficiency (Berchuck et al., 1990), and it is characterized by reduced activation of the knee extensors during walking. The results suggest that knee extensor inhibition may be a direct consequence of changed afferent input to the quadriceps muscle. Investigating the possible reflex mechanism will advance our understanding of the processes causing muscle inhibition.

Third, long-term follow-up studies in patients with knee pathologies should not only include assessment of strength and structural changes in the joint, but should also be aimed at evaluating muscle inhibition. Cross-sectional studies would allow for estimating the contribution of muscle inhibition to the long-term strength deficits observed following knee pathology, and the possible association with joint degeneration. In order to directly test the hypothesis that persistent muscle inhibition causes weakness and predisposes a joint to degeneration, intervention studies are required. Using a randomized controlled design, interventions that largely eliminate muscle inhibition following acute injury could be tested for effects on short-term and long-term strength deficits. If a rehabilitation program eliminates muscle inhibition, strength deficits should be reduced, and joint degeneration may be slowed or stopped.

In summary, the interpolated twitch technique has proven to be a useful tool for the assessment of muscular activation in normal and patient populations. By applying this technique, it has been established that muscle inhibition and weakness of the knee extensors is a long-term adjunct of knee injuries. Although it has not been demonstrated directly, there is evidence that persistent muscle weakness and muscle inhibition may contribute to or accelerate joint degeneration. Future research should be aimed at addressing this issue systematically, and focus on interventions that may eliminate or reduce muscle inhibition following knee injury.

REFERENCES

Arangio, G.A., Chen, C., Kalady, M., and Reed, J.F. (1997). Thigh muscle size and strength after anterior cruciate ligament reconstruction. *Journal of Orthopedics and Sports Physiotherapy* **26**, 238–243.

Belanger, A.Y., and McComas, A.J. (1981). Extent of motor unit activation during effort. *Journal of Applied Physiology: Environmental and Exercise Physiology* **51**, 1131–1135.

Berchuk, M., Andriacchi, T.P., Bach, B.R., and Reider, B. (1990). Gait adaptations by patients who have a deficient anterior cruciate ligament. *Journal of Bone and Joint Surgery* **72-A**, 871–877.

Brandt, K.D. (1997). Putting muscle into osteoarthritis. *Annals of Internal Medicine* **127**, 154–155.

Bülow, P.M., Norregaard, J., Danneskiold-Samsoe, B., and Mehlsen, J. (1993). Twitch interpolation technique in testing of maximal muscle strength: Influence of potentiation, force level, stimulus intensity and preload. *European Journal of Applied Physiology* **67**, 462–466.

Felson, D.T., and Zhang, Y. (1998). An update on the epidemiology of knee and hip osteoarthritis with a view to prevention. *Arthritis and Rheumatism* **41**, 1343–1355.

Gandevia, S.C., Herbert, R.D., and Leeper, J.B. (1998). Voluntary activation of human elbow flexor muscles during maximal concentric contractions. *Journal of Physiology* **512**, 595–602.

Gerber, C., Hoppeler, H., Claassen, H., Robotti, G., Zehnder, R., and Jakob, R.P. (1985). The lower-extremity musculature in chronic symptomatic instability of the anterior cruciate ligament. *Journal of Bone and Joint Surgery* **67-A**, 1034–1043.

Gillquist, J., and Messner, K. (1999). Anterior cruciate ligament reconstruction and the long term incidence of gonarthrosis. *Sports Medicine* **27**, 143–156.

Huber, A., Suter, E., and Herzog, W. (1998). Inhibition of the quadriceps muscles in elite male volleyball players. *Journal of Sports Sciences* **16**, 281–289.

Hurley, M.V., Jones, D.W., and Newham, D.J. (1994). Arthogenic quadriceps inhibition and rehabilitation of patients with extensive traumatic knee injuries. *Clinical Sciences* **86**, 305–310.

Hurley, M.V., and Newham, D.J. (1993). The influence of arthrogenous muscle inhibition on quadriceps rehabilitation of patients with early, unilateral osteoarthritic knees. *British Journal of Rheumatology* **32**, 127–131.

Iles, J.F., Stokes, M., and Young, A. (1990). Reflex actions of the knee joint afferents during contraction of the human quadriceps. *Clinical Physiology* **10**, 489–500.

Kent-Braun, J.A., and LeBlanc, R. (1996). Quantitation of central activation failure during maximal voluntary contractions in humans. *Muscle and Nerve* **19**, 861–869.

Levine, J.D., Dardick, S.J., Basbaum, A.I., and Scipio, E. (1985). Reflex neurogenic inflammation. I. Contribution of the peripheral nervous system to spatially remote inflammatory responses following injury. *Journal of Neuroscience* **5**, 1380–1386.

McAlindon, T.E., Cooper, C., Kirwan, J.R., and Dieppe, P.A. (1993). Determinants of disability in osteoarthritis of the knee. *Annals of the Rheumatic Diseases* **52**, 258–261.

Murphy, B.A., Dawson, N.J., and Slack, J.R. (1995). Sacroiliac joint manipulation decreases the H-reflex. *Electromyography and Clinical Neurophysiology* **35**, 87–94.

Natri, A., Järvinen, M., Latvala, K., and Kannus, K. (1996). Isokinetic muscle performance after anterior cruciate ligament surgery. *International Journal of Sports Medicine* **17**, 223–228.

O'Reilly, S.C., Jones, A., Muir, K.R., and Doherty, M. (1998). Quadriceps weakness in knee osteoarthritis: The effect on pain and disability. *Annals of the Rheumatic Diseases* **57**, 588–594.

Pfeifer, K., and Banzer, W. (1999). Motor performance in different dynamic tests in knee rehabilitation. *Scandinavian Journal of Medicine and Science in Sports* **9**, 19–27.

Rutherford, O.M., Jones, D.A., and Newham, D.J. (1986). Clinical and experimental application of the percutaneous twitch superimposition technique for the study of human muscle activation. *Journal of Neurology, Neurosurgery, and Psychiatry* **49**, 1288–1291.

Seto, J.L., Orofino, A.S., Morrissey, M.C., Medeiros, J.M., and Mason, W.J. (1988). Assessment of quadriceps/hamstring strength, knee ligament stability, functional and sports activity levels five years after anterior cruciate ligament reconstruction. *American Journal of Sports Medicine* **16**, 170–179.

Shakespeare, D.T., Stokes, M., Sherman, K.P., and Young, A. (1985). Reflex inhibition of the quadriceps after meniscectomy: Lack of association with pain. *Clinical Physiology* **5**, 137–144.

Shelbourne, K.D., and Gray, T. (1997). Anterior cruciate ligament reconstruction with autogenous patellar tendon graft followed by accelerated rehabilitation. *American Journal of Sports Medicine* **25**, 786–795.

Slemenda, C., Heilman, D.K., Brandt, K.D., Katz, B.P., Mazzuca, S.A., Braunstein, E.M., and Byrd, D. (1998). Reduced quadriceps strength relative to body weight. A risk factor for knee osteoarthritis in women? *Arthritis and Rheumatism* **41**, 1951–1959.

Snyder-Mackler, L., De Luca, P.F., Williams, P.R., Eastlack, M.E., and Bartolozzi, A.R. (1994). Reflex inhibition of the quadriceps femoris muscle after injury and reconstruction of the anterior cruciate ligament. *Journal of Bone and Joint Surgery* **76-A**, 555–560.

Suter, E., and Herzog, W. (1997). Extent of muscle inhibition as a function of knee angle. *Journal of Electromyography and Kinesiology* **7**, 123–130.

Suter, E., Herzog, W., and Bray, R.C. (1998a). Quadriceps inhibition following arthroscopy in patients with anterior knee pain. *Clinical Biomechanics* **13**, 314–319.

Suter, E., Herzog, W., and Bray, R.C. (1999b). Muscle inhibition and knee extensor activity in patients with ACL pathologies. In *Proceedings of the XVII Congress of the International Society of Biomechanics (ISB)* (August 8-13, 1999, Calgary, Alberta, Canada). University of Calgary, Calgary, Alberta, Canada, p. 252.

Suter, E., Herzog, W., DeSouza, K., and Bray, R.C. (1998b). Inhibition of the quadriceps muscles in patients with anterior knee pain. *Journal of Applied Biomechanics* **14**, 360–373.

Suter, E., Herzog, W., and Huber, A. (1996). Extent of motor unit activation in the quadriceps muscles of healthy subjects. *Muscle and Nerve* **19**, 1046–1048.

Suter, E., McMorland, G., Herzog, W., and Bray, R. (1999a). Decrease in quadriceps inhibition after sacroiliac joint manipulation in patients with anterior knee pain. *Journal of Manipulative and Physiological Therapeutics* **22**, 149–153.

Torry, M.R., Decker, M.J., Viola, R., O'Connor, D.D., and Steadman, J.R. (2000). Intra-articular knee joint effusion induces quadriceps avoidance gait patterns. *Clinical Biomechanics* **15**, 147-159.

Young, A. (1993). Current issues in arthrogenous inhibition. *Annals of the Rheumatic Diseases* **52**, 829–834.

22 Effects of Ageing on Eccentric and Concentric Muscle Torque Production in Lower and Upper Limbs

ANTHONY A. VANDERVOORT
Schools of Physical Therapy and Kinesiology, Faculty of Health Sciences, University of Western Ontario, London, Ontario, Canada

MICHELLE M. PORTER
Faculty of Physical Education and Recreation Studies, University of Manitoba, Winnipeg, Manitoba, Canada

DENISE M. CONNELLY
School of Rehabilitation Therapy, Physical Therapy Programme, Queen's University, Kingston, Ontario, Canada

JOHN F. KRAMER
Schools of Physical Therapy and Kinesiology, Faculty of Health Sciences, University of Western Ontario, London, Ontario, Canada

INTRODUCTION

Since the specialized, post-mitotic cells that make up the human neuromuscular system are subject to the deleterious effects of the ageing process, skeletal muscle mechanics and function in older adults reflect an overall loss of muscle mass coupled with a slowing of contractile properties. Older adults may also be negatively influenced by the effects of a sedentary lifestyle, but these effects are at least partially reversible with an appropriate exercise stimulus. This chapter begins with an overview of age-related changes in muscle size and structure. Then, the observed effects of ageing on strength and power aspects of muscle mechanics are detailed. In particular, the differential influence of ageing on the various forms of contraction is described: isometric (ISO), concentric (CONC), and eccentric (ECC), when muscle length is constant, decreases, and increases because of an external force, respectively. Since the latter is the most effective in producing high forces, a brief discussion of the adaptability of ageing muscle is also presented with an

Skeletal Muscle Mechanics: From Mechanisms to Function. Edited by W. Herzog.

an emphasis on how older adults might be able to utilize their relatively greater ECC strength for daily activities, and in effective resistance training programs.

HISTORICAL BACKGROUND

The durability of the biological motors that move the human body is remarkable, lasting for over a century in some people who live to a very old age. The well-designed post-mitotic cells that make up motor units have built-in repair and regeneration processes to maintain their highly specialized functions of electrical signalling, protein manufacturing, and force generating. However, this longevity is variable among the various nerve and muscle cells, and some of the total complement that were present at maturation will have subsequently disappeared over the life span, leaving a reduced reserve (Doherty et al., 1993; McComas, 1996).

With respect to neurons, an age-related decrease has been demonstrated throughout the nervous system, although there are regional variations in the extent (Brody and Vijayashanker, 1977). In some aspects, the existence of neuronal loss can be subtle and generally well-compensated in even the oldest individuals, while others who are less fortunate as they age may succumb to the dramatic effects of such diseases as amyotrophic lateral sclerosis (ALS) and post-polio syndrome. Studies of the numbers of motoneurons in the spinal cords of cadavers have clearly indicated a significant loss after the age of about 60 years (Tomlinson and Irving, 1977). It should also be noted that cells can still remain anatomically present in aged people, yet be dysfunctional because of biochemical changes (e.g., accumulation of lipofuscin).

One level of analysis has centred on the motor unit (MU), since it consists of a single motoneuron and its family of innervated muscle cells—cells that are uniformly Type I or II. Considerable flexibility is built into the peripheral motor pathways due to the variation between slow-twitch and fast-twitch contractile properties of these types of fibres. With ageing, the complement of motor units undergoes a process of reduction and adaptation, which in turn affects the capacity to produce forces on the joints. Using the electrophysiological technique of MU estimation, a striking decline in excitable MUs was found beginning in the seventh decade of life (Doherty et al., 1993; McComas, 1996). The ageing process also appears to cause some remodelling of synaptic connections at the neuromuscular junction, ultimately resulting in reinnervation of some of the Type II fibre complement by collateral sprouting from axons of the slower Type I MUs (Brooks and Faulkner, 1994; Lexell, 1995; McComas, 1996).

Not surprising, then, are the cadaveric and radiological observations that the thigh and leg muscles show a curvilinear pattern of reduction in size over the adult age range that corresponds to the decline in muscle strength (Hakkinen et al., 1998a; Lexell, 1995; Vandervoort and McComas, 1986). Muscles of the upper limb do not seem to show quite as much atrophy. This intriguing regional difference reflects in part the inherent variability of the effect of the ageing process on different tissues and regions of the body (Porter et al., 1995). What seems to be consistent in human muscles is an overall loss in the total number of muscle fibres with ageing in both the Type I and Type II complements, along with a significant reduction in the average fibre size of the latter.

The observed preferential atrophy of Type II fibres, and some evidence of fibre type grouping, have been interpreted as evidence of an ongoing denervation and reinnervation process, where hardy motoneurons have enlarged their own motor unit territory by capturing neighbouring fibres of failing motoneurons (Lexell, 1995; McComas, 1996). In support of this theory is the observation of extra large motor unit potentials during intramuscular needle recordings from older adults (Doherty et al., 1993; Roos et al., 1997). Since a reduction in sarcoplasmic reticulum activity also appears to occur with ageing (Delbono et al., 1997; Hunter et al., 1999), its effect, combined with the Type II fibre atrophy, produces a slowing of lower limb muscle contractile properties. Examples of this ageing effect are presented in Table 22.1, which provides *in vivo* observations of increases in the duration of evoked muscle twitches from several human muscles.

Table 22.1. Examples of muscle twitch contraction durations in young versus older adults.

Study	Muscle	Young		Older	
		Age (years)	CD (ms)	Age (years)	CD (ms)
Connelly et al., 1999	DF	20-22	185	80-85	232
Doherty et al., 1993	EF	22-38	198	60-81	203
Roos et al., 1999	VM	19-35	161	73-91	177
Vandervoort and McComas, 1986	DF	20-32	183	80-100	255
Vandervoort and McComas, 1986	PF	20-32	260	80-100	347

CD = contraction duration (time to peak tension + half relaxation time) in milliseconds (ms); DF = dorsiflexors of ankle; EF = elbow flexors; VM = vastus medialis; PF = plantarflexors of the ankle.

AGEING AND MUSCLE STRENGTH

One of the most frequently studied parameters of neuromuscular function in older people has been the loss of voluntary muscle strength that occurs with increasing age. In comparisons among adults of different ages, decreases in voluntary strength do not become apparent until after the age of about 60 (Figure 22.1). Healthy people in the seventh and eighth decades score on average about 20-40% less during tests of isometric strength than young adults, and the very old show an even greater (50% or more) reduction. Muscles in both the upper and lower limbs (including proximal and distal locations) have been examined, and the size of the age effect shows some minor variation from muscle to muscle (Porter et al., 1995; Roos et al., 1997). Males and females appear to show similar age-related trends when their values are compared on a relative basis.

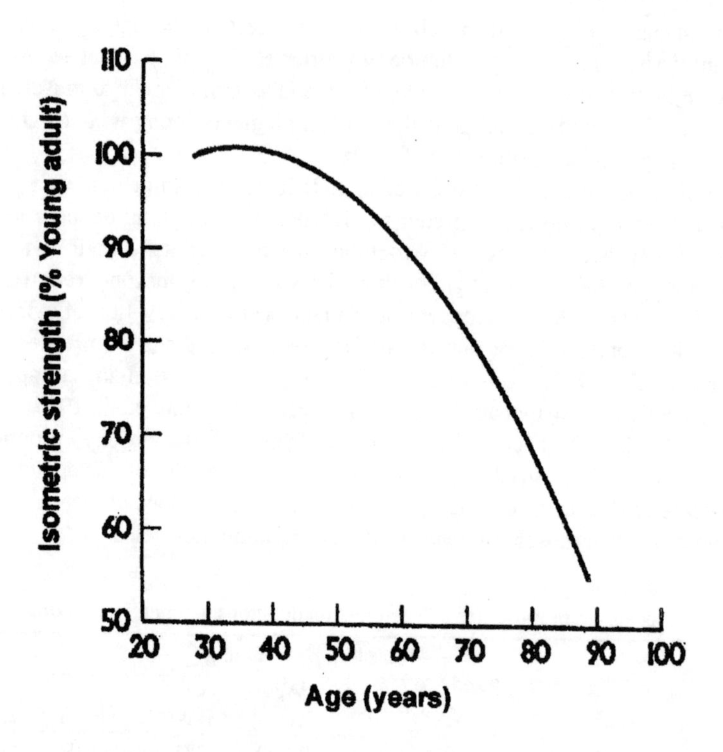

Figure 22.1. The curvilinear relationship between maximum voluntary isometric muscle strength and age in healthy adults.

The older person's strength performance will depend on both the remaining size of his or her pool of motor units and the volitional ability to turn them on completely. These parameters, in turn, are influenced by a number of systemic mechanisms (e.g., hormonal changes) and environmental factors (e.g., nutrition) that vary among individuals (Figure 22.2). Much research in the past two decades has been directed toward examining the relative effects of a sedentary lifestyle versus regular exercise on muscle metabolism, size, and function, resulting in recent publication of guidelines for older persons (American College of Sports Medicine, 1998).

Strength testing of concentric (CONC) muscle actions (in which the muscle shortens) has also revealed lower values in older people than in young adults (Porter et al., 1995). At higher velocities of movement, the age-related deficit is quite marked, and power output is also considerably reduced in older people when tested for maximal dynamic capacity (Bassey, 1997), maximal rate of evoked or voluntary isometric force development (Hakkinen et al., 1996; Vandervoort and Hayes, 1989; Winegard et al., 1996), or ability to use the stretch-shortening cycle effectively in jumping (Bosco and Komi, 1980). Since many activities of daily

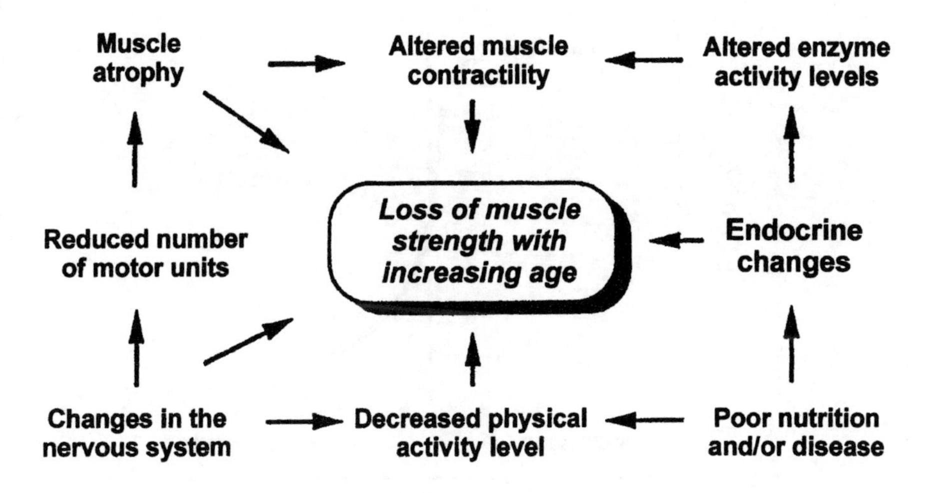

Figure 22.2. Proposed mechanisms leading to the loss of muscle strength with increasing age.

living involve dynamic movements in which power is generated by the muscles (e.g., walking, stair-climbing, sports), the functional impairments associated with low power capacity are of much current interest. It may be, for example, that very old people no longer have adequate propulsive power in their plantarflexor muscles for achieving the same gait pattern as young adults, thereby slowing them considerably, producing muscle fatigue (Faulkner and Brooks, 1995), and also putting more requirements on other leg muscles (DeVita et al., 1998).

However, muscles are also capable of generating forces while lengthening or working eccentrically against a load. Indeed, there is a radically different shape to the force-velocity curve on the lengthening side (Figure 22.3), since muscles can resist high-velocity stretching without losing force-generating capacity in the same way that rapid shortening causes decreases (Enoka, 1996; Epstein and Herzog, 1998). Part of this ECC advantage stems from the resistive contribution of passive elastic elements within the musculotendinous unit during the stretch phase, and the contribution of cross-bridges themselves (Figure 22.4). However, it has been shown in young adults that the physiologic maximum of a muscle's force-generating capacity during ECC loading (at least 1.5 times greater than ISO) is not normally reached with voluntary activation. This evidence came from experiments in which the quadriceps muscle was electrically stimulated (Westing et al., 1990), and when a quick-stretch method (Webber and Kriellaars, 1997) was used to determine total eccentric force-generating capacity. For example, in the latter study, healthy young adult subjects achieved an average voluntary ECC/CONC ratio of only 1.2, whereas testing of the muscle's full physiological potential indicated a much higher ECC/CON ratio of 2.0.

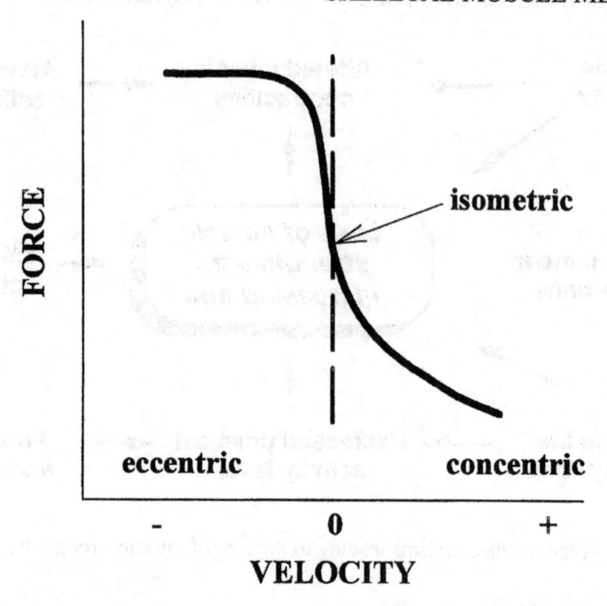

Figure 22.3. Schematic of the force-velocity relationship of muscle for both concentric (CONC) and eccentric (ECC) conditions. When muscle actively lengthens during an ECC contraction, the force-producing capabilities are higher than during a static isometric contraction (velocity zero) and concentric (CONC) contractions at varying velocities. This figure also demonstrates that force decreases with increasing shortening velocity.

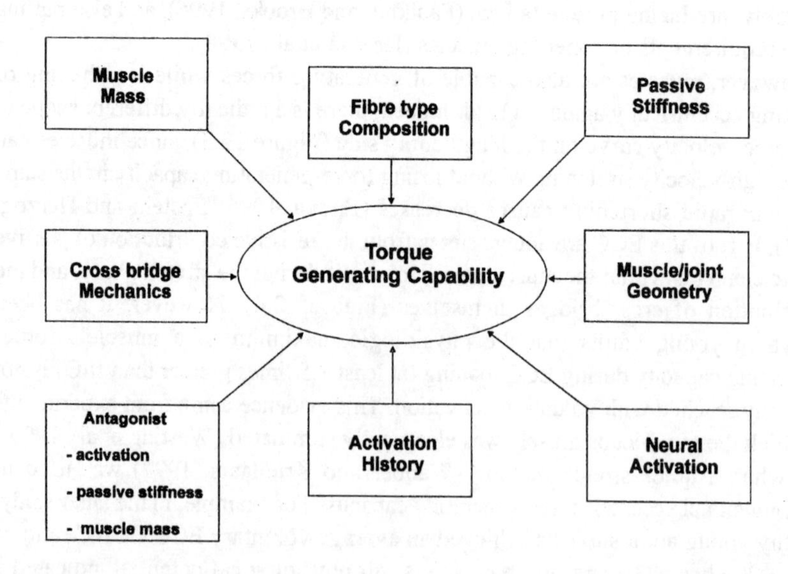

Figure 22.4. Factors influencing muscle force and torque production.

CURRENT PROBLEMS

While it has been shown that ageing causes a loss of muscle strength for isometric or concentric actions, less is known about the eccentric (ECC) performance of older persons. Therefore, the purpose of this study was to investigate the effects of ageing on ECC/CONC ratios by testing for maximum voluntary torque output of older men and women. Muscle groups at the knee, ankle, and elbow were assessed using an isokinetic protocol.

EXPERIMENTAL APPROACH

The research design used to date has involved a comparison of two different samples of men and women, representing the young and old population. Volunteer subject groups consisted of healthy and active individuals aged 65 or more, and young controls aged 20-30 years. Matching of influential variables such as height, weight, and habitual activity status was done to ensure that age groups were comparable, and highly trained athletes were not included.

Subjects came to a strength-assessment laboratory for all tests. After a warm-up ride at low intensity on a stationary bicycle, they were placed on a KINCOM isokinetic dynamometer for muscle testing. This device uses a hydraulically driven controller to regulate the angular velocity of the lever arm to which limbs are attached and, in the case of ECC loading, induces muscle lengthening against subjects' resistive efforts. Testing involved extensive practice before recording the peak torque (PT) achieved during 3-5 maximum trials for each type of muscle activation, CONC or ECC. By checking subjects' torque records for accuracy of velocity (to within 1°/s) and using consistent stabilization methods, reasonably high reproducibility of test results were achieved (I.C.C. = 0.65-0.90; see Porter et al., 1996).

The first report that older adults have significantly higher ECC/CONC ratios than young adults was published in 1990 by Vandervoort, Kramer, and Wharram, based on a comparison of knee extension strength between groups of young and old women. This higher ratio indicated that, in terms of strength, the older women were less affected by the ageing process when their muscles generated forces while lengthening versus during shortening. Indeed, for ankle dorsiflexion, although significantly lower CONC values were observed in older women compared to young (Figure 22.5), Porter et al. (1997) found no difference between age groups for the ECC condition. In several studies of various muscles since then, it has been observed that strength decreases with ageing consistently less for ECC muscle action, resulting in average ECC/CONC ratios ranging from 1.32 to 1.91 in older subjects, and 1.18 to 1.51 in young adults (Figure 22.6). For both age groups, increasing the velocity of movement had little effect on the torque generated by the ankle dorsiflexors during ECC tests, but caused significant losses in CONC strength (Figure 22.7; Connelly and Vandervoort, 2000). Thus, the CONC limb of the torque-velocity curve demonstrated a typical decrease that ended in very low values for older subjects (e.g., at 180°/s, mean PT = 6.6 Nm for CONC dorsiflexion, but ECC PT = 22.2 Nm; ECC/CONC ratio in this case = 3.38).

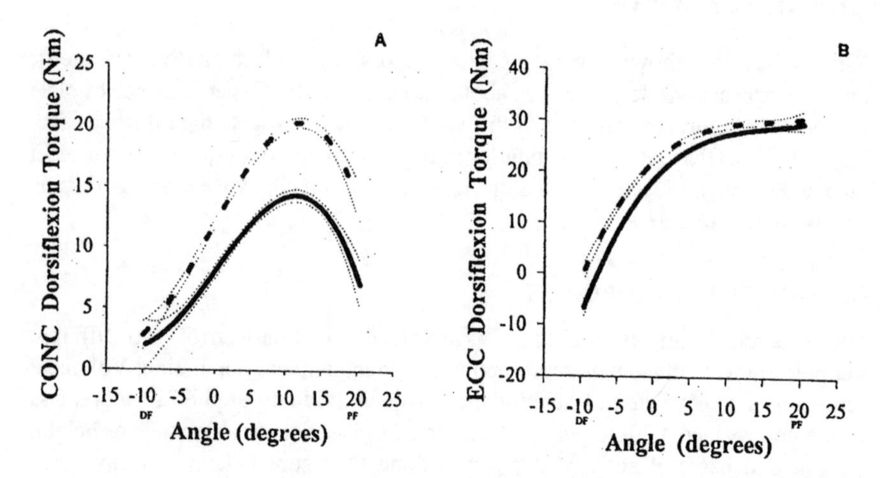

Figure 22.5. Torque-angle relationships for (A) concentric (CONC) dorsiflexion and (B) eccentric (ECC) dorsiflexion. The overlayed data for the older women (solid line) and younger women (dash), derived from polynomial curves, also include the 95% confidence intervals for each age group (dotted lines). Note the difference in scale for CONC versus ECC torque values. (Derived from Porter and Vandervoort, 1997; and Porter et al., 1997.)

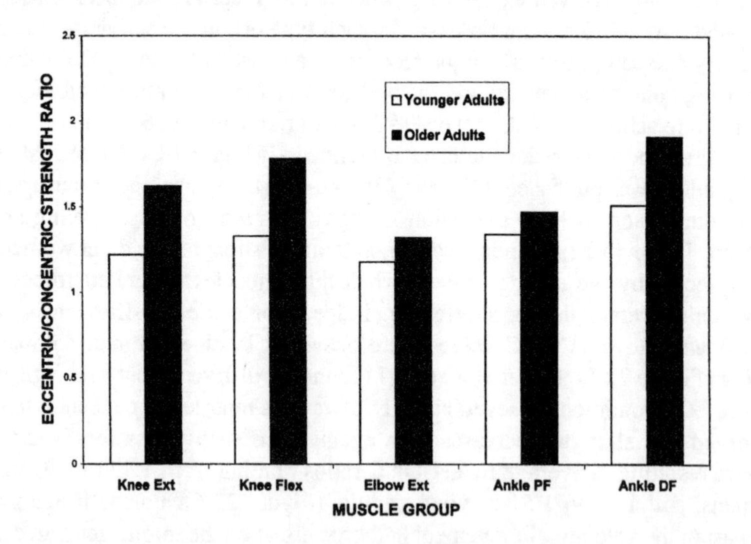

Figure 22.6. Age effect on ratios of eccentric (ECC) to concentric (CONC) muscle strength during isokinetic testing. Compiled from data collected in the authors' laboratory, based on samples of healthy young and older adults using the KINCOM isokinetic dynamometer. Moderate velocities of 90°/s for the knee and elbow tests and 30°/s for the ankle were utilized. The older adults had a mean age of 70 years.

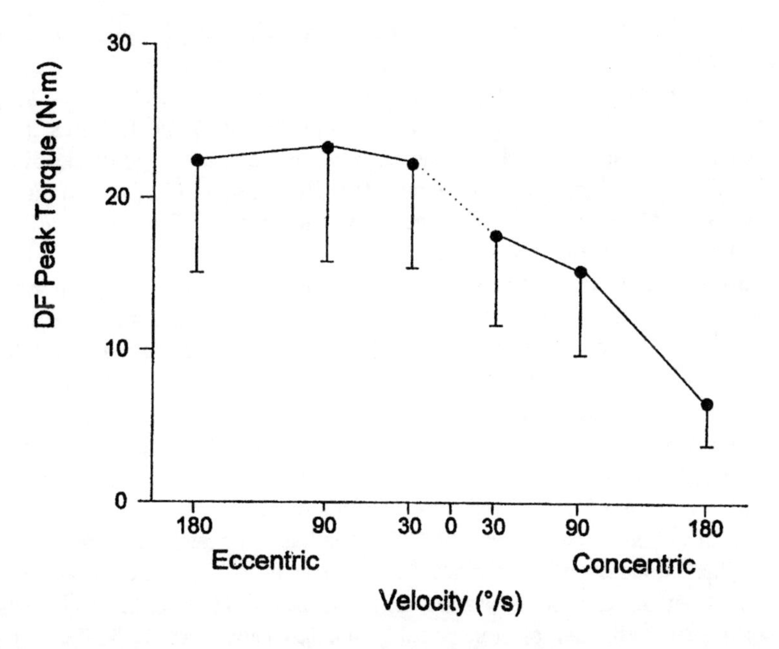

Figure 22.7. Torque velocity curve for ankle dorsiflexion strength in a group of older adults. Maximal voluntary efforts were recorded for both eccentric and concentric conditions. Note the high ECC torques, but rapid decrease in CONC output as the velocity increased. Values are mean and standard deviation. (Derived from Connelly and Vandervoort, 2000.)

FUTURE RESEARCH

Apparently, the changes in muscle mass, contraction speed, and connective tissue that cause an age-related loss of strength when muscle shortens also create a relative enhancement of performance when muscle is active under eccentric conditions (Brooks and Faulkner, 1994; Porter et al., 1995; Suominen, 1997). Other investigators have reported similar observations of higher ECC/CONC ratios for groups of old subjects compared to young adults for leg and hand muscles (Hortobyagyi et al., 1995; Lynch et al., 1999; Phillips et al., 1993). By using the skinned fibre preparations, Brooks and Faulkner (1994) concluded that the slowing of contractile properties with ageing creates a relative advantage for ECC muscle performance, presumably linked to the prolongation of the active-state duration of actin-myosin cross-bridge attachment. Therefore, some of the variation in ECC/CONC ratios for different muscles of older adults might arise from differential effects of ageing on the mixed-fibre composition in human limb musculature (Lexell, 1995). A relevant and intriguing observation by Vandervoort and McComas (1986) is that human muscle contractile properties actually become slower throughout the adult life span, although isometric strength does not decrease significantly until after the sixth decade (Doherty et al., 1993; Lexell, 1995; Porter et al., 1995). Furthermore, Trappe et al. (1995) provided evidence from a longitudinal study that there was a decreasing proportion of Type II fibres in the gastrocnemius muscle of middle-aged men, when compared to muscle biopsy samples 20 years earlier. Thus, we are currently determining ECC/CONC ratios for a sample of middle-aged men and women to see how their performance is affected by the changing muscle composition.

Another possible mechanism affecting force production in older people relates to the neural control of muscle contraction (Figure 22.4). It is possible that older adults may be able to achieve a more complete activation of their MUs during ECC efforts than the younger age group because they have fewer motor units and muscle fibres to deal with (McComas, 1996). In addition, old muscles can achieve a fused tetanic contraction at a lower firing frequency than young muscles (Connelly et al., 1999; Roos et al., 1999). Detailed electromyographical studies of young and older adults performing ECC and CONC contraction patterns will be of much value for further testing of this hypothesis of enhanced MU activation in the latter age group.

While the explanation for the relative advantage for ECC muscle contraction in older adults remains to be fully clarified, the practical applications of the phenomenon are worth noting. For example, one possible functional advantage is that for a given absolute load of muscular exercise in an older age group, Thompson et al. (1999) demonstrated that the relative intensity and hence cardiovascular response was significantly less for exercise involving eccentric contractions versus concentric. Older persons may therefore find some relative advantage in learning to use their muscles under lengthening situations. In turn, their fatigue-resistance at a given intensity of work might be enhanced during eccentric loading, as compared to concentric, because the muscle can work at a lower relative intensity of maximum capacity for a given task requirement (Enoka, 1996).

LEARNING TO UTILIZE ECCENTRIC MUSCLE CONTRACTIONS

Learning to utilize the apparent strength advantage of muscle for lengthening types of activation may be useful for older adults. We have recently begun to study this phenomenon in our laboratory within the context of concentric/eccentric strength training exercises. Use of an isokinetic, motor-driven dynamometer allows subjects to perform ECC resistance training in a controlled manner. Porter and Vandervoort (1997) found that an eight-week program of this type of exercise for the ankle dorsiflexor muscle group significantly increased both CONC and ECC strength over the whole range of movement. Thus, an additional benefit of training was increased flexibility, since the subjects found that they could dorsiflex their ankle several degrees more before reaching the equilibrium point between strength of the agonists and passive resistance of the opposing ankle structures at the back of the leg.

However, it was apparent from our preliminary observations that older adults undergo considerable learning with this task, beginning with initial movement patterns that are uncoordinated, before progressing to a more smooth and effective torque generation (Connelly et al., 2000). We noted that this effect was most pronounced at a relatively rapid isokinetic velocity for the ankle (180°/s), and was more evident in older adults than young. This observation is consistent with recent evidence that older people experience some initial difficulty in co-ordinating movement patterns if stretch-shortening cycles are used (Enoka, 1997; Greene and Williams, 1996). During the initial two weeks of training, when maximal voluntary torque production improved, there was a corresponding increase in surface EMG activity of the tibialis anterior muscle—further evidence that the older adults' nervous system adapted by learning to activate the dorsiflexor muscles at a higher intensity. Hakkinen et al. (1998a, 1998b) and other groups have also found evidence of improved intensity and co-ordination of muscle activation levels in older adults, as well as slight muscle hypertrophy, during their strength training studies of the knee extensor muscles (Porter et al., 1995; Vandervoort, 1997).

SUMMARY

As well as the inevitable ageing process, older adults can be experiencing the detrimental effects of inactivity on neuromuscular function, both leading to a reduction in reserve capacity. However, due to age-related adaptations in the connective tissue and muscle fibres, the extent of this reduction is dependent on the type of muscle action being undertaken. For example, muscle strength has been shown to decrease significantly after the sixth decade for both isometric and concentric contractions. However, older adults are relatively stronger for ECC movements in which muscles lengthen, as demonstrated by the high ECC/CONC ratios observed during tests of maximal voluntary strength. After they have learned to co-ordinate the intense ECC muscle contractions appropriately, this pattern of muscle activation can be used by older persons for effective force development in daily activities and strength training programs.

ACKNOWLEDGMENTS

This paper is based on gerontology research conducted with many colleagues and students, and supported by grants to A. A. Vandervoort from the Canadian Fitness and Lifestyle Research Institute and the Natural Sciences and Engineering Research Council of Canada.

REFERENCES

American College of Sports Medicine (1998). Exercise and physical activity for older adults. *Medicine and Science in Sports and Exercise* **30**, 992–1008.

Bassey, E.J. (1997). Measurement of muscle strength and power. *Muscle and Nerve (Supplement)* **5**, S44–S46.

Bosco, C., and Komi, P.V. (1980). Influence of aging on the mechanical behaviour of the leg extensor muscles. *European Journal of Applied Physiology* **45**, 209–219.

Brody, H., and Vijayashanker, N. (1977). Anatomical changes in the nervous system. In: *The Handbook of the Biology of Aging*. Finch, C., and Hayflick, L. (eds.), Van Nostrand Reinhold, New York.

Brooks, S.V., and Faulkner, J.A. (1994). Isometric, shortening, and lengthening contractions of muscle fiber segments from adult and old mice. *American Journal of Physiology* **267**, C507–C513.

Connelly, D.M., Carnahan, H., and Vandervoort, A.A. (2000). Motor skill learning of concentric and eccentric movements in older adults. *Experimental Aging Research* **26**, (in press).

Connelly, D.M., Rice, C.L., Roos, M.R., and Vandervoort, A.A. (1999). Motor unit firing rates and contractile properties in tibialis anterior of young and old men. *Journal of Applied Physiology* **87**, 843–852.

Connelly, D.M., and Vandervoort, A.A. (2000). Effects of isokinetic strength training on concentric and eccentric torque development in the ankle dorsiflexors of old adults. *Journals of Gerontology. Series A, Biological Sciences and Medical Sciences* **55**, (in press).

Delbono, O., Renganathan, M., and Messi, M.L. (1997). Excitation-Ca^{2+} release-contraction coupling in single aged human skeletal muscle fiber. *Muscle and Nerve (Supplement)* **5**, S88–S92.

DeVita, P., Hortobyagyi, T., Money, J., and Barrier, J. (1998). Redistribution of joint torques and powers with age. In *Proceedings of NACOB '98, Third North American Congress on Biomechanics* (August 14-18, 1998, Waterloo, Ontario, Canada). Department of Kinesiology, University of Waterloo, Waterloo, Ontario, Canada.

Doherty, T.J., Vandervoort, A.A., Taylor, A.W., and Brown W.F. (1993). Effects of motor unit losses on strength in older men and women. *Journal of Applied Physiology* **74**, 868–874.

Enoka, R.M. (1996). Eccentric contractions require unique activation by the nervous system. *Journal of Applied Physiology* **81**, 2339–2346.

Enoka, R.M. (1997). Neural strategies in the control of muscle force. *Muscle and Nerve (Supplement)* **5**, S66–S69.

Epstein, M., and Herzog, W. (1998). *Theoretical Models of Skeletal Muscle*. John Wiley & Sons, Ltd., Chichester, England.

Faulkner, J.A., and Brooks, S.V. (1995). Muscle fatigue in old animals: Unique aspects of fatigue in elderly humans. In: *Fatigue: Neural and Muscular Mechanisms*. Gandevia, S.C., Enoka, R.M., McComas, A.J., Stuart, D.G., and Thomas, C.K. (eds.), Plenum Press, New York, pp. 471–480.

Greene, L.S., and Williams, H.G. (1996). Aging and coordination from the dynamic pattern perspective. In: *Changes in Sensory Motor Behaviour in Aging*. Ferrandez, A-M., and Teasdale, N. (eds.), Elsevier, Amsterdam, pp. 89–131.

Hakkinen, K., Alen, M., Kallinen, M., et al. (1998a). Muscle CSA, force production, and activation of leg extensors during isometric and dynamic actions in middle-aged and elderly men and women. *Journal of Aging and Physical Activity* **6**, 232–247.

Hakkinen, K., Kallinen, M., Izquierdo, M., et al. (1998b). Changes in agonist-antagonist EMG, muscle CSA, and force during strength training in middle-aged and older people. *Journal of Applied Physiology* **84**, 1341–1349.

Hakkinen, K., Kraemer, W., Kallinen, M., et al. (1996). Bilateral and unilateral neuromuscular function and muscle cross-sectional area in middle-aged and elderly men and women. *Journals of Gerontology. Series A, Biological Sciences and Medical Sciences* **51**, B21–B29.

Hortobyagyi, T., Zheng, D., Weidner, M., Lambert, N.J., Westbrook, S., and Houmard, J.A. (1995). The influence of aging on muscle strength and muscle fiber characteristics with special reference to eccentric strength. *Journals of Gerontology. Series A, Biological Sciences and Medical Sciences* **50**, B399–B406.

Hunter, S.K., Thompson, M.W., Ruell, P.A., et al. (1999). Human skeletal sarcoplasmic reticulum Ca^{2+} uptake and muscle function with aging and strength training. *Journal of Applied Physiology* **86**, 1858–1865.

Lexell, J. (1995). Human aging, muscle mass and fiber type composition. *Journals of Gerontology. Series A, Biological Sciences and Medical Sciences.* **50A** (special issue), 11–16.

Lynch, N.A., Metter, E.J., Lindle, R.S., et al. (1999). Muscle quality. I. Age-associated differences between arm and leg muscle groups. *Journal of Applied Physiology* **86**, 188–194.

McComas, A.J. (1996). *Skeletal Muscle.* Human Kinetics, Champaign, Illinois.

Phillips, S.K., Rowbury, J.L., Bruce, S.A., and Woledge, R.C. (1993). Muscle force generation and age: The role of sex hormones. In: *Sensorimotor Impairment in the Elderly*. Stelmach, G.E., and Homberg, V. (eds.), Kluwer Academic Publishers, Dordrecht, The Netherlands, pp. 129–141.

Porter, M.M., and Vandervoort, A.A. (1997). Standing strength training of the ankle plantar and dorsiflexors in older women using concentric and eccentric contractions. *European Journal of Applied Physiology* **76**, 62–68.

Porter, M.M., Vandervoort, A.A., and Kramer, J.F. (1996). A method of measuring standing isokinetic plantar and dorsiflexion peak torques. *Medicine and Science in Sports and Exercise* **28**, 516–522.

Porter, M.M., Vandervoort, A.A., and Kramer, J.F. (1997). Eccentric peak torque of the plantar and dorsiflexors is maintained in older women. *Journals of Gerontology. Series A, Biological Sciences and Medical Sciences* **52**, B125–B131.

Porter, M.M., Vandervoort, A.A., and Lexell, J. (1995). Ageing of human muscle: Structure, function and adaptability. *Scandinavian Journal of Medicine and Science in Sports* **5**, 129–142.

Roos, M.R., Rice, C.L., Connelly, D.M., and Vandervoort, A.A. (1999). Quadriceps muscle strength, contractile properties, and motor unit firing rates in young and old men. *Muscle and Nerve* **22**, 1094–1103.

Roos, M.R., Rice, C.L., and Vandervoort, A.A. (1997). Age-related changes in motor unit function. *Muscle and Nerve* **20**, 679–690.

Suominen, H. (1997). Muscle collagen, aging and exercise. In *Proceedings of Fourth International Congress on Healthy Aging, Activity, and Sports* (August 27-31, 1996, Heidelberg, Germany). Health Promotion Publications, Gamburg, Germany, pp. 91–97.

Thompson, E., Versteegh, T.H., Overend, T.J., Birmingham, T.B., and Vandervoort, A.A. (1999). Cardiovascular responses to submaximal concentric and eccentric isokinetic exercise in older adults. *Journal of Aging and Physical Activity* **7**, 20–31.

Tomlinson, B.E., and Irving, D. (1977). The numbers of limb motor neurons in the human lumbosacral cord throughout life. *Journal of the Neurological Sciences* **34**, 213–219.

Trappe, S.W., Costill, D.L., Fink, W., and Pearson, D.R. (1995). Skeletal muscle characteristics among distance runners: A 20-year follow-up study. *Journal of Applied Physiology* **78**, 822–829.

Vandervoort, A.A. (1997). Neuromuscular learning and adaptations during strength training in older adults. In *Proceedings of Fourth International Congress on Healthy Aging, Activity, and Sports* (August 27-31, 1996, Heidelberg, Germany). Health Promotion Publications, Gamburg, Germany, pp. 131–138.

Vandervoort, A.A., and Hayes, K.C. (1989). Plantarflexor muscle function in young and elderly women. *European Journal of Applied Physiology* **58**, 389–394.

Vandervoort, A.A., Kramer, J.F., and Wharram, E.R. (1990). Eccentric knee strength of elderly females. *Journals of Gerontology. Series A, Biological Sciences and Medical Sciences* **45**, B125–B128.

Vandervoort, A.A., and McComas, A.J. (1986). Contractile changes in opposing muscles of the human ankle joint with aging. *Journal of Applied Physiology* **61**, 361–367.

Webber, S., and Kriellaars, D. (1997). Neuromuscular factors contributing to *in vivo* eccentric moment generation. *Journal of Applied Physiology* **83**, 40–45.

Westing, S.H., Seger, J.Y., and Thorstensson, A. (1990). Effects of electrical stimulation on eccentric and concentric torque-velocity relationships in man. *Acta Physiologica Scandinavia* **140**, 17–22.

Winegard, K.J., Hicks, A.L., Sale, D.G., and Vandervoort, A.A. (1996). A twelve-year follow-up study of ankle muscle function in older adults. *Journals of Gerontology. Series A, Biological Sciences and Medical Sciences* **51**, B202–B207.

23 Quadriceps Femoris Activation: Influence of Contraction Intensity on Neurobehaviour

D. M. PINCIVERO
Human Performance and Fatigue Laboratory, Department of Physical Therapy, Eastern Washington University, Cheney, Washington, USA

A. J. COELHO
Department of Physical Education, Health, and Recreation; Eastern Washington University, Cheney, Washington, USA

R. M. CAMPY
Department of Physical Education, Health, and Recreation; Eastern Washington University, Cheney, Washington, USA

INTRODUCTION

The quadriceps femoris (QF) is a muscle that has generated a great deal of both clinical and research interest. As a result of this muscle group's importance to physical fitness and ambulatory activities of daily living (for example, walking, and ascending and descending stairs), much work has examined QF anatomy (Hubbard et al., 1997; Nozic et al., 1997; Raimondo et al., 1998; Weinstabl et al., 1989), morphology (Edstrom and Ekblom, 1972; Pernus and Erzen, 1991; Travnik et al., 1995), biomechanics (Farahmand et al., 1998; Kouzaki et al., 1999; van Eijden et al., 1987), recruitment (Cerny, 1995; Karst and Willett, 1995; Powers et al., 1996), and fatigue thresholds (Housh et al., 1995, 1996; Vaz et al., 1996). Less effort has been made to address the role of contraction intensity on the neurobehaviour relationship between the different components of the QF muscle. Although it has been surmised that the different components of the QF muscle act in a co-ordinated fashion to generate knee extensor torque, their relative contribution to this function remains in question. Experimental evidence demonstrates that variations in tension of the different parts of the QF muscle affect patello-femoral kinematics and forces (Farahmand et al., 1998). Therefore, alterations in QF neuroactivation may have significant implications on knee-joint function. From a clinical perspective, 'selective VMO atrophy' has often been considered a persistent complication following knee injury or surgery

Skeletal Muscle Mechanics: From Mechanisms to Function. Edited by W. Herzog.

(Grabiner et al., 1994). Enhancing the understanding of QF neurobehaviour under voluntary contraction conditions will advance further inference towards 1) rehabilitation activities aimed at restoring what may be considered 'normal' QF function, and 2) a reduction in the incidence of overuse injuries, such as anterior knee pain, through preventive strengthening.

The overall objective of this chapter is to enhance the understanding of QF function during voluntary muscle contractions. The specific objectives of this chapter are as follows:

1. to present a concise understanding of QF function based upon anatomical structure,
2. to illustrate how electromyography (EMG) can be used effectively as a tool to measure muscle activation,
3. to identify current problems, controversies, and equivocal aspects of QF function,
4. to demonstrate the findings of an experiment to investigate activation parallelism of the different components of the QF muscle,
5. to provide an outlook on how activation parallelism can be applied clinically and how future research can address this issue under different conditions.

It is anticipated that this approach to examining the QF muscle will not only help guide future research, but will further narrow the gap between experimental findings and clinical applicability.

BACKGROUND

The following sections will focus on the anatomical structure of the QF muscle, the use of EMG to measure activation of different portions of the muscle, as well as highlighting current problems associated with the understanding of QF muscle function.

QUADRICEPS FEMORIS STRUCTURE

The quadriceps femoris (QF) consists of four muscles, one bi-articular and three uni-articular, although an accessory quadricep muscle has been rarely identified as either an additional head or a bi-laminar structure associated with the vastus lateralis or vastus intermedius (Hollinshead and Rosse, 1985; Nwoha and Adebisi, 1994). The QF muscle receives its nerve supply via diverging medial and lateral branches of the femoral nerve (from lumbar spinal nerve roots 2 to 4), which supply the upper lateral, and medial and lower portions of the muscle, respectively (Thiranagama, 1990). The bi-articular muscle, known as the rectus femoris (RF), has an origin on the anterior iliac spine and on the upper border of the acetabulum (Hollinshead and Rosse, 1985; Lewandowsky, 1994). The three uni-articular muscles, grouped as the vasti, are the vastus lateralis (VL), vastus intermedius (VI), and vastus medialis (VM). The VL, which spans the entire lateral aspect of the thigh, originates from the lateral lip of the linea aspera, the lower border of the greater trochanter, and the lateral intermuscular septum of the VI (Hollinshead and Rosse, 1985). The VI is the deepest of the quadriceps and is covered entirely by the RF, VL, and the VM. The origin of the VI is on the anterior and lateral surfaces of the body of the femur, and develops into a superficial tendinous lamina approximately midway long the thigh (Lewandowsky, 1994). The VM originates from the lower anterior portion of the intertrochanteric line, and then proceeds inferiorly down the medial lip of the linea aspera, where it inserts into the medial border of the QF tendon and the medial rim and medial base of the patella (Hollinshead and Rosse, 1985; Weinstabl et al., 1989). The components of the QF muscle blend into the quadriceps tendon, which attaches to the upper portion of the patella. The patella, which is the largest sesamoid bone in the body, is embedded within this tendinous tissue as it continues distally to insert onto the tibial tuberosity by way of the patellar tendon (Hollinshead and Rosse, 1985).

The QF functions in a manner that generates knee extensor torque during isometric, concentric, or eccentric contractions. It has been reported that the VI is the purest knee extensor, as its line of pull lies parallel to the shaft of the femur (Lieb and Perry, 1968). Through a cadaveric examination of the insertion angle, the proximal portion of the VL has been shown to exert lateral tension at an average of 12.5° from the long axis of the femur, while the distal portion exerts a lateral tension of approximately 32° (Weinstabl et al., 1989). Perhaps the most widely investigated component of the QF muscle, the VM has been characterized as a patella stabilizer rather than a knee extensor (Norkin and Levangie, 1992). However, empirical evidence has demonstrated that the proximal and distal portions of this muscle exert tensions at different angles, similarly to the VL. The upper fibres of the VM have demonstrated an angle of insertion onto the patella of 11.46° (±2.96 SD) (Nozic et al., 1997), 14.6° (±0.2 SEM) (Hubbard et al., 1997), 16.8° (Weinstabl et al., 1989), and 18° (± 2.0 SD) (Raimondo et al., 1998). The angle of insertion of the distal portion of the VM is much greater than the angle displayed by the distal portion of the VL. Specifically, this angle has been shown to be 48.9° (Weinstabl et al., 1989), 52.0° (±6.20 SD) (Nozic et al., 1997), 52.0° (±2.0 SD) (Raimondo et al., 1998), and 53.2° (±0.36 SEM) (Hubbard et al., 1997). These different angles of the VM have led many researchers and clinicians to refer to the upper fibres as the vastus medialis longus (VML), and to the lower fibres as the vastus medialis oblique (VMO) (Lieb and Perry, 1968; Speakman and Weisberg, 1977; Bose et al., 1980; Weinstabl et al., 1989). Anatomical investigations appear to support the notion that the VM is one composite muscle rather than two distinct parts. It should be noted that this conclusion is based on studies that relied on relatively large sample sizes and the criteria of a distinct fibrofascial plane separating the two portions of the muscle (Hubbard et al., 1997; Weinstabl et al., 1989; Nozic et al., 1997).

EMG AS A TOOL FOR MEASURING NEUROACTIVATION

Electromyography can be defined as the recording and interpreting of the electrical signals generated by muscle fibres following motor neuron activation and preceding contraction. The well-documented usage of this non-invasive technique has provided researchers and clinicians with a valid tool for evaluating activation characteristics of muscles during voluntary contractions. Typically recorded with a bipolar electrode configuration, the surface EMG signal can be considered, in some instances, a representative estimate of muscle recruitment (Enoka, 1994; Kamen and Caldwell, 1996). The generation of the raw EMG signal stems from a temporal and spatial distribution of muscle fibre action potentials (MFAP) within the detection range of the recording electrodes (Enoka and Fuglevand, 1993). A differential triphasic waveform that is constructed from the recording of a single MFAP (Φ) is largely a function of the distance of the recording electrode to the source (r_1 - *depolarization*) and sink (r_2 - *repolarization*) currents (Equation 23.1, McComas, 1996; Winter, 1990).

$$\Phi = \frac{I}{4\Pi\sigma}\left(\frac{1}{r_1} - \frac{1}{r_2}\right) \tag{23.1}$$

The summation of all MFAPs of a single motor unit—the motor unit action potential (MUAP)—subsequently gives rise to the generation of a number of MUAPs when the motor unit is repeatedly activated: the motor unit action potential train (MUAPT) (Basmajian and DeLuca, 1985). A frequency-modulated MUAPT from a number of activated motor units during a given muscle contraction is algebraically summed to form the raw EMG signal (also referred to as the interference pattern), where Nm = number of active motor units, $MUAPTm$ = the mth motor unit action potential train, and $n(t)$ = background or instrument noise (Equation 23.2, Basmajian and DeLuca, 1985; Sanders et al., 1996; Stashuk, 1998).

$$EMG = \sum_{m=1}^{Nm} MUAPTm(t) + n(t) \tag{23.2}$$

Through various types of processing techniques (full wave rectification, root mean square, etc.), the raw surface EMG signal can be quantified to yield inferential results regarding overall muscle activation (Figure 23.1).

Although many factors exert an effect on the outcome of the raw EMG signal, alterations in voluntary neural drive to the recipient muscle will elicit the largest response. During low-force muscle contractions where input to the alpha-motoneuron pool is relatively weak, low threshold motor units are first activated (Ghez and Gordon, 1995). As more motor units are recruited to achieve higher

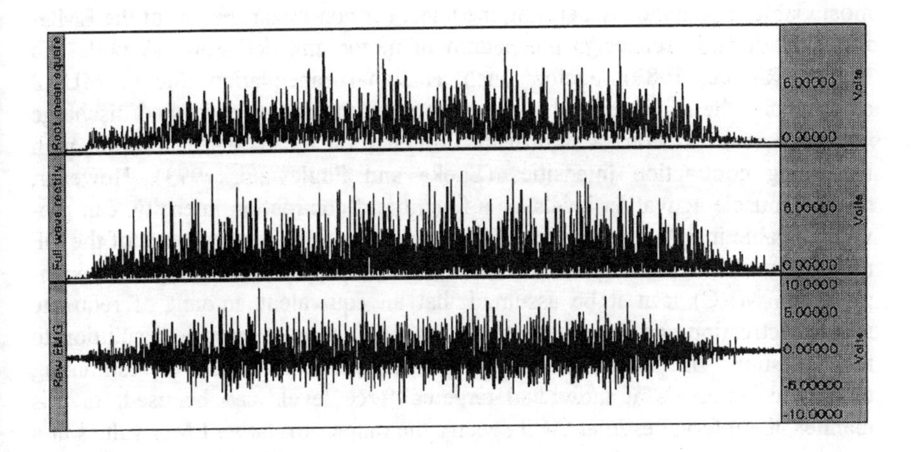

Figure 23.1. Illustration of the raw and processed (full wave rectification and root mean square) surface EMG signal of the quadriceps during a sustained, submaximal isometric contraction.

target force levels (orderly recruitment pattern, Ghez and Gordon, 1995; Henneman, 1979), firing rate modulation occurs in order to sustain these force levels (De Luca et al., 1982a, 1982b; Ghez and Gordon, 1995). As a result, a steady increase in the recorded EMG signal will tend to follow linearly increasing contraction intensities up to near-maximal levels. The use of surface EMG can prove to be valuable in this instance, as an overall examination of muscle neuroactivation is possible. However, due to factors such as electrode placement (Webster, 1984), electrode orientation (Fuglevand et al. 1992), or subcutaneous tissue (De la Berra and Milner, 1994), inter-muscle comparisons can be limiting. Normalizing processed raw EMG signals to a maximal voluntary contraction (MVC) under isometric conditions has been deemed an acceptable method for assessing neuroactivation differences between muscles. As this method is predicated on the notion that motor units are maximally activated during the MVC, component muscles of a larger group (i.e., the QF muscles) receiving the same nerve supply (femoral nerve) can be examined and compared for varying recruitment characteristics.

A popular approach for studying the effect of contraction intensity on muscle activation is through an examination of the EMG-force relationship. This method has been carried out repeatedly to assess the linearity of this relationship. Linear (Woods and Bigland-Ritchie, 1983; Milner-Brown and Stein, 1975) and non-linear (Woods and Bigland-Ritchie, 1983; Kuroda et al., 1970; Lawrence and De Luca, 1983) EMG-force relationships have been demonstrated in numerous studies. As a result, it has been stated that 'the qualitative relation between these two parameters is largely unknown; even for the relatively "simple" cases involving isometric contractions' (Herzog, 1996). The most viable explanation underlying the linear (or non-linear) nature of the EMG-force relationship relates to the pattern of motor unit activation (Woods and Bigland-Ritchie, 1983). Factors such as signal cancellation due to MUAP overlap, and the non-linear relationship between motor unit force and discharge rate are known to exert a significant influence on the increase in EMG with increasing contraction intensities (Enoka and Fuglevand, 1993). However, relative muscle activation levels, as a function of contraction intensity, can provide valuable insight to potential differences between the components of the QF muscle. In order to generate a muscle force of a particular intensity (for example, 50% MVC), can it be assumed that an equivalent amount of requisite muscle activation occurs concomitantly? In an attempt to provide a solution to this question, the simplistic yet very descriptive approach of estimating confidence intervals at individual targeted force levels can be used. In this manner, the interval estimate will specify the range of relative EMG values at a given contraction intensity about a sample average within which the population mean can be expected with a chosen level of confidence (Gardner and Altman, 1989; Sim and Reid, 1999). Although traditional hypothesis testing will tend to arrive at only a binary decision (accept or reject) between individual muscles, the confidence interval (CI) will provide further descriptive value.

CURRENT PROBLEMS

As presented, the EMG-force relationship demonstrates how activation of a muscle drives its force-generating capabilities. It has also been shown that the components of the QF muscle are very distinctive in terms of structure, morphology, and EMG characteristics. To further complicate the understanding of QF neuroactivation, the distinctive muscles of this complex receive their nerve supply from lumbar spinal roots L2-L4, via the femoral nerve (Hollinshead and Rosse, 1985). Based upon the demonstrated differences between the individual parts of the QF muscle, it may be speculated that individual activation will display distinctive trends. As such, inferences derived from surface EMG signals alone may not suitably answer the question of whether activation differences are due to specialized recruitment strategies, or simply manifestations of the structure itself (i.e., muscle fibre type differences). Only when such information is put in context of the muscles' physiological function and structure can further hypotheses be tested that bear direct clinical relevance. To isolate one aspect of QF neurobehaviour, the present experiment was conducted to investigate the influence of contraction intensity on activation of the different portions of the QF muscle. If the assumption is made that efferent impulses are transmitted equally (spatially and temporally) along the diverging branches of the femoral nerve to the components of the QF muscle, then it can be expected that activation of these muscles across the contraction intensity spectrum will be parallel. Therefore, the purpose of this study was to examine activation parallelism between the superficial components of the QF muscle during linearly increasing isometric intensities.

METHODS

The following sections detail an experimental procedure used to evaluate activation parallelism of the different components of the QF muscle.

SUBJECTS

Subjects for this study consisted of 30 healthy male and female volunteers (mean±SD age = 23.1±2.2 years, mean±SD height = 170.8±9.8 cm, mean±SD mass = 68.8±13.2 kg). All subjects were physically active but had not actively taken part in an intensive resistance training program for the lower extremity at least six months prior to the study. Individuals with a history of cardiovascular disease, hypertension, or orthopaedic pathology were excluded from participating in this study. All subjects provided written informed consent as approved by the Institutional Review Board at Eastern Washington University.

MEASUREMENT OF ISOMETRIC TORQUE

Prior to the measurement of isometric torque, all subjects completed a warm-up period that consisted of submaximal cycling for three to five minutes. Isometric torque was measured on the Biodex System II Isokinetic Dynamometer (Biodex Medical, Inc., Shirley, NY). Subjects were placed in a comfortable, upright, seated position on the Biodex Accessory Chair and were secured using thigh, pelvic, and torso straps in order to minimize extraneous body movements. The lateral femoral epicondyle was used as the bony landmark for matching the axis of rotation of the knee joint with the axis of rotation of the dynamometer resistance adapter. Gravity correction was obtained by measuring the torque exerted on the dynamometer resistance adapter with the knee in a relaxed state at full extension. Values for isometric torque were automatically adjusted for gravity by the Biodex Advantage software program. During the measurement of isometric torque, subjects were required to fold their arms across their chests and were given verbal encouragement as well as visual feedback from the Biodex computer monitor in an attempt to achieve a maximal voluntary effort level (Hald and Bottjen, 1987; Kim and Kramer, 1997; McNair et al., 1996). All procedures and verbal encouragement were administered by the same investigator for all subjects. Calibration of the Biodex dynamometer was performed according to the manufacturer's specifications by hanging a 10-kg mass on the resistance adapter in the horizontal position, prior to every testing session.

Once the subjects were seated in the chair, the knee was fixed at an angle of 60° of flexion, the angle of maximal isometric force generation (Thorstensson et al., 1976; Tihanyi et al., 1982). Following two to three submaximal isometric contractions, followed by two to three maximal isometric contractions for familiarization purposes, subjects were asked to contract their quadriceps as hard as they could (MVC) and to hold this contraction for five seconds. This contraction was repeated two more times with a minimal rest of two minutes between contractions. The average peak torque of the three MVCs was calculated to yield a representative estimate of an individual's maximal voluntary effort. Subjects were then asked to perform a voluntary isometric contraction of the quadriceps at the following intensities: 10%, 20%, 30%, 40%, 50%, 60%, 70%, 80%, and 90% of MVC. All subjects performed each contraction for five seconds with a minimal rest period of two minutes between contractions. Subjects were asked to match a horizontal line on the Biodex computer monitor that corresponded to the torque level at each intensity. The order of exercise intensity was randomized. During all testing, the subjects were blinded to the absolute torque values they were generating.

MEASUREMENT OF NEUROMUSCULAR ACTIVATION

Neuromuscular activation was assessed through surface electromyography for the vastus medialis (VM), vastus lateralis (VL), and rectus femoris (RF) muscles. Preamplified bipolar circular surface electrodes (Ag/AgCl – 0.8 cm diameter) were placed on each muscle with a fixed inter-electrode distance (centre-to-centre) of 2 cm. Prior to electrode placement, the skin area was shaved, cleaned with isopropyl alcohol, and abraded with coarse gauze in order to reduce skin impedance and to ensure good adhesion of the electrodes. Electrode placement for the VM was 20% of the distance from the medial joint line of the knee to the anterior superior iliac spine (ASIS) (Zipp, 1982). The VM electrode was placed at an approximate 45° angle between the anatomical horizontal and frontal planes in order to be oriented along the direction of the muscle fibres. Electrode placement for the VL was the midpoint between the head of the greater trochanter and the lateral femoral epicondyle (Housh et al., 1995), and electrode placement for the RF was 50% of the distance from the ASIS to the superior pole of the patella (Zipp, 1982). The reference electrode was placed over the medial shaft of the tibia, approximately 6-8 cm below the inferior pole of the patella. EMG activity was collected by a four-channel unit (Therapeutics Unlimited, Iowa City, IA) at a rate of 1000 Hz for each muscle. The common mode rejection of the current system was 87 dB at 60 Hz with an input impedance of >25 Mohms d.c. The gain range used in this study was 10K and signals were bandpass filtered between 20-500 Hz. Raw EMG signals were digitized and stored on computer disks for subsequent analysis by the Acknowledge software program, version 3.2.6 (Biopac Systems Inc., Santa Barbara, CA). The signals collected within the first and last second of each 5-second isometric contraction were not used for analysis because of knee movement that may have occurred at the initiation and completion of the test. Therefore, a 3-second window of EMG signals was used for analysis. The 3-second window of raw EMG activity at each intensity was full-wave rectified and integrated (area: $mV \cdot s$). The integration algorithm in the software is as follows:

$$f_{output}(n) = \sum_{k=1}^{n-1} f_{input}(k) + \left[\left[f_{input}(n-1) + f_{input}(n)\right]/2\right] \cdot \Delta t \qquad (23.3)$$

In this integration algorithm, $f(n)$ represents the data values ($mV \cdot s$), and the Δt represents the horizontal sampling interval (seconds) (MP100 Systems Guide, Biopac, Inc.). The EMG signal at each submaximal intensity was then normalized as a percentage of the integrated activity during the middle three seconds of the averaged MVCs for each muscle. Previous measurements of integrated surface EMG during isometric contractions have demonstrated moderate to high reliability coefficients for the quadriceps muscles ranging from $r = 0.71$ to $r = 0.98$ (Kollmitzer et al., 1999; Poyhonen et al., 1999; Sleivert and Wenger, 1994; Viitasalo and Komi, 1975).

STATISTICAL ANALYSIS

In order to determine the variability and reliability of the three MVCs, descriptive statistics, the intra-class correlation (ICC) coefficient (2,1) and the corresponding standard error of measurement, and 95% confidence intervals were calculated (Shrout and Fleiss, 1979; Stratford and Goldsmith, 1997). One-sample t-tests were performed to test the null hypothesis (H_0) that the normalized EMG at each contraction intensity was not significantly different from the equivalent percentage of MVC. In this manner, the observation of normalized EMG at each intensity was tested against the H_0: μ = intensity level (for example, 40% MVC). In the case where the normalized EMG at a given intensity level was significantly different from the tested H_0 (i.e., μ = intensity level), H_0 was rejected, and it was concluded that the activation level of that muscle at a given intensity was different than the relative torque output (e.g., 40% MVC) (Glass and Hopkins, 1996). For example, the normalized EMG at 10% MVC for each muscle was evaluated for a difference from the value of 10. As this analysis did not evaluate differences between subsets of means (i.e., utilize a multiple comparison procedure), the risk of alpha inflation and, hence, the need for applying the Bonferonni inequality correction factor was not warranted (Glass and Hopkins, 1996). Confidence intervals (95%) were subsequently calculated to estimate the relative range of neuromuscular activation at each contraction intensity for the VM, VL, and RF muscles. A two-factor (muscle-by-intensity) ANOVA with repeated measures was calculated on the integrated EMG across all levels of contraction intensity to detect significant main effects and interactions. All tests of significance were carried out at a preset alpha level of $p<0.05$.

RESULTS

The mean absolute torque level (±standard deviation) for each of the three MVCs were as follows: MVC 1 – 214.25±63.26 N·m, MVC 2 – 218.96±62.64 N·m, and MVC 3 – 218.77±63.40 N·m. The calculated ICC between the three MVCs was 0.98, and the standard error of measurement (SEM) was 8.92 N·m, or 4.1% of the mean value. The 95% confidence interval for the SEM value was found to be 3.4% to 4.8%.

The normalized EMG (% MVC) values (means, standard deviations, and 95% confidence intervals) for each muscle are summarized in Table 23.1. These results demonstrate that activation of the vastus medialis was significantly lower than the expected % MVC between 20% to 70% MVC. Activation of the vastus lateralis and rectus femoris muscles were significantly higher than the expected values at 10% MVC. The rectus femoris muscle was also shown to be activated significantly less than the expected values at 40% to 70% MVC.

Table 23.1. Means, standard deviations, and 95% confidence intervals (CI) for normalized EMG of the vastus medialis (VM), vastus lateralis (VL), and rectus femoris (RF) at contraction intensities from 10% to 90% MVC.

MVC	Vastus medialis		Vastus lateralis		Rectus femoris	
(%)	Mean±SD	95% CI	Mean±SD	95% CI	Mean±SD	95% CI
10	8.9±4.8	7.1 – 10.7	14.6±4.1*	13.1 – 16.2	14.7±5.4*	12.7 – 16.8
20	14.1±6.2*	11.8 – 16.4	22.0±5.5	20.0 – 24.1	20.7±6.0	18.5 – 23.0
30	20.6±7.7*	17.7 – 23.5	29.8±6.3	27.5 – 32.2	27.6±6.8	25.1 – 30.2
40	27.1±7.7*	24.3 – 30.0	38.6±7.2	35.9 – 41.3	35.1±8.8*	31.9 – 38.4
50	36.8±11.7*	32.4 – 41.1	48.9±8.6	45.7 – 52.1	43.5±10.3*	39.6 – 47.3
60	49.6±14.4*	44.2 – 55.0	64.0±19.2	56.9 – 71.2	54.9±11.2*	50.7 – 59.1
70	57.7±11.9*	53.2 – 62.1	71.2±16.1	65.2 – 77.2	62.9±10.4*	59.1 – 66.8
80	72.5±17.2	66.1 – 78.1	81.3±10.6	77.3 – 85.3	79.5±9.7	75.9 – 83.1
90	93.9±20.4	86.3 – 101.5	91.5±10.5	87.6 – 95.4	90.6±9.7	87.0 – 94.2

* Indicates statistically different from the % MVC.

The results from the two-factor ANOVA demonstrated significant main effects for both intensity ($F_{8,232} = 553.6$, $p < 0.05$) and muscle ($F_{2,58} = 19.50$, $p<0.05$). These findings (depicted in Table 23.1) illustrate that activation of the vastus medialis was significantly lower than the rectus femoris ($F_{1,29} = 41.38$, $p<0.05$), while activation of the rectus femoris muscle was lower than the vastus lateralis ($F_{1,29} = 9.63$, $p = 0.015$). A significant muscle-by-intensity interaction ($F_{16,464} = 3.9$, $p<0.05$) was also noted. Specifically, one interaction was detected between the vastus medialis and vastus lateralis at 10% and 20% MVC ($F_{1,29} = 10.60$, $p<0.03$), where the latter muscle showed a significantly higher increase in activation from 10% MVC to 20% MVC. A significant interaction was also detected between the vastus lateralis and rectus femoris muscles between intensity levels of 70% and 80% MVC ($F_{1,29}=4.49$, $p=0.043$). Activation of the vastus medialis increased significantly more from 80% to 90% MVC than the vastus lateralis ($F_{1,29}=14.83$, $p=0.001$) and the rectus femoris muscle ($F_{1,29}=9.63$, $p=0.004$). The lack of a significant muscle-by-intensity interaction between the levels of 20% MVC and 70% MVC indicates a parallel increase in activation of the three muscles.

DISCUSSION

The major findings of this study demonstrate that VM activation was lower than a preset level of torque within the mid-range of contraction intensities (20% to 70% MVC). Such was also the case with the RF muscle within the 40% to 70% MVC range. Within the contraction intensity mid-range, VL activation was shown to be significantly greater than the RF and VM muscles. The most notable finding, however, was the presence of activation non-parallelism at near maximal intensities, as VM activation increased significantly more from 80% to 90% MVC than activation of the other two muscles.

The use of inferential statistics in the present investigation sheds light on QF activation. As it is well known that sample size exerts a significant effect on the outcome of many statistical analyses, through a reduction in sample variance (Glass and Hopkins, 1996), exploring individual muscle activation via more 'descriptive' measures becomes important. The significant deviation of RF and VM activation from a preset level of mechanical output (i.e., torque) within the mid-range of contraction intensities suggests that recruitment characteristics of muscles are altered. We assumed that the EMG-force relationship is essentially linear (Woods and Bigland-Ritchie, 1983; Milner-Brown and Stein, 1975) and that balanced increases in contraction intensity result in balanced increases in muscle activation. For VM and RF this was not the case. As shown in Table 23.1, the EMG activity at each contraction intensity can be compared to the 'amount' of EMG generated during maximal voluntary efforts. The term 'muscle recruitment' should be differentiated from the more widely accepted term 'motor unit recruitment', as the former alludes to a gross combination of various motor unit recruitment strategies (recruitment and rate coding) (Enoka and Fuglevand, 1993). As a result of the width of the confidence interval, the results of the present study suggest VM and RF contribute less to the knee extensor torque at mid-contraction intensities than VL.

The capability of muscles to generate force is in direct proportion to the physiological cross-sectional area (PCSA). Of the QF muscles, the VL has the largest PCSA, followed by the VI and RF, and finally VM (Farahmand et al., 1998; Wickiewicz et al., 1983). Based on PCSA measurements, Farahmand et al. (1998) estimated a 35% contribution to quadriceps force by the combined RF and VI, 40% from VL, and 25% from VM. These contributions are similar to those presented by Wickiewicz et al. (1983) for VL (38%), VM (25%), VI (22%), and RF (15%). It is interesting to note that according to these measurements, the force-generating capability of RF is lower than that of VM. However, the results of the present investigation show that RF was activated significantly more than VM within the contraction intensity mid-range. This result may demonstrate a unique method of neuromuscular control, as it might be expected that a large muscle would be relied upon preferentially for force generation (Ebenbilcher et al., 1998). The recruitment pattern of QF may be explained by the bi-articular nature of RF and the uni-articular nature of VM and VL. This speculation was proposed by Ebenbilcher et al. (1998), who examined surface EMG of the superficial QF components during isometric submaximal fatiguing contractions in 18 healthy males. A reduction in the root mean square (RMS) EMG of RF was observed during a sustained 70% MVC, as compared to increases in VM and VL activation. During a series (*n=50*) of brief isometric MVCs, Kouzaki et al. (1999) observed a greater reduction in the median frequency of RF EMG and mechanomyographic (MMG) signal than VM and VL. It was suggested that, as a function of muscle fatigue, 'a divergence of mechanical activity within the quadriceps muscles' occurred during maximal voluntary efforts (Kouzaki et al., 1999). The results of the present investigation appear to support the possible presence of non-uniform mechanical and electrical activity of QF as presented by Kouzaki et al. (1999) and Ebenbilcher et al. (1998), respectively. Although the neurophysiological processes differ considerably between the experimental protocols of these two published reports and those of the present investigation, the non-parallel activation of the QF components at near-maximal contraction intensities further supports the potential of altering recruitment characteristics. The similar activation patterns of the three muscles at 90% MVC suggests that excitation of the motoneuron pool to maximal levels results in maximal activation (recruitment and rate coding) of all motor units to a convergent level (De Luca et al., 1982a, 1982b). This finding, however, contradicts the phenomenon of common drive, which suggests that an increase in surface EMG across the intensity spectrum should occur in a parallel manner between the three muscles in this study. Contrary results to activation non-parallelism were found by Housh et al. (1996) for the QF muscles during an incremental cycle ergometer protocol that was performed to volitional exhaustion. The onset of neuromuscular fatigue during this test was shown to occur at a similar rate between VM, VL, and RF. Similar findings were also presented between VL and RF during a sustained isometric 70% MVC (Vaz et al.,1996), and between VM and VL during sustained isometric 30% and 60% MVCs (Grabiner et al., 1991). In a separate investigation, Housh et al. (1995) demonstrated that the neuromuscular fatigue threshold of RF was significantly lower than that of the VL and VM during cycling. Although these studies may appear to support the idea of the QF functioning as a unit, their protocols involved a significant fatigue component. In the present investigation, it was attempted to provide a precise understanding of the neuromuscular control of the three superficial quadriceps muscles in the non-fatigued state. The findings from this study support the notion of convergence at near-maximal force levels. In considering the anatomical location of the more distal VM compared to the VL and RF, the distance through which action potentials are travelling along the femoral nerve should receive future examination. Increased amplitudes of muscle compound motor-unit action potentials occurred with decreased distance of nerve stimulation in the biceps brachii (Kereshi et al., 1983; McComas, 1996). Although the present study does not specifically address this notion, the relative position of the three superficial quadriceps should warrant future investigation regarding the degree of motoneuron activation and distal muscle recruitment.

FUTURE DIRECTIONS

The QF muscle contributes a very integral part to lower extremity function. Disruption of the 'normal' function of this muscle through injury, disease, or surgical procedures produces disability for an individual. It has been demonstrated that experimentally altering the forces generated by the different components of the QF muscle, different patello-femoral kinematics and forces result (Farahmand et al., 1998; Goh et al., 1995). There is little empirical evidence about the neuroactivation of these muscles as a function of contraction intensity. This is an important consideration as it pertains directly to rehabilitation activities, particularly in the case of patello-femoral dysfunction. Although it has been suggested that differential activation of the QF components may be the cause of patello-femoral dysfunction (Biedert and Gruhl, 1997; Koskinen and Kujala, 1992; Morrish and Woledge, 1997; Voight and Wieder, 1991), other findings tend to refute this relationship (Hubbard et al., 1998; Karst and Willett, 1995; Powers et al., 1996; Taskiran et al., 1998). Based upon the results of the present investigation, the efficacy of prescribing rehabilitation exercises should be considered from a 'contraction intensity' perspective. It is tempting to speculate that the more distally located VM muscle that is innervated by progressively smaller, diverging branches of the femoral nerve may necessitate a stronger drive from the central nervous system for optimal recruitment. If this is the case, isolated quadriceps contraction at relatively high intensities may be an appropriate prescription to help re-establish normal patello-femoral function. Unfortunately, justification for this presumption presents conflicting results in the scientific literature. The implications of this line of inquiry, however, will have a profound effect on rehabilitation procedures and, ultimately, patient outcome.

ACKNOWLEDGEMENTS

This work was supported by a Summer Faculty Research Grant awarded by the Office of Grants and Research Development at Eastern Washington University.

REFERENCES

Basmajian, J.V., and DeLuca, C.J. (1985). *Muscles Alive: Their Functions Revealed by Electromyography* (5th edition). Baltimore, Maryland, pp. 19–64.

Biedert, R.M., and Gruhl, C. (1997). Axial computed tomography of the patellofemoral joint with and without quadriceps contraction. *Archives of Orthopaedic and Trauma Surgery* **116**, 77–82.

Bose, K., Kanagasuntheram, R., and Osman, M.B.H. (1980). Vastus medialis oblique: An anatomic and physiologic study. *Orthopedics* **3**, 880–883.

Cerny, K. (1995). Vastus medialis oblique/vastus lateralis muscle activity ratios for selected exercises in persons with and without patellofemoral pain syndrome. *Physical Therapy* **75**(8), 672–683.

De la Berra, E., and Milner, T. (1994). The effects of skinfold thickness on the selectivity of surface EMG. *Electroencephalography and Clinical Neurophysiology* **93**, 91–99.

DeLuca, C.J., LeFever, R.S., McCue, M.P., and Xenakis, A.P. (1982a). Behavior of human motor units in different muscles during linearly varying contractions. *Journal of Physiology* **329**, 113–128.

DeLuca, C.J., LeFever, R.S., McCue, M.P., and Xenakis, A.P. (1982b). Control scheme governing concurrently active human motor units during voluntary contractions. *Journal of Physiology* **329**, 129–142.

Ebenbilcher, G., Kollmitzer, J., Quittan, M., Uhl, F., Kirtley, C., and Fialka, V. (1998). EMG fatigue patterns accompanying isometric fatiguing knee-extensions are different in mono- and bi-articular muscles. *Electroencephalography and Clinical Neurophysiology* **109**, 256–262.

Edstrom, L., and Ekblom, B. (1972). Differences in sizes of red and white muscle fibres in vastus lateralis of musculus quadriceps femoris of normal individuals and athletes: Relation to physical performance. *Scandinavian Journal of Clinical Laboratory Investigations* **30**, 175–181.

Enoka, R.M. (1994). *Neuromechanical Basis of Kinesiology* (2nd edition). Champaign, Illinois, pp. 166–200.

Enoka, R.M., and Fuglevand, A.J. (1993). Neuromuscular basis of the maximum voluntary force capacity of muscle. In: *Current Issues in Biomechanics.* Grabiner, M.D. (ed.), Champaign, Illinois, pp. 215–236.

Farahmand, F., Tahmasbi, M.N., and Amis, A.A. (1998). Lateral force-displacement behaviour of the human patella and its variation with knee flexion: A biomechanical study *in vitro*. *Journal of Biomechanics* **31**, 1147–1152.

Fuglevand, A., Winter, D., Patla, A., and Stashuk, D. (1992). Detection of motor unit action potentials with surface electrodes: Influence of electrode size and spacing. *Biological Cybernetics* **67**, 143–153.

Gardner, M.J., and Altman, D.G. (1989). Estimation rather than hypothesis testing: Confidence intervals rather than p values. In: *Statistics With Confidence: Confidence Intervals and Statistical Guidelines.* Gardner, M.J., and Altman, D.G. (eds.), British Medical Association, London, England, pp. 6–19.

Ghez, C., and Gordon, J. (1995). An introduction to movement. In: *Essentials of Neural Science and Behavior.* Kandel, E.R., Schwartz, J.H., and Jessell, T.M. (eds.), Stamford, Connecticut, pp. 489–500.

Glass, G.V., and Hopkins, K.D. (1996). *Statistical Methods in Education and Psychology* (3rd edition). Needham Heights, Massachusetts, pp. 444–480.

Goh, J.C.H., Lee, P.Y.C., and Bose, K. (1995). A cadaver study of the function of the oblique part of the vastus medialis. *Journal of Bone and Joint Surgery* **77-B**, 225–231.

Grabiner, M.D., Koh, T.J., and Draganich, L.F. (1994). Neuromechanics of the patellofemoral joint. *Medicine and Science in Sports and Exercise* **26**(1), 10–21.

Grabiner, M.D., Koh, T.J., and Miller, G.F. (1991). Fatigue rates of vastus medialis oblique and vastus lateralis during static and dynamic knee extension. *Journal of Orthopaedic Research* **9**, 391–397.

Hald, R.D., and Bottjen, E.J. (1987). Effect of visual feedback on maximal and submaximal isokinetic test measurements of normal quadriceps and hamstrings. *Journal of Orthopedic and Sports Physical Therapy* **9**, 86–93.

Henneman, E. (1979). Functional organization of motoneuron pools: The size-principle. In: *Integration in the Nervous System.* Asanuma, H., and Wilson, V.J. (eds.), Igaku-Shoin, Tokyo, pp. 13–25.

Herzog, W. (1996). Force-sharing among synergistic muscles: Theoretical considerations and experimental approaches. In: *Exercise and Sport Science Reviews*. Holloszy, J.O. (ed.), Baltimore, Maryland, **24**, 173–202.

Hollinshead, W.H., and Rosse, C. (1985). *Textbook of Anatomy* (4th Edition). Philadelphia, Pennsylvania, pp. 373–389.

Housh, T.J., deVries, H.A., Johnson, G.O., Evans, S.A., Housh, D.J., Stout, J.R., Bradway, R.M., and Evetovich, T.K. (1996). Neuromuscular fatigue thresholds of the vastus lateralis, vastus medialis and rectus femoris muscles. *Electromyography and Clinical Neurophysiology* **36**, 247–256.

Housh, T.J., deVries, H.A., Johnson, G.O., Housh, D.J., Evans, S.A., Stout, J.R., Evetovich, T.K., and Bradway, R.M. (1995). Electromyographic fatigue thresholds of the superficial muscles of the quadriceps femoris. *European Journal of Applied Physiology* **71**, 131–136.

Hubbard, J.K., Sampson, H.W., and Elledge, J.R. (1997). Prevalence and morphology of the vastus medialis oblique muscle in human cadavers. *Anatomical Record* **249**, 135–142.

Hubbard, J.K., Sampson, H.W., and Elledge, J.R. (1998). The vastus medialis oblique muscle and its relationship to patellofemoral joint deterioration in human cadavers. *Journal of Orthopedic and Sports Physical Therapy* **28**, 384–391.

Kamen, G., and Caldwell, G. (1996). Physiology and interpretation of the electromyogram. *Journal of Clinical Neurophysiology* **13(5)**, 366–384.

Karst, G.M., and Willett, G.M. (1995). Onset timing of electromyographic activity in the vastus medialis oblique and vastus lateralis muscles in subjects with and without patellofemoral pain syndrome. *Physical Therapy* **75(9)**, 813–823.

Kereshi, S., Manzano, G., and McComas, A.J. (1983). Impulse conduction velocities in human biceps brachii muscles. *Experimental Neurology* **80**, 652–662.

Kim, H.J., and Kramer, J.F. (1997). Effectiveness of visual feedback during isokinetic exercise. *Journal of Orthopedic and Sports Physical Therapy* **26(6)**, 318–323.

Kollmitzer, J., Ebenbichler, G.R., and Kopf, A. (1999). Reliability of surface electromyographic measurements. *Clinical Neurophysiology* **110(4)**, 725–734.

Komi, P.V., and Buskirk, E.R. (1970). Reproducibility in electromyographic measurements with inserted wire electrodes and surface electrodes. *Electromyography* **4**, 357 367.

Koskinen, S.K., and Kujala, U.M. (1992). Patellofemoral relationships and distal insertion of the vastus medialis muscle: A magnetic resonance imaging study in nonsymptomatic subjects and in patients with patellar dislocation. *Journal of Arthroscopic and Related Surgery* **8**, 465–468.

Kouzaki, M., Shinohara, M., and Fukunaga, T. (1999). Non-uniform mechanical activity of quadriceps muscle during fatigue by repeated maximal voluntary contraction in humans. *European Journal of Applied Physiology* **80**, 9–15.

Kuroda, E., Klissouras, V., and Milsum, J.H. (1970). Electrical and metabolic activities and fatigue in human isometric contraction. *Journal of Applied Physiology* **29**, 358–367.

Lawrence, J.H., and DeLuca, C.J. (1983). Myoelectric signal versus force relationship in different human muscles. *Journal of Applied Physiology* **54**, 1653–1659.

Lewandowski, J. (1994). Variations in quadriceps femoris muscle in human fetuses. *Folia Morphologiica (Warsz.)* **53**(2), 117–125.

Lieb, F.J., and Perry, J. (1968). Quadriceps function: An anatomical and mechanical study using amputated limbs. *Journal of Bone and Joint Surgery (American)* **50**, 1535–1548.

McComas, A.J. (1996). *Skeletal Muscle: Form and Function*. Champaign, Illinois, pp. 127–146.

McNair, P.J., Depledge, J., Brettkelly, M., and Stanley, S.N. (1996). Verbal encouragement: Effects of maximum effort voluntary muscle activation. *British Journal of Sports Medicine* **30**, 243–245.

Milner-Brown, H.S., and Stein, R.B. (1975). The relation between the surface electromyogram and muscular force. *Journal of Physiology* **246**, 549–569.

Morrish, G.M., and Woledge, R.C. (1997). A comparison of the activation of muscles moving the patella in normal subjects and in patients with chronic patellofemoral problems. *Scandinavian Journal of Rehabilitation Medicine* **29**, 43–48.

Norkin C.C., and Levangie, P.K. (1992). *Joint Structure and Function: A Comprehensive Analysis* (2nd edition). Philadelphia, Pennsylvania, pp. 337–378.

Nozic, M., Mitchell, J., and de Klerk, D. (1997). A comparison of the proximal and distal parts of the vastus medialis muscle. *Australian Journal of Physiotherapy* **43**(4), 277–281.

Nwoha, P.U., and Adebisi, S. (1994). An accessory quadriceps femoris muscle in Nigerians. *Acta Anat. Nippon* **69**, 175–177.

Pernus, F., and Erzen, I. (1991). Arrangement of fiber types within fascicles of human vastus lateralis muscle. *Muscle and Nerve* **14**, 304–309.

Powers, C.M., Landel, R., and Perry, J. (1996). Timing and intensity of vastus muscle activity during functional activities in subjects with and without patellofemoral pain. *Physical Therapy* **76**(9), 946–955.

Poyhonen, T., Keskinen, K.L., Hautala, A., Savolainen, J., and Malkia, E. (1999). Human isometric force production and electromyogram activity of knee extensor muscles in water and on dry land. *European Journal of Applied Physiology* **80**, 52–56.

Raimondo, R.A., Ahmad, C.S., Blankevoort, L., April, E.W., Grelsamer, R.P., and Henry, J.H. (1998). Patellar stabilization: A quantitative evaluation of the vastus medialis obliquus muscle. *Orthopedics* **21**(7), 791–795.

Sanders, D., Stalber, E., and Nandedkar, S. (1996). Analysis of the electromyographic interference pattern. *Journal of Clinical Neurophysiology* **13**(5), 385–400.

Shrout, P.E., and Fleiss, J.L. (1979). Intraclass correlations: Uses in assessing rater reliability. *Psychological Bulletin* **86**, 420–428.

Sim, J., and Reid, N. (1999). Statistical inference by confidence intervals: Issues of interpretation and utilization. *Physical Therapy* **79**(2), 186–195.

Sleivert, G.G., and Wenger, H.A. (1994). Reliability of measuring isometric and isokinetic peak torque, rate of torque development, integrated electromyography and tibial nerve conduction velocity. *Archives of Physical Medicine and Rehabilitation* **75**, 1315–1321.

Speakman, H.G., and Weisberg, J. (1977). The vastus medialis controversy. *Physiotherapy* **63**, 249–254.

Stashuk, D. (1998). EMG signal decomposition: How it performed and how can it be used? *North American Congress on Biomechanics*, pre-event workshop. August 14–18, 1998. Waterloo, Ontario, Canada.

Stratford, P.W., and Goldsmith, C.H. (1997). Use of the standard error as reliability index of interest: An applied example using elbow flexor strength data. *Physical Therapy* **77**, 745–750.

Taskiran, E., Dinedurga, Z., Yagiz, A., Uludag, B., Ertekin, C., and Lok, V. (1998). Effect of the vastus medialis obliquus on the patellofemoral joint. *Knee Surgery and Sports Traumatology and Arthroscopy* **6**, 173–180.

Thiranagama, R. (1990). Nerve supply of the human vastus medialis muscle. *Journal of Anatomy* **170**, 193–198.

Thorstensson, A., Grimby, G., and Karlsson, J. (1976). Force-velocity relations and fiber composition in human knee extensor muscles. *Journal of Applied Physiology* **40**, 12–16.

Tihanyi, J., Apor, P., and Fekete, G. (1982). Force-velocity-power characteristics and fiber composition in human knee extensor muscles. *European Journal of Applied Physiology* **48**, 331–343.

Travnik, L., Pernus, F., and Erzen, I. (1995). Histochemical and morphometric characteristics of the normal human vastus medialis longus and vastus medialis obliquus muscles. *Journal of Anatomy* **187**, 403–411.

van Eijden, T.M.G.J., Weijs, W.A., Kouwenhoven, E., and Verburg, J. (1987). Forces acting on the patella during maximal voluntary contraction of the quadriceps femoris muscle at different knee flexion/extension angles. *Acta Anatomica* **129**, 310–314.

Vaz, M.A., Zhang, Y., Herzog, W., Guimaraes, A.C.S., and MacIntosh, B.R. (1996). The behavior of rectus femoris and vastus lateralis during fatigue and recovery: An electromyographic and vibromyographic study. *Electromyography and Clinical Neurophysiology* **36**, 221–230.

Viitasalo, J.H.T., and Komi, P.V. (1975). Signal characteristics of EMG with special reference to reproducibility of measurements. *Acta Physiologica Scandinavica* **93**, 531–539.

Voight, M.L., and Wieder, D.L. (1991). Comparative reflex response times of vastus medialis obliquus and vastus lateralis in normal subjects and subjects with extensor mechanism dysfunction. *American Journal of Sports Medicine* **19**, 131–137.

Webster, J. (1984). Reducing motion artifacts and interference in biopotential recording. *IEEE Transactions in Biomedical Engineering* **31**, 823–826.

Weinstabl, R., Scharf, W., and Firbas, W. (1989). The extensor apparatus of the knee joint and its peripheral vasti: Anatomic investigation and clinical relevance. *Surgical and Radiologic Anatomy* **11**, 17–22.

Wickiewicz, T.L., Roy, R.R., Powell, P.L., and Edgerton, V.R. (1983). Muscle architecture of the human lower limb. *Clinical Orthopaedics* **179**, 275–283.

Winter, D.A. (1990). *Biomechanics and Motor Control of Human Movement* (2nd Edition). New York, pp. 191–212.

Woods, J.J., and Bigland-Ritchie, B. (1983). Linear and non-linear surface EMG/force relationships in human muscles. *American Journal of Physical Medicine* **62**(6), 287–299.

Zipp, P. (1982). Recommendations for the standardization of lead positions in surface electromyography. *European Journal of Applied Physiology* **50**, 41–54.

24 Adaptation of Human Ankle-joint Stiffness to Changes in Functional Demand

DANIEL LAMBERTZ, CHRISTOPHE CORNU, CHANTAL PÉROT, AND FRANCIS GOUBEL
Biomécanique et Génie Biomédical, Université de Technologie de Compiègne, France

INTRODUCTION

The present study was devoted to changes in musculo-articular stiffness of the human ankle plantarflexors by using sinusoidal perturbation techniques. The purpose was not to describe new methods or modelling techniques in musculo-articular biomechanics, but rather to present some results of changes in bio-mechanical parameters caused by changes in muscle functional demand. One special topic will be the influence of space flight on human ankle joint stiffness. Changes in musculo-articular stiffness when using a special training technique will also be discussed.

HISTORICAL BACKGROUND IN JOINT DYNAMICS

A variety of ergometer devices have been constructed to facilitate the quantitative study of the dynamic mechanical response characteristics of human joints like the ankle, elbow, and wrist. Among these joints, the ankle joint is of special interest and a great number of devices have been used to rotate the ankle (Agarwal and Gottlieb, 1985; Kearney and Hunter, 1990).

The dynamics of human joints deal with the relation between the angular position of a joint and the torque/force acting about it. Joint dynamics define the interaction between a joint, its associated muscles and limbs, and the environment (e.g., external forces like gravitational forces or ground reaction forces). They determine the displacements evoked by perturbing forces during postural control and the forces that must be generated to perform a voluntary movement. The relationship between force and joint displacement is described by joint stiffness (see below), which can be influenced by changes in the functional demands (e.g., disuse or 'overuse'). In general, a large joint stiffness provides good position control and favours the application of additional forces to counterbalance position instability. The control of joint stiffness can be influenced

Skeletal Muscle Mechanics: From Mechanisms to Function. Edited by W. Herzog.

by at least three muscle components: 1) the passive and intrinsic muscle mechanical properties, 2) the reflex excitability, and 3) the supraspinal control (Sinkjaer, 1997). As shown in Figure 24.1, the mechanisms underlying joint dynamics can be seen as an input-output system including limb dynamics, external torques, articular mechanics, and muscle mechanics (i.e., contractile mechanics and activation mechanics). Further parameters that can be included in the input-output systems are reflex dynamics and higher central commands.

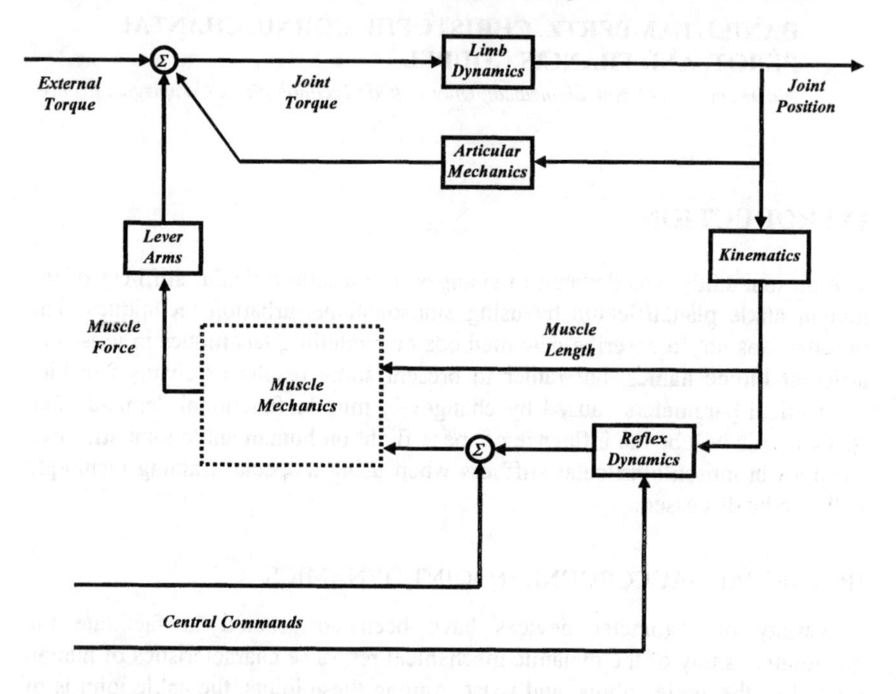

Figure 24.1. The musculoskeletal control system designed as an input-output system with possible sites for adaptation due to changes in the functional demands. (Adapted from Kearney and Hunter, 1990. Reproduced by permission of Begell House, Inc.)

So, joint dynamics arise from the interaction of a number of subsystems, and numerous possibilities of adaptation exist when there are changes in the functional demands. However, position and torque are the most important signals that can be observed and manipulated in order to describe joint dynamics. In addition, EMGs from the surrounding muscles provide an indirect measure of the neural input. The characteristics of the other subsystems are generally difficult to determine independently using simple input-output models, but should be considered when changes in the functional demand occur. Nevertheless, measuring joint dynamics can be regarded as a classical problem in system identification through the analysis of an appropriate transfer function to describe the relation between the input (position or torque) and the output (torque or position) records.

A common linear non-parametric model, with no assumptions about system structure or order, may use frequency response functions to describe joint dynamics. The relation between position and torque (or the transfer function) is often formulated in terms of a compliance frequency function $C(j\omega)$ or a stiffness frequency function $S(j\omega)$, as the inverse formulation in which position is regarded as the input and torque as the output (Figure 24.2).

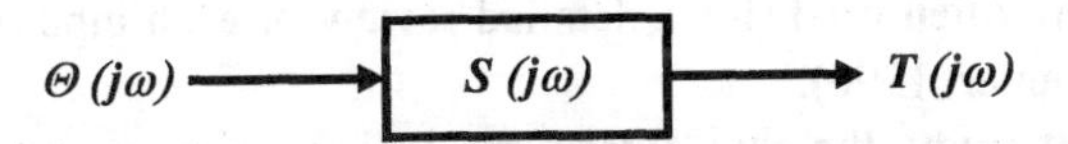

Figure 24.2. Reduction of the musculoskeletal control system to a simple input-output system can serve as a non-parametric model. $\Theta\ (j\omega)$ is the Fourier transform of the angular displacement, $T\ (j\omega)$ is the Fourier transform of the joint torque, and $S\ (j\omega)$ is the stiffness frequency response function.

The compliance and stiffness functions are related to each other by:

$$C(j\omega) = \frac{1}{S(j\omega)} \tag{24.1}$$

It might appear that there would be no particular advantage to use one or the other function when the joint system is considered *a priori* as a linear input-output system. However, the use of angular displacement perturbations has the advantage that the muscle length and its rate of change are directly controlled. This facilitates the analysis, because the response of the muscle system will be easier to separate from the passive joint dynamics. Thus, the adjusted torque response will then reflect only the muscle dynamics, and this will be useful when considering the human stretch reflex induced by displacement perturbations.

An alternative description of joint dynamics is the use of parametric models that are successful if an analytic expression or model structure is selected appropriately. This may be done using *a priori* knowledge about the dynamics of the system components. The selection of a model structure can be made on the basis of the results of non-parametric experiments. For example, the form of a transfer function may be estimated from an examination of the gain portion (i.e., the input-output amplitude ratio) of a system frequency response. The model structure is then usually chosen to provide a concise description of behaviour without necessarily assigning any physical meaning to model parameters. Although the simple second-order model of joint dynamics was derived in this manner, the underlying physiology is often too complex to be appropriately described by such simple models and some confusion exists when assigning the different parameters to muscle mechanical properties (Desplantez et al., 1999). However that may be, many studies are based on this model and the results often give a good approximation of the underlying functional relation.

It remains, then, to define the input waveform to characterize joint dynamics. The selection of an appropriate perturbation is another methodological consideration but should, in principle, be independent to obtain estimates of joint dynamics. In practice, each class of inputs is associated with particular experimental requirements, analysis techniques, and underlying assumptions that influence the results. Among the different classes of input waveforms, sinusoidal perturbations are often used (for a detailed review of each input waveform, see Kearney and Hunter, 1990).

In the present study, the sinusoidal perturbation technique was chosen for a number of reasons. Sinusoidal perturbations are easy to generate and they involve no sudden impulsive movements that might synchronize a large number of sensory receptors in an artificial way. Furthermore, the use of sinusoids is a useful and well established technique in engineering and the methods for analysis are well understood. A common representation of frequency-dependent changes in gain and phase is the use of a Bode diagram. To construct a Bode diagram, it has been shown that the use of complex values allows for easy handling of sinusoidal or cosine functions. Thus, a sinusoidal input and output signal can be described as follows:

$$\begin{aligned} \Theta(t) &= \hat{\Theta}\sin(\omega t) \quad \rightarrow \Theta(j\omega) = \hat{\Theta}e^{jwt} \\ T(t) &= \hat{T}\sin(\omega t + \varphi) \rightarrow T(j\omega) = \hat{T}e^{j(\omega t+\varphi)} \end{aligned} \tag{24.2}$$

where $\hat{\Theta}$ = input amplitude
$\hat{T}$ = output amplitude
φ = phase lag
ω = imposed frequency

Thus, as shown above in Figure 24.2, the stiffness frequency function can be described by a gain value and a phase lag such as:

$$S(j\omega) = \frac{T(j\omega)}{\Theta(j\omega)} = \frac{\hat{T}}{\hat{\Theta}} \cdot e^{j\varphi} \tag{24.3}$$

Then, for each frequency, the gain value and the phase lag are plotted in a gain diagram and a phase diagram respectively, using a logarithmic scale. Typical Bode diagrams and their associated models are well known in engineering. In Figure 24.3, the compliance frequency response displays a number of characteristic features. The gain is constant at low frequencies, there is a resonant volley at intermediate frequencies, and the gain decreases with a slope of –40 dB/decade at high frequencies. The phase shift starts at 0° at low frequencies, decreases to –90°, and approaches –180° at high frequencies.

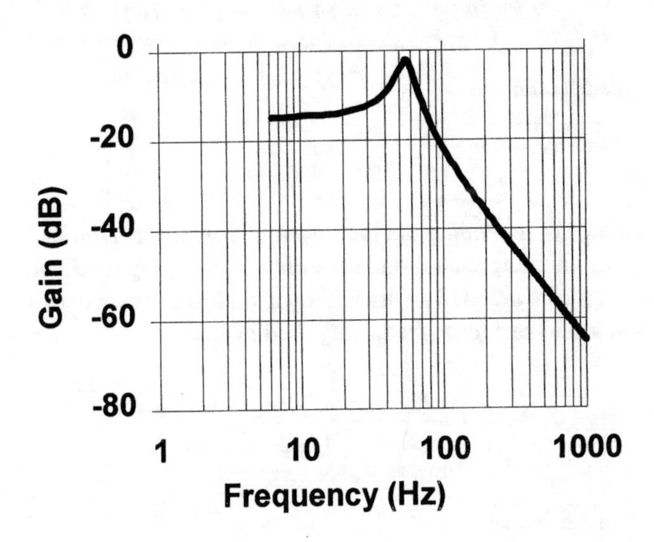

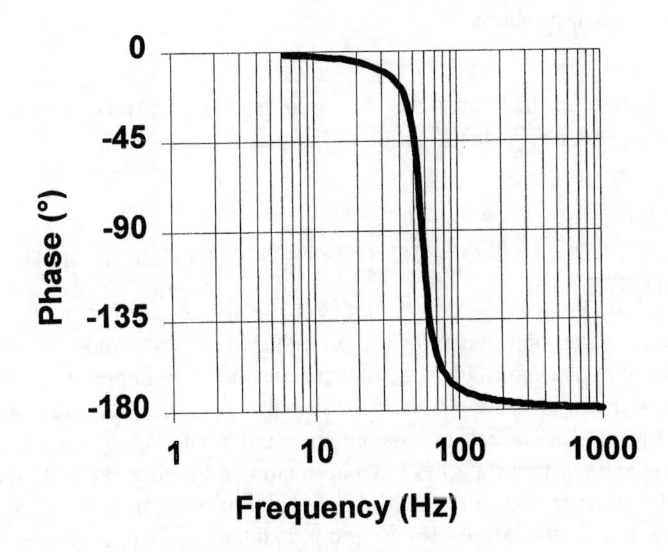

Figure 24.3. The Bode diagram shows the typical behaviour of a second-order model including an inertia (*I*), a viscosity (*B*), and an elastic parameter (*K*). Gain and phase diagrams were achieved from a parametric model simulation.

These features correspond to the response of a second-order system, and the joint dynamics have been modelled frequently using a parametric model of the form:

$$I \cdot \ddot{\Theta}(j\omega) + B \cdot \dot{\Theta}(j\omega) + K\Theta(j\omega) = T(j\omega) \tag{24.4}$$

where I = inertial parameter
B = viscous parameter
K = elastic parameter

An advantage of modelling the joint dynamics with Equation 24.4 is that each parameter has an interpretation in terms of the underlying mechanics. However, Equation 24.4 can be Laplace-transformed and manipulated to express joint dynamics in a standard form, frequently used in electronics:

$$\frac{\Theta(s)}{T(s)} = \frac{A\omega_0^2}{s^2 + 2 \cdot \xi \cdot \omega_0 \cdot s + \omega_0} \tag{24.5}$$

where s = Laplace operator
A = static gain
ω_0 = natural frequency
ξ = damping parameter

This expression facilitates the determination of the different mechanical parameters when using identification techniques.

CURRENT PROBLEMS OF INTERPRETATION OF THE SOLICITED STRUCTURES

Knowledge of the joint dynamics can be useful in understanding the underlying mechanisms in neuromuscular disease, but they are also important in aspects of rehabilitation engineering when restoring the function of paralyzed and/or injured limbs. The effect of adaptation on the characteristics of muscular contractile and elastic properties have been studied because skeletal muscles are known to present dynamic entities, capable of changing their phenotypic properties under the influence of various exogenous factors. These measurements are often accompanied by biochemical, histochemical, and immunohistochemical studies in order to assess muscle fibre-type transition due to changes in muscle functional demand. The involved structures can be, among others, elements constituting the components of Hill's model (Hill, 1938) such as the series elastic component (SEC) and the parallel elastic component (PEC). The SEC is classically separated in two fractions (Shorten, 1987): an active fraction (muscle fibres) and a passive fraction (tendon). In humans, under voluntary activation, the elastic characteristics of the SEC can be obtained by means of a quick-release technique derived from classical methods in isolated muscles, and may be interpreted in terms of musculo-tendinous elastic characteristics (Goubel and Pertuzon, 1973; Pousson et al., 1990). The PEC is composed of the sarcolemma, connective tissue, and residual acto-myosin bridges (Hill, 1968).

The physiological interpretation of contributing structures when adaptations occur in a more complex system is somewhat delicate considering the number of structures that are potentially modifiable (see Figure 24.1), and the global measurements performed during such experiments. The structures involved when using sinusoidal perturbations can be the musculo-tendinous complex, articular structures, and musculo-articular structures as a whole, without any possibility to distinguish the effective contribution of each of them. Thus, care must be taken when discussing adaptations with regard to animal studies, where results are obtained on isolated muscles. The adaptations that occur in isolated muscles focus on only a fraction of the musculo-articular system, and an extrapolation to the whole system can lead to questionable interpretations.

METHODS

As microgravity preferentially affects postural muscles, the apparatus was initially designed to test muscle groups crossing the ankle joint. To improve knowledge about adaptation of the human ankle-joint system, the concept of mechanical impedance was used to express the intrinsic elastic characteristics of the musculo-articular system. The proposed protocol—to test the mechanical impedance of the musculo-articular system—was evaluated in preliminary tests (Tognella et al., 1997) when subjects exerted a constant plantarflexion torque. The elaboration of this protocol using different levels of tonic activation (at least three levels under active condition) allowed us to propose a normalized representation of active musculo-articular stiffness.

MATERIALS

The ankle ergometer used for these studies has been developed in our laboratory. The technical support of this ergometer has been described elsewhere (Tognella et al., 1997) but will be reviewed briefly. The ergometer consists of two main units: (1) a power unit that contains the actuator, its power supply unit, a position and torque transducer, and its associated electronics, and (2) a driving unit composed of a PC-type computer equipped with a specific 12-bit A/D converter and a timer board. Angular displacement is measured with an optical digital sensor and angular velocity is captured from a resolver bound to the rotor (except for velocities greater than 15.7 rad·s^{-1}, which required a tachometer). Angular torque was obtained using a strain-gauge torque transducer. A specific menu-driven software controlled all procedures and recorded all mechanical variables (1 kHz sampling frequency) for later analysis. A dual-beam oscilloscope gave the subject a visual-feedback about the procedure in progress.

EXPERIMENTAL METHODS

All subjects were placed on an adjustable seat with their left feet attached rigidly to the actuator of the ankle ergometer. The horizontal bi-malleolar axis coincided with the axis of rotation of the actuator. The knee was extended to 110° and the ankle was placed at 90° (neutral position; Figure 24.4).

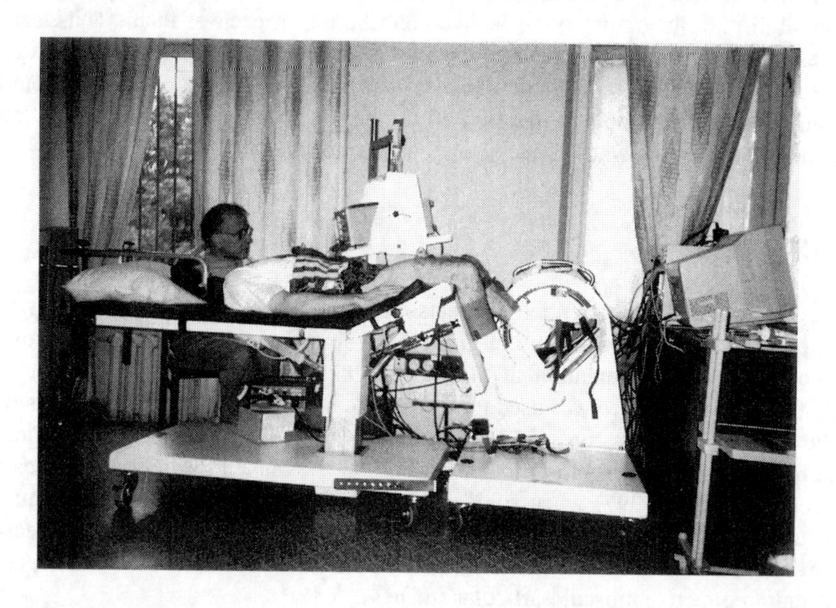

Figure 24.4. The ankle ergometer device in an experimental situation at the Y.G. Gagarin Cosmonauts Training Centre, Star City, Moscow.

In a first test, maximal voluntary contraction (MVC) was determined in plantarflexion under isometric conditions. The subject was asked to develop an MVC against the actuator. Three trials were carried out and the best performance was considered as the true MVC of the day. Then, sinusoidal oscillations were imposed on the joint to characterize the mechanical impedance of the musculo-articular system in plantarflexion. The duration of the sinusoidal oscillations was 4 seconds and the displacement amplitude was 3° peak-to-peak. Frequencies of 4-16 Hz were successively presented using steps of 1 Hz. During such experiments, the subject had to maintain a constant level of torque equal to a percentage of his MVC. In order to mask unavoidable oscillations in torque, a filtering of the torque signal was performed and the filtered torque was provided as feedback. Thus, all subjects were able to keep the level of force at the target value. Sinusoidal perturbations with no participation of the subject (i.e., 0% of MVC) were performed in most cases. Resting periods between the different trials were observed to prevent muscle fatigue.

DATA PROCESSING AND PRESENTATION OF RESULTS

The parameters used to characterize the mechanical properties of the musculo-articular system during sinusoidal oscillations were angular displacement and torque. The following analysis only considered torque that was modulated at the driving frequency. In doing so, non-linearities were neglected (Kearney and Hunter, 1990). Averaged displacement-to-torque amplitude ratios (i.e., compliance) and displacement-to-torque phases were plotted on a Bode diagram. As in other studies (Kearney and Hunter, 1990), the gain diagram showed a peak for the resonant frequency, followed by a linearly decreasing slope of –40 dB/decade. The phase plot showed a –90° phase lag for the resonant frequency. Frequency-dependent gain changes and phase shifts reflect the mixed mechanical contribution from inertia (I), viscosity (B), and elasticity (K), of the musculo-articular system. Using identification techniques (Levy, 1959), a second-order model including such parameters was adjusted to the Bode diagram. This led to the expression Z, the mechanical impedance of the musculo-articular system:

$$Z(s) = I \cdot s + B \cdot (\alpha) + K(\alpha) \cdot s^{-1} \tag{24.6}$$

where s = Laplace operator
α = level of tonic activation

Figure 24.5 shows classical gain diagrams and the typical behaviour for two different levels of voluntary activation for a given subject: one can distinguish differences in gain increase at low frequencies and a shift of the resonant frequency to higher values by increasing the level of voluntary activation. Fitting by a second-order model was always satisfactory.

Musculo-articular stiffness was investigated in passive and active conditions. When using different activation levels, musculo-articular stiffness can be linearly related to torque. This means that the greater the number of activated

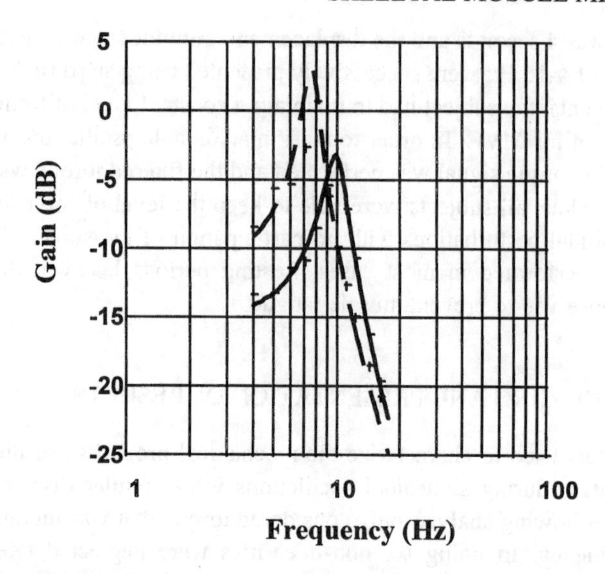

Figure 24.5. Typical example of a gain diagram of the musculo-articular system at 35% MVC (dashed line) and 50% MVC (solid line), adjusted by a second-order model. The gain is expressed in terms of a compliance.

fibres, the stiffer the musculo-articular system. This kind of relationship was also described in the review of Kearney and Hunter (1990). The stiffness-torque relationships were fitted by a linear model,

$$K = a \cdot T + b \tag{24.7}$$

where the slope (a) is defined as the stiffness index (SI_{MA}) for the musculo-articular system, T is the torque, and b describes the intercept.

This representation is commonly used in the literature, and taking the slope of the curve for characterizing the musculo-articular stiffness has several advantages:

- The SI_{MA} value is independent of the torque level, so it can be readily compared with results from other studies.
- MVC measurements for normalizing musculo-articular stiffness data are not necessary.

It was of interest to describe changes in musculo-articular stiffness by using SI_{MA} and b. The stiffness index parameter (SI_{MA}) identifies musculo-articular stiffness under active conditions. This parameter is assumed to be similar to an index of SEC stiffness (SI_{MT}), described by the slope of the linear stiffness-torque relationship achieved by means of a quick-release test.

The intercept b describes the musculo-articular system under passive conditions and should be close to the passive elasticity values found at 0% of

MVC. This parameter is a passive resistance of the ankle joint to the imposed movement when muscles are relaxed. This passive elastic stiffness reflects the combined effects of passive structures including skin, muscle, tendon, ligaments, and articular surface (Weiss et al., 1988) and can be attributed in part to the PEC.

APPLICATION IN MUSCULAR DISUSE

A number of studies have documented that the absence of a weight-bearing environment encountered during a period of real or simulated microgravity induces alterations in skeletal muscle function. In humans, the loss in muscle mass and force has been reported to be a prominent factor after space flight (Edgerton and Roy, 1996). Nevertheless, it seems that no one has raised the question of investigating changes in musculo-articular stiffness after exposure to microgravity. The clinical literature indicates that disuse increases muscle and joint stiffness, which might have consequences in terms of musculoskeletal dysfunction and control of movement (Akeson et al., 1987). If such changes also occur during space flight, this may alter neuromuscular performance since stiffness governs the mechanics of the interactions between the musculoskeletal system and the external environment. Therefore, the aim of the present work was to determine whether stiffness properties of the human plantarflexors and the ankle joint were modified after a long-term space flight.

The experiments were performed on 14 cosmonauts (C1 to C14) twice before flight (baseline data collection [BDC]; BDC1, BDC2) and two to three days after space flight (return [R]; R+2 / R+3) of 180-day duration, except for C2 and C3 whose flight duration was 90 days. Elasticity in the passive condition (Kp: 0% MVC for cosmonauts C4 to C14) and in an active condition (K_a: 50% MVC for cosmonauts C1 to C14) were investigated. Then, all K values were pooled to characterize global changes in SI_{MA} after exposure to microgravity.

RESULTS

Firstly, musculo-articular stiffness at 0% and 50% MVC was investigated for the population as mentioned above. Concerning the passive sinusoidal perturbation test (0% MVC), eight cosmonauts had a decrease in Kp from 3.25% to 36.82%, whereas three cosmonauts had an increase from 5.00% to 21.90%. A Wilcoxon-Signed-Rank test ($P<0.05$), taking each subject into account, indicated that post-flight data were significantly lower than pre-flight data. Mean Kp changed from 39.12±3.97 N·m·rad^{-1} to 34.78±4.20 N·m·rad^{-1}, which corresponded to a decrease of 11.1% (Cochran t-test, $P<0.05$). Concerning the results obtained in the active condition, it appeared that because of a strong loss in force production, some subjects were not able to maintain the demanded pre-flight reference value. On the other hand, changes in reference torque between

BDC1 and BDC2 led to differences in pre-flight stiffness. So, by using the BDC1 and BDC2 stiffness it was possible to calculate, by extrapolation, a pre-flight musculo-articular stiffness corresponding to the torque developed in post-flight condition. Then, pre-flight and post-flight stiffnesses (K_a) were comparable for the same torque value. The use of extrapolated pre-flight K_a appeared valid in light of the existing linear relationship. It was found that, after the flight, 11 cosmonauts had a decrease in their musculo-articular stiffness by values ranging from 1.75 N·m·rad^{-1} to 54.24 N·m·rad^{-1}, which corresponds to a decrease of 0.75% to 31.86%. Three cosmonauts had an increase in their musculo-articular stiffness of 0.5% to 10.75%. The Wilcoxon-Signed-Rank test ($P<0.05$), taking each subject into account, revealed that K_a was significantly lower in post-flight tests than in pre-flight tests. Mean K_a changed from 293.45±47.86 N·m·rad^{-1} (pre-flight) to 275.48±62.88 N·m·rad^{-1} (post-flight). This corresponds to a non-significant decrease of 6.11%, as determined by a Cochran t-test ($P>0.05$).

Finally, changes in SI_{MA} after exposure to microgravity were characterized for the population by pooling all individual data (Figure 24.6). The global SI_{MA} changed from 4.94±0.14 rad^{-1} to 4.86±0.10 rad^{-1}, i.e., it remained nearly unchanged (−1.62%). The intercept value b was found to change from 41.96±6.93 N·m·rad^{-1} to 33.61±4.28 N·m·rad^{-1}, corresponding to a significant decrease of 19.89% (Cochran t-test, $P<0.05$). As mentioned above, the intercept values were close to the mean passive stiffness, Kp, which was 39.12±3.97 N·m·rad^{-1} and 34.78±4.20 N·m·rad^{-1} for pre- and post-flight, respectively.

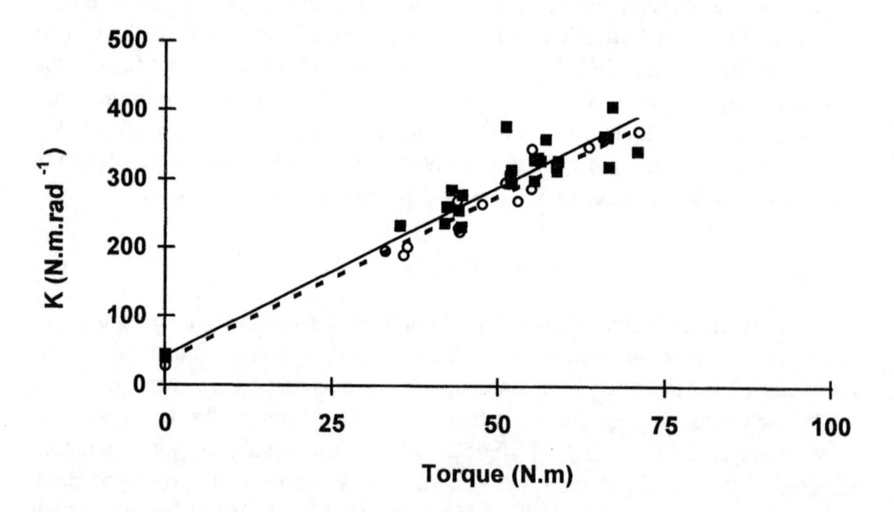

Figure 24.6. Stiffness-torque relationships from sinusoidal perturbation tests BDC 1/2 (■, solid line) data and R+2 / R+3 (○, dashed line) data for the population. The slopes indicate changes in SI_{MA}, and the intercepts represent changes in passive stiffness caused by space flight. The correlation coefficient r was 0.983 (BDC) and 0.994 (R+).

DISCUSSION

The musculo-articular stiffness values from pre-flight tests found in this study in passive (Kp) or active (K_a) conditions, as well as the global SI_{MA} or the intercept b, are in good agreement with values reported by other investigators (Shorten, 1987; Kearney and Hunter, 1990; Toft et al., 1991; Tognella et al., 1997).

Data about detailed changes in mechanical properties of human muscles due to exposure to microgravity remain scarce. A review of space-flight induced changes in force production under isometric and isokinetic conditions indicated a decrease in muscular force and an increase in maximum shortening velocity (Edgerton and Roy, 1996). The same decrease in force was also observed after a long-term bed-rest, a classical human model of simulated microgravity (Edgerton and Roy, 1996).

Considering animal studies, soleus mechanical properties have been studied using a model of simulated weightlessness, the hindlimb suspension (Thomason and Booth, 1990). Changes in elastic properties in isolated soleus muscle of hindlimb suspended rats have been reported by Canon and Goubel (1995); these changes were a decrease in SEC stiffness, as determined by tension-extension curves. This decrease was partly attributed to a fibre-type transition phenomenon from slow to fast fibres, since fast fibres are less stiff than slow fibres (Kovanen et al., 1984; Goubel and Marini, 1987). On the other hand, changes in SEC stiffness are not only related to the active but also to the passive part of the SEC. Alterations in tendon composition during hindlimb suspension were reported by Nakagawa et al. (1989). They found a decrease in tendon stiffness, and so, the decrease in SEC stiffness might have two origins (Canon and Goubel, 1995; Almeida-Silveira et al., 2000). The effect of simulated microgravity on *in vivo* passive structures using hindlimb suspended rats was described as an increase in hindlimb passive tension which was attributed to musculo-tendinous units, but also to the joint (Gillette and Fell, 1996). Furthermore, clinical studies showed an increase in human knee-joint stiffness (Heerkens et al., 1986) and ankle-joint stiffness (Namba et al., 1991) in rabbits as a result of immobilization. In addition, changes in posture can alter the mechanical properties of the musculo-tendinous system and musculo-articular system. Gillette and Fell (1996) observed increased plantarflexion during hindlimb suspension, leading to an increase in joint stiffness when retested in a more dorsiflexed position. However, for Gillette and Fell (1996), the increase in stiffness did not appear to be due to a shorter muscle, but was caused by changes in muscle architecture, cytoskeletal proteins, or visco-elastic properties of the muscle and its connective tissue.

Considering changes in musculo-articular stiffness after long-term space flight, these changes indicated a significant decrease in Kp. Kp reflects the combined effect of passive elastic structures including skin, muscle, tendon, ligament, and the articular surface. The influence of disuse-induced atrophy might be evident, since modifications in morphological and biochemical characteristics for the Achilles tendon were observed (Nakagawa et al., 1989). These authors reported a decrease in surface area of collagen fibres and this could be a factor related to the decrease in tendon stiffness (Almeida-Silveira et al., 2000). However, Riemersma and Schamhardt (1985) showed that within a tendon, the loading components (collagen fibres) are approximately constant, whereas the non-loading components (e.g., fat, glucosaminoglycans) determine the difference in cross-sectional area (CSA). Thus, the decrease in tendon stiffness reported by Almeida-Silveira et al. (2000) could be due to a re-modelling of collagen fibres, since no changes in CSA were observed. The effects of immobilization on morphological and biochemical characteristics of articular structures are well known (Akeson et al., 1987), leading to a proliferation of connective tissue concentration within the joint space, and thus, the increased passive stiffness may be explained. However, concerning the changes in human musculo-articular structures after space flight, any hypothesis will be speculative because no detailed data of morphological and biochemical properties are available. Furthermore, the influence of countermeasures, e.g., physical exercise during the space flight, should be mentioned. Namba et al. (1991) reported that continuous passive motion during immobilization can maintain joint function and thus limit the increase in joint stiffness. However, there is no consensus on the type of exercise or the duration of exercises that should be performed during long-term space flight (Edgerton and Roy, 1996). In general, it is assumed that countermeasures can limit the effects of microgravity, but not alleviate them.

Concerning the musculo-articular stiffness in the active condition (K_a), no data from disuse-induced changes are available. However, K_a is assumed to reflect the combined effects of musculo-tendinous stiffness and passive ankle-joint stiffness. Goubel (1997) reported in a preliminary study an increase in SEC stiffness as a result of space flight when using a musculo-tendinous stiffness index (SI_{MT}) for characterizing human plantarflexors. Thus, as a result of space flight, musculo-tendinous stiffness increases whereas passive ankle-joint stiffness decreases. Such an interaction between these two stiffnesses was proposed by Farley and Morgenroth (1999) for explaining leg stiffness adjustment when hopping, taking into account ankle-joint stiffness. From this point of view, the decrease in K_a might be attributed to a greater decrease in Kp in order to compensate for the increase in musculo-tendinous stiffness. Thus, by using the pooled data for SI_{MT} and the intercept b, it was found that SI_{MT} and b altered by about the same amount, but in the opposite direction. Consequently, SI_{MA} should not change much, which was confirmed for the pooled data.

APPLICATION IN TRAINING

The use of specific training methods enables athletes to improve performance during competition. One training method is 'plyometric training' in which athletes perform repeated stretch-shortening cycles of activated muscles. The intent of this type of exercise is to increase the instantaneous explosive power output as required during short-term efforts like sprinting or jumping. Using plyometric training, athletes may enhance mechanical efficiency because of the stretch-shortening contractions compared to training using shortening contractions exclusively (Bosco et al., 1982). The aim of this study was to quantify the effects of plyometric training on ankle musculo-articular structures in humans using the above described ankle ergometer, and to give another example of how changes in functional demand can affect the mechanical properties of the musculo-articular system.

The plyometric training program and the detailed experimental protocol has been published elsewhere (Cornu et al., 1997), but will be reviewed briefly here. Fourteen college students volunteered for this study. Two methods were employed to increase the functional demand: increases in the number of exercises specifically involving a stretch-shortening cycle, and increases in the overloading during the eccentric phase of the movement (a series of plyometric jumps from platforms of varying heights). The training period lasted for seven weeks and comprised two sessions of one hour per week. A progressive evolution in terms of the number and execution speed of the jumps was imposed so as to improve from 300 jumps per session to more than 900 jumps at the end of the training period. Musculo-articular stiffness was investigated at three levels of voluntary activation (35%, 50%, and 75% of MVC) and in the passive condition (i.e., 0% of MVC).

RESULTS

The slopes of the stiffness-torque relationships were used to describe changes in the activated musculo-articular system, according to Equation 24.7. Furthermore, the intercept value b was extrapolated to study the musculo-articular system under passive conditions. This was done for each subject and for the whole population.

Twelve of the 14 subjects showed a significant change in their SI_{MA} as a result of the training. For two subjects, there was no change. The SI_{MA} for the whole group changed from 7.00±0.23 rad^{-1} to 5.25±0.23 rad^{-1}. This corresponds to a significant decrease of 32.69%. For the intercept value, 11 of the 14 subjects showed an increase. The global intercept of the subjects changed from 85.05±9.29 N·m·rad^{-1} to 134.69±10.28 N·m·rad^{-1}, for a significant increase of 58.36%. Further results from this experiment, such as the musculo-articular viscous properties, have been reported elsewhere (Cornu et al., 1997).

DISCUSSION

The results indicate that plyometric training modifies the elastic properties of the musculo-articular structures crossing the ankle.

Regarding changes in stiffness of the musculo-articular system after plyometric training, the role of the musculo-tendinous complex will be considered first. In humans, by using a quick-release technique, Pousson et al. (1995) reported an increase in the SEC stiffness of the plantarflexors following plyometric training. A decrease in SEC stiffness was reported by Almeida-Silveira et al. (1994) when analyzing the elastic behaviour of the isolated rat soleus muscle. This discrepancy could originate from the heterogeneity of the SEC. In the rat soleus, plyometric training induces an increase in type II fibres (Almeida-Silveira et al., 1994) which are less stiff than type I fibres (Goubel and Marini, 1987). This adaptation of the active part of the SEC may cause the decrease in stiffness despite an increase of stiffness in the passive parts (Woo et al., 1981). Moreover, it is certain that a part of the passive structure (free tendon) is removed in an isolated muscle. Thus, in human plantarflexors tested *in situ*, the observed increase in musculo-tendinous stiffness could arise from an adaptation of the passive part of the SEC. This increase in stiffness might be more pronounced than the decrease in stiffness caused by a transition in fibre types. In the present experiment, SI_{MA} was found to decrease with plyometric training. It may be hypothesized that, as in other experiments (Pousson et al., 1995), plyometric training increased the stiffness of the musculo-tendinous complex. The observed decrease in stiffness of the musculo-articular system may be attributed to an adaptation of the articular structures. However, there is no published evidence of such an adaptive response to plyometric training. The idea that plyometric training may lead to changes in the mechanical properties that oppose those found in disuse situations (e.g., Heerkens et al., 1986) remains an attractive hypothesis (first formulated by Woo et al., 1982). On the other hand, it may be speculated that in the present experiment, the training was so intensive that the transition of fibre types from slow to fast was more pronounced than in other experiments. Therefore, the balance between factors leading to an increase or a decrease in the stiffness of the musculo-articular system might have been different than in other studies. However, no data about fibre-type distribution in the human plantarflexor muscle group before and after training are available to support this hypothesis.

Concerning changes in the intercept *b*, they correspond to changes in the passive stiffness of the musculo-articular system. The results indicate an increase in *b* with training. In terms of muscle contribution, this means that there should be an increase in the stiffness of the PEC. Since the PEC corresponds to collagenous structures, an adaptive response similar to that observed in tendons might occur (i.e., an increase in stiffness as postulated above). A slow to fast fibre-type transition phenomenon may limit the increase in passive stiffness, since fast muscle fibres are less stiff than slow ones when tested passively (Kovanen et al., 1984). According to Woo et al. (1982), articular structures may also contribute to the decrease in joint stiffness.

CONCLUSION AND FUTURE DIRECTIONS

An important result from the space flight experiment was the decrease in passive musculo-articular stiffness. This result indicates that disuse caused by microgravity affects joint stiffness differently than disuse caused by immobilization in humans and animals. It was attempted to explain these results by considering that alterations in musculo-articular stiffness that occur during space flight may include active musculo-tendinous stiffness. The decrease in passive musculo-articular stiffness may affect overall leg stiffness, and thus, may contribute to adaptation in postural control.

Musculo-articular stiffness was adapted by plyometric training. Such adaptations are compatible with published results and hypotheses. The fact that a mechanical system, such as the plantarflexors and the ankle joint, can have one part (tendon) that exhibits an increase in stiffness while another part (muscle fibres) exhibits a decrease in stiffness is attractive from a functional point of view. This mode of adaptation to overuse may satisfy two contradictory requirements: a decrease in stiffness to allow for storage of additional potential energy, and an increase in stiffness for better force transmission to the periphery (Cavagna et al., 1981).

The two examples of disuse and overuse adaptations of ankle stiffness showed that changes in the functional requirements cause adaptations that depend on the amount of loading. In the following, changes in passive musculo-articular stiffness are considered, assuming that a global adaptive behaviour can be deduced from the experiments. In both cases, the changes in joint stiffness can be attributed to alterations in collagenous structures that might be similar to changes observed in tendons. Concerning the 'overuse' model, it is speculated that tendon stiffness increases in humans as it does in corresponding experiments using isolated tendons of rats. There might be a link between *in vivo* human-tendon adaptations and *in vitro* rat-tendon adaptation that might be used to explain the observed increase in passive musculo-articular stiffness following plyometric training. Concerning the disuse model, the literature indicates that after hindlimb suspension, isolated muscles and tendons show a decrease in stiffness. The observed decrease in passive musculo-articular stiffness after microgravity exposure of cosmonauts may then be interpreted using the results found in the hindlimb suspension studies. Thus, adaptation has its origin in the passive musculo-tendinous system. Summarizing, the passive elements adapt in an opposite manner after disuse caused by either space flight or 'overuse'. Nevertheless, in both cases, the adaptations appear to be consistent with the functional demands of the musculo-articular structures.

Concerning musculo-articular stiffness under active conditions, the interpretation of adaptations caused by changes in the functional demands is more difficult because of all the different structures that might be involved. Thus, further experiments including mechanical (e.g., the role played by muscle fibres) and reflexological approaches are needed to elucidate the precise origins in musculo-articular stiffness.

In conclusion, the knowledge gained from animal experiments may help to explain the mechanisms of *in vivo* joint stiffness adaptations. However, alterations in musculo-articular stiffness appear to have multiple origins, and the consequences of these adaptations for satisfying changing functional demands are not known. Thus, the interpretation of *in vivo* changes in joint stiffness following a change in the functional demands is more complex than in animal experiments, because no clear distinction between the relative contribution of different structures to joint stiffness can be made.

ACKNOWLEDGMENTS

The authors express their gratitude to the students of the University of Compiègne who participated in the training program, and to the crews, their back-ups, and the medical staff of Y.A. Gagarin Cosmonauts Training Centre at Star City, Moscow. The technical assistance of Alain Mainar and Clotilde Vanhoutte for performing the experiments is appreciated. This study was supported by the Centre National d'Etudes Spatiales (CNES).

REFERENCES

Agarwal, G.C., and Gottlieb, G.L. (1985). Mathematical modeling and simulation of the postural control loop, part III. *Critical Reviews in Biomedical Engineering* **13**, 49–93.

Akeson, W.H., Amiel, D., Abel, M.F., Garfin, S.R., and Woo, S.L. (1987). Effects of immobilization on joints. *Clinical Orthopaedics and Related Research* **219**, 28–37.

Almeida-Silveira, M.I., Lambertz, D., Pérot, C., and Goubel, F. (2000). Changes in stiffness induced by hindlimb suspension in rat Achilles tendon. *European Journal of Applied Physiology* **81**, 252–257.

Almeida-Silveira, M.I., Pérot, C., Pousson, M., and Goubel, F. (1994). Effects of stretch-shortening cycle training on mechanical properties and fibre type transition in the rat soleus muscle. *Pflügers Archiv* **427**, 289–294.

Bosco, C., Tarkka, I., and Komi, P.V. (1982). Effects of elastic energy and myoelectric potentiation of triceps surae during stretch-shortening cycle exercise. *International Journal of Sports Medicine* **3**, 137–140.

Canon, F., and Goubel, F. (1995). Changes in stiffness induced by hindlimb suspension in rat soleus muscle. *Pflügers Archiv* **429**, 332–337.

Cavagna, G.A., Citterio, G., and Jacini, P. (1981). Effects of speed and extent of stretching on the elastic properties of active frog muscle. *Journal of Experimental Biology* **91**, 131–143.

Cornu, C., Almeida-Silveira, M.I., and Goubel, F. (1997). Influences of plyometric training on the mechanical impedance of the human ankle joint. *European Journal of Applied Physiology* **76**, 282–288.

Desplantez, A., Cornu, C., and Goubel, F. (1999). Viscous properties of human muscle during contraction. *Journal of Biomechanics* **32**, 555–562.

Edgerton, V.R., and Roy, R.R. (1996). Neuromuscular adaptations to space flight. Chapter 32 in *Handbook of Physiology: Volume 1, Section 4, Environmental Physiology.* American Physiological Society, Bethesda, Maryland. pp. 721–763.

Farley, C.T., and Morgenroth, D.C. (1999). Leg stiffness primarily depends on ankle stiffness during hopping. *Journal of Biomechanics* **32**, 267–273.

Gillette, P.D., and Fell, R.D. (1996). Passive tension in rat hindlimb during suspension unloading and recovery: Muscle/joint contributions. *Journal of Applied Physiology* **81**, 724–730.

Goubel, F. (1997). Changes in mechanical properties of human muscle as a result of space flight. *International Journal of Sports Medicine* **18**, S285–S287.

Goubel, F., and Marini, J.F. (1987). Fibre type transition and stiffness modification of soleus muscle of trained rats. *Pflügers Archiv* **410**, 321–325.

Goubel, F., and Pertuzon, E. (1973). Evaluation de l'élasticité du muscle *in situ* par une méthode de quick-release. *Archives Internationales de Physiologie et Biochimie* **81**, 697–707.

Heerkens, Y.F., Woittiez, R.D., Huijing, P.A., Huson, A., Van Ingen Schenau, G.J., and Rozendal, R.H. (1986). Passive resistance of the human knee: The effect of immobilization. *Journal of Biomedical Engineering* **8**, 95–104.

Hill, A.V. (1938). The heat of shortening and the dynamic constants of muscle. *Proceedings of the Royal Society of London B* **216**, 136–195.

Hill, D.K. (1968). Tension due to interaction between the sliding filaments in resting striated muscle: The effect of stimulation. *Journal of Physiology* **199**, 637–683.

Kearney, R.E., and Hunter, I.W. (1990). System identification of human joint dynamics. *Critical Reviews in Biomedical Engineering* **18**, 55–87.

Kovanen, V., Suominen, H., and Heikkinen, E. (1984). Mechanical properties of fast and slow skeletal muscle with reference to collagen and endurance training. *Journal of Biomechanics* **17**, 725–735.

Levy, E.C. (1959). Complex curve fitting. *IEEE Transactions on Automatic Control* **4**, 37–43.

Nakagawa, Y., Totuska, M., Sato, T., Fukuda, Y., and Hirota, K. (1989). Effect of disuse on the ultrastructure of the Achilles tendon in rats. *European Journal of Applied Physiology* **59**, 45–49.

Namba, R.S., Kabo, J.M., Dorey, F.J., and Meals, R.A. (1991). Continuous passive motion versus immobilization. The effect of post-traumatic joint stiffness. *Clinical Orthopaedics and Related Research* **267**, 218–223.

Pousson, M., Legrand, J., Berjaud, S., and Van Hoecke, J. (1995). Détente et élasticité: effets d'un entraînement plyométrique. *Science et Motricité* **25**, 19–26.

Pousson, M., Van Hoecke, J., and Goubel, F. (1990). Changes in elastic properties of human muscle induced by eccentric exercise. *Journal of Biomechanics* **23**, 343–348.

Riemersma, D.J., and Schamhardt, H.C. (1985). *In vitro* mechanical properties of equine tendons in relation to cross-sectional area and collagen content. *Research in Veterinary Science* **39**, 263–270.

Shorten, M.R. (1987). Muscle elasticity and human performance. In: Karger, Basel (ed.), *Medicine and Sport Sciences* **25**, 1–18.

Sinkjaer, T. (1997). Muscle, reflex and central components in the control of joint in healthy and spastic man. *Acta Neurologica Scandinavica* **96**, Supplement No. 170.

Thomason, D.B., and Booth, F.W. (1990). Atrophy of the soleus muscle by hindlimb unweighting. *Journal of Applied Physiology* **68**, 1–12.

Toft, E., Sinkjaer, T., Andreaseen, S., and Larsen, K. (1991). Mechanical and electromyographic responses to stretch of the human ankle extensors. *Journal of Neurophysiology* **68**, 1402–1410.

Tognella, F., Mainar, A., Vanhoutte, C., and Goubel, F. (1997). A mechanical device for studying mechanical properties of human muscles *in vivo*. *Journal of Biomechanics* **30**, 1077–1080.

Weiss, P.L., Hunter, I.W., and Kearney, R.E. (1990). Human ankle joint stiffness over the full range of muscle activation levels. *Journal of Biomechanics* **21**, 539–544.

Woo, S.L., Gomez, M.A., Amiel, D., Ritter, D.A., Gelberman, R.H., and Akeson, W.H. (1981). The effects of exercise on the biomechanical and biochemical properties of swine digital flexor tendon. *Journal of Biomedical Engineering* **103**, 51–56.

Woo, S.L., Gomez, M.A., Woo, Y.K., and Akeson, W.H. (1982). Mechanical properties of tendons and ligaments. *Biorheology* **19**, 397–408.

25 Measuring Human Finger Flexor Muscle Force *In Vivo*: Revealing Exposure and Function

JACK TIGH DENNERLEIN
Harvard School of Public Health, Boston, Massachusetts, USA

Throughout this book, the authors are examining the microcosm of the internal workings of skeletal muscle and the systems with which they interact. This chapter, which is more macro in nature, will present some of the benefits of measuring *in vivo* muscle force of the finger flexor in adult humans. The benefits are mainly that the measurements further validate models predicting the exposure of internal tissues to forces and the possible roles of muscles in human movement. During the open carpal tunnel release surgery of nine adults, the force of the flexor digitorum superficialis (FDS) for the long finger was measured. The patients completed two tasks: an isometric pinch, and tapping on a keycap and keyswitch of a computer keyboard. The objectives of these experiments were: (1) to validate current musculoskeletal models that predict muscle and tendon forces based on the external load and the finger posture, and (2) to observe and describe the tendon force during a dynamic activity such as tapping on a keyboard. The observations made through these measurements revealed that assumptions used in musculoskeletal models have limitations. During the dynamic activity of the tapping task, the tension in the muscle remains elevated long after the fingertip has been removed from the keycap. Finally, the measurements also revealed that passive muscle force might be used to propel the finger during dynamic activities, such as typing.

BACKGROUND

As humans, our primary interface with the physical world is our hands. We use our hands every day to explore and manipulate the world around us, and even more so with the advent of the computer and the Internet, we communicate with one another through our hands. For example, I sit here in my office typing this text by impacting the keyboard with my fingertips at least once for each character you see. Through these movements, the ideas of an author are communicated to you, the reader. In essence, the desktop computer has brought manual labour to

Skeletal Muscle Mechanics: From Mechanisms to Function. Edited by W. Herzog.

the office environment and knowledge-based workers, not traditionally known for manual types of work. Hence, computer work has been added to other jobs where injury to the hands and wrists hinders performance at work, not to mention the fundamental activities of daily living.

In the last 10 years, injuries to the tendons of the hand at the wrist and adjacent tissues associated with repetitive work of the upper extremity have increased. In 1994, these injuries were reported to be as high as 55% of all work-related repetitive motion disorders (Bureau of Labor Statistics, 1995). But the injury mechanisms are not well understood. Epidemiological studies (Armstrong et al., 1987; Moore and Garg, 1994; Silverstein et al, 1986) have identified several risk factors for tendon-related disorders, which, along with motion, posture, and vibration, include the force exerted during a repetitive task. Understanding how force is transmitted from the site of external application (fingertip) to the internal tissue (finger flexor tendons) is critical to understanding the mechanics of repetitive motion disorders. This understanding determines how the internal biological tissues support an externally applied force, and such knowledge identifies tendons that are exposed to high forces.

Currently, clinicians and researchers use models predicting tendon force (for example, Harding et al., 1993; Chao et al., 1989) to assess the loads applied to the internal tissues of the upper extremity. These models are based on the equations of static equilibrium at each joint of the finger and a set of tendon force constraints to evaluate the supporting loads. However the musculoskeletal system is redundant with many muscles, tendons, and ligaments contributing to the balance of static equilibrium, and for sub-maximal forces there are infinite combinations of muscle forces that balance these equations. Solution of this indeterminate problem requires assumption of additional constraint relationships, usually constraints on how the forces are distributed over the muscles and tendons. There are several different sets of assumptions, which are described in the Methods and Materials section (for example, An et al., 1984). These assumptions have been validated to a certain degree by experimental data, mainly electromyographic (EMG) and cadaver studies (Valero-Cuevas et al., 1998). Measuring *in vivo* muscle force aids in the process of validating these methods that predict muscle forces.

The knowledge of *in vivo* muscle force also guides techniques of tendon repair, procedures for rehabilitation (Komi, 1990; Schuind et al., 1992), and the design of joint replacements (Weightman and Amis, 1982). Furthermore, this knowledge provides a direct measure of motor control through muscle force not revealed by electromyography (Komi et al., 1987).

The aims of these *in vivo* muscle force studies were to evaluate the relationship between fingertip force and finger flexor muscle force by measuring the *in vivo* force in the finger flexor digitorum superficialis tendon during two tasks, isometric pinch and rapid finger motion of a keystroke. The goals also included evaluating models that predict muscle force based on the participants' joint thickness, hand length, finger joint posture, and the contact force at the fingertip.

The specific research questions included: (1) Are the forces, as reflected in the force of the FDS, distributed across the tendons of the finger consistent with the patterns predicted in the literature? (2) For the rapid finger motion of a keystroke, is the relationship between force of the finger flexor tendon and force at the fingertip proportional? More specifically, do existing isometric models based on quasi-static equilibrium accurately predict the observed ratios of tendon-to-tip force? And, is the relationship between tendon and tip force, as defined by the slope of lines fitted to the data, different during the loading and unloading portions of a keystroke?

Results obtained while investigating these questions provided additional goals and questions. First, the results suggest that current models do not predict all possible force distributions for the finger. Therefore, a modified model is presented that better predicts the wide range of force distribution. Second, muscle force remains elevated for some time before and after the fingertip is in contact with the keycap. Possible mechanism and functions for the elevated muscle force, including motor control issues, are discussed.

METHODS AND MATERIALS

Measuring *in vivo* muscle force provides several challenges, including the obvious of entering inside the human body in a safe and efficient manner. Open carpal tunnel release surgery provided access to the extrinsic finger flexor tendons.

BUCKLE TRANSDUCER

Muscle and tendon force transducers, commonly referred to as buckle transducers, have been used to measure tensions in the human finger flexor muscles (Schuind et al., 1992), the human Achilles tendon (Komi et al., 1987; and Komi, 1990), the primate extraocular muscle tendon (Miller and Robins, 1992), the cat soleus and gastrocnemius tendons (Gregor et al., 1988; Sherif et al., 1983; Whiting et al., 1984), and the horse digital extensor tendon (Barnes and Pinder; 1974). By displacing the path of the tendon around a frame and fulcrum, the buckle converts tendon force into a bending load on the buckle frame. Strain gauges placed on the frame measure the resulting strain. The output of the strain gauges are calibrated to known tensions of *in vitro* tendons in a uniaxial materials-testing machine (Dennerlein et al., 1997, An et al., 1990).

The transducer used in this study consists of a 9 × 16 × 4.5 mm stainless steel frame and a removable stainless steel fulcrum (Figure 25.1), and fits tendons up to 5 mm wide, 3 mm thick. The tendon lies and self-aligns in the semicircular arches ($r = 2.5$ mm) in the frame and fulcrum. The cylindrical design and arrangement of the wires allows the transducer to slide between other tendons with minimal interference. It does not twist the tendon. Lengthening of the muscle-tendon unit by passing the tendon over the fulcrum is 1.2 mm for a 3 mm thick tendon. Material and transducer dimensions were designed for loads between 0 and 50 N, the physiological range of interest. The transducer has 6% accuracy (Dennerlein et al., 1997) for quasi-static loads and *in vitro* dynamic tests of the tendon-transducer, where a tapping-like load (Rempel et al., 1994) was applied to the tendon, indicating that the bandwidth of the system is above 50 Hz.

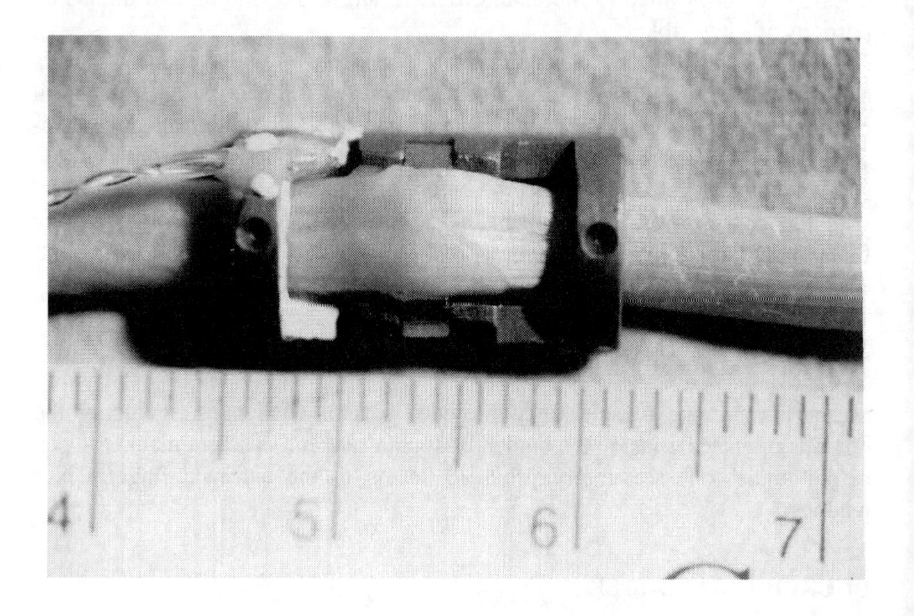

Figure 25.1. The buckle transducer with an *in vitro* finger flexor tendon. The removable fulcrum allows for *in vivo* installation of the transducer. The scale is in centimetres with millimetre markings.

IN VIVO FORCE MEASUREMENTS

Nine subjects (eight females and one male, mean age 48 ± 21 years) undergoing open carpal tunnel release surgery at the University of California, San Francisco (UCSF) participated in the study. The University of California San Francisco Committee on Human Research and the University of California Berkeley Committee for the Protection of Human Subjects approved the procedures. Prior to surgery, the hand length and the metacarpal phalangeal joint (MP), the proximal interphalangeal joint (PIP), and the distal interphalangeal joint (DIP) thicknesses were measured for each subject. Hand lengths, measured from the distal wrist crease to the tip of the long finger, ranged from 5 to 57 percentile of the population, and the average joint thicknesses ranged from 60 to 93 percentile (Garret, 1970 a, b). The subjects practised the pinch tasks to be performed during the experiments prior to their surgery.

Surgery was performed, as usual, under local anaesthesia at the incision site, and each subject retained motor control of the forearm musculature throughout the procedure. The subjects were prone, with the shoulder abducted to 90° and the palm rotated upward resting on the operating table. The carpal tunnel contents were exposed through a 5 cm longitudinal carpal tunnel incision. The flexor digitorum superficialis (FDS) tendon of the long finger was identified and the synovium removed. A gas-sterilized tendon force transducer was then mounted onto the FDS tendon (Figure 25.2).

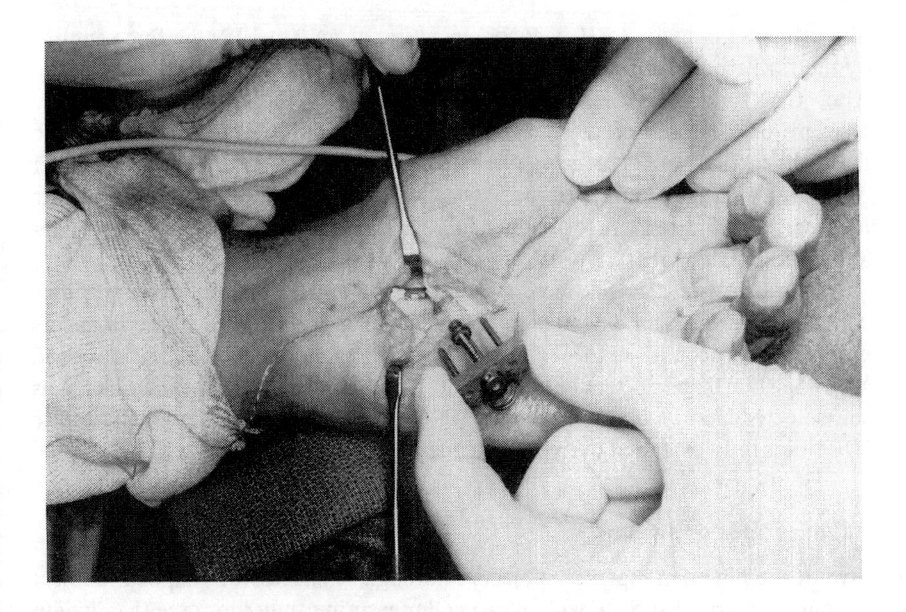

Figure 25.2. The buckle transducer mounted on the FDS inside the carpal tunnel during surgery. The three-prong device was used for calibration purposes.

After the subject flexed the long finger 10 times to seat the transducer onto the tendon, the tendon thickness was measured *in situ* for use in the transducer calibration factor (An et al., 1990; Dennerlein et al., 1997). The subject's forearm was rotated 90° from full supination toward a neutral forearm posture, with the thumb upward and with the palm towards the feet. Both tasks were performed so that the finger movement was in the horizontal plane. The tourniquet was released prior to data collection, relieving any ischemia, and the forearm and wrist were manually stabilized during the tapping tasks with the wrist straight. The surgeon removed all transducers once the tasks were completed, and surgery continued. The procedure extended the time of surgery by approximately 20 minutes.

A video camera mounted above the surgical field recorded the sagittal view of the finger posture and load cell alignment during the isometric pinch tasks for all but the first subject. For the first subject, the video viewed the hand from the side, and thus specific posture data from this subject were not acquired and ratios not predicted. The angle of intersection of lines aligned with the dorsal surface of adjacent finger segments defined the approximate joint angle, which is later used in the models (Long et al., 1960).

Isometric pinch

The subjects gradually (> 10 seconds) increased from 0 to 10 N the force applied by the fingertip to a single-axis load cell (GreenLeaf Medical Pinch Meter, Palo Alto, CA), and then decreased it monotonically to 0 N while observing a measure of the force on a visual monitor. These ramps were repeated two to three times. The force range is typical of that applied in keyboard and other occupational tasks (Rempel et al., 1994; Johnson et al., 1993). Data from the load cell and the tendon force transducer were recorded on a computer data-acquisition system at 50 samples per second.

The gradual pinch tasks were conducted at three different finger postures, which ranged from an extended to a flexed pinch posture. Pre-set goniometers guided the surgeon to set the angle of the MP joint, and the load cell was aligned such that the fingertip would not slip off the load cell. During pinch, subjects self-selected either a tip pinch posture (DIP joint flexed) or a pulp pinch posture (DIP joint fully extended or hyper-extended, DIP angle $\leq 0°$).

Tapping on a keyswitch

Five of the nine participants, Subjects 5 through 9, successfully tapped. Subject 5 tapped on a rigid surface with a load cell measuring fingertip force (bandwidth of 1000 Hz, GreenLeaf Medical Pinch Meter, Palo Alto, CA). The remaining four subjects tapped on a keycap and keyswitch assembly. The keycap contained a custom-designed load cell with a 3% of full-scale accuracy and 1000 Hz bandwidth (Smutz et al., 1994). Subjects were unable to view their hands, however, a buzzer connected to the switch in the keyswitch assembly provided audio feedback of contact. Subjects were asked to tap at one keystroke per second and minimize fingertip contact time with the keycap.

Data from the fingertip load cell and tendon force transducers were recorded on a computer data-acquisition system at 2000 samples per second. The data were digitally low-pass filtered with a cut-off frequency of 300 Hz. A video camera mounted above the surgical field recorded the sagittal view of the finger posture, fingertip position, and load cell alignment at 30 frames per second.

MUSCULOSKELETAL MODEL

A model incorporating the three observed postures and subjects' anthropometry was used to predict the muscle force and tendon tension to tip force ratio for the eight subjects with posture measurements. The three predictions of the force ratio were averaged for each subject, because the variations of the muscle tensions and the ratio predictions within subjects were small. The tendon tension model represents the finger as three movable rigid bodies (the proximal, middle, and distal phalanxes), linked by free hinges at the phalangeal joints and moving in the sagittal plane. The tendons of the six muscles that move the long finger span these joints (Table 25.1). The tendons facilitating extension through the contraction of the extensor and intrinsic (ulnar interosseous (UI), radial interosseous (RI), lumbrical (LU)) muscles of the long finger are: extensor slip (ES), radial band (RB), ulnar band (UB), and terminal extensor (TE) (Chao et al., 1989). The distribution of forces across the muscles to the components of the extensor mechanism is given by Equations 25.1.

$$\begin{aligned} TE &= RB + UB \\ RB &= 0.133\ RI + 0.167\ EDC + 0.667\ LU \\ UB &= 0.313\ UI + 0.167\ EDC \\ ES &= 0.133\ RI + 0.313\ UI + 0.167\ EDC + 0.333\ LU \end{aligned} \qquad (25.1)$$

Table 25.1. Tendons of the long finger that cross each joint and their contribution to either flexion or extension (Dennerlein et al., 1998a, Chao et al., 1989).

Joint	Flexion	Extension
MP	Flexor Digitorum Superficialis (FDS) Flexor Digitorum (FDP) Ulnar Interosseous (UI) Radial Interosseous (RI) Lumbrical (LU)	Extensor Digitorum Communis (EDC)
PIP	FDS, FDP	Extensor Slip (ES) Radial Band (RB) Ulnar Band (UB)
DIP	FDP	Terminal Extensor (TE)

Tension in the tendon acting at a distance from the axis of rotation (the moment arm) and a load applied to the fingertip create moments at the joints. Hence, the static equilibrium balance for the *j*th joint is:

$$\sum_i F_i r_{ij} + F_{tip} r_{j,tip} = 0 \qquad (25.2)$$

where F_i is the tension in the *i*th tendon, r_{ij} is the moment arm for the *i*th tendon at the *j*th joint, and $F_{tip}\ r_{j,tip}$ is the moment at the *j*th joint of the force at the fingertip. The joint thickness and the posture measurements are used to predict the moment arms (Armstrong and Chaffin, 1978; An et al., 1983). Hand length measurements determined the distances between the axes of rotation of adjacent joints (Buchholz and Armstrong, 1992). Equations 25.1 and 25.2, plus the equilibrium moment balances at the MP joint under abduction and adduction axis yield four constraint balances involving the six unknown muscle-tendon tensions. Additional constraints maintain the muscle-tendon forces as tensile only (force ≥ 0).

Three methods of solution were applied to the indeterminate problem (Table 25.2). Method 1 prescribes that only the extrinsic flexor tendon forces act about the PIP and DIP joints, providing two equations and two unknowns (FDS and FDP; Johnson et al., 1995). Method 2 assumes that the three intrinsic muscles act in synchronization and that the tension in the extensor (EDC) is zero, giving three unknowns and three balances (Harding et al., 1993; Weightman and Amis, 1982; Smith et al., 1964). Method 3 assumes that the muscle contraction balances the tension to minimize the sum of the squares of the muscle stress (An et al., 1984; Pedotti et al., 1978):

$$J = \sum_i \left(F_i^2 \Big/ PCSA_i^2 \right) \tag{25.3}$$

where $PCSA_i$ is the physiological cross-sectional area of the ith muscle from Chao et al. (1989). Quadratic programming techniques provide a solution for the six muscle-tendon forces (Matlab, Mathworks Inc, Natick, MA). Equation 25.3 defines a so-called cost function, which is minimized through the quadratic programming technique. Many different cost functions exist in the literature. For example, one cost is the summation of all the muscle forces, while another is the summation of the muscle stresses. There is little physiology supporting any cost function. However, preliminary results, as well as results found in the literature, suggest that the square of the muscle stress provides a good correlation with electromyographic studies.

Table 25.2. Three solution methods for the indeterminate problem (Dennerlein et al., 1998a).

Method	Reference	DIP Constraint
1.) Set tension in RB, UB, TE, and EDC = 0	Johnson et al. (1995)	Set tension in FDP = 0 when DIP ≤ 0°.
2.) Lump RI, UI, and LU into INT and EDC = 0 INT = RI+UI+LU	Harding et al. (1993), Weightman and Amis (1982), Smith et al. (1964)	Set tension in FDP = 0 when DIP ≤ 0°.
3.) Optimization methods; muscle stress $J = \sum_{tend} F_{tend}^2 \Big/ PCSA^2$	Chao et al. (1989)	Add constraint moment at the DIP joint when DIP ≤ 0°.

Parametric statistical methods and graphical methods were employed to address the specific research questions. The *in vivo* force ratios were compared to predicted force ratios using paired t-tests. The suspected non-proportional relationship and the dissipation of the high-frequency fingertip force components were observed by graphically plotting the two forces against each other. The non-proportional relationship was evaluated by comparing slopes of lines fitted to the loading and unloading regions using paired t-tests.

RESULTS

The *in vivo* experiments provided relationships for both the isometric condition (Table 25.3) and the dynamic keystroke (Figure 25.3). The models did not predict the wide range of tendon-to-tip ratios observed in both cases, however, these discrepancies are explained with a simple modification to the model.

IN VIVO MEASUREMENTS

The ratios of FDS tendon-to-tip force during isometric pinch varied substantially between subjects (Table 25.3). For each subject the tension in the FDS tendon was proportional to the force applied at the fingertip (Dennerlein et al., 1998a). Subject tendon-to-fingertip force ratios averaged across the three postures ranged from 1.7 to 5.8 (mean=3.3, SD=1.4). Within each subject, the variation (standard deviation) of the ratios across the three postures ranged from 0.1 to 1.9.

Table 25.3. *In vivo* and predicted finger tendon force ratios during fingertip loading. A) The measured FDS tendon force from this study. B) *In vivo* tendon forces reported by Schuind et al. (1992). C) Predicted forces using measured joint postures and the three different solutions methods. D) Predicted forces using the three different solution methods with DIP constraint added to the model. E) Predicted tendon forces reported in the literature. (Reprinted from Dennerlein et al., 1998a, with permission from Elsevier Science.)

Source	FDS*	FDP*	Intrinsic*	EDC	$r^{2\dagger}$	*RSME*††
A *In vivo*	3.3 (1.4)	–	–	–	–	–
B Schuind et al. 1992 (*in vivo*)	1.7 (1.5)	7.9 (6.3)	–	–	–	–
C Method 1	1.1 (0.4)	3.1 (0.7)	–	–	0.05	1.3
Method 2	1.2 (0.4)	3.9 (0.7)	2.8 (1.0)	–	0.02	1.3
Method 3	1.2 (0.4)	3.8 (0.8)	2.8 (0.6)	0	0.04	1.3
D Method 1 (DIP Constraint)	3.1 (1.8)	1.5 (1.4)	–	–	0.62	0.8
Method 2 (DIP Constraint)	3.7 (2.0)	1.6 (1.7)	1.7 (1.1)	–	0.71	0.7
Method 3 (DIP Constraint)	2.9 (1.7)	2.2 (1.4)	2.3 (0.6)	0	0.85	0.5
E Chao et al., (1989)	0.3–2.1	1.9–3.1	2.4–3.9	0	–	–
Harding et al., (1993)	0.8–2.7	1.2–3.2	0.5–3.2	–	–	–
Weightman and Amis (1982)	1.6–2.8	2.1–2.6	1.3–2.6	–	–	–

* Mean (standard deviation) tendon force in units of applied tip force.
† Correlation (r^2) between the predicted and the measured force.
†† RSME is the root square mean error.

The ratio of FDS tendon-to-tip force depended upon the DIP posture selected by the subjects during the isometric pinch task. Five subjects applied a tip pinch posture (DIP joint flexed) and four subjects applied a pulp pinch posture (DIP joint fully to hyper-extended, DIP angle ≤ 0; Chao et al., 1989). The tendon-to-tip force ratios measured for the four pulp-pinch subjects (mean = 4.4, SD = 1.4) were significantly higher (two-sample t-test performed and $p = 0.022$) than the ratios observed during tip pinch postures (mean=2.4, SD=0.6). This trend contradicts those predicted by the implemented model and the predictions of Chao et al. (1989).

For the five participants that successfully tapped on a keyswitch, 17 to 35 keystrokes were collected per subject. Summary measures for tendon force for each subject are presented in Table 25.4.

Table 25.4. Flexor digitorum superficialis (FDS) muscle forces and parameters relating muscle-to-fingertip force during a keystroke. Results presented are subject mean and (standard deviation). (Reprinted with permission from the *Journal of Orthopaedic Research*, Dennerlein et al., 1999.)

	Subjects					
	1	2	3	4	5	Mean
Keystroke Duration (ms)	136 (33)	277 (70)	158 (33)	210 (36)	224 (37)	219 (53)
Max. Tendon Tension (N)	11.2 (3.4)	14.0 (6.6)	14.6 (3.6)	8.3 (2.0)	16.6 (6.8)	12.9 (3.3)
Mean Tendon Tension (N)	6.5 (1.8)	8.0 (4.2)	7.9 (2.0)	5.0 (1.4)	8.6 (3.4)	7.2 (1.4)
Ratio of Max. Forces[1]	3.5 (1.0)	4.4 (0.3)	6.6 (1.0)	5.5 (1.7)	6.8 (1.5)	5.4 (1.4)
Ratio of Mean Forces[1]	3.4 (1.2)	4.5 (1.2)	6.5 (1.0)	6.2 (1.8)	5.6 (1.3)	5.2 (1.3)
Slope Loading[1]	1.9 (0.6)	3.7 (0.8)	5.0 (1.4)	3.0 (1.0)	5.5 (1.4)	3.8 (1.5)
Intercept Loading (N)	4.7 (1.1)	1.7 (2.2)	2.8 (1.3)	3.1 (1.3)	2.0 (1.5)	2.9 (1.2)
Slope Unloading[1]	0.6 (0.6)	2.1 (1.3)	3.2 (1.5)	1.5 (1.9)	4.7 (1.3)	2.4 (1.6)
Intercept Unloading (N)	8.8 (2.7)	7.1 (4.1)	7.9 (4.0)	5.9 (2.4)	5.3 (3.4)	7.0 (1.4)
Predicted Ratio I[1,2]	1.9 (0.6)	1.4 (0.1)	0.9 (0.3)	0.8 (0.1)	1.0 (0.2)	1.2 (0.5)
Predicted Ratio II[1,2]	1.9 (0.6)	5.9 (0.2)	3.8 (0.9)	4.5 (0.7)	4.2 (0.4)	4.1 (1.5)
Isometric Ratio[1,3]	2.7 (0.9)	4.8 (0.9)	5.8 (0.6)	4.5 (0.4)	2.8 (0.8)	4.1 (1.4)

1 The ratios are in terms of units of force at the fingertip.
2 Ratio of FDS tendon-to-tip force predicted from tendon force model (I: (7), II: DIP passive torque added, Dennerlein et al., 1998a).
3 Ratio of tendon-to-tip force measured during an isometric task reported in reference Dennerlein et al. (1998a).

The relationship between tendon and tip forces was described by four different multiplier (tendon-to-tip gain) parameters: the ratio of tendon-to-tip maximum forces, the ratio of tendon-to-tip mean forces during contact, the slope of a line fitted to the force data during the loading region, and the slope of a line fitted to the data during the unloading region. When the fingertip was in contact with the keycap, the relationship between the tension of the FDS tendon and the fingertip force was not simply proportional (Figure 25.3 and Figure 25.4). First, while the fingertip force patterns (Figure 25.3) contained the three phases observed by

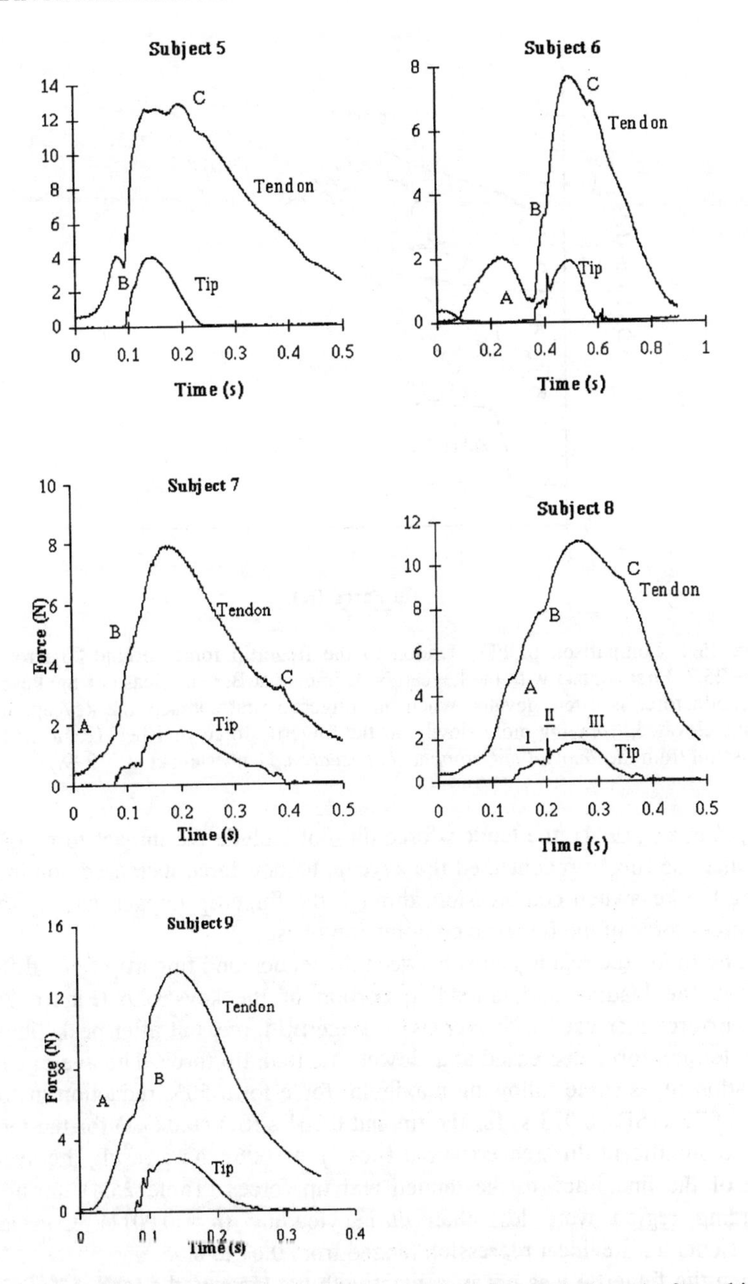

Figure 25.3. Flexor digitorum superficialis (FDS) tendon and fingertip force histories during a keystroke for the five participants. Abrupt changes in tendon force slope are denoted at Points A, B, and C, and correspond to a change in muscle contraction-state. (Reprinted with permission from the *Journal of Orthopaedic Research*, Dennerlein et al., 1999.)

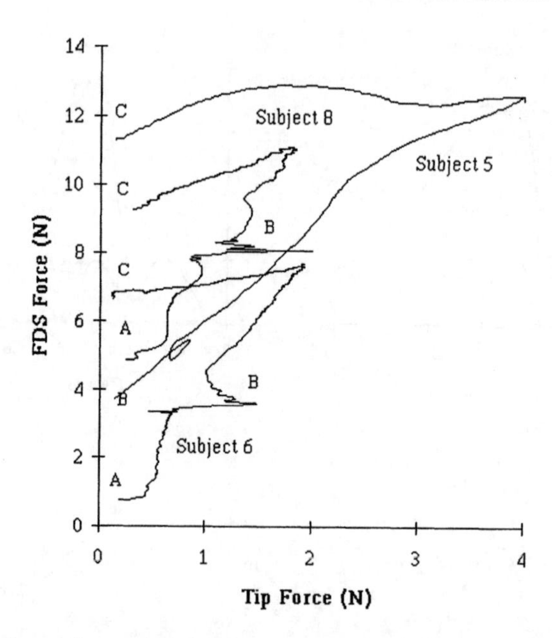

Figure 25.4. Comparison of FDS tendon to the fingertip force for the keystrokes of Figure 25.3. First contact with the keycap is A, impact at B, and release of the keycap at C. Tendon force is often elevated when the fingertip first contacts the keycap. It also remains elevated, decaying more slowly, as the fingertip force vanishes. (Reprinted with permission from the *Journal of Orthopaedic Research*, Dennerlein et al., 1999.)

Rempel et al. (1994), the tendon force did not include the impact force (Phase II). Once the fingertip contacted the keycap, tendon force increased continually during the keyswitch compression, through the fingertip impact, and up to the maximum force of the fingertip compression phase.

Furthermore, the relationships between the tendon and fingertip force differed between the loading and unloading portion of the keystroke (Figure 25.3). Tendon force increased with increasing fingertip force, but after peak fingertip force, tendon force decreased at a slower rate than tip force. The average half relaxation times (time following maximum force for a 50% reduction in force) were 0.071 s (SD=0.013 s) for the tip and 0.151 s (SD=0.023 s) for the tendon force, a significant difference (paired t-test: $p=0.006$). As a result, the average slope of the line fitted to the tendon and tip forces (Table 25.4) during the unloading region were less than during loading ($p = 0.0011$). Correlation coefficients for the linear regression ranged from 0.67 to 0.95.

When the fingertip was not in contact with the keycap, the tendon force was often elevated (Figure 25.3 and Figure 25.4). Tendon force increased before contact of the fingertip with the keycap. The tendon was preloaded at the point of contact. As contact ended, the tendon force was elevated approximately 57% of the maximum value averaged across subjects (Figure 25.4).

MODEL PREDICTIONS

While the postures varied widely between subjects and covered a large range of pinch scenarios, the predicted ratios from the implemented tendon force models varied little and were smaller than the measured values for the isometric pinch data (Figure 25.5). The most extended posture involved joint angles of 32°, 11°, and −15° for the MP, PIP, and DIP, and 72° for the angle between the distal phalanx and the applied tip force. The most flexed posture involved joint angles of 30°, 95°, 0°, and 60°. Joint angles ranged from 0° to 47° for the MP, from 11° to 95° for the PIP, and from −15° to 44° for the DIP, and the angle between the distal phalanx and the applied tip force ranged from 24° to 86°. The predicted (solution Method 3) tendon-to-tip force averaged across the three postures ranged from 0.7 to 1.9 (mean = 1.2, SD = 0.4) across all subjects. The within-subject variation ranged from 0.0 to 0.3. Table 25.3, C presents the results of the two other solution methods. The mean predicted ratio of 1.2 is less than half the value of the measured ratio, and the largest average predicted ratio of 1.9 is only slightly larger than the smallest measured ratio of 1.7. Furthermore, the subject-averaged predicted ratios poorly correlated ($r^2 \leq 0.04$) with the measured ratios (Figure 25.5).

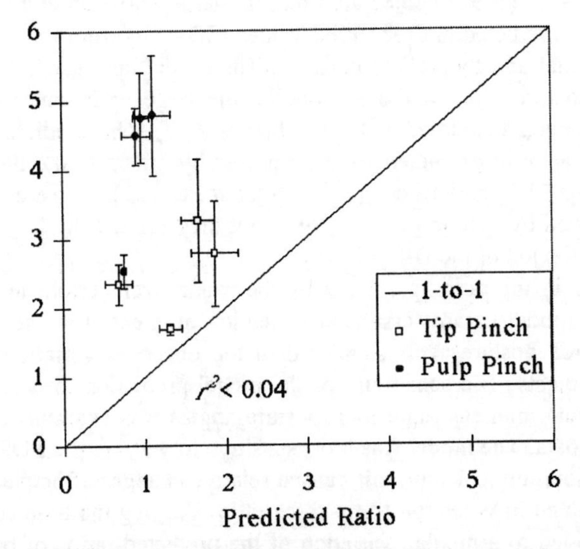

Figure 25.5. Mean measured FDS tendon-to-tip force ratios and the predicted isometric force ratios using solution Method 3. The error bars represent one standard deviation of the three measurements (vertical) and of the three predictions (horizontal) for the three postures. (Reprinted from Dennerlein et al., 1998a, with permission from Elsevier Science.)

For the dynamic activity of tapping on a keyswitch, the model estimates of the tendon-to-tip force ratio (Predicted Ratio I) were significantly ($p < 0.03$) less than the four multiplier or ratio parameters calculated from the tendon force histories (Table 25.4).

DISCUSSION

The direct measures of muscle force revealed the exposure of the tendons to force for both isometric and dynamic activities. The measurements also revealed aspects of the motor control of finger motion and provided a test bed to validate and discuss the limitations of inverse biomechanical models predicting muscle force.

REVEALING EXPOSURE THROUGH PREDICTIONS OF MUSCLE FORCE

Direct measurement of tension of a muscle provides a measure of forces within the human musculoskeletal system that have been predicted by isometric force models. These models make assumptions about the distribution of forces across the different muscles. By directly measuring the tension in the FDS tendon, it was observed that the FDS muscle force varied widely across subjects, more so than models predict. It was proposed that the force supported by different tendons varied across subjects and can be predicted based on the observation of extreme joint postures.

Schuind et al. (1992) conjectured that the large variation and the differences between the predicted and observed values (SD = 1.4) were due to a difference between actual and theoretical postures. The model here used a large range of observed postures; yet, it did not predict the large variations (SD = 0.4) seen between subjects (Table 25.3, C; and Figure 25.5). The predicted values were within the range of predicted values reported by others in the literature (Table 25.3, E). The FDS tendon-to-tip force ratios measured here were larger than the ratios reported by Schuind et al. (1992), but they limited the DIP posture to tip pinch with flexion of the DIP joint.

The error in the ratios predicted by the model from errors in estimation of anatomical model parameters, such as tendon moment arms, segment lengths, and measured postures was smaller than the observed variation of the ratios between subjects. The sensitivity of the model prediction to errors in the anatomical input parameters and joint posture angles was examined by perturbing the subject data. The model was most sensitive to varying the FDP moment arm at the DIP joint up to 1 mm; this caused relative changes of between 0.25 to 1.0 in the predicted FDS tendon-to-tip force ratio. Varying the moment arm for the other joints led to a median variation of the predicted ratios of between 0.1 to 0.6. Variation of the segment lengths and joint angles of 10% changed the predicted force ratio less than 0.4 and 0.3, respectively.

The measured difference between the FDS force ratios of the pulp and tip pinch postures results from the mechanics of the DIP joint in those postures. In the models, the DIP acts as an unrestrained hinge joint and so causes FDS force ratios to decrease in the pulp pinch posture. As the DIP extends, the angle of the applied load, and hence the mechanical advantage of the applied fingertip load about the DIP joint, increases. The tension of the FDP, the only flexor tendon at the DIP, increases to maintain equilibrium at the DIP joint. The increased FDP tension allows a decreased FDS tension required for equilibrium at the PIP joint. The model predicts this balance of FDP and FDS forces at equilibrium. Yet, the predicted decrease of FDS tension was not observed between the tip and pulp pinch subjects.

The observed *in vivo* FDS tension of the pulp pinch subjects was significantly higher than the tension measured in the tip pinch subjects. It was proposed that a passive joint constraint moment, produced by soft, connective tissues of the DIP joint, contributes to static equilibrium at the DIP joint near full extension, hence the assumption of an unrestrained hinge joint is unrealistic in this circumstance.

A new DIP constraint model, which selects one of two model structures based upon observed posture, was developed to explain the higher FDS tendon force. In the tip pinch model, the DIP is an unrestrained hinge joint where tendons alone maintain equilibrium. In the pulp pinch posture, the new model includes a constraint moment at the DIP joint that assists the FDP by supporting the moment of the load at the fingertip. The implementation of the constraint moment for the three solution methods is presented in Table 25.2. For Methods 1 and 2, the new model assumes that the FDP force vanishes, and the DIP joint is rigid (distal and middle phalanxes are a single rigid body). Method 3 adds a joint constraint moment at the DIP joint to the set of variables in Equation 25.3. The moment required is predicted by optimization.

The predicted tendon-to-tip force ratios using this new DIP joint constraint model (Table 25.3, D) correlated well ($r^2 > 0.62$) with the ratio of the measured tendon-to-tip forces (Figure 25.6). The variance of the subject data accounted for by this new model (r^2=85%) and the root mean square error (0.51) were best with Method 3. The new model predicts a large increase in FDS tendon force during pulp pinch, whereas the old model did not.

For the ratios obtained during the dynamic activity of tapping on a keyswitch, the data indicated that in predicting simple relationships, such as the ratio of maximum forces, the models underestimated the measured ratios. The estimates from the models were in the same range as those reported by Harding et al. (1993)—0.8 to 2.7 for five postures emulating the hand during piano playing. Some of this discrepancy may be attributed to the DIP hyper-extension (pulp pinch) posture selected during a keystroke, as proposed above. Adding a DIP constraint torque, which aids the FDP tendon to counteract the torque created by the tip force at the DIP joint, appears to accommodate the different strategies related to DIP pulp-pinch posture. Although the DIP constraint provides some improvement in model prediction, none of the models include the important bias of muscle pre-load or co-contraction, and thus they under-predict the ratios of the mean forces (mean error = 1.2).

As mentioned above, a limitation of the models is that they do not include the effects of co-contraction of antagonist muscles, and therefore, under-predict tendon force ratios. The mean difference between the measured and predicted force ratios for the pulp pinch postures is 0.07, yet in the tip pinch postures, the

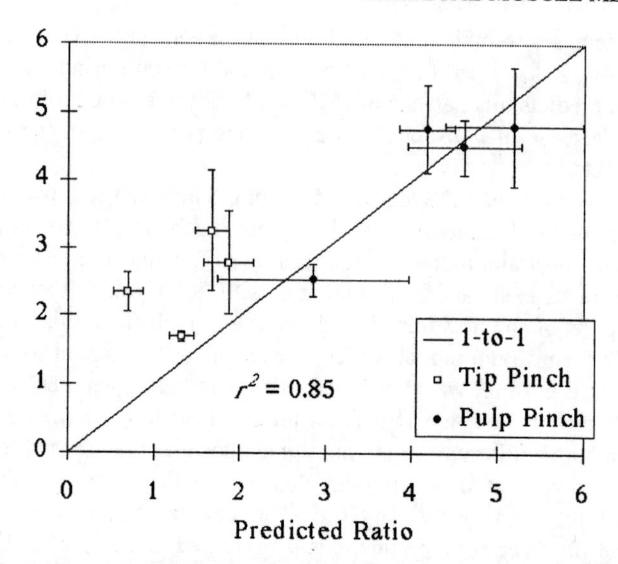

Figure 25.6. Mean measured FDS tendon-to-tip force ratios and the predicted isometric force ratios from the new DIP constraint model using solution Method 3. The error bars represent one standard deviation of the three measurements (vertical) and of the three predictions (horizontal) for the three postures. (Reprinted from Dennerlein et al., 1998a, with permission from Elsevier Science.)

model under-predicts the measured ratio by approximately 1 (Figure 25.6). While model sensitivity may explain some of this difference, antagonist co-contraction may exist. The tip pinch posture requires active stabilization of the DIP joint by co-contraction, whereas the pulp pinch posture can rely on the passive joint constraint to stabilize the joint. Therefore, increased co-contraction is expected in the tip pinch postures.

Debate exists over which cost function should be used in the optimization routine that provides a solution to the statically indeterminate system. Pedotti et al. (1978) evaluated several different methodologies for the lower extremity during human locomotion. Pedotti et al. compared the predictions with EMG recordings during gait, and concluded that little difference existed between the different predictions and the measured activity. Similarly, the three different solution methods for the indeterminate problem here did not yield significantly different predictions for tendon force (Table 25.3). The question arises: which function or method is best? For presentation purposes, Method 3 was chosen because it provided the highest correlation and the least error. These models, as discussed, often do not include co-contraction but there is usually only conjecture on the existence of co-contraction to stabilize the joints during a task. Questions, therefore, remain on how to quantify co-contraction and predict it based on some mechanism, such as joint stability. Valero-Cuevas et al. (1998) discuss the existence of interossei activity during lateral pinch in order to stabilize the joint, but again the predictions do not match the *in vivo* EMG data.

REVEALING DYNAMIC FUNCTION WHILE TAPPING ON A KEYSWITCH

In some settings, the relationship between muscle force and externally measured forces is assumed to be proportional, for example, Harding et al. (1993). The *in vivo* measurement of forces shows that for tapping on a keyswitch, the relationship was *not* proportional. Most inverse muscle force models assume an inverse causal relationship—that is, the applied load at the fingertip creates the muscle-tendon tension. For isometric models, this assumption is acceptable because the relationship is proportional (Dennerlein et al, 1998a) and causality is a moot issue. However, the true causality is reversed. The muscle, with the effects of dynamics and the forces of several muscles superimposed, creates the tip force. The result is that the force histories of the flexor digitorum super-ficialis (FDS) during a keystroke are a complicated function of fingertip loading and muscle mechanics associated with different contraction states.

Change in the muscle contraction states coincides with abrupt transitions in the slope of the tendon force with respect to time at Points A, B, and C in Figure 25.3. Just prior to fingertip-keycap contact, the muscle shortens as the finger flexes—an unloaded concentric contraction. When the fingertip first contacts the keycap (Point A), the inertia of the keycap and the force of the keyswitch spring are added to the fingertip, increasing the load of the concentric contraction. When the fingertip stops moving at the end of keyswitch travel (Point B), an isometric contraction begins and the tendon force continues to increase. As the release motion begins (Point C), the isometric contraction ends and the flexor muscle lengthens, initiating an eccentric contraction.

The properties of the muscle mechanics may contribute to the tendon force remaining elevated for some time after the tip force vanishes. The speed associated with the co-ordinated movement of a keystroke requires that the force applied by the fingertip is removed quickly. However, the decay of the FDS muscle force is set by muscle physiology and is relatively slow, on the order of 0.1 seconds (Lieber, 1992). EMG studies of typing indicate that the beginning of extensor activity overlaps with the end of the flexor activity (Dennerlein et al, 1998b; Weiss et al., 1996). The flexor muscle contraction is near the maximum value as the release movement begins and its contraction is eccentric. This type of contraction produces higher forces than an isometric or shortening muscle (Joyce et al., 1969). Superimposed on the residual eccentric contraction is the muscle's passive force. As the fingertip moves away from the keycap, the flexor muscle lengthens past its free length, and passive muscle force increases.

REVEALING MOTOR CONTROL FUNCTION

During the isometric pinch, tendon force was directly proportional to the force applied by the fingertip, as expected, because the mechanical advantage of the tendon to the tip remains constant. The first order linearity also suggests, however, that recruitment of synergist muscles, like the flexor digitorum profundus, follows the flexor digitorum superficialis for these submaximal voluntary efforts. If the recruitment did not follow the FDS, then a more complicated relationship would exist.

EMG studies of the extrinsic muscles of the finger during touch-typing (Dennerlein et al, 1998b) indicate that the flexor digitorum superficialis is not actively producing muscle force until the fingertip first contacts the keycap. However, for Subject 6, whose movement pattern followed that of touch-typing, muscle force was present prior to contact between the finger and the key. In fact, there was a large increase and then decrease of muscle force before contact. Passive muscle force may explain this observed increase and decrease of FDS force prior to contact with the keycap. Figure 25.7 shows the position of the fingertip relative to the keycap, as observed from the video system for five keystrokes just prior to contact. The muscle force followed a repeatable pattern with respect to fingertip position prior to contact. The data displayed the same shape and magnitude as the muscle's passive force properties, which were measured on Subjects 8 and 9 (Dennerlein et al., 1999). Although this passive muscle force prior to the keystroke was observed with only one subject, the data of this *in vivo* muscle force and previous EMG studies suggest that passive force is present during the motion of a keystroke.

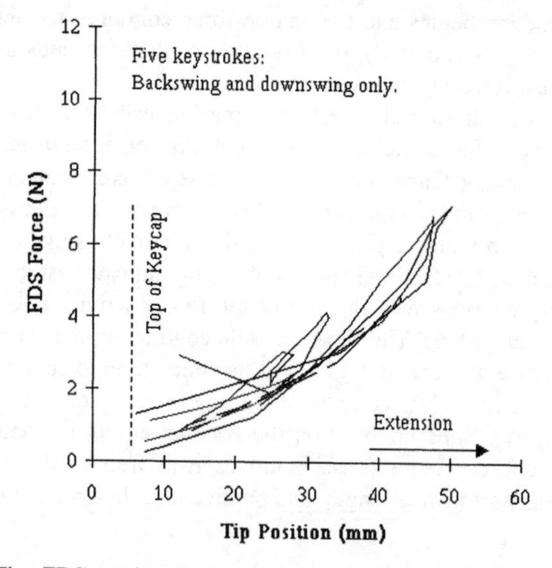

Figure 25.7. The FDS tendon forces with respect to fingertip position during the backswing and downswing for five keystrokes of Subject 6. Comparing tendon force with fingertip position and the repeatable pattern suggests that the forces during the backswing and downswing are passive force-length properties of the FDS muscle. (Reprinted with permission from the *Journal of Orthopaedic Research*, Dennerlein et al., 1999.)

LIMITATIONS OF *IN VIVO* MEASUREMENTS

The present study has several limitations. First, only one of the six muscles of the long finger was measured in this study, providing a limited view of all the internal forces. The forces in the other finger muscles, specifically the flexor digitorum profundus (FDP), remain unknown. Ideally, simultaneous measurement of both muscle forces would elucidate how the FDS and FDP work passively and actively together during a keystroke. Second, tapping on a single keyswitch while prone on one's back in a surgical setting is a different task than touch-typing. The posture of the upper extremity and the slight sedation of the subjects during surgery may have compromised the motor performance during a keystroke. Indeed, the duration of the keystroke was longer, contact velocities were slower, and the motion patterns, except for one subject, were different than touch-typing (Dennerlein et al., 1998b; Rempel et al., 1994). For the pinch procedures, these factors may have contributed to the observed hyper-extension of the DIP joint in the pulp pinch posture, however, it does occur during work tasks such as touch-typing (Dennerlein et al., 1993) and writing. Third, the subjects had carpal tunnel syndrome, and their associated sensory deficiencies may have affected the motor control of the keystroke. These limitations suggest that the reported maximum and mean tendon forces and the observed fingertip forces may be different than what occurs during touch-typing. However, the conclusions about the ratios and the four parameters relating fingertip force and tendon force rely on the mechanics of the musculoskeletal system, rather than the motor-control, and therefore remain valid.

FUTURE WORK

The discussion identifies three directions for future work. First, more transducers on more tendons would address the limitation of the single force measurement and the conclusions about the complete distribution of forces on other muscles. The next step would be to measure additionally the force of the flexor digitorum profundus (FDP) to validate the predicted lower force in the FDP for the pulp pinch postures. The physical limitations and invasiveness of the procedures, though, make observations of all the muscles involved difficult. Other experimental improvements would include combining other model validation techniques, such as electromyography with the *in vivo* measurement techniques. Such work would allow for more robust conclusions further validating the hypotheses proposed here.

In terms of predicting muscle force, there is an obvious need to include the effects of co-contraction in the solution methodologies. The question, though, is how? Joint stability is often proposed as a mechanism or motivation for co-contraction, but what metric does one use to describe stability? And then, what motor control strategies are used to achieve this joint stability? Mechanical stability is a complex issue, but with so much conjecture surrounding co-

contraction, it could be a significant contribution to the musculoskeletal modelling community.

A third point for future directions relates to the causality of the muscle force models. As discussed, models often work their way from the outside, predicting muscle force based on the external measurements. The true causality is reverse—it is the muscle force that is creating the external load. This becomes evident for dynamic activities. Forward-type models (Zajac, 1993) and muscle models can be developed to help model the dynamic activity of the keystroke or other types of impact activities.

Finally, one note with regard to understanding tendon injury mechanism: *in vivo* force measurements provide a better understanding of the exposure-dosage relationship (that is, the relationship between external force, exposure, and the loads on the internal tissues), specifically for those known to be at risk for injury. However, the dosage-tissue response is not addressed by *in vivo* studies. A next logical step would be to build a basis of knowledge for the tissue response in a physiological sense to the internal loads measured here or predicted by the models. This step is necessary for improvements in treatment, rehabilitation, and prevention programs that work with musculoskeletal disorders.

CONCLUSIONS

- The findings of this *in vivo* study of muscle force provided a view inside the musculoskeletal system during isometric pinch and tapping on a keyswitch that, to date, has only been conjectured.
- Muscle forces of the flexor digitorum superficialis were higher than estimates by models for isometric pinch, because these models did not include passive properties of the joints.
- The flexor digitorum superficialis (FDS) tendon force histories during tapping on a keyswitch are a complicated function of fingertip loading, fingertip motion, and muscle contraction-state. FDS tendon force remained elevated after the force at the fingertip vanished.
- The measurements revealed that passive muscle force might play a role in the control of the finger during a keystroke.

ACKNOWLEDGMENTS

The author would like to acknowledge the patients and staff of the University of California, San Francisco Ambulatory Care Surgery Center. Gratitude is also expressed to the staff of the UCSF Ergonomics Program for its support and cooperation. Special thanks go out to Drs. David Rempel, Steve Lehman, Edward Diao, and C.D. Mote, Jr., who were the author's mentors and collaborators throughout these projects during his doctoral dissertation at the University of California, Berkeley.

REFERENCES

An, K.N., Berglung, L., Cooney, W.P., Chao, E.Y.S., and Kovacevic, N. (1990). Direct *in vivo* tendon force measurement system. *Journal of Biomechanics* **23**(12), 1269–1271.

An K.N., Kwak B.M., Chao E.Y., and Morrey B.F. (1984). Determination of muscle and joint forces: A new technique to solve the indeterminate problem. *Journal of Biomechanical Engineering* **106**, 364–367.

An, K.N., Ueba, Y., Chao, E.Y., Cooney, W.P. III, and Linscheid, R.L. (1983). Tendon excursion and moment arm of index finger muscles. *Journal of Biomechanics* **16**(6), 419–425.

Armstrong, T.J., and Chaffin D.B. (1978). An investigation of the relationship between displacements of the finger and wrist joints and the extrinsic finger flexor tendons. *Journal of Biomechanics* **11**, 119–128.

Armstrong, T.J., Fine, L.J., Goldstein, S.A., Lifshitz, Y.R., and Silberstein, B.A. (1987). Ergonomics considerations in hand and wrist tendonitis. *Journal of Hand Surgery* **12A**, 830–837.

Barnes, G.R.G., and Pinder D.N. (1974). *In vivo* tendon tension and bone strain measurement and correlation. *Journal of Biomechanics* **7**, 35–42.

Buchholz, B., and Armstrong, T.J. (1992). A kinematic model of the human hand to evaluate its pretensile capabilities. *Journal of Biomechanics* **25**(2) 149–162.

Bureau of Labor Statistics (BLS) (1995). Reports on survey of occupational injuries and illnesses in 1977–1994. Washington, DC: Bureau of Labor Statistics, U.S. Department of Labor.

Chao, E.Y., An, K.N., Cooney, W.P., and Linscheid, R.L. (1989). *Biomechanics of the Hand: A Basic Research Study*. World Scientific, Singapore.

Dennerlein, J.T., Diao, E., Mote, C.D. Jr., and Rempel, D. (1998a). Tensions of the flexor digitorum superficialis are higher than a current model predicts. *Journal of Biomechanics* **31**(4), 295–301.

Dennerlein, J.T., Diao, E., Mote, C.D. Jr., and Rempel, D. (1999). *In vivo* finger flexor tendon forces while tapping on a keyswitch. *Journal of Orthopaedic Research* **17**(2) 178–184.

Dennerlein, J.T., Miller, J., Mote, C.D. Jr., and Rempel, D. (1997). A low profile human tendon force transducer, the influence of tendon thickness on calibration. *Journal of Biomechanics* **30**(4), 395–398.

Dennerlein, J.T., Mote, C.D. Jr., and Rempel, D. (1998b). Control strategies for finger movement during touch-typing: The role of the extrinsic muscles during a keystroke. *Experimental Brain Research* **121**, 1–6.

Dennerlein, J.T., Serina, E.R., Mote, C.D. Jr., and Rempel, D. (1993). Fingertip kinematics and forces during typing. In *Proceedings of 17th American Society of Biomechanics*. Iowa City, Iowa.

Garrett, J.W. (1970a). Anthropometry of the hands of female air force flight personnel (Report AMRL-TR-69-26). Wright-Patterson Air Force Base, Ohio, Aerospace Medical Research Laboratory, Aerospace Medical Division, Air Force Systems Command.

Garrett, J.W. (1970b). Anthropometry of the hands of male air force flight personnel (Report AMRL-TR-69-42). Wright-Patterson Air Force Base, Ohio, Aerospace Medical Research Laboratory, Aerospace Medical Division, Air Force Systems Command.

Gregor, R.J., Roy, R.R., Whiting, W.C., Lovely, R.G., Hodgson, J.A., and Edgerton, V.R. (1988). Mechanical output of the cat soleus during treadmill locomotion, *in vivo* vs. *in situ* characteristics. *Journal of Biomechanics* **21**(9), 721–732.

Harding, D.D., Brandt, K.D., and Hillberry, B.M. (1993). Finger joint force minimization in pianists using optimization techniques. *Journal of Biomechanics* **26**(12), 1403–1412.

Johnson, P.W., Dennerlein, J.T., Armstrong, T.A., and Rempel, D.M. (1995). A graphical computer model simulating forces in the extrinsic finger flexor tendons during static work. In *Proceedings of the 5th International Symposium on Computer Simulations in Biomechanics*. Jyväskylä, Finland.

Johnson, P.W., Tal, R., Smutz, W.P., and Rempel, D.M. (1993). Computer mouse designed to measure finger forces during operation. In *Proceedings of the IEEE EMBS*. San Diego, California.

Joyce, G.D., Rack, P.M.H., and Westbury, D.R. (1969). The mechanical properties of cat soleus muscle during controlled lengthening and shortening movements. *Journal of Physiology* **204**, 461–474.

Komi, P.V., 1990. Relevance of *in vivo* force measurements to human biomechanics. *Journal of Biomechanics* **23**(1), 23–34.

Komi, P.V., Salonen, M., Jarvinen, M., and Kokko, O. (1987). *In vivo* registration of Achilles tendon forces in man, I. Methodological development. *International Journal of Sports Medicine* **8**, 3–8.

Lieber, R.L. (1992). *Skeletal Muscle Structure and Function, Implications for Rehabilitation and Sports Medicine*. Williams & Wilkins, Baltimore.

Long, C., Brown, M.E., and Weiss, G. (1960). An electromyographic study of the extrinsic-intrinsic kinesiology of the hand, preliminary report. *Archives of Physical Medicine and Rehabilitation* **41**, 175–181.

Miller, J.M., and Robins, D. (1992). Extraocular muscle forces in alert monkey. *Vision Research* **32**(6), 1099–1113.

Moore, J.S., and Garg, A. (1994). Upper extremity disorders in a pork processing plant, relationships between job risk factors and morbidity. *American Industrial Hygiene Association Journal* **55**(8), 703–715.

Pedotti, A., Krishnan, V.V., and Stark, L. (1978). Optimization of muscle-force sequencing in human locomotion. *Mathematical Biosciences* **38**, 57–76.

Rempel, D., Dennerlein, J., Mote, C.D. Jr., and Armstrong, T. (1994). A method of measuring fingertip loading during keyboard use. *Journal of Biomechanics* **27**(8), 1101–1104.

Schuind, F., Garcia-Elias, M., Cooney, W.P., and An, K.N. (1992). Flexor tendon forces, *in vivo* measurements. *Journal of Hand Surgery* **17A**(2), 291–298.

Sherif, M.H., Gregor, R.J., Liu, L.M., Roy, R.R., and Hager, C.L. (1983). Correlation of myoelectric activity and muscle force during selected cat treadmill locomotion. *Journal of Biomechanics* **16**(9), 691–701.

Silverstein, B.A., Fine, L.J., and Armstrong, T.J. (1986). Hand wrist cumulative trauma disorders in industry. *British Journal of Industrial Medicine* **43**, 779–784.

Smith, E.M., Juvinall, R.C., Bender, L.F., and Pearson, J.R. (1964). Role of the finger flexors in rheumatoid deformities of the metacarpophalangeal joints. *Arthritis and Rheumatism* **7**, 467–480.

Smutz, P., Serina, E., and Rempel, D. (1994). A system for evaluating the effect of keyboard design on force, posture, comfort, and productivity. *Ergonomics* **37**(10), 1649–1660.

Valero-Cuevas, F.J., Zajac, F.E., and Burgar, C.G. (1998). Large index-fingertip forces are produced by subject-independent patterns of muscle excitation. *Journal of Biomechanics* **31**, 693–703.

Weightman, B., and Amis, A.A. (1982). Finger joint force predictions related to design of joint replacements. *Journal of Biomedical Engineering* **4**, 197–205.

Weiss, J., Helmut, K., and Thomas, L. (1996). Repetitive finger movements and antagonistic muscle activity. *Advances in Occupational Ergonomics and Safety I* **2**, 508–510.

Whiting, W.C., Gregor, R.J., Roy, R.R., and Edgerton, V.R. (1984). A technique for estimating mechanical work of individual muscles in the cat during treadmill locomotion. *Journal of Biomechanics* **17**(9), 685–695.

Zajac, F.E. (1993). Muscle coordination of movement, a perspective. *Journal of Biomechanics* **26** (Supplement 1), 109–124.

IV *In Vivo* Muscle Function (Animals)

26 Sarcomere Length Non-uniformities and Stability on the Descending Limb of the Force-length Relation of Mouse Skeletal Muscle

T. L. ALLINGER
Department of Mechanical Engineering, University of Calgary, Calgary, Alberta, Canada

W. HERZOG
Human Performance Laboratory, University of Calgary, Calgary, Alberta, Canada

H. E. D. J. TER KEURS
Faculty of Medicine, University of Calgary, Calgary, Alberta, Canada

M. EPSTEIN
Department of Mechanical Engineering, University of Calgary, Calgary, Alberta, Canada

INTRODUCTION

Materials with softening properties are unstable. This is because the force-elongation relation for a softening material has a negative slope. Here, stability is defined to exist if the potential energy function has a minimum. In muscle, the descending limb of the force-length (F-L) relation has a negative slope under isometric conditions and maximal stimulation (Gordon et al., 1966b). As a result, the potential energy function does not have a minimum for lengths on the descending limb. Because of this, sarcomeres are assumed to have unstable length behaviour on the descending limb of the F-L relation.

Hill (1953) was the first to note that sarcomeres may be unstable on the descending limb of the F-L relation, because of the negative slope of this part of the F-L relation. The argument for instability follows this reasoning. Muscle fibres are composed of a large number of sarcomeres in series. An inequality in force production between sarcomeres occurs because of differences in the initial sarcomere length, differences in sarcomere cross-sectional area, and/or differences in sarcomere activation (Edman and Reggiani, 1984). Any difference in

Skeletal Muscle Mechanics: From Mechanisms to Function. Edited by W. Herzog.

force production by the sarcomeres results in some sarcomeres shortening and some sarcomeres lengthening. If the sarcomeres in a fibre follow the F-L relation and are operating at lengths on the descending limb of the F-L relation, unstable behaviour will result (Hill, 1953): the weak (long) sarcomeres are stretched by the strong (short) sarcomeres (open circles, Figure 26.1). Thus, for fibres operating on the descending limb of the F-L relation, some sarcomeres will lengthen and some sarcomeres will shorten during a contraction until they reach a stable position (positively sloped F-L relation). Thus, these sarcomere length changes should continue until some sarcomeres are on the resting and others on the ascending limb of the F-L relation (Figure 26.1).

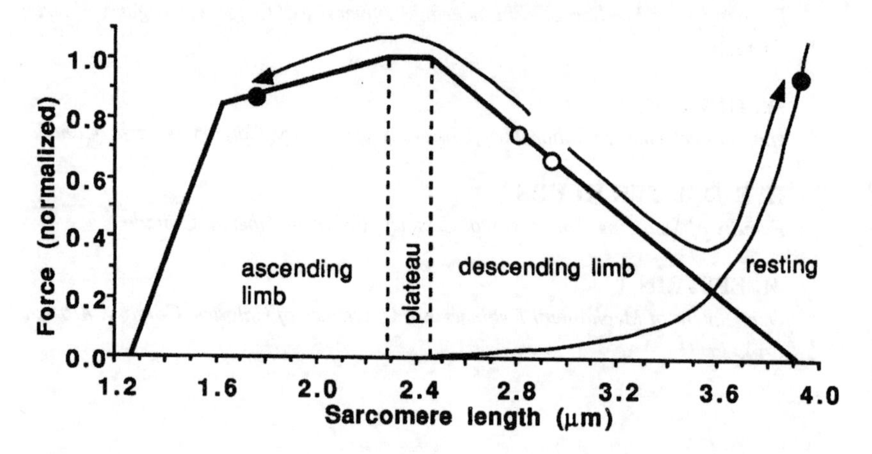

Figure 26.1. Theoretical force-length relation for mouse skeletal muscle and the behaviour of unstable sarcomeres during a fixed-end contraction of a muscle fibre. On the descending limb, sarcomeres that differ in length and force at rest (open circles) will change non-uniformly in length during contraction. Short (strong) sarcomeres will shorten onto the ascending limb, while long (weak) sarcomeres will lengthen onto the resting limb of the F-L relation.

Instability of sarcomere length on the descending limb of the F-L relation has been associated with sarcomere length non-uniformity. Force 'creep' (slow rise in force) is believed to occur because of non-uniform changes in sarcomere length in a fibre during a fixed-end contraction (Huxley and Peachey, 1961; Gordon et al., 1966a; Julian et al., 1978; Julian and Morgan, 1979a, 1979b; Lieber and Baskin, 1983; Edman and Reggiani, 1984). These slow changes in sarcomere length during a contraction motivated Gordon, Huxley, and Julian (1966a) to devise the 'segment clamp' method in order to reduce or eliminate sarcomere lengthening (central region of frog single fibres) during an isometric contraction. Non-uniform changes in sarcomere length along the length of a fibre were attributed to unstable F-L properties on the descending limb of the F-L relation (Hill, 1953; Huxley and Peachey, 1961; Gordon et al., 1966b;

Julian et al., 1978; Julian and Morgan, 1979a; Lieber and Baskin, 1983; Edman and Reggiani, 1984).

Theoretical models using sarcomeres with unstable F-L properties have also spawned interesting results. Fibres modelled as a number of sarcomeres in series illustrated how force creep is associated with non-uniform changes in sarcomere length (Morgan et al., 1982; Edman and Reggiani, 1984). Similar models appear to demonstrate non-uniform changes in sarcomere length that result in the force-depression phenomenon (Edman et al., 1993). Morgan (1990) predicted that fibres undergoing a stretch possess sarcomeres that 'pop' or are rapidly stretched onto the resting limb of the F-L relation. This model demonstrated how non-uniform changes in sarcomere length could explain the force enhancement after stretch (i.e., increased isometric force following stretch). Although non-uniform changes in sarcomere length may explain many phenomena, the hypothesis of instability is not proven.

The idea that sarcomeres are unstable over a large portion of their operating range is not appealing intuitively (Allinger et al., 1996a). Instability would result in:

1. gross differences in sarcomere lengths along a fibre,
2. sarcomere extension being limited only by passive structures where force production cannot be regulated,
3. fibres that are incapable of regulating their length over extended contractions,
4. large sarcomere length changes resulting in increased cross-bridge cycling (ATP splitting) and increased energy requirements, compared to a more stable behaviour, and
5. unstable torque-angle relations at joints caused by the unstable muscle properties, leading to problems in movement control.

In fact, stability may be an inherent property of the active force-producing process within a sarcomere, and the instability theory may be a misinterpretation of the descending limb of the F-L relation. Some investigators have suggested that sarcomeres in muscle fibres are stable on the descending limb of the F-L relation, but they did not provide rigorous theoretical or experimental evidence for their statements (Deleze, 1961; Hill, 1977; ter Keurs et al., 1978; Pollack, 1983, 1990). The F-L relation is derived from separate contractions. However, during a single contraction, the force rises during elongation and decreases during shortening. Thus, in one contraction the force-elongation curve has a positive slope and could be considered stable. Recently it has been demonstrated analytically that a sarcomere with mechanical properties, which have been observed experimentally, can be stable and yet possess a negatively sloped F-L region (Allinger et al., 1996a). One indication of stable sarcomere length behaviour has not been investigated: the shape of the sarcomere length-time traces.

The purpose of this study was to measure the non-uniformities in sarcomere length in mouse skeletal muscle, and to determine if sarcomeres exhibit stable length behaviour at lengths on the descending limb of the F-L relation during fixed-end contractions. Although sarcomere length non-uniformities have been measured in many studies using single-fibre preparations of frog skeletal muscle, little work is available involving the sarcomere length behaviour in whole-muscle preparations (amphibian skeletal muscle studies, ter Keurs et al., 1978; Paolini and Roos, 1975; Paolini et al., 1976), and, in particular, mammalian skeletal muscle. It is not known if the non-uniform sarcomere length behaviour observed in single fibres also occurs in whole muscle preparations.

METHODS

Sarcomere dynamics were measured on the serratus anterior muscle of the mouse, *in vitro*. Laser diffraction was used to measure sarcomere length at different locations on the muscle during fixed-end contractions. These sarcomere length measures were used to evaluate sarcomere length non-uniformity and stability.

PREPARATION

Twelve muscle preparations were used from the distal head of the mouse serratus anterior (14.9±4.0 g body weight). The distal head of the serratus anterior originates on the vertebral border of the scapula and inserts into rib 7 (Figure 26.2). The fibres in this preparation are fusiform and run from the scapula to the rib with no visible tendon. Little series elasticity existed in the preparation because of the lack of tendon and rigid bony attachment sites. The muscle preparation forms a flat band of fibres 11.1±1.2 mm long (scapula to rib), 0.7±0.2 mm wide, and 0.3±0.1 mm thick.

EXPERIMENTAL APPARATUS

Muscle preparations were mounted in a Plexiglas experimental chamber (Figure 26.3). The rib was attached to a force transducer (strain gauge) and the scapula was attached to a motor. An oxygenated, physiological solution flowed through the experimental chamber (rate 6 ml/min) at room temperature (20°C). The muscle was stimulated through platinum electrodes lying parallel to the muscle fibres in the bath. Tetanic stimulation was achieved using a square wave pulse of 0.5 ms duration at 200 Hz, and a train duration of 250 or 300 ms at a supramaximal voltage. Sarcomere length was measured using a laser diffraction system with a resolution of 5 nm (ter Keurs et al., 1978).

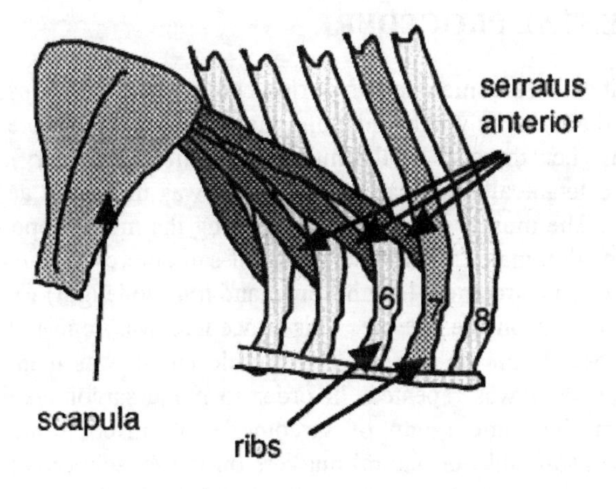

Figure 26.2. Diagram of the serratus anterior muscle originating on the scapula and inserting into the ribs. The muscle bundle attached to rib 7 was dissected away from the other bundles.

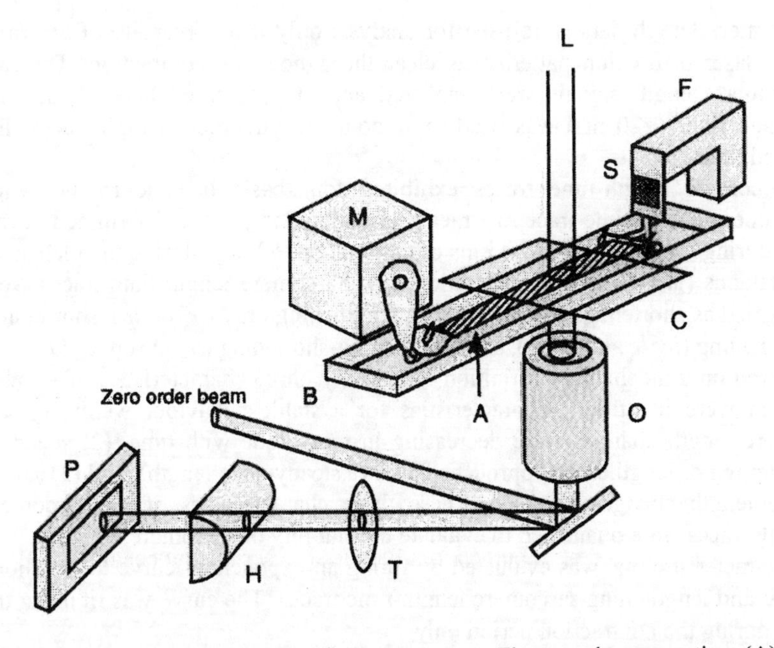

Figure 26.3. Schematic diagram of test apparatus. The muscle preparation (A) was connected to the motor arm (M) and a force transducer (S). Solution flowed past the muscle in a Plexiglas chamber (B) covered by a cover slip (C). Sarcomere length was measured using a laser beam (L) directed through the muscle. The first-order diffraction pattern was focused (O, T) and concentrated (H) by lenses onto a photosensitive diode array (P).

EXPERIMENTAL PROCEDURE

Non-uniformities in sarcomere length during contraction on the descending limb were measured across the width and along the length of the muscle under fixed-end conditions (i.e., the ends of the muscle were held stationary). The muscle was stimulated tetanically and sarcomere length was measured at one location on the muscle. The muscle was then moved using the microscope stage so that the laser beam illuminated a different group of sarcomeres. Tetanic stimulation and data collection (sarcomere length, force, and muscle length) were performed at this new location on the muscle. This procedure was repeated on as many locations as possible on the muscle. The muscle length was then changed and the above procedure was repeated. In order to make sarcomere length measurements from the same group of sarcomeres at different muscle lengths, surface markers (6-0 silk) or natural markers on the muscle (connective tissue) were used to reposition the laser beam. A total of 143 contractions were collected at 26 locations along the length of 12 muscles.

DATA ANALYSIS

Sarcomere length data were used for analysis only if the intensity of the first-order laser diffraction pattern was clear throughout the contraction. The raw sarcomere length signals were analyzed and are presented here. A moving average filter (±20 ms) was used to smooth the sarcomere length traces displayed in the results.

Sarcomere length-time traces exhibited four basic behaviours. For each sarcomere length-time trace, the mean sarcomere length was determined at rest (lr), during early tetanus (25-45 ms of contraction, le), and during the plateau of the tetanus (last 40 ms of contraction, lp). Sarcomere length-time traces were classified as shortening only ($lr>le>lp$), lengthening only ($lr<le<lp$), shortening-lengthening ($lr>le$ and $le<lp$), and lengthening-shortening ($lr<le$ and $le>lp$).

Based on a mechanical definition of stability, three characteristics of a stable system were identified. Characteristics for a stable behaviour were: (1) sarcomere length changes were decreasing in magnitude with time, (2) speed of shortening or lengthening approached zero (a steady-state length), and (3) sarcomere length changes reversed. These three characteristics of the sarcomere length traces were quantified to evaluate the stability of sarcomere lengths.

Characteristic one was evaluated by fitting an exponential curve to the shortening and lengthening sarcomere length-time traces. The curve was fit using the data during the contraction period only,

$$SL(t) = a + b\,e^{-c\,t} \tag{26.1}$$

where: a, b, c = constants. The steady-state sarcomere length was represented by the constant a (μm), and the time constant for the decaying portion of the function was determined as $1/c$ (s). The Levenberg-Marquardt method was used to fit the exponential function to the data.

Characteristic two was evaluated by calculating the velocity of the sarcomere length data at the end of contraction. A straight line was fit to the data during the tetanic plateau,

$$SL(t) = u + v\,t \tag{26.2}$$

where: SL=sarcomere length (μm), v=sarcomere length velocity (μm/s), t= time (s), u and v = constants. The chi-square merit function was used to explicitly solve for the constants u and v.

Characteristic three was evaluated by determining the number of sarcomere length traces showing a reversal in length change. The number of traces that demonstrated a shortening-to-lengthening or lengthening-to-shortening behaviour was counted. More details about the experiments can be found in Allinger (1995).

EXPERIMENTAL RESULTS

The experimental results were used to evaluate the non-uniform changes in sarcomere length during contraction. In addition, three criteria were used to evaluate whether sarcomere length is stable on the descending limb of the force-length relation.

SARCOMERE LENGTH NON-UNIFORMITIES

Non-uniform sarcomere length behaviour was observed along the length of the muscles; some sarcomeres shortened while others lengthened (Figure 26.4). For example, in muscle C, the difference in sarcomere length between two groups of sarcomeres increased about 10X, from 0.02 μm (rest) to 0.2 μm (end of contraction). This increase in sarcomere non-uniformity from rest to the tetanic plateau was observed in all muscles where more than one group of sarcomeres was measured along the length of the muscle.

Based on the four classifications of sarcomere length, muscles B and C (Figure 26.4) illustrated lengthening (a24, a7) and shortening-lengthening (a23, a8) sarcomere-length behaviour in the two regions shown. Muscle A had shortening (a23) and shortening-lengthening (a22) curves in the two regions shown, and muscle D contained a sarcomere length trace (a20) of the lengthening-shortening type in the region of the muscle illuminated by the laser. A total of 143 measurements in 26 regions of 12 muscles were made at different muscle lengths along the length of the muscles. Of the 143 sarcomere length traces collected, 31 were of the shortening, 60 of the lengthening, 49 of the shortening-lengthening, and three of the lengthening-shortening type. These results do not necessarily represent the proportion of sarcomeres that follow these four types of sarcomere length-time behaviour in the serratus anterior of the mouse, because sarcomere length measurements could be made only from selected regions of the muscle.

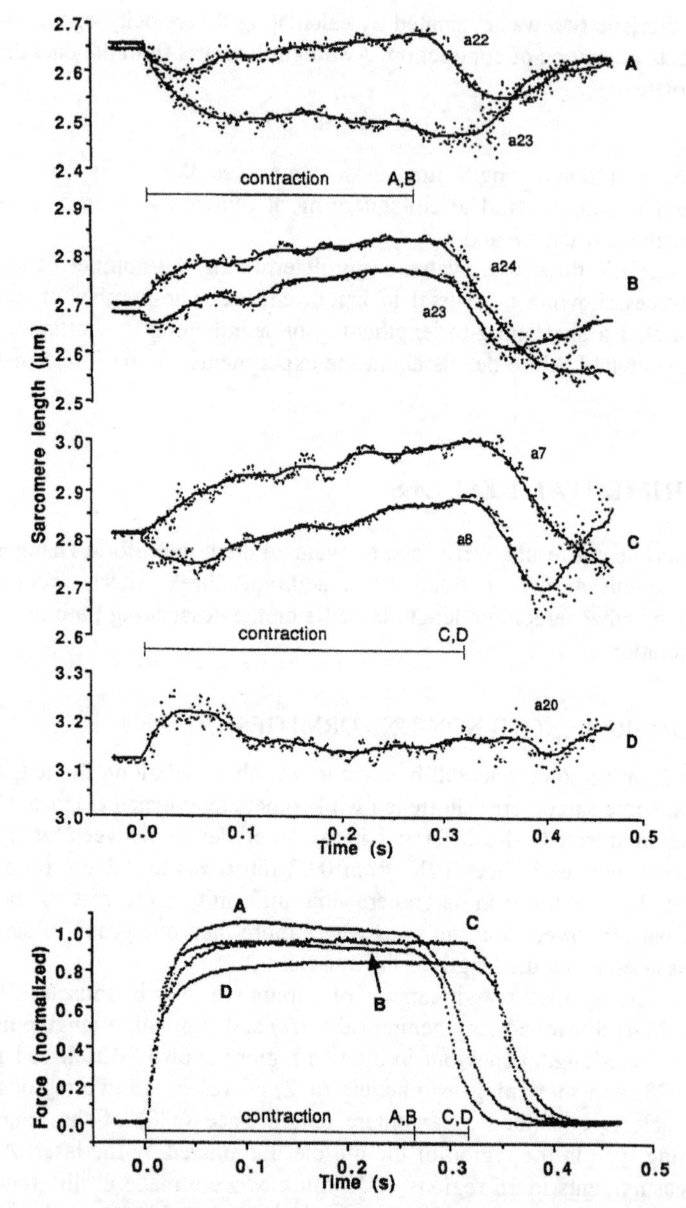

Figure 26.4. Typical sarcomere length-time traces at different locations along the length of four muscles. Traces for a given muscle are at one muscle length. Raw data were represented by dots and smoothed data by the solid lines. Force-time traces for the corresponding muscles are shown. Forces were normalized to the peak force during the plateau at optimal length. Muscles A, B, C, and D had plateau forces at optimal lengths of 86, 75, 490, and 100 mN/mm^2, and cross-sectional areas of 0.18, 0.08, 0.32, and 0.35 mm^2, respectively.

STABILITY TESTS

All of the 143 sarcomere length-time traces collected from the 12 muscles in this study demonstrated stable sarcomere length behaviour by satisfying at least two of the stability criteria. All three characteristics of the stability criteria were met by 52 of the sarcomere length-time traces.

First stability criteria

All of the shortening and lengthening types of sarcomere length-time traces exhibited a decrease in the rate of change during the contraction (first stability criteria). Upon contraction, sarcomere lengths changed rapidly during force development and then decreased in speed as the contraction progressed (Figure 26.5). The exponential curve-fitting routine converged successfully for 42 of the 60 lengthening and 15 of the 31 shortening types of sarcomere length traces. The root mean variance between the actual data and the fitted curves (±0.019 μm, n = 57) was similar to that during the tetanic plateau for all contractions combined (±0.021 μm, n = 143). The reason why the exponential curves could not be fitted successfully to all sarcomere length traces was because the changes in sarcomere length during the contractions were too small in those cases.

From the fitted equations for each sarcomere length trace, the time constant (*1/c*) for the logarithmic decrement was calculated. For the lengthening traces, the time constant was 105±8 ms (mean±standard error), and for the shortening traces the time constant was 55±13 ms (mean±standard error). Thus, the shortening regions of the muscle had sarcomeres that approached a steady-state length at a faster rate than the lengthening regions of the muscle.

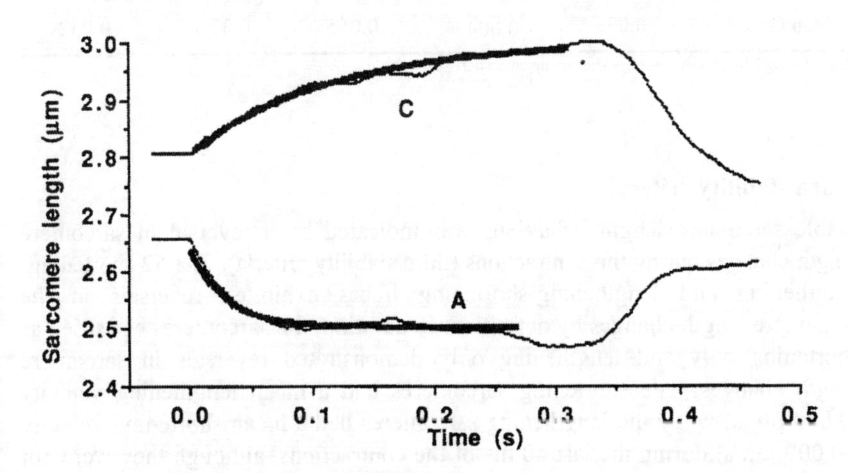

Figure 26.5. Example of the exponential ($SL(t) = a + b\,e^{-ct}$) curve fit during contraction (thick line) for shortening (muscle A) and lengthening (muscle C) sarcomere length-time traces for muscles shown in Figure 26.4.

Second stability criteria

The sarcomere length behaviour for the four types of sarcomere length-time traces approached a steady-state value at the end of the contractions (second stability criteria). The velocity of sarcomere length change at the end of the contraction was determined by fitting a straight line to the last 40 ms of each of the 143 contractions. The mean sarcomere length velocity for the shortening, lengthening, shortening-lengthening, and all types of contractions grouped together was 0.132 μm/s (0.055 Lo/s) or less (Table 26.1). The lengthening-shortening type traces had a larger speed of lengthening during the plateau than the other sarcomere length types; however, only three observations of this type of sarcomere length behaviour were found. The mean velocity for all types of sarcomere length traces and for all contractions grouped together was not significantly different from zero, indicating that a steady-state or nearly steady-state length was achieved by the sarcomeres near the end of the 250 or 300 ms contraction.

Table 26.1. Rate of change in sarcomere length (velocity) for the last 40 ms of a fixed-end contraction (plateau).

	Type of Behaviour				
Velocity	Shorten Only	Lengthen Only	Shorten-lengthen	Lengthen-shorten	All Grouped
Mean (μm/s)	0.080	–0.009	0.132	0.899	0.078
S.E. (μm/s)	0.129	0.091	0.090	0.925	0.059
n	31	60	49	3	143
Mean (Lo/s)	0.033	–0.004	0.055	0.37	0.032

Note: + lengthening sarcomere, – shortening sarcomere, Lo = 2.4 μm

Third stability criteria

Stable sarcomere length behaviour was indicated by a reversal in sarcomere length changes during the contractions (third stability criteria). The 52 shortening-lengthening and lengthening-shortening traces exhibited reversals in the sarcomere length changes by definition. In addition, the sarcomeres classified as shortening only and lengthening only demonstrated reversals in sarcomere length changes. The shortening sarcomeres had a mean lengthening velocity (0.080 μm/s) while the lengthening sarcomeres had a mean shortening velocity (-0.009 μm/s) during the last 40 ms of the contractions, although they were not statistically different from zero. The raw data show small oscillations about the smoothed curve that may be real or measurement artefact (Burton and Huxley, 1995). These reversals in sarcomere length changes are characteristic of a stable system with a damped oscillatory motion.

DISCUSSION

Sarcomere length changes in muscle were compared to those in single fibres from another study, and arguments supporting sarcomere stability on the descending limb of the force-length relation were made. Also, statements supporting the instability hypothesis are argued.

MUSCLE VERSUS SINGLE FIBRES

We found that the non-uniform sarcomere length behaviour observed in mammalian muscle was similar to that of amphibian single fibres during fixed-end contractions. Edman and Reggiani (1984) measured sarcomere length changes using small segments (0.5 mm long) identified by surface markers in six single-fibre preparations of the frog during fixed-end contractions. They demonstrated that different segments possess varying amounts of length change and reversals of length change during contractions. Their results (Figure 6 and Figure 7 in Edman and Reggiani, 1984) were classified into the four types of sarcomere length behaviours that were used in this study. Comparisons of the results from this study and the study of Edman and Reggiani (1984) show similar findings in the distribution of the four types of sarcomere length behaviours along the length of muscles and fibres (Table 26.2). Their study probably represents the actual distribution of sarcomere length behaviours better than our study, since they made measurements along the entire length of the same fibre. In our study, measurements could only be made from selected locations along the length of the muscle.

Table 26.2. Comparisons of the four types of sarcomere length behaviours between muscle (our study) and single fibres (Edman and Reggiani, 1984, Figure 6 and Figure 7).

	Muscle *		Single Fibre **	
Type	Number	Per Cent	Number	Per Cent
Shorten only	31	22%	47	37%
Lengthen only	60	42%	18	14%
Shorten-lengthen	49	34%	62	48%
Lengthen-shorten	3	2%	1	1%
Total	143	100%	128	100%

* 12 fibres, 26 regions along muscles, various muscle lengths
** 6 fibres, 64 segments along fibres, 2 fibre lengths

STABILITY TESTS

Three characteristics indicating stable sarcomere length behaviour were found in the sarcomere length-time results of mouse serratus anterior muscle. (1) Sarcomere length changes decreased in magnitude with time, as exhibited by the exponential decay in length changes for the shortening-only and lengthening-only types. (2) Sarcomere lengths approached a steady-state length as demonstrated by the small velocities during the tetanic plateau. (3) Reversals in the sign of sarcomere length changes occurred during contractions. All three of these characteristics of a stable system were found in this study.

Other studies also demonstrated stable sarcomere length behaviour if interpreted using the characteristics described in our study. Table 26.3 lists the investigations and the number of sarcomere length-time traces shown in those studies that exhibit a decrease in lengthening or shortening speed during isometric contractions (labelled stable). In the investigations shown in Table 26.3, frog single fibres and different methods of measuring sarcomere length were used. Also, sarcomere behaviour was measured at various locations along the length of the fibres. Based on our stability criteria, almost unanimous support for stable sarcomere length behaviour was found in these investigations. The four observations listed as unstable (Table 26.3) may in fact be displaying stable behaviour, because the contraction time was too short to declare if the sarcomere speed was decreasing or increasing over time. A reinterpretation of the sarcomere length-time data found in the literature indicates stable sarcomere behaviour as defined here on the descending limb of the F-L relation.

Table 26.3. List of the number of sarcomere length-time traces demonstrating stable and unstable behaviour.

Reference	Figure	Stable	Unstable	Region of Muscle	Method
Edman and Reggiani (1984)	4	20	0	Whole length	Segment
	8	8	0	Ends	Segment
Edman et al. (1993)	2b	9	0	Whole length	Segment
Granzier and Pollack (1990)	2, 3, 4, 8, 9	10	0	Central	Segment and diffraction
Huxley and Peachey (1961)	4	1	0	End	Micrographs
Julian et al. (1978)	2	2	0	Central	Segment
Julian and Morgan (1979a)	7, 9	8	0	Central	Segment
Lieber and Baskin (1983)	2, 5b, 6a, 7	11	0	End and central	Diffraction
	4a, 4b, 5a, 6b	0	4	End	Diffraction
Sugi and Tsuchiya (1988)	6	10	0	Whole length	Segment
ter Keurs et al. (1978)	5, 6	7	0	Central	Diffraction
ter Keurs et al. (1984)	1	4	0	Central	Diffraction
Total:		90	4		

CONSIDERATIONS ON INSTABILITY

Several arguments have been used to explain sarcomere behaviour under the hypothesis that sarcomere length is unstable on the descending limb of the F-L relation: (1) force-velocity properties, (2) non-uniform sarcomere length changes, and (3) negative slope of the force-length relation.

Force-velocity properties

It has been argued that the force-velocity properties of sarcomeres provide stability or decrease instability of the sarcomeres. Force-velocity properties refer to the characteristic of sarcomeres to produce more force during stretch and less force during shortening than the isometric-force value at the corresponding sarcomere length. On the descending limb of the F-L relation, the hypothesis of instability predicts that the long sarcomeres are continuously stretched and can produce more force than would be predicted by the F-L relation because of the force-velocity properties. Likewise, the short sarcomeres are shortening and, therefore, produce less force than the isometric value for the same length.

However, force-velocity properties cannot produce a stable sarcomere length behaviour since velocity properties are not part of the potential energy function that determines sarcomere stability (Allinger et al., 1996a). Force-velocity properties may slow the lengthening of weak sarcomeres and the shortening of strong sarcomeres, but they cannot make the system stable, for example, by reversing the direction of the length changes in sarcomeres.

Non-uniform sarcomere length changes

How can sarcomeres be stable when there are non-uniform sarcomere length changes during contraction? Stability can be demonstrated by taking a system at equilibrium, perturbing it, and observing whether it returns to the equilibrium state. The question of stability and non-uniformity can be explained through a thought experiment.

Tie a strong and a weak rubber band together and stretch them between two stationary pegs (Figure 26.6). The system is now in equilibrium. Perturb the system by grabbing the point connecting the two rubber bands and pulling it to one side. Now release your fingers and the system returns to the equilibrium condition. If you plot the length-time curves for the two rubber bands, you will observe non-uniform changes in length. However, we know the system is stable because it always returns to its equilibrium position.

This thought experiment might be analogous to how sarcomeres behave in a fibre during contraction. Think of the sarcomere length during the tetanic plateau as the equilibrium position of the sarcomeres. The perturbation occurs when the muscle is stimulated and the contraction begins. Although the sarcomeres are at a stable equilibrium while at rest, once the muscle is stimulated, the sarcomeres assume a new equilibrium condition. The system is

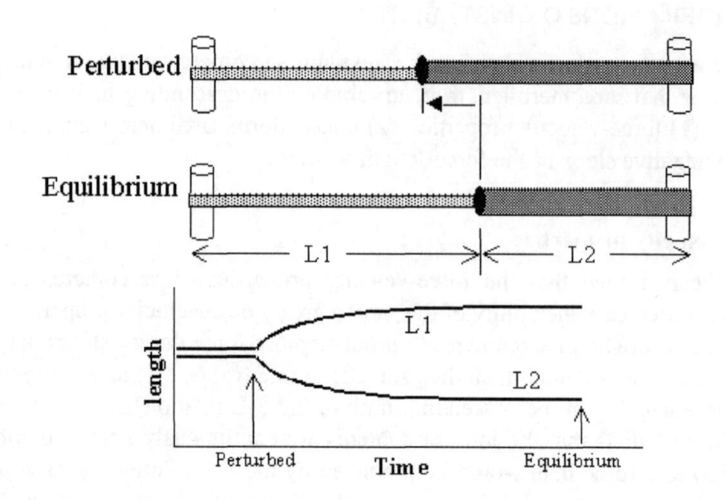

Figure 26.6. Illustration of how a stable system of two rubber bands can have non-uniform length changes. The system is perturbed from equilibrium. Band one (L1) lengthens while band two (L2) shortens once released from the perturbed condition. The perturbed condition is analogous to the moment of stimulation, and the equilibrium condition is near the tetanic plateau in a muscle fibre under fixed-end conditions.

perturbed upon activation because of differences in initial sarcomere length, force-producing potential (i.e., cross-sectional area), and/or the time and rate of activation of individual sarcomeres. Thus, sarcomeres behave non-uniformly upon activation, just like the rubber band system.

If sarcomeres were unstable, large non-uniformities in sarcomere length (>1.6 μm) should not only develop along the length of a fibre, but also within a small segment (Morgan, 1990). It is possible that in this study a few sarcomeres may have been overstretched and were not detected by the laser diffraction system. However, no secondary peaks were evident in the laser diffraction patterns that might have indicated large differences in sarcomere lengths. Others have found that non-uniformities within the test regions do not increase during tetanic contractions, but reach a steady state in myofibrils (Fabiato and Fabiato, 1978; Iwazumi, 1987), in frog single fibres (Cleworth and Edman, 1972; Hill, 1977; ter Keurs et al., 1978; Haskell and Carlson, 1981; Tameyasu et al., 1982), and in whole muscle (ter Keurs et al., 1978). Sarcomere length non-uniformities in the test region have been reported to be maximally 0.130 μm (Tameyasu et al., 1982, laser diffraction) in single fibres. In whole muscle, non-uniformities in sarcomere length increase slightly from rest to contraction (2.6% or 0.068 μm at 2.6 μm sarcomere length) but not enough to demonstrate unstable sarcomere behaviour (Paolini and Roos, 1975). The uniformity of adjacent sarcomeres in series may not be the result of an interaction between the sarcomeres, but may be related to stable sarcomere properties (Allinger et al., 1996a).

Negative slope of the F-L relation

The negative slope of the descending limb of the F-L relation does not necessarily mean sarcomeres are unstable. This may be illustrated by the following thought experiment (Allinger et al, 1996b; Allinger, 1995). Imagine that two combs are placed together, bristle to bristle, to represent cross-bridges in a sarcomere. If the combs are pulled apart perpendicularly to the bristles, the force is proportional to the number of bristles engaged before the pull. With the same displacement between combs, more force is developed at a short length (combined length of two combs) than at a long length. Plotting the force and length of the two-comb system reveals a negatively sloped F-L relation, similar to the descending limb of the F-L relation for sarcomeres. However, the two-comb system is stable because the force-elongation curve for each bristle has a positive slope.

The comb example illustrates how differences in force-elongation and force-length relationships may be interpreted incorrectly. The F-L relation in this comb example has a negative slope because the 'forces' of independent 'contractions' were connected to produce the relationship. This is identical to the way in which F-L relationships are determined. If the force-elongation relation is drawn for the comb example, it will have a positive slope because the more the bristles are bent, the higher the force. Similarly, positively sloped force-elongation curves have been demonstrated in skeletal muscle.

Many studies have demonstrated changes in the force production of sarcomeres following shortening or stretching that are not related to the F-L relation. After small sarcomere shortening and lengthening, the immediate and short-term force production is depressed and enhanced, respectively. Ford et al. (1977) demonstrated this by measuring the short-range stiffness of sarcomeres, and showed that the force-elongation relation has a positive slope (T1 and T2 curves). On a longer time scale, sarcomeres produce more isometric force after a stretch than during a purely isometric contraction at the same length (Hill, 1977; Julian and Morgan, 1979b; Edman et al., 1978, 1982). Similarly, sarcomeres produce less force after shortening than isometrically at the same length (Abbott and Aubert, 1952; Edman, 1980; Granzier and Pollack, 1989; Sugi and Tsuchiya, 1988; Edman et al., 1993). These results demonstrate that sarcomeres possess a positively sloped force-elongation curve, both in the short and long term, on the descending limb of the force-length relation. This positive slope is a necessary requirement for sarcomere length stability.

Other studies have demonstrated results contradictory to those mentioned above. Most scientists will agree that the short-range stiffness of sarcomeres has a positive slope. However, Morgan (1990) proposed that the force enhancement after stretch is not caused by increased force production in sarcomeres but by popping sarcomeres. Edman and Reggiani (1984) hypothesized that force depression after shortening is not a property of the sarcomere, but is caused by non-uniform changes in sarcomere length after the shortening. Using a segment-clamp technique, they demonstrated that after shortening from the descending limb to the plateau of the F-L relation, no force deficit existed. Furthermore, Horowitz et al. (1992) demonstrated a force enhancement of isometric tension after very small shortening, thereby implying a negative sarcomere force-length behaviour that is not consistent with sarcomere length stability.

FUTURE RESEARCH

Further research is necessary to determine if our hypothesis is valid: sarcomere length is stable on the descending limb of the force-length relation. One experiment that would directly test this hypothesis is a perturbation experiment. When a fibre is at rest, the passive elements keep the system in a stable equilibrium. That is, if the length of a sarcomere is perturbed, the sarcomere will return to its equilibrium length. When a fibre is stimulated, sarcomeres are expected to move toward a new equilibrium length because of differences in initial lengths, cross-sectional areas, and times of activation along the length of the fibre. This new equilibrium length for sarcomeres is likely history-dependent. In other words, the length changes of a sarcomere during contraction will change its potential energy and so affect its equilibrium length.

In order to uniquely determine whether sarcomere length is stable on the descending limb, a perturbation experiment should be performed. Such an experiment has not been done to date, but is technically feasible and could be performed readily in many laboratories around the world. It is based on the idea that if sarcomere lengths on the descending limb of the force-length relationship are stable, then sarcomere lengths will return to the stable equilibrium position following small perturbations.

Based on these background considerations, the following experiment is proposed: Take two isolated muscle fibres, attach them in series, allow for a motor to attach to the two fibres at the place they meet, and attach the free ends of each fibre to a force transducer (Figure 26.7). In the initial configuration, both fibres are at a similar mean sarcomere length on the descending limb of the force-length relation. In a first test (control), the two fibres are stimulated simultaneously and it is observed whether one fibre tears the other fibre apart or if the fibres reach an equilibrium position (i.e., the junction of the two fibres does not move). If the fibres reach an equilibrium position, a second experiment is performed. In this second experiment, the two fibres are stimulated simultaneously and allowed to reach the equilibrium position. Once they are in equilibrium (i.e., the junction of the two fibres does not move), the site of junction of the two fibres is grabbed by the motor (Figure 26.7), and the junction is moved forcibly to one side, thereby stretching one fibre and shortening the other. Once perturbed in this way, the junction is released, and if the system is stable, the junction site should move back to the initial equilibrium position. We hope that this experiment may be performed in the near future to resolve the question of (in-)stability of sarcomere length and force production on the descending limb of the force-length relationship once and for all.

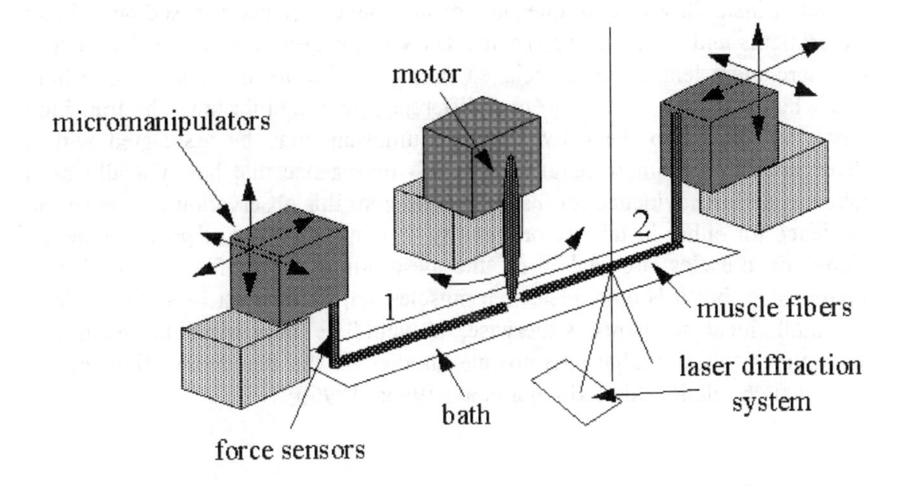

Figure 26.7. Test apparatus for stability tests. Two single frog-muscle fibres are attached to force transducers at each end of the bath. The two fibres are also attached to each other and a motor. Sarcomere length is measured with either a laser-diffraction or a segment-length method. Micromanipulators change the fixed-end lengths of the fibres and centre the fibre in the bath. Platinum electrodes (not shown) lie parallel to the fibre in the bath for muscle stimulation.

CONCLUSIONS

Since Hill (1953) proposed that sarcomeres are unstable on the descending limb of the force-length relationship, instability has become an accepted 'fact' that is not questioned. However, Hill's (1953) interpretation of the descending limb of the force-length relationship was incorrect, and so his conclusions must be questioned. In the meantime, it is well accepted and generally known that the descending limb of the force-length relationship represents a static property. It is obtained by stimulating a muscle at one length, recording the corresponding force, deactivating the muscle, setting it to a new length, stimulating it again, etc., that is, the force-length relationship is a series of isolated data points, obtained from isolated and independent tests. It is absolutely impossible to stretch or shorten a muscle on the descending limb of the force-length relationship with a constant activation such that the force follows the data points of the force-length relationship. Therefore, Hill's (1953) interpretations were incorrect. However, despite the relative ease of performing a perturbation experiment as proposed above, to establish uniquely whether or not sarcomere lengths are stable on the descending limb of the force-length relationship, such an experiment has not been performed. Rather scientists have either accepted the notion of instability without question, or they have accepted it based on indirect experiments and inconclusive results. The current experimental results suggest that sarcomere lengths in the relaxed fibre (muscle) are in stable equilibrium, and when activated tend to go to a different stable equilibrium. The transition from the relaxed to the active stable equilibrium may be associated with a divergence in sarcomere lengths. It is this divergence that has typically been taken as definite evidence for instability, whereas this observation by itself is no evidence for either instability or stability. It is not resolved at present whether fibres or muscles are stable on the descending limb of the force-length relationship, but it is quite clear that muscles and fibres must be stable, at least in small, local segments. Otherwise, it would be impossible for even two sarcomeres to exist adjacently to one another on the descending limb of the force-length relationship (Allinger et al., 1996a, 1996b).

REFERENCES

Abbott, B.C., and Aubert, X.M. (1952). The force exerted by active striated muscle during and after change of length. *Journal of Physiology* **117**, 77–86.

Allinger, T.L. (1995). Stability and the descending limb of the force-length relation in mouse skeletal muscle: A theoretical and experimental examination. Ph.D. thesis, University of Calgary, Calgary, Alberta, Canada.

Allinger, T.L., Epstein, M., and Herzog W. (1996a). Stability of muscle fibers on the descending limb of the force-length relation: A theoretical consideration. *Journal of Biomechanics* **29**(5), 627–633.

Allinger, T.L., Herzog, W., and Epstein, M. (1996b). Force-length properties in stable skeletal muscle fibers: Theoretical considerations. *Journal of Biomechanics* **29**(9), 1235–1240.

Burton, K., and Huxley, A.F. (1995). Identification of source oscillations in apparent sarcomere length measured by laser diffraction. *Biophysical Journal* **168**, 2429–2443.

Cleworth, D.R., and Edman, K.A.P. (1972). Changes in sarcomere length during isometric tension development in frog skeletal muscle. *Journal of Physiology* **227**, 1–17.

Deleze, J.B. (1961). The mechanical properties of the semitendinosus muscle at lengths greater than its length in the body. *Journal of Physiology* **158**, 154–164.

Edman, K.A.P. (1980). Depression of mechanical performance by active shortening during twitch and tetanus of vertebrate muscle fibers. *Acta Physiologica Scandinavica* **109**, 15–26.

Edman, K.A.P., Caputo, C., and Lou, F. (1993). Depression of tetanic force induced by loaded shortening of frog muscle fibres. *Journal of Physiology* **466**, 535–552.

Edman, K.A.P., Elzinga, G., and Noble, M.I.M. (1978). Enhancement of mechanical performance by stretch during tetanic contractions of vertebrate skeletal muscle fibers. *Journal of Physiology* **281**, 139–155.

Edman, K.A.P., Elzinga, G., and Noble, M.I.M. (1982). Residual force enhancement after stretch of contracting frog single muscle fibers. *Journal of General Physiology* **80**, 769–784.

Edman, K.A.P., and Reggiani, C. (1984). Redistribution of sarcomere length during isometric contraction of frog muscle fibres and its relation to tension creep. *Journal of Physiology* **351**, 169–198.

Fabiato, A., and Fabiato, F. (1978). Myofilament-generated tension oscillations during partial calcium activation and activation dependence of the sarcomere length-tension relation of skinned cardiac cells. *Journal of General Physiology* **72**, 667–699.

Ford, L.E., Huxley, A.F., and Simmons, R.M. (1977). Tension responses to sudden length change in stimulated frog muscle fibres near slack length. *Journal of Physiology* **269**, 441–515.

Gordon, A.M., Huxley, A.F., and Julian, F.J. (1966a). Tension development in highly stretched vertebrate muscle fibres. *Journal of Physiology* **184**, 143–169.

Gordon, A.M., Huxley, A.F., and Julian, F.J. (1966b). The variation in isometric tension with sarcomere length in vertebrate muscle fibres. *Journal of Physiology* **184**, 170–192.

Granzier, H.L.M., and Pollack, G.H. (1989). Effect of active pre-shortening on isometric and isotonic performance of single frog muscle fibres. *Journal of Physiology* **415**, 299–327.

Granzier, H.L.M., and Pollack, G.H. (1990). The descending limb of the force-sarcomere length relation of the frog revisited. *Journal of Physiology* **421**, 595–615.

Haskell, R.C., and Carlson, F.D. (1981). Quasi-elastic light-scattering studies of single skeletal muscle fibers. *Biophysical Journal* **33**, 39–62.

Hill, A.V. (1953). The mechanics of active muscle. *Proceedings of the Royal Society of London B* **141**, 104–117.

Hill, L. (1977). A-band length, striation spacing and tension change on stretch of active muscle. *Journal of Physiology* **266**, 677–685.

Horowitz, A., Wussling, H.P.M., and Pollack, G.H. (1992). Effect of small release on force during sarcomere isometric tetani in frog muscle fibres. *Biophysical Journal* **63**, 3–17.

Huxley, A.F., and Peachey, L.D. (1961). The maximum length for contraction in vertebrate striated muscle. *Journal of Physiology* **165**, 150–165.

Iwazumi, T. (1987). Mechanics of the myofibril. In: *Mechanics of the Circulation.* ter Keurs, H.E.D.J., and Tyberg, J.V. (eds.), Martinus Nijhoff Publishers, Dordrecht.

Julian, F.J., and Morgan, D.L. (1979a). Intersarcomere dynamics during fixed-end tetanic contractions of frog muscle fibres. *Journal of Physiology* **293**, 365–378.

Julian, F.J., and Morgan, D.L. (1979b). The effect on tension of non-uniform distribution of length changes applied to frog muscle fibres. *Journal of Physiology* **293**, 379–392.

Julian, F.J., Sollins, M.T., and Moss, R.L. (1978). Sarcomere length non-uniformity in relation to tetanic responses of stretched skeletal muscle fibres. *Proceedings of the Royal Society of London B* **200**, 109–116.

Lieber, R.L., and Baskin, R.J. (1983). Intersarcomere dynamics of single muscle fibers during fixed-end tetani. *Journal of General Physiology* **82**, 347–364.

Morgan, D.L. (1990). New insights into the behaviour of muscle during active lengthening. *Biophysical Journal* **57**, 209–221.

Morgan, D.L., Mochon, S., and Julian, F.J. (1982). A quantitative model of intersarcomere dynamics during fixed-end contractions of single frog muscle fibers. *Biophysical Journal* **39**, 189–196.

Paolini, P.J., and Roos, K.P. (1975). Length-dependent optical diffraction pattern changes in frog sartorius muscle. *Physiological Chemistry and Physics* **7**, 235–254.

Paolini, P.J., Sabbadini, R., Roos, K.P., and Baskin, R.J. (1976). Sarcomere length dispersion in single skeletal muscle fibers and fiber bundles. *Biophysical Journal* **16**, 919–930.

Pollack, G.H. (1983). The cross-bridge theory. *Physiological Reviews* **63**(3), 1049–1113.

Pollack, G.H. (1990). *Muscles and Molecules.* Ebner and Sons, Seattle, pp. 221–222.

Sugi, H., and Tsuchiya, T. (1988). Stiffness changes during enhancement and deficit of isometric force by slow length changes in frog skeletal muscle fibres. *Journal of Physiology* **407**, 215–229.

Tameyasu, T., Ishide, N., and Pollack, G.H. (1982). Discrete sarcomere length distribution in skeletal muscle. *Biophysical Journal* **37**, 489–492.

ter Keurs, H.E.D.J., Iwazumi, T., and Pollack, G.H. (1978). The sarcomere length-tension relation in skeletal muscle. *Journal of General Physiology* **72**, 565–592.

ter Keurs, H.E.D.J., Luff, A.R., and Luff, S.E. (1984). Force-sarcomere-length relation and filament length in rat extensor digitorum muscle. In: *Contractile Mechanisms in Muscle*. Pollack, G.H., and Sugi, H. (eds.), Plenum Press, New York, pp. 511–525.

27 *In Vivo* Function and Functional Design in Steady Swimming Fish Muscle

STEPHEN L. KATZ
Zoology Department, Duke University, Durham, North Carolina, USA

ROBERT E. SHADWICK
Scripps Institute of Oceanography, La Jolla, California, USA

INTRODUCTION

When my (SLK) mother asks me what I do, I tell her, 'I try to figure out how a fish fillet works.' And she says, 'Don't we know that?' I think she is very rational.

We wanted this anecdote to introduce two things. Primarily, that this paper is about how axial muscle works in propelling fish. In other words, what we know about *in vivo* muscle function. In particular, we want to talk about the structural design of the muscle and how it powers swimming via the production of undulations of the axial skeleton. We will not discuss the use of median or paired fins as oars or lifting foils. The second thing the anecdote suggests is that our knowledge about how muscle powers fish swimming is really rather limited. In other words, what we do not know, and what things are useful to do at this point.

Design itself is a difficult term to use in biology, since its use as a verb is loaded with teleology. So it is worth being clear from the outset. In this paper, 'design' is a noun, and it refers to the actual mechanical structure of the subject of study—in this case the axial, musculoskeletal structures in fish. The word design was also chosen to help introduce the idea that we can use analytical tools to evaluate the structural design of the fish fillet, and gain insight into its performance and quality. In the course of this paper, we will introduce the anatomical design of fish myotomes and try to point out that the anatomy has consequences for the mechanics of the muscle. Then we will try to present what we know about the biomechanics and force-generating physiology of fish muscle. This will lead into a description of how we must rely on *in vitro* studies of muscle to infer what is happening *in vivo*. Finally, we would like to comment on the synthesis that is emerging about how fish power swimming and what is still missing in our understanding of fish muscle function.

Skeletal Muscle Mechanics: From Mechanisms to Function. Edited by W. Herzog.

A critical key to understanding how the axial muscle of fish works *in vivo* is to incorporate an appreciation of the coupled nature of internal and external forces. In swimming, two things happen that are an expression of force development. The first is the deformation of the body to produce a propulsive wave that travels from anterior to posterior. The material of the body has a non-zero stiffness, and this deformation therefore requires the application of internal forces. At the same time, the undulations of the body push the surrounding fluid around, imparting momentum to the wake and producing thrust. This momentum transfer generates external, reaction forces that act on the undulating body. These internal and external forces acting on the swimming fish comprise a coupled mechanical system, the components of which are not independent. Achieving a complete understanding of *in vivo* muscle function is critically dependent on incorporating both these internal and external forces (Jordan, 1996; Katz and Jordan, 1997). As this paper proceeds, we hope that the reader will appreciate that this incorporation is not a common component in research on fish muscle biomechanics.

A CONTEXTUAL FRAMEWORK

There are numerous recent reviews on the physiology and biomechanics of fish axial muscle (Johnston, 1983; Wainwright, 1983; Videler, 1993; Wardle et al., 1995). To distinguish this paper, and to help add some structure to what follows, we have included a flow diagram (Figure 27.1) that expresses some of the levels on which design can be expressed. It also provides a simple design context for the material that follows. It summarizes a sequence of events, or indeed flow of information, that must occur in any animal locomotion—swimming or otherwise. Obviously, in this paper we will be addressing specific issues that relate to fish swimming. At the top of the diagram, there is a decision to swim, which might involve cognition or not, that results in activation of brain stem central pattern generators (Grillner and Kashin, 1976). This, in turn, is followed by the transmission of neural activity along the spinal chord and out the motor nerves to the axial muscle. This activation is revealed by electromyography (EMG). When the EMG arrives at the muscle, there is some time taken up in the transfer of information through the excitation-contraction coupling mechanism. This time delay may have functional consequences. Once the muscle has been activated, it generates force and, for a given loading, attempts to shorten. These force-generating and shortening events within the muscle are manifest in gross movement and, if co-ordinated appropriately, locomotion. At this level, external forces can feed back into muscle force production by providing mechanical afterload, or at a higher level, by influencing activation through proprioreceptor feedback. This is obviously a very simple framework and experienced readers will undoubtedly chafe at its simplification of what in detail are very complex physiological events. However, it is still useful in pointing out that locomotion is

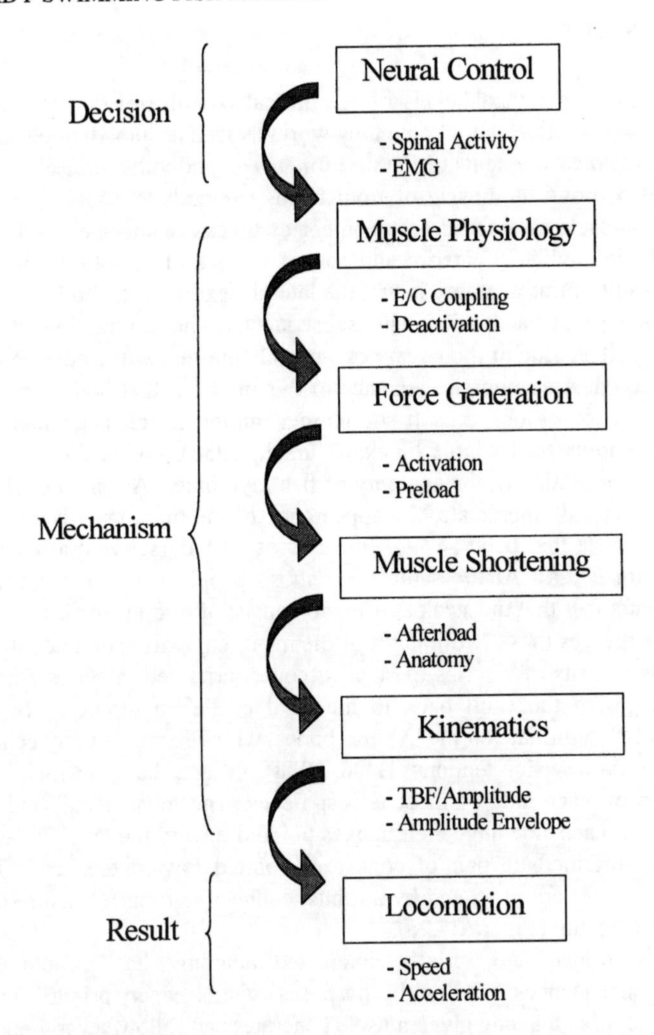

Figure 27.1. Flow diagram indicating the sequence of events in a motor control process, or motor control cascade. Headings indicate generic elements in the cascade. Subheadings indicate manifestations that are subject to experimental examination. See the text above for further explanation.

a consequence of a cascade of events, from nerve to outside world. Any constraint operating on one part of this motor control cascade will feed back up the chain, placing constraints on upstream elements. We will start this paper at the point of activation and proceed down the cascade. At the end of the paper, we hope to be able to show how what we are now learning about *in vivo* muscle function places important constraints on the design of the muscles all the way back up the chain.

THE ANATOMY

Fish myotomes have several levels of anatomical complexity that impact our ability to make predictions about how they work, as well as impeding our ability to make discrete measurements of the performance of individual muscles. In his early account of how the muscle of trout bends the body, Sir George Cayley described the muscle as a serial arrangement of blocks of muscle. Each block attaches to the immediately anterior and posterior block of muscle (Bone et al., 1995). One result of this anatomy is that the lateral flexure of the body is simply the progression of the activation, and subsequently, the contraction of these muscles along either side of the body—one muscle pulling on the next posterior muscle, and so on. A second consequence of this model is that body bending is simply the reflection of local muscle shortening. But the muscle is geometrically much more complex than simple blocks (Nursall, 1956). Figure 27.2 is a diagrammatic representation of the anatomy of fish myotomes. As anyone who has eaten a fish fillet will appreciate, the appearance of the myotomes is as folded chevrons of muscle that bend several times across the dorsal-ventral extent of the fish (Figure 27.2a). At the same time, those people who have eaten fish steaks will remember that the steak appears to consist of a series of rings (Figure 27.2b). If one merges those two images of the meat, one will appreciate that the fillet actually consists of a series of stacked cones, arranged in series from the posterior margin of the skull back to the final caudal myotomes. The final myotomes of fish terminate on the hypural bones, which form the bony centre of the caudal fin via insertion tendons (Bone, 1978). In fact, there are four sets of stacked cones on each side of most teleost fishes. The most dorsal and most ventral cones on each side have their apexes pointed toward the tail. The apexes of the larger, intermediate pair of cones are pointed toward the head. These middle sets of cones are separated by a robust connective tissue structure called the mid-lateral septum (Figure 27.2a).

This highly folded and serially connected anatomy has a number of mechanical consequences, but for the purposes of this paper, primary among these is that the force that any given muscle generates can follow several possible trajectories (Johnsrude and Webb, 1985). Each muscle fibre pulls on the next posterior muscle fibre across the collagenous, connective tissue myoseptum that divides adjacent myotomes (Willemse, 1966). Therefore, some force is transmitted across from muscle to muscle. Force is also transmitted along the myoseptum to adjacent skin and spinal column and can act to bend the body locally (Wainwright, 1983). Westneat et al. (1993) have also documented the highly ordered structure of connective tissue within the mid-lateral septum of Scombrid fishes (tunas). They have made an excellent case for force being transmitted along the body from muscle in one location to a point of application on the spinal column at a more posterior location via the tendons contained in the mid-lateral septum. The same structure exists in other fishes, and it seems reasonable to expect that the mechanical function attributed to this septum in

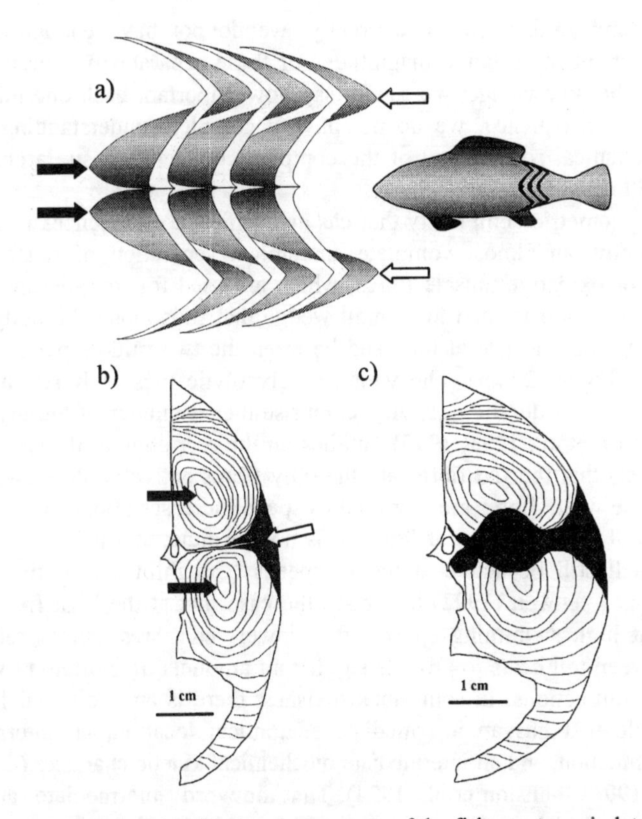

Figure 27.2. a) Diagram indicating the anatomy of the fish myotome in lateral view. The diagram shows three myotomes that are separated by the removal of two intervening myotomes. This view indicates the folded and nested nature of myotomes. The silhouette of the fish provides the orientation of the myotomes within the fillet. The four sets of cones within each myotome are indicated by arrows: black arrows indicate the anterior-pointing cones, and white indicate the posterior-pointing cones. See the text for further description. b) Diagram showing the anatomy of the fish myotomes in cross-section or steak view. This view shows the generalized condition for bony fish where the white muscle makes up the bulk of the myotome cones, visible in this view as rings. The red muscle, shaded grey, is restricted to a small wedge of parallel-fibred muscle in close apposition to the skin, and nested into the large anterior-pointing cones. To provide orientation, the centres of the anterior-pointing cones are indicated by black arrows. The position of the mid-lateral septum is indicated by a white arrow. See the text for further description. c) Diagram showing the anatomy of tuna myotomes in cross-section or steak view. This view shows the condition that is special to tunas and is similar to the condition seen in some Alopid and Lamnid sharks. In these fish, a large portion of the cross-section that in other fish would be composed of white fibres is expressing the red phenotype. Tunas have this red muscle in addition to the wedge of red muscle seen in other fish.

tunas will also exist for the mid-lateral septum of other fishes. It is also possible that forces can be transmitted along the body for a longer distance through the skin acting as a structural exotendon (Hebrank, 1980; Hebrank and Hebrank

1986; Katz and Jordan, 1997). Currently, we do not have enough discrete information about the relative magnitudes of these separate force trajectories, and without that knowledge, we cannot say how important each one might be. As a further consequence, we do not have a complete understanding of the specific mechanical role of any of the complex connective-tissue architecture within the fillet.

A second geometric complexity that has mechanical consequences is that fish myotomes show an almost complete anatomical separation of muscle fibre types. Red, or oxidative muscle fibres, which are used for long-duration, low-intensity activity, are located in a small wedge that runs along the body under the skin along the mid-lateral line, and between the two middle pairs of myotomal cones (Figure 27.2b). The white, or glycolytic muscle fibres, used for high-intensity, short-duration activity, comprise the remainder of the myotomal cones (Johnston et al., 1993, 1995). Studies on the recruitment of these muscle fibres suggests that the red fibres are the only fibres active at slow swimming speeds and the white fibres are recruited as speed increases (Bone, 1966; Rome et al., 1984). This separation of function is not absolute, and it is possible that red fibres will still be active at higher speeds (Coughlin and Rome, 1996). However, van Leeuwen (1992) has made the case that at the high frequencies employed at high swimming speeds, the slower red fibres cannot relax fast enough between tail beats to provide significant amounts of shortening work in subsequent contractions. In some teleost fishes, there is an additional layer of 'pink' muscle that has an intermediate anatomical location, an intermediate mechanical function, and an intermediate biochemical kinetic character (Coughlin and Rome, 1996; Johnston et al., 1974). That the word 'intermediate' appeared so many times in that sentence is an impressive expression of the structure-function paradigm in the field of biomechanics. In several Scombrid fishes (tunas), and some Lamnid (e.g., Mako) and Alopid (e.g., Threasher) sharks, there is an additional loin of red muscle fibres that runs along the body in a medial position within the myotome cones in a location that in other fish is composed exclusively of white fibres (Figure 27.2c). This additional loin of red muscle has a number of mechanical consequences that make those fish a special case (Knower et al., 1999; Shadwick et al., 1999). Among the mechanical consequences of this separation of red and white muscle fibres is that in slow swimming, when only the superficial red muscle fibres are powering body bending, most of the myotome is inactive and undergoing passive strain. This strain will consume some unknown fraction of the force generated by the red fibres, and therefore, constitutes an additional force trajectory that needs to be incorporated into the total force budget of the swimming muscle. Since the white muscle is heavily invested with connective tissue myosepta, it is clearly not homogeneous. This might lead one to expect that the passive shearing that occurs during slow swimming may not be like that in a simple beam constructed of isotropic material (see below).

A third complexity is that muscle fibre orientation within the myotome is not uniform. In the red muscle, the fibres do appear to be oriented primarily along the body axis, but in the white muscle this is not the case. Alexander (1969) performed serial sections on fixed specimens and demonstrated that as one looks from the inside to the outside of the fish fillet, the muscle fibre angles change in a progressive manner. At the apex of the myotome cone, the muscle fibres are approximately co-linear with the axis of the cone. However, as one looks at fibres that are farther from the apex of the cone, the angle of the fibre with respect to the axis of the cone increases both in the horizontal and sagittal planes. It has been said that this results in a net helical orientation to the fibres with the centre of the helix co-linear with the axis of the myotome cones. One should keep in mind, however, that each muscle fibre is straight as it spans the distance from one myoseptum to the next, but as one follows the trajectory of fibres across many myosepta the net trajectory is helical along the body. It was proposed that this arrangement allows all the muscle fibres within the cone to work on an equivalent portion of the force-length relationship, and to shorten by a very small amount for a given amount of body bending

The bottom line is that this anatomical complexity makes discrete biomechanical measurements technically very difficult. Indeed, in the absence of a complete and discrete force budget, we are not able to describe the exact biomechanical action of any specific muscle fibre within the myotome. As a result, we must rely on indirect measures of the biomechanical performance of the muscle in fish myotomes.

INDIRECT INDICATORS OF MUSCLE BIOMECHANICS

Since the complicated anatomy and serial arrangement of the myotomes make discrete *in vivo* measurements of muscle function difficult, we have to rely on indirect indicators of muscle performance to describe the *in vivo* muscle function in fish swimming. The strategy has been to:

1. Describe the *in vitro* performance of small pieces of muscles.
2. Then map these results onto the kinematics and electromyography determined from *in vivo* studies of fish.
3. From this mapping, predict the actual mechanical behaviour of the muscles *in vivo*.

Making these mechanical measures in muscles that cyclically shorten is difficult because the shortening velocity continually changes during the power stroke. In addition, force is often developed as the muscle is being lengthened or is shortening. Estimates of power output using *in vitro* force-velocity curves derived from isotonic shortening experiments are inappropriate for modelling fish swimming muscle because these measurements only consider force produced during shortening, and will overestimate the *in vivo* power output (Swoap et al., 1993).

The oscillatory work-loop technique addresses some of these limitations (Josephson, 1985). In this technique, isolated muscle fibres are subjected to sinusoidal length changes and phasic stimulation is applied in an attempt to mimic the *in vivo* conditions (Rome et al., 1990, 1992a,b). Thus, the work-loop technique is appropriate for determining values of power output from fish muscle (Altringham and Johnston, 1990a,b; Johnson and Johnston, 1991; Anderson and Johnston, 1992; Johnson et al., 1991; Moon et al., 1991; Rome and Swank, 1992; Altringham et al, 1993; Rome et al., 1993). The performance of muscle in these experiments is dependent on numerous parameters, including but not limited to cycle frequency, amplitude, and temperature (Johnson et al., 1991 and 1994; Rome et al., 1992a). As a general statement, however, these studies indicate that activation that occurs before peak length will produce force while the muscle shortens. This results in a loop in the force-extension plane with a counter-clockwise trajectory, indicating positive work done by the muscle on the outside world. If the activation occurs approximately half-way between mean length and peak length, then the area enclosed by this work-loop, and the net work done by the muscle, will be maximal. On the other hand, if the muscle is only active while lengthening, or active late in shortening, then the work-loop will have a clockwise direction (Johnson and Johnston, 1991). A clockwise work-loop indicates a negative work contraction, and indicates that the muscle is having work done on it.

In what follows, it is important to distinguish net negative work from instantaneous negative work. In those contractions that produce a large counter-clockwise trajectory in the force-extension plane (indicating maximal net positive work), there is usually a portion of the curve at long length, but lengthening, where activation and force generation have already begun. This small amount of force generation while being lengthened is critical to maximizing net force output of the muscle (Altringham and Johnston, 1990a; Altringham et al., 1993). By the same token, it is also critical to the production of net positive work that the EMG termination be appropriately matched to the relaxation characteristics of the muscle. EMG termination that is too early will result in force only being produced while lengthening (net negative work). Too long an EMG will prevent adequate relaxation and lengthening between contractions by inhibiting stretch-induced deactivation of the muscle (Rome et al., 1993; Coughlin and Rome, 1996). In general practice, the distinction between *net* positive and *net* negative work is made based on the timing of EMG onset relative to the shortening cycle of the muscle. We would ask readers, however, to put these other issues in a corner of their brain while they read the rest of this paper and remember that the reality is somewhat more complex than simply EMG onset timing. It is also critically important to remember that in these experiments, the muscle strain is imposed by the experimental apparatus and the force generated by the muscle in response to activation is measured. *In vivo*, the muscle generates force and the shortening emerges from a coupling of the internal muscle forces and external forces imposed by the outside world.

Once the work-loop experiments have produced a library of data on how the muscles behave *in vitro* when subjected to a wide range of strain and excitation combinations, *in vivo* behaviour of the muscle is estimated by mapping this library onto the actual strain and excitation. Therefore, having accurate measurements of *in vivo* excitation and strain are critical for understanding the *in vivo* performance of the muscle. Until recently, this mapping of *in vitro* experiments on isolated muscle samples onto whole animals has relied on measures of body bending to estimate local muscle strain patterns and EMG to estimate local patterns of muscle activation.

PATTERNS OF MUSCLE ACTIVATION

Studies of EMG in muscle have been numerous (Bone, 1966; Rayner and Keenan, 1967; Brill and Dizon, 1979; Wardle and Videler, 1993; Jayne and Lauder, 1995a,b; Shadwick et al., 1998; Hammond et al., 1998; Gillis, 1998). However, there are few simple statements that can be made that apply to fish in general. In part, this results from the diverse morphologies and kinematics of the fish species looked at, the choice of recording from red versus white muscle, and the way the scientific questions were posed within each study. Also, please keep in mind that in the past, the high variability in EMG data within a single fish has not allowed the various experimentalists to discriminate a constant progression of EMG activity along the body from a changing rate of EMG arrival. As a consequence, most people treat the progression rate of EMG onset as constant. In the future, it may become functionally important to make sufficiently detailed measurements to discriminate these two cases (see below).

What one can say is that the EMG activity arrives at myotomes in an anterior to posterior sequence. It seems inappropriate to say that the EMG has a wave velocity along the body since the activity is not transmitted through the muscle directly, but EMG onset arrivals do proceed along the body at a rate. In general, the rate of onset progression exceeds the rate of offset progression along the body. As a result, the duty cycle (time the muscle is active as a fraction of a complete cycle) decreases from anterior to posterior. In eels, which may pass one or several wavelengths of undulation along their body during swimming (= low relative wavelength) (D'Aout and Aerts, 1999), the progression of offset is almost the same as onset, and the duty cycle is close to constant along the body (Gillis, 1998). At the other end of the performance spectrum, tunas, which pass only one wavelength on the body while swimming (= high relative wavelength), display a red muscle offset that occurs simultaneously along the body (Knower et al., 1999). Also, as one proceeds from fish with low relative wavelength to high, one observes less time with simultaneous contralateral muscle activity (Knower et al., 1999).

These statements about the patterns of muscle activity are almost exclusively based on red muscle. It is worth pointing out that these measurements are technically difficult to make since one is generally trying to record very small, high-frequency, electrical activity within a large saline body of water (for marine fish), which is a particularly noisy environment. Commonly, fish are instrumented with long masses of fine wire, since the muscles of interest are locomotion muscles and the fish will move. In the case of performing these experiments in a water tunnel treadmill, the seawater is moving, making noise suppression even more difficult. In addition, the limits on the lengths of wire and the ability to alter the speed of water tunnel treadmills makes recording sprinting activity from white muscle particularly difficult. Recently, techniques have been developed for the remote recording of EMG from swimming fish (Rogers and Weatherley, 1983; Dewar et al., 1999), and one might expect to see more information on white muscle become available soon.

It is also important to keep in mind that EMG is not an explicit surrogate for force development. There is a finite time delay built into the excitation-contraction coupling machinery. In addition, the muscle continues to generate force after EMG offset. It is also possible that these delays are a varying function of location on the body. The functional role of these time differences, or their location dependencies, is a subject of some debate. On the one hand, He et al. (1990) reported for the mackerel *Scomber scombrus*, that there is no significant difference in the contraction timing of red muscle sampled from rostral and caudal locations. Van Leeuwen et al. (1990) also treated the time delay between the arrival of EMG onset and the start of force generation as the same at all points along the body. However, Rome et al. (1993) have shown that important characteristics of the swimming muscle of scup are body-position dependent. They observed the force production from posterior red muscle to decay more slowly than anterior muscle following the termination of EMG activity. They concluded that the most posterior muscle was still producing positive work in spite of the early EMG with respect to the muscle strain cycle. However, Johnston et al. (1993) working with the sculpin (*Myxocephalus scorpius*) did not find similar body-location dependence to intrinsic white-muscle activation or deactivation characteristics. Van Leeuwen (1992) incorporated the idea that posterior muscles are slower in the decay of force than anterior muscles in an update of their model. Importantly, this later model predicts dynamic properties of the muscle in swimming carp that demonstrated behaviour close to that observed in live fish (see below).

PATTERNS OF MUSCLE STRAIN

The other important thing to know in mapping *in vitro* onto *in vivo* is the pattern of muscle strain. Generally, this is estimated by measuring the amount of local curvature during body bending. It is important to keep in mind that using curvature as an index of muscle strain assumes that the body of the fish is bending, as would a homogeneous beam. In a bending beam, the strain of any small piece is the product of the local curvature and the distance of that small piece from the neutral axis of the beam. The neutral axis is a line, usually near the mid-line of the beam, which neither stretches nor compresses as the beam bends. In fish, this has been shown to be equivalent to the position of the backbone when viewed from above (Shadwick et al., 1998 and 1999). Clearly, this is a critical assumption, because if local body curvature does not specifically define the phase and amplitude of muscle strain at the same anatomical location, then the mapping of *in vitro* muscle experiments to *in vivo* dynamics will not accurately predict muscle work output during swimming.

Recent work on the mackerel (*Scomber japonicus*) using X-ray videography indicates that in these fish, local curvature is, in fact, a very reliable indicator of local red-muscle strain (Shadwick et al., 1998). Radio-opaque, gold beads were surgically introduced into the muscle of the fish, which were then allowed to swim in a small flume that was placed in the beam of a video-radiography apparatus. Local changes in muscle length, indicated by the distance between gold beads, were correlated in time with estimates of muscle strain deduced from local body curvature and thickness.

In our laboratory, we have recently concluded a series of experiments to validate this assumption using sonomicrometry and simultaneous video image analysis on the milkfish (*Chanos chanos*). Briefly, swimming fish were filmed and the amount of body curvature and thickness measured from the images. From these data, strain in the red muscle and in the approximate centre of the fillet was estimated. In each of these fish, sonomicrometry transducers had been surgically placed in the muscle at the same locations for which strain had been estimated from video.

Sonomicrometry is a technique where pairs of small (~2 mm diameter) piezo-electric crystals are implanted in the muscle. One of the crystals is an ultrasound (5 MHz) emitter and the other is a receiver. The time it takes for an ultrasound pulse to travel between the crystals when multiplied by the speed of sound in muscle gives a high-quality measure of the distance between the crystals (Ohmens et al., 1993). If the crystals are carefully aligned with the shortening axis of the muscle, then, and only then, sonomicrometry provides a high-quality measure of muscle shortening. Using these techniques, Coughlin et al. (1996) validated the assumption that strain in superficial red muscle of scup is accurately predicted by local body bending. However, Covell et al. (1991), working on fast starts in trout, found that local body bending was not an accurate predictor of the phase of local white-muscle strain.

Working with the milkfish (*Chanos*), we were able to show that muscle strain estimated from local body bending accurately predicted muscle strain measured with sonomicrometry (Figure 27.3). In all the fish looked at, and over a variety of speeds, strain measured with sonomicrometry validated those estimated from body bending (Katz et al., 1999). This study also suggested that both the phase and amplitude of muscle strain within the myotome cone was accurately predicted by external images of body bending. This validated the assumption that the fillet undergoes strain as would be expected in a homogeneous beam. This finding is somewhat surprising given the complex organization of the connective tissue. Recently, Wakeling and Johnston (1999), working on carp, indicated that muscle strain within the myotome cone is distributed in a non-uniform manner. Their findings were intermediate between the prediction of a homogeneous beam (described above), and that of Alexander (1969), who predicted that all muscle strain would be the same and independent of position within the cone. Working on the red muscle distribution pattern in skipjack tunas, Shadwick et al., (1999) have shown that the red-muscle strain is not accurately predicted by local body bending. The additional loin of red fibres located deep in the cones of these fish actually shortens later in time than local body bending predicts. In fact, it contracts in phase with bending of the body at a location as much as 20% of body length more posterior. It seems clear that this pattern of muscle strain is allowed by the tunas' distinctive anatomy, and it is unlikely to be demonstrated by other fish that lack this distribution of red muscle.

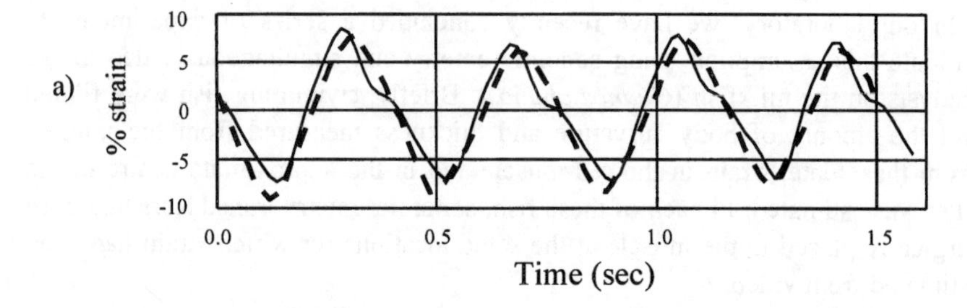

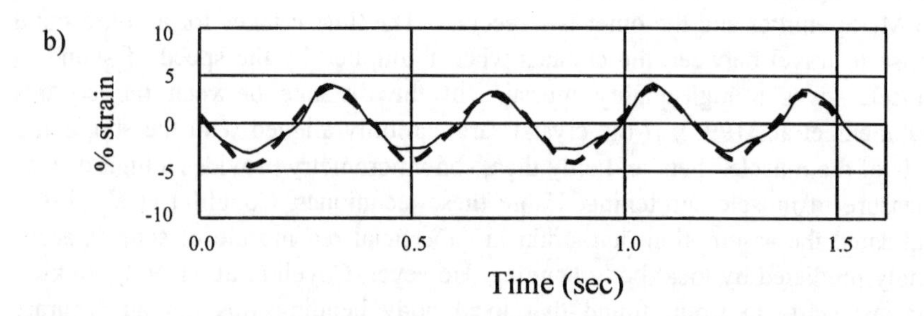

Figure 27.3. a) Plot comparing muscle strain calculated from video images of body bending and from sonomicrometry at a location of 0.53 FL (FL = fork length) on the body located in close apposition to the skin in a milkfish swimming at 2.34 L/s. At this speed, the fish is using a steady gait. In all parts of this figure, the estimates from video images are presented as dashed lines, while the estimates based on sonomicrometry are presented as solid lines. The average muscle strain amplitude measured with sonomicrometry was 7.5%, and the maximum cross-correlation coefficient was 0.95 with no phase difference. b) Comparison of muscle strain calculated at 0.53 L for white muscle located within the conic myotome. At this swimming speed, the white muscle is not expected to be recruited. The average muscle strain amplitude from sonomicrometry is 4.25%, and the maximum cross-correlation coefficient was 0.98 with zero phase difference.

MAPPING *IN VITRO* ONTO *IN VIVO*

One might reasonably expect that with this information in hand, a successful mapping of *in vitro* onto *in vivo* would have produced a consensus regarding the *in vivo* power generation in fish muscle. However, no such consensus has yet emerged. The work that has been done has produced two very different conclusions concerning the role of swimming muscle in different regions of the animal. As mentioned above, mapping *in vitro* experiments of isolated muscle samples onto whole animals has used body curvature, or surrogates for curvature, to estimate local muscle-strain histories and EMG to estimate local patterns of muscle activation. If a fish body were a simple sine wave, the maximum value of the lateral movement of the body would coincide with the maximum in curvature on the other side of the fish. In the past, this intuition has lead to use of lateral movement or undulation of the body as a surrogate for curvature. But fish do not adopt simple harmonic waveforms. As a result, lateral undulation of the body is an imprecise surrogate for body curvature, and has contributed to the conclusion that the progression of EMG onset proceeds along the body faster than does the wave of muscle strain. Thus, the posterior red muscle is thought to be activated too early in the shortening history of the muscle to produce significant amounts of positive work (Williams et al., 1989; Wardle et al., 1995). Using an explicit form of curvature as an index of muscle strain, studies on carp and saithe white muscle (Hess and Videler, 1984; van Leeuwen et al., 1990; Altringham et al., 1993) suggest that in these fish the anterior red muscles are activated with the appropriate phase to produce maximal positive work. The most caudal muscles are predicted to do predominately negative work, thus functioning essentially as tendons rather than actuators.

Conversely, several reports have suggested that while all the muscle is activated in a manner to produce positive work, the anterior swimming muscle has such a small strain that the amount of work available to power swimming is minimal. Performing work-loop experiments with red muscle from scup (*Stenotomus chrysops*, Sparidae [= Porgies]), Rome et al. (1993) reached this conclusion. They further suggested that the small strains seen in anterior red muscle inhibit the natural tendency of the muscle to relax after experiencing a large stretch, or stretch-induced relaxation. As a consequence, the anterior muscles do not lengthen adequately between activations to allow significant shortening work upon subsequent activation (Rome et al., 1993; Coughlin and Rome, 1996). Thus, posterior muscles generate the majority of the total swimming power, since they undergo a larger amplitude of strain. These measurements of muscle strain were confirmed by Coughlin et al. (1996). Subsequent research has not effectively reconciled these apparently incompatible conclusions regarding regional differences in work output from muscle.

We feel that a consensus is emerging between these different theories, a consensus in which all the red muscle used in steady swimming does similar amounts of positive work and no region of the fish is devoted to negative work production. This simple view of muscle function emerges from recognizing two things: one, the small strains seen in anterior myotomes of scup are not a general result for all fish, and two, curvature must be specified very carefully.

Sonomicrometry measurements of red-muscle strain in milkfish indicate that while red-muscle strain amplitude increases from rostral to caudal myotomes, this increase is relatively modest (Katz et al., 1999). The muscle undergoes a strain in excess of ±5%, with maxima approaching ±10% (values of strain are per cent change away from resting length). This should allow all the muscle to produce significant mechanical work while shortening. Indeed, the magnitude of strain seen in milkfish at slow swimming speeds is almost the same as seen in mackerel (Shadwick et al., 1998). That study, in fact, predicted that net positive work would occur all along the length of red muscle. In any case, neither the milkfish nor the mackerel shows an increase in muscle strain amplitude that is similar to that reported for the scup. There are two potential sources for this difference; either there has been technical error in the estimate of muscle strain, or there are real differences in the manner that these different species of fish use their muscles. It seems unlikely that a technical error in the estimate of strain is the reason for this difference, since so many separate technologies seem to show the same results. Specifically, good agreement is achieved in each paired comparison of the three techniques mentioned (sonomicrometry, video radiography, and external image analysis). Thus, one concludes that there are real differences in the way different species use their red muscle to power swimming. Since similar large magnitudes of muscle shortening are observed in the anterior myotomes of mackerel (Shadwick et al., 1998), trout (Hammond et al., 1998), and milkfish (Katz et al., 1999), we must conclude that these fish are capable of generating a good deal of positive work from these muscles during swimming. Scup are distinct from these other species in having laterally compressed fillets. It is possible that they use their muscles in a manner different from other fish and thus represent a special case. This suggests that in the future, an appreciation of the potential differences displayed between species in their use of swimming muscles should be incorporated into conclusions reached about how various fish power swimming. As a generality, it seems at least inappropriate to suggest that anterior myotomes are unable to perform significant work.

By more carefully specifying the history of muscle shortening, Katz and Shadwick (1998) showed that the wave of contraction in red muscle proceeds along the body in a manner that keeps pace with the progression of activation, rather than lagging behind. As a result, the red muscle is likely activated in a manner to produce primarily positive work and negligible negative work in fish such as mackerel. In that study, fish were videotaped while swimming, and the position of the body mid-line was digitized. Polynomial curves were fit to the position of the body mid-line, and the curvature of the fitted function was calculated. It is important to remember that when the position of the body mid-line is defined in an orthogonal co-ordinate system, x and z perhaps, curvature is a rather specific function:

$$\kappa(x) = \frac{z''(x)}{(1 + z'(x)^2)^{3/2}} \tag{27.1}$$

where $\kappa(x)$ is curvature as a function of position on the body, x, $z(x)$ is the position of the body in space, and $z'(x)$ and $z''(x)$ indicate the first and second derivatives of the body position with respect to x (Shanks and Gambill, 1969) (Figure 27.4). By using a simple analytical geometry treatment, Katz and Shadwick (1998) were able to show that when the amplitude of body undulation is not constant along the body (i.e., in all fish), using surrogates for curvature can give misleading results. In fact, curvature proceeds along the body more quickly than does lateral movement of the body. It proceeds quickly enough to proceed at the same rate as EMG arrivals, and suggests that all the red muscle is activated in a similar manner with respect to strain cycle.

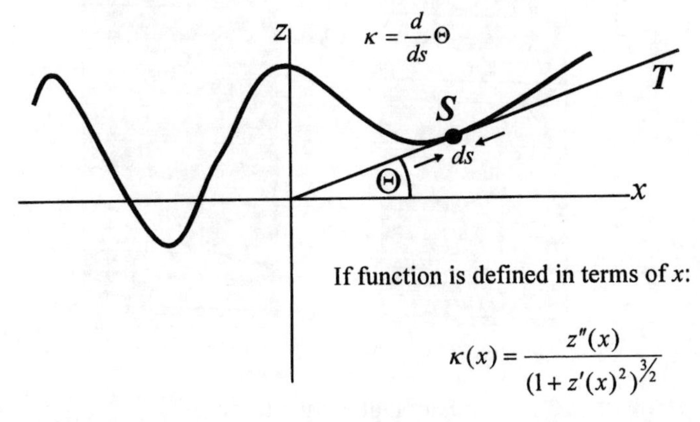

Figure 27.4. Cartoon depicting the relevant terms in the calculation of curvature. The curvature, κ, of a line segment at a point, S, is the rate of change of the slope of the tangent, Θ, with respect to arc length, ds. The specific formula for the calculation of curvature will depend on the choice of parameterization.

That study was criticized because the conclusions were based on a limited data set. The authors anticipated that the use of analytical geometry that was independent of species-specific anatomy would avoid that criticism, but they were wrong. Since that time, similar data have been collected for other species (milkfish, skipjack, and yellowfin tuna), and in all cases the same results were obtained. Figure 27.5 is an example of this for milkfish. Figure 27.5a is a surface generated by taking each polynomial curve that is fit to the body mid-line in each video field. The video is sampled at 60 Hz, and so we can append each sequential fitted polynomial to create this surface, which shows the position of the body in space from head to tail (going from left to right), and through time (going from bottom to top). It is immediately apparent that the waves of lateral undulation, which are indicated by shades of grey, proceed in a straight line. A heavy dashed line indicates one crest in the lowest cell. Since the surface describes space and time, the fact that the line is straight indicates that the wave of lateral undulation travels at a constant velocity. This is the same result as that

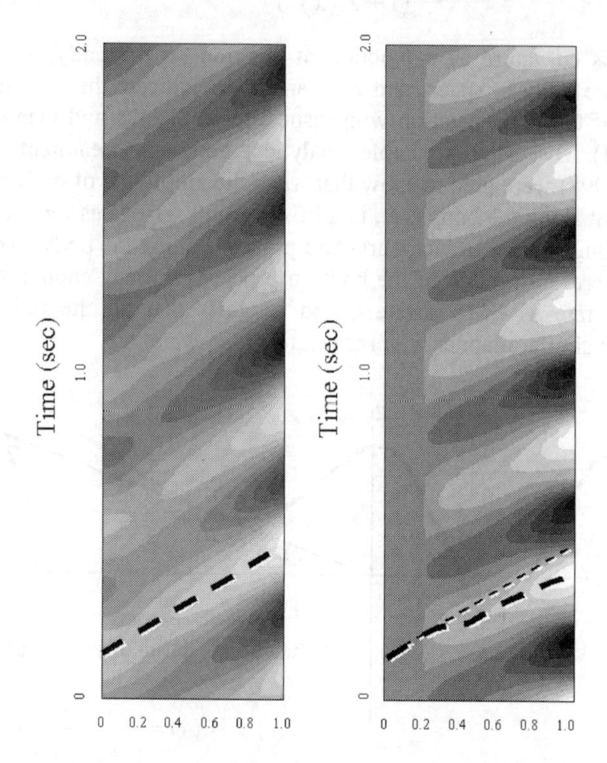

Figure 27.5. a) Contour plot of body position, *z(x)*, indicating the lateral undulation of the body. These data are for a milkfish swimming at 2.53 FL/s (FL=fork length). This sequence lasts 2.0 seconds and covered almost five tail beats. Time increases from bottom to top, and position on the body (reported as a fraction of FL) increases from left to right. Light shades indicate deflection of the body mid-line to the right, and dark shades indicate deflection of the body to the left. The dashed line indicates the progression of a peak in lateral deflection as it moves along the body through time. Significantly, the straight line indicates that the wave of lateral undulation proceeds along the body at a constant velocity. b) Contour plot of body mid-line curvature for the same locations and times as in Figure 27.5a. The waves in curvature progress caudally from left to right, and through time from bottom to top. Therefore, this panel maps directly onto the panel in Figure 27.5a. Light shades indicate concavity to the right side of the fish and convexity to the left (i.e., muscle shortening on the right). Dark shades indicate the converse. The thick dashed line indicates the progression of a peak in curvature as it moves along the body through time. Also plotted is a thin dashed line that is the same as in Figure 27.5a, indicating lateral deflection. The result, that the wave in curvature travels at a velocity different from lateral undulation, is indicated by the divergence in these two lines.

obtained by Videler and Hess (1984) for saithe (*Pollachius virens*) and mackerel (*Scomber scombrus*), and Katz and Shadwick (1998) for mackerel (*Scomber japonicus*), and Müller et al. (1997) for mullet (*Chelon labrosus*). Figure 27.5b shows a surface that describes body curvature for the same moments in time as the surface in Figure 27.5a. Superimposed on the curvature surface is a heavy dashed line, indicating the travelling crest of curvature that correlates with the travelling wave of lateral undulation marked in Figure 27.5a. What is apparent is that the travelling wave of curvature proceeds more quickly over the posterior two-thirds of the body than does lateral undulation. (Curvature in the anterior portion of the fish is zero, since the bending of the skull is thought to be negligible.) This indicates that the result in Katz and Shadwick (1998) for mackerel was not specific to that species alone, but rather looks to be general.

So far, this observation is a feature of relatively simple geometry. It becomes functionally significant when we map an *in vivo* pattern of excitation onto our prediction for the phase of muscle strain based on curvature. Figure 27.6a is a contour plot that combines the information of EMG measurements and phase of local shortening from *Scomber japonicus* (Shadwick et al., 1998). Wardle et al. (1995) defined the temporal cycle in body undulations using a degree scale. An anatomical segment at mean length, but lengthening, is labelled 0° or 360°; a segment at maximum length is 90°; a segment at mean length, but shortening, is labelled 180°; and a segment at its shortest length is at 270°. In this framework, the time is normalized by the tail-beat period (T) and the body location by the fork length (FL). The straight lines from the lower left to the upper right in Figure 27.6a are the progressions of lateral deflection along the body from head to tail and through time. The average onset and offset of EMG activity as a function of body position and time is indicated by vertical, dark solid lines. EMG was recorded in lateral swimming muscle at three locations: 0.4 L, 0.68 L, and 0.75 L. Shadwick et al. (1998) indicated that the anterior muscle was active at a point in the muscle strain cycle, based on lateral deflection of the body, corresponding to 37° and inactive at 209°, while the posterior muscle was active at 349° and inactive at 81°. If lateral deflection were an accurate reflection of muscle strain in the posterior location, this pattern of activation would not produce any activity while the muscle was shortening (i.e., no positive work).

If the muscle-strain estimates are based on curvature, a different conclusion is reached. Figure 27.6b is homologous to Figure 27.6a, except that the phase of muscle strain is correlated to curvature. Since the wave of curvature arrives at a point on the body earlier than the wave of lateral deflection, the corrected lines that indicate relative segmental length based on curvature appear as curved lines immediately below each corresponding straight line. In this scheme, lateral muscle at a location of 0.4 L has an EMG onset of 64°, while at a longitudinal location of 0.75 L, the EMG onset occurred at 41°. These onsets correspond to muscle that is lengthening, but not quite at maximal segment length. At the anterior location, the EMG offset occurs at 233°, or just after the segment has passed mean length while shortening. The posterior location EMG offset occurs

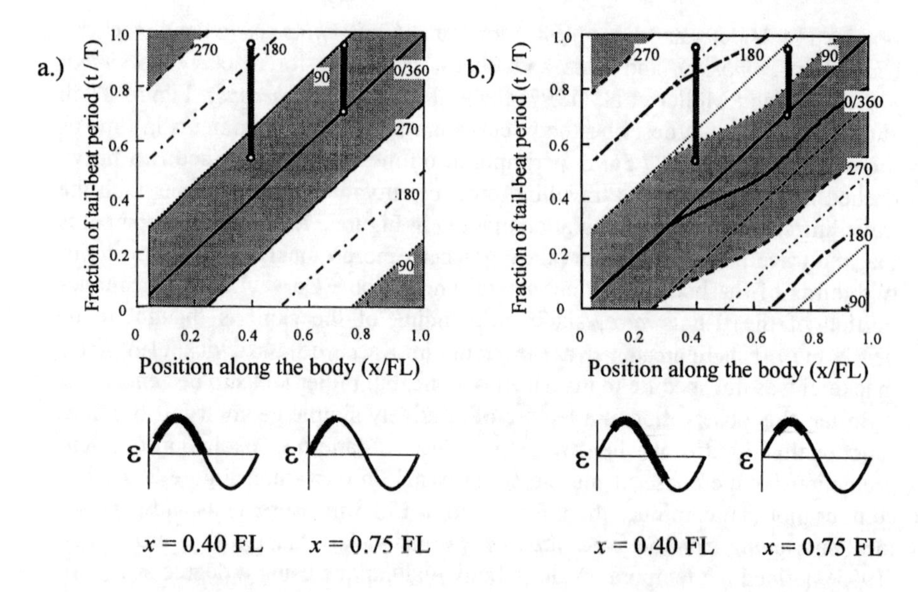

Figure 27.6. a) Phase diagram of local muscle length changes as a function of location on the body and through time. The cycle of segment length is defined as 0° or 360° when the muscle is at mean length, but lengthening; 90° at maximum length; 180° when at mean length, but shortening; and 270° when at minimum length. The time axis is normalized by tail-beat period (T) and the body location axis is normalized by fork length (FL). The straight lines proceeding from left to right and from bottom to top are lines of equivalent phase of segment length based on lateral deflection as in Figure 27.5a, and also represented in Wardle and Videler (1993). These data are based on a mackerel (*Scomber japonicus*) swimming at 3.25 L/s. In an effort to simplify this plot, the regions that correspond to segment lengthening are shaded grey, and the regions that correspond to segment shortening are blank. The average onset and offset of EMG are indicated by the dark grey, vertical lines located at 0.4 L and 0.75 L—the range reported in Shadwick et al. (1998). The sinusoids below the figure express the results from the phase diagram onto a pattern of muscle length change. The thick portion of the lines corresponds to EMG activity mapped onto phase of strain for these two locations on the body. Using this surrogate for curvature, the muscle at 0.4 L is active before peak length and continues to be active for a good deal of shortening. This pattern is thought to produce maximal positive work contractions. The muscle at 0.75 L is active at approximately mean length and is inactive at peak length—a pattern of excitation that is thought to produce essentially negative work. b) The same phase diagram as in Figure 27.6a, with the same axes and representation. In this panel, however, the phase of local segment shortening is based explicitly on curvature. In this panel, local lengthening is shaded grey and lengthening is blank. It is apparent that in this plot, muscle shortening proceeds along the body at a changing rate. When the EMG activity is plotted over the phase diagram, a different result is obtained. Now the muscle at 0.4 L is active before peak length and while shortening—similar to Figure 27.6a, but the muscle at 0.75 L is now active at about the same point and continues to be active well after peak length. This is a pattern of excitation and strain that is thought to produce positive work contractions.

at 148°, or approximately half-way between maximal length and mean length. Therefore, lateral muscle in both locations is becoming active at approximately the same time with respect to local segmental length changes. Because the activation time of about 45-60° is close to what has been found to be optimal for net positive work production (Johnson and Johnston, 1991; and see above), we can predict that the muscle in both anterior and posterior locations is producing net positive work. However, the larger duty cycle (EMG-on as a fraction of the entire cycle) at the anterior location means that the anterior muscle remains active until the muscle has shortened almost to its minimum length, while the posterior muscle becomes inactive well before the muscle reaches minimum length. While there are some remaining unknowns concerning the physiology of the red muscle, in particular excitation-contraction coupling (see above), we feel confident proposing that the muscle is activated in a manner to produce net positive work.

Significantly, curvature proceeds along the body at a changing velocity. If we propose that all of the muscle is active at a similar point in the strain cycle, and we incorporate the observation of He et al. (1990) that the temporal relationship between EMG onset and force generation in red muscle is the same at all points on the body, then we would predict that a careful examination of EMG timing would reveal that EMG onset proceeds with a non-constant rate. In fact, there are data to suggest that this is the case (Williams et al., 1989; Jayne and Lauder, 1995a; Hammond et al., 1998). However, it does not appear that the data were collected or analyzed in the context of this hypothesis, and we are not in a position to accept the hypothesis on this basis. On the other hand, we cannot reject the hypothesis, and it would be really neat to see someone do the experiment that would allow this idea to be examined critically. Such an examination is exciting in providing a test of functional design on this level of the motor control cascade.

WORK PRODUCTION AND DESIGN

The conclusion that muscle is performing net positive work in anterior myotomes and net negative work in posterior myotomes (Videler, 1993; Wardle et al., 1995) is an exciting result. This is true in as much as the muscle appears to be similar and continuous from the front to the back of the fish, and, in spite of this, has the mechanical function of a tendon-like force transmitter in posterior muscles. We feel that recent experimental evidence suggests that all of the muscle is in fact doing the same type of mechanical work, and in particular, that the posterior myotomes are not designed to perform net negative work. However, is there a logical basis for evaluating the quality of these design ideas that allows us to do more than just compare data sets? Can we ask the following question: Is having the muscle perform significant amounts of negative work a high-quality design?

In order to evaluate this question, it must be remembered that muscles do negative work when an outside force is applied. And since the amplitude of strain in the posterior myotomes can be as high as ±10%, strains are in excess of what we may expect from elasticity residing entirely within the myosin cross-bridge. Thus, the negative work attributed to the posterior myotomes is active, large-scale lengthening of the whole muscle, and this has consequences for evaluating muscle design. While much has been made of the increased ability of muscle to generate force while being lengthened (Hill, 1938), it must be remembered that this only occurs while the muscle is being lengthened. Muscles do not actively lengthen by themselves. The other thing to keep in mind is that muscle has a very poor energy storage capacity. While collagenous tendon is capable of storing 5000 J kg^{-1}, muscle is only able to store 10 J kg^{-1} (Alexander and Bennet-Clark, 1977). This discrepancy is even larger when one considers that tendon has a much larger energetic resilience than muscle, and thus is capable of recovering a much larger fraction of this stored energy in subsequently doing useful work. In fact, the work done to extend the posterior muscles in fish by 10% is far in excess of what can be recovered elastically. Negative work, therefore, is lost, and the force spent to overcome the resistance of the active muscle is completely analogous to friction. Negative work dissipates energy; it does not store energy elastically. This makes the use of terms like 'stiffness' to describe the force-extension characteristics of muscle doing negative work particularly problematic.

While negative work has important and useful consequences for terrestrial musculoskeletal designs, the characteristics of actively lengthening muscle are less useful for aquatic designs. In terrestrial systems, muscles doing negative work are absorbing energy produced by muscles at some other location or at some other moment in time. In running over ground, the absorbed energy could be kinetic energy of the centre of mass directed downward at the end of the aerial phase. Many terrestrial animals have evolved sophisticated mechanisms for absorbing that energy, storing it, and then releasing it at appropriate times. Since muscles doing negative work can generate almost twice the force of a muscle undergoing a positive work contraction (Katz, 1939), one might expect that a gram of posterior muscle doing negative work is absorbing and dissipating the positive work produced by roughly twice as much anterior muscle. Therefore, for each gram of muscle doing negative work in antagonism of two grams of anterior muscle doing positive work (and consuming metabolic fuel), the net result is that three grams of muscle are not doing any work on the fluid. It would be hard to characterize this as a high-quality biomechanical design. A number of authors have made this or similar arguments in trying to evaluate the use of negative work in swimming (Rome et al., 1993; van Leeuwen, 1992; Katz and Shadwick, 1998). On the other hand, in situations where there are large transfers of kinetic energy that require sophisticated control mechanisms—such as abrupt manœuvring in high Reynolds number locomotion (Wardle et al.,

1995)—one might then expect the muscle to perform significant amounts of negative work. This seems less likely in steady swimming of fish in general.

WHAT'S NEXT?

There are still a couple of big holes in our understanding of muscle function in fish. One of them is assembling a complete force budget for single muscles within the fillet. The other one is getting an integrated understanding of coupled internal and external forces.

Readers will have noticed that direct measures of force production are absent from this paper. It was mentioned above that the anatomical complexities of fish swimming muscle have made this technically difficult. Just as important, the fluid character of the water also makes direct measurement difficult. A good deal of the momentum imparted to the water by a swimming fish is dissipated by the water's viscosity and is difficult to measure in any meaningful way. A horse can be made to run over a ground-reaction force plate and the instantaneous loads and moments present in the musculoskeletal system can be measured. In a fish there is no simple analogue to the ground-reaction force plate because of the complex relationship between the material properties of water and the dynamical structure of the wake. It is possible to make some estimates of work and power in the wake using flow visualization techniques (Müller et al., 1997; Wolfgang et al., 1999). However, this approach relies on a large number of assumptions, different combinations of which are made in different papers. In practice, this approach can produce widely different predictions of power output and efficiency of swimming fish, depending on the choice of assumptions (Müller et al., 1997). It remains to be seen how precise these techniques can be made when proper validation tools are available and brought to bear.

Recently, attempts have been made to measure the actual tension developed in the caudal tendons of some species of tunas. Knower et al. (1999) have placed tendon-buckle force transducers onto some of the tendons in the caudal peduncle of skipjack and yellowfin tuna. Tendon-buckle force transducers are essentially small (3 to 7 mm), three-tined forks, with a small strain gauge on the middle tine. A tendon is woven through the tines of the fork, and as tension is developed within the tendon, the middle tine is displaced. The displacement is transformed into tension with appropriate calibration of the strain gauge transducer. These transducers have successfully been used to measure the forces in the tendons of a wide variety of terrestrial animals (e.g., Biewener et al., 1988), but the technology has not been exploited in the study of fish. In part, this results because the geometrically complex fish myotomes do not usually have easily identified tendons, and if they do, the tendons are nested deep within the muscle making a surgical approach difficult. Tunas are somewhat unique in having large, robust collagenous tendons in the caudal peduncle. The most caudal myotomes are increasingly reduced and their aponeuroses (= posterior myosepta) collect into four major tendons at the caudal termination of each fillet. Knower et al. (1999) have shown that tension begins to rise in the caudal tendons when only the most anterior myotomes are active as indicated by EMG. This is significant because it indicates that force is transmitted along the body of tunas over almost the entire length of the fillet, while all of the intervening muscle is inactive and presumably not generating force. It is important to keep in mind that tuna have a unique distribution of red muscle within the fillet (see above) and that these measures of force represent the net contribution from whatever muscles are anterior of the transducer. Therefore, in fish, tendon buckles are still not describing the force output of single muscle fibres. Progress in our understanding of *in vivo* muscle performance will come with the development of more refined technologies to make discrete measurements of force production in individual muscles. With this type of discrete measurement in hand we will then be able to assemble a complete force budget for individual muscle components of the myotome, and thus be able to make more definitive statements about *in vivo* function.

The other issue that remains is assembling all this diverse information into a synthetic model of muscle function. It has often been observed that the wave of lateral undulation of the body travels at a constant velocity (Videler and Hess, 1984; Müller et al., 1997; Katz and Shadwick, 1998; Figure 27.6). Katz and Shadwick (1998) have also shown that for animals that increase the amplitude of undulation as it proceeds caudally (i.e., all fish of which we are aware), curvature travels at a non-constant velocity. The use of curvature as a surrogate for red-muscle strain in non-tunas has also been validated (Coughlin et al., 1996; Wakeling and Johnston, 1999; Katz et al., 1999). If all this is true, and if muscle were designed to produce similar positive work in contraction, then one would expect that the EMG would be constrained to follow a single temporal pattern to ensure co-ordinated locomotion. Implicit here is the idea that the activation would have to account for possible regional variation in excitation-contraction coupling characteristics in the muscles. Now return to Figure 27.1 in the context of this information. The result is a body wave of constant velocity (Figure 27.5a), and this limitation will feed back up the motor control cascade in Figure 27.1. This feedback will place very narrow constraints or limits on the design at each step. Achieving a quantitative understanding of these limits would represent a significant synthesis in our appreciation of design.

As exciting as that is, it is possible that there is another level to integrate into this framework. We have made the case that the constant velocity of the body wave has placed constraints on the design of the locomotion machinery, but is there something that constrains the body wave velocity itself? If there were some feature of the fluid that makes it undesirable to alter the body wave velocity, this would act to constrain the wave to a constant velocity. In turn, this would constrain the motor control cascade as described above. It would be intolerably naïve to suggest that there is no physical property of water that impacts the specifics of fish locomotion. At this point, however, we do not know enough about the system to identify with confidence which mechanical feature (or features) of water might be constraining body wave velocity. However, the generality of the observation that the velocity is constant along the body is strongly suggestive. At the very least, it seems worth investigating the alternative hypothesis that there is NO important characteristic of the fluid that places this constraint on body kinematics and motor control of movement in fish as a strategy of identifying this feature. If it turns out that there are good mechanical reasons to pass that wave form along the body at a constant rate, this would be a powerful connection between external and internal forces and a major step toward assembling a synthetic understanding of fish muscle design.

ACKNOWLEDGEMENTS

This research was funded by a National Science Foundation grant (IBN95-14203). The authors would like to thank Kobe Tai, Charles A. Pell, Taylor Hayes, H. Scott Rapoport, Torre Knower, and Lene Hefner for their help in preparing this material.

REFERENCES

Alexander, R.M. (1969). Orientation of muscle fibres in the myomeres of fishes. *Journal of the Marine Biological Association of the UK* **49**, 263–290.

Alexander, R.M., and Bennet-Clark, H.C. (1977). Storage of elastic strain energy in muscle and other tissues. *Nature* **265**, 114–117.

Altringham, J.D., and Johnston, I.A. (1990a). Modeling muscle power output in a swimming fish. *Journal of Experimental Biology* **148**, 395–402.

Altringham, J.D., and Johnston, I.A. (1990b). Scaling effects on muscle function: Power output of isolated fish muscle fibers performing oscillatory work. *Journal of Experimental Biology* **151**, 453–467.

Altringham, J.D., Wardle, C.S., and Smith, C.I. (1993). Myotomal muscle function at different locations in the body of a swimming fish. *Journal of Experimental Biology* **182**, 191–206.

Anderson, M.E., and Johnston, I.A. (1992). Scaling of power output in fast muscle fibers of the Atlantic cod during cyclical contractions. *Journal of Experimental Biology* **170**, 143–154.

Biewener, A.A., Blickhan, R., Perry, A.K., Heglund, N.C., and Taylor, C.R. (1988). Muscle forces during locomotion in kangaroo rats: Force platform and tendon buckle measurements compared. *Journal of Experimental Biology* **137**(1), 191–206.

Bone, Q. (1966). On the function of the two types of myotomal muscle fibre in elasmobranch fish. *Journal of the Marine Biological Association of the UK* **46**, 321–349.

Bone, Q. (1978). Locomotor muscle. In: *Fish Physiology*, Volume 7. Hoar, W.S., Randall, D.J. (eds.), Academic Press, New York, London, pp. 361–424.

Bone, Q., Marshall, N.B., and Blaxter, J.H.S. (1995). *Biology of Fishes* (2nd edition). Blackie Academic & Professional, Glasgow, Scotland, p. 332.

Brill R.W., and Dizon, A.E. (1979). Red and white muscle fiber activity in swimming skipjack tuna, *Katsuwonus pelamus* (L.). *Journal of Fish Biology* **15**, 679–685.

Coughlin, D.J., and Rome, L.C. (1996). The roles of pink and red muscle in powering steady swimming in scup, *Stenotomus chrysops*. *American Zoologist* **36**, 666–677.

Coughlin, D.J., Valdes, L., and Rome, L.C. (1996). Muscle length changes during swimming in scup: Sonomicrometry verifies the anatomical high-speed cine technique. *Journal of Experimental Biology* **199**, 459–463.

Covell, J.W., Smith, M., Harper, D.G., and Blake, R.W. (1991). Skeletal muscle deformation in the lateral muscle of the intact rainbow trout *Oncorhynchus mykiss* during fast start manoeuvres. *Journal of Experimental Biology* **156**, 453–466.

D'Aout, K., and Aerts, P. (1999). A kinematic comparison of forward and backward swimming in the eel *Anguilla anguilla*. *Journal of Experimental Biology* **202**, 1511–1521.

Dewar, H., Deffenbaugh, M., Thurmond G., Lashkari, K., and Block B.A. (1999). Development of an acoustic telemetry tag for monitoring electromyograms in free swimming fish. *Journal of Experimental Biology* **202**, 2693–2699.

Gillis, G.B. (1998). Neuromuscular control of anguilliform locomotion: patterns of red and white muscle activity during swimming in the American eel *Anguilla rostrata*. *Journal of Experimental Biology* **201**, 3245–3256.

Grillner S., and Kashin, S. (1976). On the generation and performance of swimming in fish. In: *Neural Control of Locomotion*. Herman, R.M., Grillner, S., Stein, P.S.G., and Stuart, D.G. (eds.), Plenum Press, New York, London, pp. 181–201.

Hammond, L., Altringham, J.D., and Wardle, C.S. (1998). Myotomal slow muscle function of rainbow trout *Oncorhynchus mykiss* during steady swimming. *Journal of Experimental Biology* **201**, 1659–1671.

He, P., Wardle, C.S., and Arimoto, T. (1990). Electrophysiology of the red muscle of mackerel *Scomber scombrus* L. and its relation to swimming at low speeds. In: *The 2nd Asian Fisheries Forum*. Hirano, R., and Hanyu, I. (eds.), Asian Fisheries Society, Manila, pp. 469–472.

Hebrank, M.R. (1980). Mechanical properties and locomotor functions of eel skin. *Biological Bulletin* **158**, 58–68.

Hebrank, M.R., and Hebrank, J.H. (1986). The mechanics of fish skin: Lack of an 'external tendon' role in two teleosts. *Biological Bulletin* **171**, 236–247.

Hess, F., and Videler, J.J. (1984). Fast continuous swimming of saithe (*Pollachius virens*): A dynamical analysis of bending moments and muscle power. *Journal of Experimental Biology* **109**, 229–251.

Hill, A.V. (1938). The heat of shortening and the dynamic constants of muscle. *Proceedings of the Royal Society of London B* **126**, 136–195.

Jayne, B.C., and Lauder, G.V. (1995a). Speed effects on mid-line kinematics during steady undulatory swimming of largemouth bass, *Micropterus salmoides*. *Journal of Experimental Biology* **198**, 585–602.

Jayne, B.C., and Lauder, G.V. (1995b). Are muscle fibers within fish myotomes activated synchronously? Patterns of recruitment within deep myomeric musculature during swimming in largemouth bass. *Journal of Experimental Biology* **198**, 805–815.

Johnson, T.P., and Johnston, I.A. (1991). Power output of fish muscle fibers performing oscillatory work: Effect of seasonal temperature change. *Journal of Experimental Biology* **157**, 409–423.

Johnson, T.P., Johnston, I.A., and Moon, T.W. (1991). Temperature and the energy cost of oscillatory work in teleost fast muscle fibers. *Pflügers Archiv* **419**, 177–183.

Johnson, T.P., Syme, D.A., Jayne, B.C., Lauder, G.V., and Bennett, A.F. (1994). Modeling red muscle power output during steady and unsteady swimming in largemouth bass. *American Journal of Physiology* **267**, R481–R488.

Johnsrude, C.L., and Webb, P.W. (1985). Mechanical properties of the myotomal musculo-skeletal system of rainbow trout *Salmo gairdineri*. *Journal of Experimental Biology* **119**, 71–83.

Johnston, I.A. (1983). Dynamic properties of fish muscle. In: *Fish Biomechanics*. Webb, P.W., and Weihs, D. (eds.), Praeger, New York, pp. 36–67.

Johnston, I.A., Franklin, C.E., and Johnson, T.P. (1993). Recruitment patterns and contractile properties of fast muscle fibers isolated from rostral and caudal myotomes of the short-horned sculpin. *Journal of Experimental Biology* **185**, 251–265.

Johnston, I.A., Patterson, S., Ward, P.S., and Goldspink, G. (1974). The histochemical demonstration of myofibrillar adenosine triphosphatase activity in fish muscle. *Canadian Journal of Zoology* **52**, 871–877.

Johnston, I.A., van Leeuwen, J.L., Davies, M.L.F., and Beddow, T. (1995). How fish power predation fast-starts. *Journal of Experimental Biology* **198**, 1851–1861.

Jordan, C.J. (1996). Coupling internal and external mechanics to predict swimming behaviour: A general approach? *American Zoologist* **36**, 710–722.

Josephson, R.K. (1985). Mechanical power output from a striated muscle during cyclic contraction. *Journal of Experimental Biology* **114**, 493–512.

Katz, B. (1939). The relation between force and speed in muscular contraction. *Journal of Physiology* **96**, 45–64.

Katz, S.L., and Jordan, C.E. (1997). A case for building integrated models of aquatic locomotion that couple internal and external forces. In *Proceedings of the 10th International Symposium on Unmanned, Untethered Submersibles (Biological Propulsions Supplement)*. Autonomous Undersea Systems Institute, pp. 135–152.

Katz, S.L., and Shadwick, R.E. (1998). Curvature of swimming fish mid-lines as an index of muscle strain suggests swimming muscle produces net positive work. *Journal of Theoretical Biology* **193**, 243–256.

Katz, S.L., Shadwick, R.E., and Rappoport, H.S. (1999). Muscle strain histories in swimming milkfish in steady and sprinting gaits. *Journal of Experimental Biology* **202**, 529–541.

Knower, T., Shadwick, R.E., Katz, S.L., Graham, J.B., and Wardle, C.S. (1999). Red muscle activation patterns in yellowfin (*Thunnus albacares*) and skipjack (*Katsuwonus pelamis*) tunas during steady swimming. *Journal of Experimental Biology* **202**, 2127–2138.

Moon, T.W., Altringham, J.D., and Johnston, I.A. (1991). Muscle energetics and power output of isolated fish fast muscle performing oscillatory work. *Journal of Experimental Biology* **158**, 261–273.

Müller, U.K., van den Heuvel, B.L.E., Stamhuis, E.J., and Videler, J.J. (1997). Fish foot prints: Morphology and energetics of the wake behind a continuously swimming mullet (*Chelon labrosus* Risso). *Journal of Experimental Biology* **200**, 2893–2906.

Nursall, J.R. (1956). The lateral musculature and the swimming of fish. *Proceedings of the Zoological Society of London B* **126**, 127–143.

Ohmens, J.H., MacKenna, D.A., and McCulloch, A.D. (1993). Measurement of strain and analysis of stress in resting rat left ventricular myocardium. *Journal of Biomechanics* **26**, 665–676.

Rayner, M.D., and Keenan, M.J. (1967). Role of red and white muscles in the swimming of the skipjack tuna. *Nature* **214**, 392–393.

Rogers, S.C., and Weatherley, A.H. (1983). The use of opercular muscle electromyograms as an indicator of the metabolic costs of fish activity in rainbow trout, *Salmo gairdineri*, Richardson, as determined by radiotelemetry. *Journal of Fish Biology* **23**, 535–547.

Rome, L.C., Choi, I.H., Lutz, G., and Sosnicki, A. (1992b). The influence of temperature on muscle function in the fast swimming scup. 1. Shortening velocity and muscle recruitment during swimming. *Journal of Experimental Biology* **163**, 259–279.

Rome, L.C., Funke, R.P., and Alexander, R. McN. (1990). The influence of temperature on muscle velocity and sustained performance in swimming carp. *Journal of Experimental Biology* **154**, 163–178.

Rome, L.C., Loughna, P.T., and Goldspink, G. (1984). Muscle fiber recruitment as a function of swim speed and muscle temperature in carp. *American Journal of Physiology* **247**, R272–R279.

Rome, L.C., Sosnicki, A., and Choi, I.H. (1992a). The influence of temperature on muscle function in the fast swimming scup. 2. The mechanics of red muscle. *Journal of Experimental Biology* **163**, 281–295.

Rome, L.C., and Swank, D. (1992). The influence of temperature on power output of scup red muscle during cyclical length changes. *Journal of Experimental Biology* **171**, 261–281.

Rome, L.C., Swank, D., and Corda, D. (1993). How fish power swimming. *Science* **261**, 340–342.

Shadwick, R.E., Katz, S.L., Korsmeyer, K.E., Knower, T., and Covell, J.W. (1999). Muscle dynamics in skipjack tuna *Katsuwonus pelamus*: Timing of red muscle shortening and body curvature during steady swimming. *Journal of Experimental Biology* **202**, 2139–2150.

Shadwick, R.E., Steffensen, J.F., Katz, S.L., and Knower, T. (1998). Muscle dynamics in fish during steady swimming. *American Zoologist* **38**, 755–770.

Shanks, M.E., and Gambill, R. (1969). *Calculus of the Elementary Functions*. Holt, Rinehart & Winston, New York, 545 pp.

Swoap, S.J., Johnson, T.P., Josephson, R.K., and Bennett, A.F. (1993). Temperature, muscle power output and limitations on burst locomotor performance of the lizard *Dipsosaurus dorsalis*. *Journal of Experimental Biology* **174**, 185–197.

van Leeuwen, J.L. (1992). Muscle function in locomotion. In: *Mechanics of Animal Locomotion*. Volume 11 of Advances in Comparative and Environmental Physiology. Alexander, R. McN. (ed.), Springer-Verlag, Heidelberg, pp. 191–250.

van Leeuwen, J.L., Lankheet, M.J.M., Asker, H.A., and Osse, J.W.M. (1990). Function of red axial muscles of carp (*Cyprinus carpio*): Recruitment and normalized power output during swimming in different modes. *Journal of Zoology, London* **220**, 123–145.

Videler, J.J. (1993). *Fish Swimming*. Chapman and Hall, London.

Videler, J.J., and Hess, F. (1984). Fast continuous swimming of two pelagic predators, saithe (*Pollachius virens*) and mackerel (*Scomber scombrus*): A kinematic analysis. *Journal of Experimental Biology* **109**, 209–228.

Wainwright, S.A. (1983). To bend a fish. In: *Fish Biomechanics*. Webb, P.W., and Weihs, D. (eds.), Praeger, New York, pp. 68–91.

Wakeling, J.M., and Johnston, I.A. (1999). White muscle strain in the common carp and red to white muscle gearing ratios in fish. *Journal of Experimental Biology* **202**, 521–528.

Wardle, C.S., Videler, J.J., and Altringham, J.D. (1995). Tuning into fish swimming waves: Body form, swimming mode and muscle function. *Journal of Experimental Biology* **198**, 1629–1636.

Wardle, J.M., and Videler, J.J. (1993). The timing of the EMG in the lateral myotomes of mackerel and saithe at different swimming speeds. *Journal of Fish Biology* **42**, 347–359.

Westneat, M.W., Hoese, W., Pell, C.A., and Wainwright, S.A. (1993). The horizontal septum: Mechanisms of force transfer in locomotion of scombrid fishes (*Scombridae perciformes*). *Journal of Morphology* **217**, 183–204.

Willemse, J.J. (1966). Functional anatomy of the myosepta in fishes. *Proceedings of the Akad. Wetensch. Kon. Nederlands* **69C**, 58–63.

Williams T.L., Grillner, S., Smoljaninov, V.V., Wallen, P., Kashin, S., and Rossignol, S. (1989). Locomotion in lamprey and trout: The relative timing of activation and movement. *Journal of Experimental Biology* **143**, 559–566.

Wolfgang, M.J., Anderson, J.M., Grosenbaugh, M.A., Yue, D.K.P., and Triantafyllou, M.S. (1999). Near-body flow dynamics in swimming fish. *Journal of Experimental Biology* **202**, 2303–2327.

28 Visco-elastic Properties of Cardiac Trabeculae: Re-examination of Diastole

BRUNO D. STUYVERS
Faculty of Medicine, University of Calgary, Calgary, Alberta, Canada

MASAHITO MIURA
1st Department of Medicine, Tohoku University School of Medicine, Sendai, Japan

HENK E. D. J. TER KEURS
Faculty of Medicine, University of Calgary, Calgary, Alberta, Canada

Diastole is commonly described as a period of cardiac cycle where blood entering the ventricle stretches the myocardial tissue passively, while intracellular ionic concentrations are in a steady state and cellular processes of contraction are silent. During this period, myocardial fibres elongate and generate passive tension on ventricular content through the classical tension-length relation.

The shape of the relation between passive tension and level of stretch is determined by the visco-elastic properties of cardiac tissue. It is commonly admitted that these properties are steady and, at rest, the myocardium behaves like a passive visco-elastic system. Based on this concept of passive system, the myocardial fibres should respond to the same stretch with identical visco-elastic resistance as long as the muscle stays unstimulated. Actually, recent studies on rat cardiac trabeculae revealed that visco-elastic resistance of cardiac muscle varies during resting periods separating two twitches (Stuyvers et al., 1997). We found that such changes resulted from two phenomena: 1) intracellular free-Ca^{2+} concentration ($[Ca^{2+}]_i$) varies continuously during the resting period (Stuyvers et al., 1997), and 2) the visco-elastic properties are affected by small variations of Ca^{2+} through a mechanism other than normal Ca^{2+} activation of contractile filaments (Stuyvers et al., 1998). We review in this chapter the major experimental features leading to the novel and challenging idea that an important factor of the 'active' properties of cardiac muscle, such as calcium ions, may regulate the 'passive' properties as well.

Skeletal Muscle Mechanics: From Mechanisms to Function. Edited by W. Herzog.

DIASTOLIC LENGTHENING OF CARDIAC SARCOMERES

Trabeculae constitute a well-known model of properties of cardiac muscle. Continuous measurement of the sarcomere length (SL) by laser diffraction technique permits monitoring of the active shortening of contractile units during contraction. When the preparation is stimulated repetitively, resting periods (wherein normally no major contractile event should be detected) separate active shortenings of the sarcomeres. Contrary to this expectation, we noticed systematically a slow and small increase of SL trace during the time interval between two twitches. We decided to examine whether this SL increase reflected a physiological event or was simply a technical artefact.

PREPARATION

Trabeculae (n = 17) were dissected from right ventricles of rat hearts and mounted in an experimental chamber as described previously (Stuyvers et al., 1997; De Tombe and ter Keurs, 1992). The preparation was stimulated every two seconds and left to equilibrate for one hour at 25°C in the perfusion medium. The standard solution was a modified Krebs-Henseleit buffer solution composed of (mM) 112 NaCl, 1.2 $MgCl_2$, 5 KCl, 2.4 Na_2SO_4, 2 NaH_2PO_4, 1 $CaCl_2$, 10 Glucose, 19 $NaHCO_3$, and equilibrated with 95% O_2 and 5% CO_2; pH 7.4 (adjusted with KOH 1 M).

FORCE, MUSCLE LENGTH, AND SARCOMERE LENGTH MEASUREMENTS

Methods used to measure force (F), muscle length (ML), and sarcomere length (SL) have previously been described in detail (Stuyvers et al., 1997; De Tombe and ter Keurs, 1992; Daniels et al., 1984). Variations of muscle length were determined from the displacement of a motor arm controlled by a dual servo-amplifier (model 300s, Cambridge Technology, Watertown, MA, USA). Force was measured by a silicon strain gauge (model X17625, SensoNor, Horten, Norway) with a resolution of 0.63 μN (Stuyvers et al., 1997). SL was measured by laser diffraction techniques as previously described (Stuyvers et al., 1997; Daniels et al., 1984). The resolution of SL measurements was 2-3 nm.

RESULTS

Figure 28.1 shows the typical SL changes following a twitch. Complete relaxation of twitch force was accompanied by a sarcomere shortening transient followed by a gradual lengthening (ΔSL) (Figure 28.1A). The course of ΔSL was individually fitted by a monoexponential function (Figure 28.1B). The variation of SL during the resting period was reproducible from muscle to muscle, but the amplitude varied under otherwise similar conditions (ΔSL =25±9 nm, n=15).

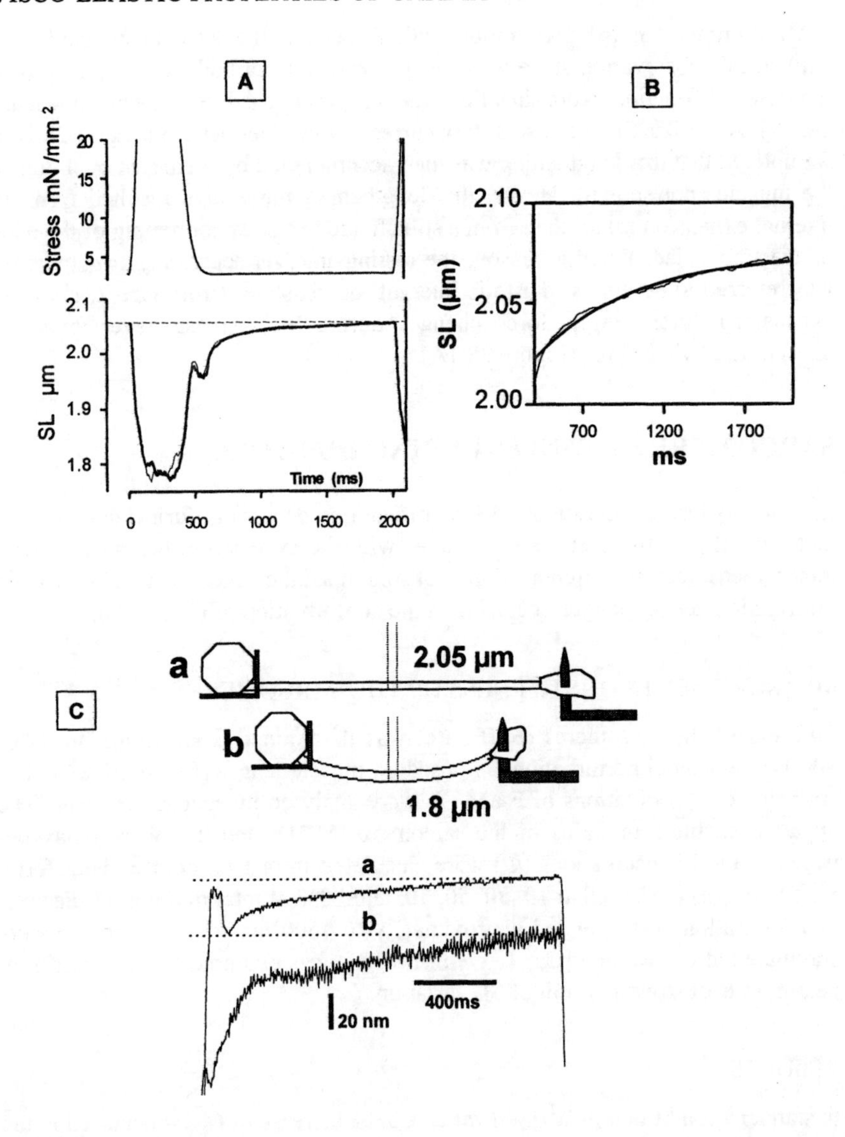

Figure 28.1. SL change during diastolic interval in intact rat cardiac trabeculae. A: Typical time course of stress and sarcomere length (SL) during and in between twitches of a rat cardiac trabecula. B: Diastolic sarcomere lengthening from panel A represented on a larger scale; the superimposed continuous line represents the non-linear regression fit to the data: $SL = -0.02 \cdot e^{(-t/618)} + 2.1$; $(r = 0.92)$; temperature: 25°C; pH: 7.4; $[Ca^{2+}]_o = 1$ mM). C: Lengthening of the sarcomere in the central segment of a trabecula at two different sarcomere lengths. Note that diastolic lengthening observed at SL: 2.05 µm (a) still occurred below slack length (b).

Misalignment of the preparation with respect to the laser beam could not explain this observation since SL changes derived from both first order lines of the laser diffraction were identical (see superimposed SL traces in Figure 28.1A). A similar SL increase still occurred below slack length (Figure 28.1C); we noticed that this lengthening was then accompanied by increased buckling of the muscle (not shown). Hence, this lengthening must have resulted from an internal expansion rather than from a stretch due to a slow shortening of the ends of the trabeculae. For this reason, the resting interval separating two twitches was referred to herein as 'diastolic interval' or 'diastole' (from Greek, *diastole*: expansion). Interestingly, force changes during SL increase never exceeded 0.3% of the twitch force (Figure 28.1A).

SARCOMERE STIFFNESS DURING DIASTOLE

In order to further investigate the modifications occurring during diastole, we measured the stiffness of the sarcomeres with the expectation that such a parameter measured at different times of the diastolic interval would provide information about changes occurring in the conformation of the system.

MEASUREMENTS OF STIFFNESS OF THE SARCOMERES

Stiffness of the sarcomeres (Stiff-Sarc) was determined from 30-ms bursts of 500 Hz sinusoidal perturbations imposed on muscle length (Figure 28.2A). The corresponding oscillations of F and SL were analysed by spectral analysis. The apparent stiffness modulus of the sarcomere (MOD) and the phase delay between F and SL oscillations (Φ) were calculated from Fast Fourier Transform. Stiff-Sarc was measured at 10, 30, 50, 70, and 90% of total duration of diastole. 100% duration corresponded approximately to 1500 ms when trabeculae were stimulated at 0.5 Hz and $[Ca^{2+}]_o = 1$ mM. MOD was expressed in mN/mm^2/µm, i.e., in units of stress per unit of SL variation.

RESULTS

In standard conditions ($[Ca^{2+}]_O = 1$ mM), a 30% increase of MOD occurred in the initial 450 ms of the diastole (MOD: 9.3±0.6 at 10% versus MOD: 12.2±0.5 mN/mm^2/µm at 50%, $n = 158$; $p < 0.05$, 18 muscles, Figure 28.2B, left panel). MOD decreased slightly at the end of the diastole to 11.6±0.65 mN/mm^2/µm. On average, MOD increased during diastole following a monoexponential function (see Figure 28.2B, left panel, dotted line) with a time constant of ~300 ms.

During the same period, Φ decreased significantly from 84±3° to 73±4° ($n = 158$; $p<0.05$; 18 muscles; Figure 28.2B, right panel). At $[Ca^{2+}]_O = 1$ mM, 15 of 18 trabeculae studied showed discrete sarcomere spontaneous activity with a minimal effect on force and restricted to a brief period at the end of diastole (see the example of

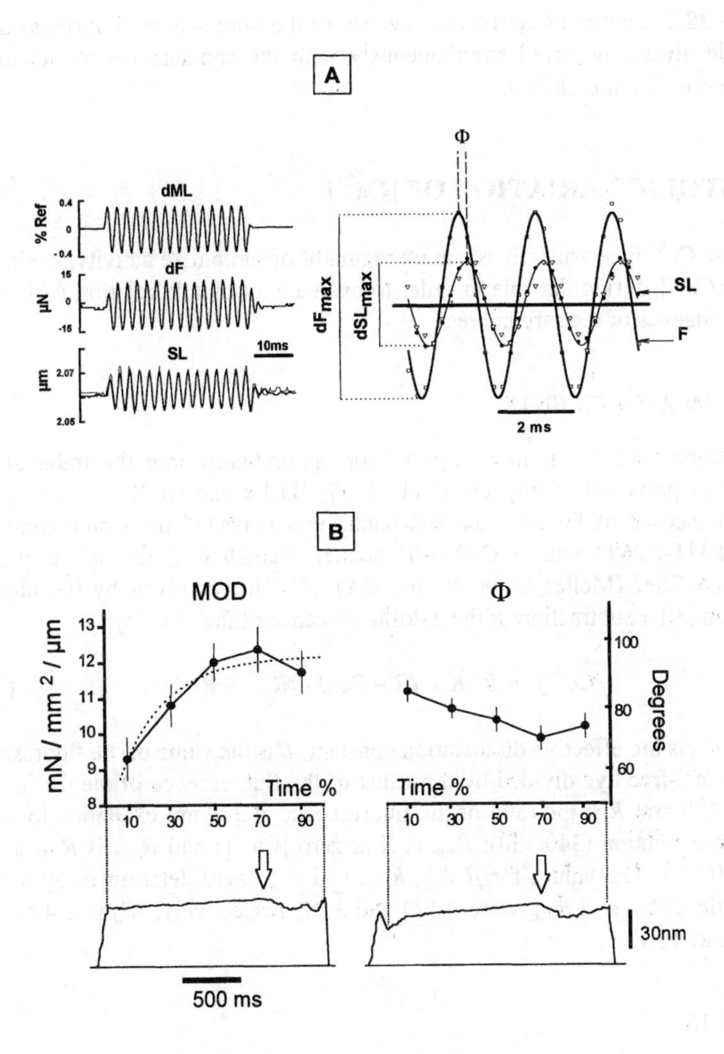

Figure 28.2. Sarcomere stiffness during diastole. A, left panel: Typical records of response of sarcomere length (SL) and force (dF) during a 30-ms burst of 500 Hz sinusoidal perturbations of muscle length (dML). SL was recorded from two detectors as described in Daniels et al. (1984). A, right panel: SL and F sinusoids represented on a slower time base. B: Time course of stiffness during diastole at $[Ca^{2+}]_o = 1$ mM. Data are expressed as mean±SEM ($n=18$). The time scale was expressed in % of the diastolic interval (see text). The time course of MOD was fitted with a monoexponential function (see dotted line): $MOD = -9 \cdot e^{(-t/308)} + 12$; $(r = 0.92)$. Lower traces show an example of time course of sarcomere length obtained from a representative trabeculae showing slight spontaneous fluctuations of SL starting at 70% of diastole (see arrows).

Figure 28.2B, bottom traces). The reversal of the time course of stiffness during diastole always occurred simultaneously with the spontaneous motion of the sarcomeres (Figure 28.2B).

DIASTOLIC VARIATION OF $[Ca^{2+}]_i$

Because Ca^{2+} is considered as a main regulator of sarcomere activity, we investigated $[Ca^{2+}]_i$ during diastole in order to evaluate a possible relation with mechanical changes of the sarcomeres.

$[Ca^{2+}]_i$ MEASUREMENTS

Ca^{2+} probe Fura-2 was microinjected iontophoretically into the trabeculae as described previously (Stuyvers et al., 1997; Backx and ter Keurs, 1993). The epifluorescence of Fura-2 from trabeculae was collected by a photomultiplier tube (PMT-R2693 with a C1053-01 socket, Hamamatsu) through a 500 nm bandpass filter (Melles Griot, Irvine, CA). $[Ca^{2+}]_i$ was given by the classical equation (after subtraction of the autofluorescence of the muscle):

$$[Ca^{2+}]_i = \beta \cdot K'_d \cdot (R - R_{min}) / (R_{max} - R) \tag{28.1}$$

where K'_d is the effective dissociation constant; β is the value of the fluorescence of the Ca^{2+}-free dye divided by the value of the fluorescence of the Ca^{2+}-bound dye at 380 nm; R is the ratio of the fluorescence at 340 nm excitation to that at 380 nm excitation (340/380); R_{min} is R at zero $[Ca^{2+}]_i$; and R_{max} is R at a saturating $[Ca^{2+}]$. The values for β, K'_d, R_{min}, and R_{max} were determined by *in vitro* calibration. R_{min} and R_{max} were 0.129 and 3.71, respectively; K'_d was 4.57 mM, and β was 12.1.

RESULTS

As illustrated in Figure 28.3, relaxation of the Ca^{2+} transient did not end with the active force development. Instead, $[Ca^{2+}]_i$ continued to decrease during diastole. When $[Ca^{2+}]_o = 1$ mM, the amplitude of the latter decline was approximately 120 nM. Decay of $[Ca^{2+}]_i$ was completed approximately 200 ms before the end of the diastolic interval, i.e., at 90% of total duration. The time constant of the decay during the diastole was calculated from the fit of a monoexponential function to the decline of $[Ca^{2+}]_i$. The time constant (t_{Ca}) was 210-325 ms, i.e., similar to the time constant of increase of sarcomere stiffness during the same period (see Figure 28.2B). Similarity between time courses of both parameters suggested that visco-elastic properties of cardiac sarcomeres were influenced by variations of $[Ca^{2+}]_i$ in the submicromolar range.

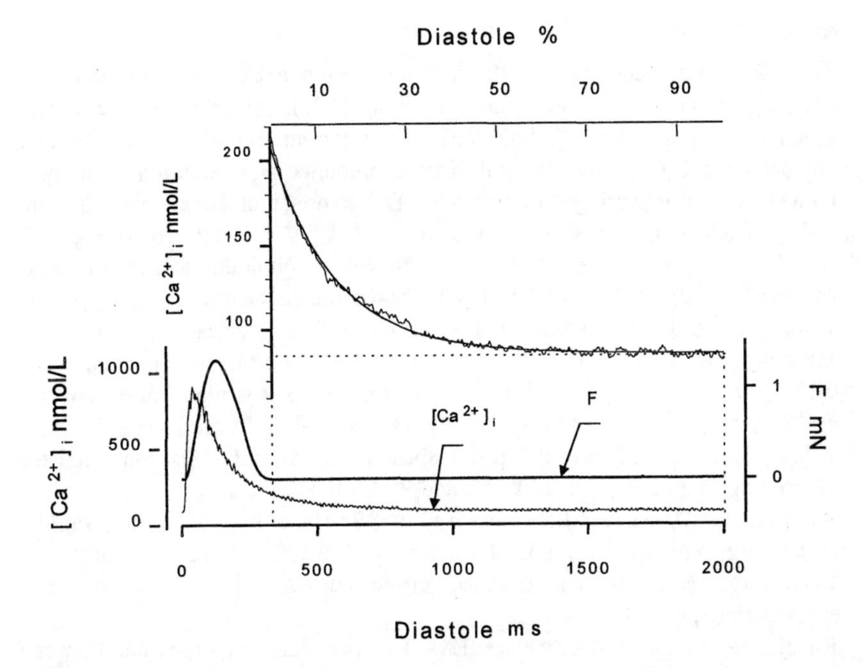

Figure 28.3. $[Ca^{2+}]_i$ during the diastolic interval: assessment of $[Ca^{2+}]_i$ simultaneously with F showed an exponential decline of $[Ca^{2+}]_i$ from the peak systolic to the end-diastolic level. The diastolic part of $[Ca^{2+}]_i$ decay is represented on a larger scale on the upper part of the diagram. Amplitude of the decay during diastole was 123 nM; data were fitted through a monoexponential function (see superimposed continuous line): $[Ca^{2+}]_i = 738 \cdot e^{(-t/224)} + 86$; $(r = 0.98)$. The time scale was expressed in ms (lower part) and in % of diastole (upper part).

EVIDENCE FOR Ca^{2+}-DEPENDENCE OF PASSIVE PROPERTIES OF CARDIAC SARCOMERES

The above results suggested the existence of a relationship between visco-elastic properties of the sarcomeres and Ca^{2+} in a range of concentrations wherein no contractile process was expected to occur. In order to further investigate this possibility, we used the same approach to sarcomere stiffness in trabeculae made permeable with detergent.

Ca^{2+} AND SARCOMERE STIFFNESS IN SKINNED TRABECULAE

The use of skinned fibres enabled us to control $[Ca^{2+}]$ surrounding the myofibrils. We focussed our study on $[Ca^{2+}]$-concentrations measured during diastolic decrease of $[Ca^{2+}]_i$ in intact trabeculae, and we calculated the corresponding stiffness of the sarcomeres using the same method of sinusoidal perturbations.

Protocol

Trabeculae were incubated for 30-45 minutes with a Ca^{2+}-free solution (RS) containing saponin (50 μg/mL) (Stuyvers et al., 1998). Then the preparation was washed four times with RS. Solutions with different free Ca^{2+}-concentrations were prepared by mixing RS with varied amounts of a solution containing 10 mM of Ca^{2+} ('activating solution,' AS). The composition (total concentrations in mM) of RS was: 100 BES^-, 10 $EGTA^{4-}$, 6.8 ATP^{4-}, 10 CP^{3-}, 6.30 Mg^{2+}, 55 $CH_3CH_2COO^-$ (propionate), 12.6 Cl^-, 33.6 Na^+, 1 Dithiothreitol; temperature: 25°C; pH 7.1 (adjusted with KOH 1 M). Total ionic concentrations ($[\,]_{total}$) were calculated in order to obtain final solutions with appropriate free ion concentrations ($[\,]_{free}$), with net charge = 0, ionic strength = 0.2 M, and ionic equivalent = 0.17 M at 25°C and pH 7.1. The composition of AS was identical except that $[Mg^{2+}]_{total}$ and $[Ca^{2+}]_{total}$ were respectively 5.85 and 10 mM (equivalent $[Mg^{2+}]_{free}$ and $[Ca^{2+}]_{free}$: 3.1 mM and 400 μM respectively). All solutions contained the protease inhibitors Leupeptin (40 μM) and P.M.S.F. (0.5 mM).

Sarcomere stiffness was first measured in RS. The solution was then rapidly removed and replaced by 1 mL of solution with a different $[Ca^{2+}]$. Free Ca^{2+}-concentrations between 0 and 430 nM were tested randomly. All measurements were performed at 25°C.

Results were expressed as mean±SEM. The significance of relations between stiffness and $[Ca^{2+}]$ was judged on the basis of ANOVA and the correlation coefficients of the regressions. Student's t-test was used for comparisons of two means; the difference was considered significant when $p<0.05$. Data were fitted using a Marquardt-Levenberg algorithm.

Results

MOD was first measured in intact muscle at 1350 ms after the stimulation, i.e., approximately at half of diastole, and was taken as the reference (MOD_{ref}): MOD_{ref} was 36.6±4.8 mN/mm²/μm ($n=17$). In Fura-2 micro-injected trabeculae, $[Ca^{2+}]_i$ had then reached about 100 nM. We found that MOD of skinned fibres at $[Ca^{2+}] = 100$ nM was 36.8±6.5 mN/mm²/μm ($n = 17$), i.e., identical to MOD_{ref} measured under equivalent conditions in skinning experiments while the muscle was still intact. The phase delay between SL and F oscillations (Φ) measured at $[Ca^{2+}] = 100$ nM was similar to Φ measured in skinned fibres before skinning (Φ_{ref}): $\Phi = 160±16°$ and $\Phi_{ref} = 170±27°$ ($n = 17$). Below 450 nM, MOD varied with $[Ca^{2+}]$ following a triphasic relation (17 muscles; Figure 28.4A). At $[Ca^{2+}]$ below 70 nM, MOD was independent of $[Ca^{2+}]$. Between 70 and 200 nM, MOD decreased with increase of $[Ca^{2+}]$. Above 200 nM, a steep increase in MOD was observed with an increase of $[Ca^{2+}]$. Over the total range of $[Ca^{2+}]$ tested, the data could be fitted with the sum of two sigmoidal relationships. Figure 28.4A shows that values of MOD obtained below 200 nM of $[Ca^{2+}]$ fitted well with a sigmoidal relationship with a negative Hill coefficient. Polynomials fitted through the data between 90 and 200 nM (see insert Figure 28.4A) permitted a model-independent estimate of the maximal variation of MOD versus [Ca^{2+}], and showed that MOD decreased by 50% over an increase in [Ca^{2+}] of 100 nM. Under identical conditions, similar polynomials showed that Φ increased by 80° (see insert Figure 28.4B). Between 1 and 200 nM, changes of Φ with [Ca^{2+}] were opposite to variations of MOD (Figure 28.4A and Figure 28.4B). Above 200 nM, Φ decreased again with further increase of [Ca^{2+}], and reached the same values as measured below 70 nM.

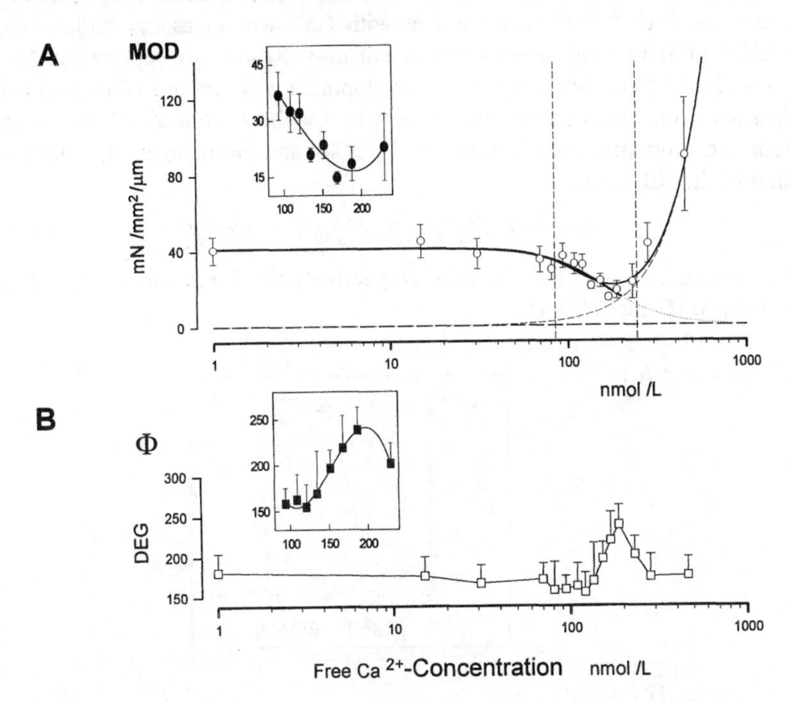

Figure 28.4. Sarcomere stiffness-[Ca^{2+}] relationship in skinned trabeculae: MOD and Φ were plotted against free Ca^{2+}-concentration; data were expressed as mean±SEM (n=17). Between 1 and 200 nM, MOD fitted through the following Hill function (function A; see thick continuous line): $MOD = MOD_{max} \cdot [Ca^{2+}]^{n_H} / ([Ca^{2+}]_{50}^{n_H} + [Ca^{2+}]^{n_H})$ where MOD_{max} is the maximal value of MOD obtained between 1 and 200 nM (41±2 mN/mm²/µm); n_H is the Hill coefficient (−2.6±0.7); $[Ca^{2+}]_{50}$ is [Ca^{2+}] at half amplitude of the sigmoidal decay between 1 and 200 nM ($[Ca^{2+}]_{50}$ = 160±13 nM). It was assumed that the lower part of the sigmoid (dotted part of the curve at [Ca^{2+}] > 200 nM) tends to reach the stiffness of weakly attached cross-bridges (wXb stiffness: horizontal dashed line; see text); stiffness modulus of wXb (MOD_{wXb}) was 0.06 mN/mm²/µm (see text). The 'active stiffness' (function B; see ascending dashed line) was calculated from the sum of wXb stiffness and sXb stiffness. Development of sXb stiffness with [Ca^{2+}] was predicted from the typical active force-[Ca^{2+}] relation described in Figure 28.5 (see text for details). Data obtained between 1 and 450 nM were fitted with the sum of function A, function B, and wXb stiffness. The lower panel shows variations of Φ in the range of [Ca^{2+}] = 1-430 nM. Inserts in both upper and lower panels show on a linear horizontal scale data between 90-230 nM; third-order polynomial functions have been

used to fit the data (thick continuous line) and to accurately estimate (in a model-independent fashion) variations of MOD and Φ with $[Ca^{2+}]$.

ACTIVE FORCE DEVELOPMENT

In order to verify that our protocol did not alter the contractile properties of the sarcomeres, active force development with Ca^{2+} was measured under experimental conditions used in our study of stiffness. Results are reported in Figure 28.5, and they show that active force development was similar to data previously published from skinned cardiac trabeculae (without control of the internal sarcomere shortening: see Kentish et al., 1986, and Harrison et al., 1988), and followed the Hill equation:

$$F = Fmax \cdot [Ca^{2+}]^{n}{}_{H} / ([Ca^{2+}]_{50}{}^{n}{}_{H} + [Ca^{2+}]^{n}{}_{H}) \qquad (28.2)$$

where *Fmax*, n_H, and $[Ca^{2+}]_{50}$ were respectively 56±2 mN/mm^2, 2.1±0.2, and 3.4±0.2 μM (Figure 28.5A).

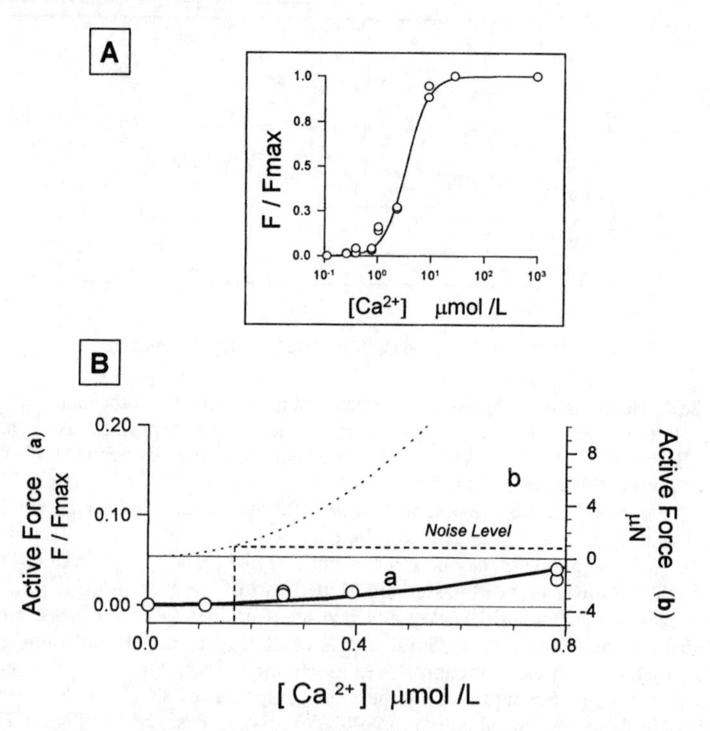

Figure 28.5. The active force-$[Ca^{2+}]$ relationship. A: Active force developed by two saponin skinned trabeculae under experimental conditions used for study of stiffness. Active force follows a typical Hill function when $[Ca^{2+}]$ is increased up to 1 mM: Fmax= 56±2 mN/mm^2, Hill coefficient (n_H)=2.1±0.2, $[Ca^{2+}]_{50}$=3.4±0.3 μM (see text for details);

force was normalized to Fmax. B: Data of panel A at $[Ca^{2+}]$ below 800 nM represented on a linear scale. Force was expressed in μN (trace b) in order to facilitate the comparison with the noise level of the system (dashed line as indicated). Horizontal thin lines indicate the force zero level. The detection of force development at submicromolar $[Ca^{2+}]$ was limited by the noise in our system (0.63 μN). Given the coefficients of the Hill regression for these muscles (cross-sectional area: 0.01 mm^2), we would have expected that active force equalled the noise level at $[Ca^{2+}]=140$ nM, and exceeded the noise level by a factor of 4-5 at $[Ca^{2+}]=300$ nM, as it was observed experimentally (see Figure 28.5B).

THEORETICAL ANALYSIS OF THE STIFFNESS-$[Ca^{2+}]$ RELATIONSHIP

The stiffness-$[Ca^{2+}]$ relationship involved three steps, depending on the range of concentrations tested. Figure 28.4A shows that data could be accounted for if we assumed that the relationship between stiffness and $[Ca^{2+}]$ resulted from the sum of two sigmoidal functions. One function (A) showed a sigmoidal decrease of MOD with a Kd of 160±13 nM and a slope (n_A) of −2 to −3 (−0.6±0.7). The second function (B) reflected a sigmoidal increase of MOD with increasing $[Ca^{2+}]$. Function B follows directly from the active force-$[Ca^{2+}]$ relationship (Figure 28.5) if one assumes that stiffness of activated cross-bridges increased in proportion to the force. It can be shown (Stuyvers et al., 1997; Brenner et al., 1982) that the contribution of cross-bridges in weakly attached states (wXb) to stiffness at 500 Hz is 0.06 $mN/mm^2/\mu m$, i.e., ~600 times lower than the values measured in our study (see horizontal dashed line in Figure 28.4A). Stiffness of Ca^{2+}-dependent cross-bridges (in strongly attached states, sXb) followed from instantaneous release experiments. It has been demonstrated for skeletal muscle (Huxley and Simmons, 1971) and cardiac muscle (Backx and ter Keurs, 1988) that a near instantaneous release (of ~12 nm in cardiac trabeculae) of the fully activated sarcomere causes force to drop to zero. Considering that maximal force developed under our experimental conditions was approximately 60 mN/mm^2, stiffness of uniformly activated trabecula would have been ~5000 $mN/mm^2/\mu m$. Hence, the relationship between $[Ca^{2+}]$ and stiffness of sXb followed directly from the parameters of the Hill function fitting the data of Figure 28.4A. Development of 'active stiffness,' as it is represented in Figure 28.4A, resulted from the sum of wXb stiffness and sXb stiffness. The excellent fit between experimental data and the above model suggested that the stiffness-$[Ca^{2+}]$ relationship measured in our study between 1 and 450 nM resulted from the sum of the following two different Ca^{2+}-dependent mechanisms: 1) a decrease with increase of $[Ca^{2+}]$ of the stiffness generated by a system which dominates at low $[Ca^{2+}]$ (below 200 nM), i.e., at $[Ca^{2+}]$ in which intact muscle operates during diastole, and 2) at $[Ca^{2+}] > 200$ nM, an increase of stiffness due to the formation of cross-bridges in the strongly attached state.

COMPARISON BETWEEN INTACT AND SKINNED TRABECULAE

We measured the diastolic stiffness of intact trabeculae before skinning, at 1350 ms after stimulation. We have found that, at this time, $[Ca^{2+}]$ was ~100 nM. At $[Ca^{2+}]=100$ nM, the stiffness moduli of skinned trabeculae were nearly identical to those measured before skinning, suggesting that saponin at steady $[Ca^{2+}]$ did not affect significantly the apparent stiffness of the sarcomeres. The variation of MOD and Φ over a range of $[Ca^{2+}]$ encountered in intact muscle during diastole (230-90 nM) was also the same in skinned muscles as compared to intact trabeculae: MOD was inversely related to $[Ca^{2+}]$ and the variations of MOD and Φ with $[Ca^{2+}]$ were opposite. It is clear that the stiffness-$[Ca^{2+}]$ relation of the sarcomeres in skinned fibres can explain the increase of sarcomere stiffness, while $[Ca^{2+}]_i$ gradually declined from 200 to 90 nM during diastole of intact stimulated muscles.

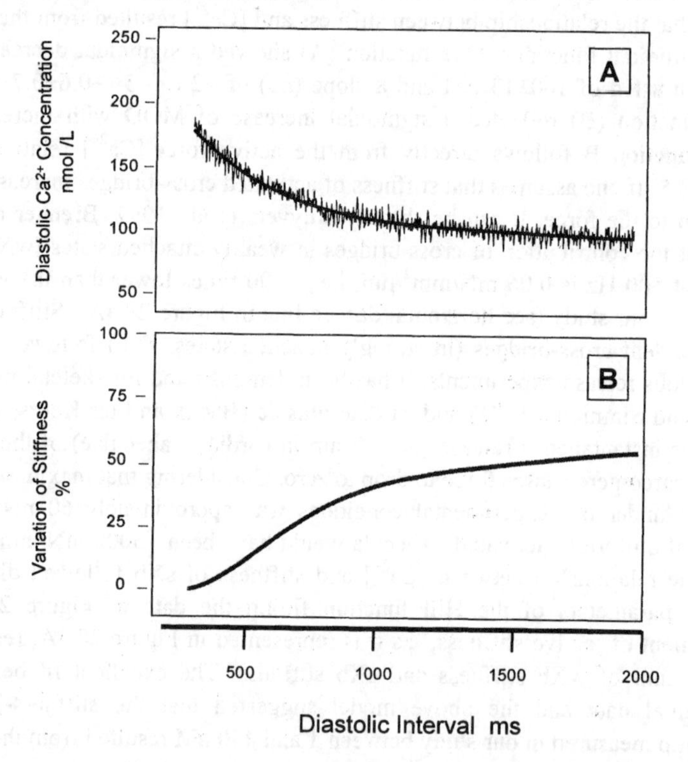

Figure 28.6. Modelling of the time course of stiffness during diastole from the stiffness-$[Ca^{2+}]$ relation. A: Variation of $[Ca^{2+}]_i$ observed experimentally during diastole as represented in Figure 28.3. B: Values of MOD calculated from values of $[Ca^{2+}]_i$ of panel A using the sum of functions A and B of Figure 28.4.

Interestingly, we could faithfully reproduce the time course of stiffness observed in intact trabeculae during diastole from the stiffness-[Ca^{2+}] relation determined in skinned trabeculae (see Figure 28.6): values of [Ca^{2+}] measured during the diastolic decline of $[Ca^{2+}]_i$ in intact trabeculae (Figure 28.3) were incorporated in the function based on the sum of two sigmoids and fitting the data of Figure 28.4. As shown in Figure 28.6B, the calculated course of the stiffness modulus matched well the increase seen experimentally in intact trabeculae (see Figure 28.2B).

CONCLUSION

Our work shows that resting properties of intact cardiac sarcomeres are sensitive to [Ca^{2+}] in the submicromolar range. Further analysis of these results revealed that such a Ca^{2+} dependence originates in a visco-elastic system that is independent of the formation of cross-bridges and that generates stiffness in an inverse Ca^{2+}-dependent manner. Importantly, this system operates in the range of Ca^{2+} concentrations wherein the myocytes regulate their cytosolic Ca^{2+} level during the diastole of cardiac muscle. Our present finding is critical because it suggests that pharmacological and pathological conditions affecting Ca^{2+}-homeostasis in the myocytes are potentially responsible for alterations of diastolic properties of the myocardium as well.

Preliminary results obtained on skinned fibres in our laboratory strongly suggest that the large endo-sarcomeric protein titin is involved in the Ca^{2+}-dependence of visco-elastic properties of cardiac sarcomeres.

REFERENCES

Backx, P.H., and ter Keurs, H.E.D.J. (1988). Restoring forces in rat cardiac trabeculae. *Circulation* **78**, 11–68.

Backx, P.H., and ter Keurs, H.E.D.J. (1993). Fluorescent properties of rat cardiac trabeculae microinjected with fura-2 salt. *American Journal of Physiology* **264**, H1098–H1110.

Brenner, B., Schoenberg, M., Chalovich, J.M., Greene, L.E., and Eisenberg, E. (1982). Evidences for cross-bridge attachment in relaxed muscle at low ionic strength. *Proceedings of the National Academy of Sciences USA* **79**, 7288–7291.

Daniels, M.C.G., Noble, M.I.M., ter Keurs, H.E.D.J., and Wohlfart, B. (1984). Velocity of sarcomere shortening in rat cardiac muscle: Relationship to force, sarcomere length, calcium, and time. *Journal of Physiology (London)* **355**, 367–381.

De Tombe, P.P., and ter Keurs, H.E.D.J. (1992). An internal viscous element limits unloaded velocity of sarcomere shortening in rat myocardium. *Journal of Physiology (London)* **454**, 619–642.

Harrison, S.M., Lamont, C., and Miller, D.J. (1998). Hysteresis and the length dependence of calcium sensitivity in chemically skinned rat cardiac muscle. *Journal of Physiology (London)* **401**, 115–143.

Huxley, A.F., and Simmons, R.M. (1971). Proposed mechanism of force generation in striated muscle. *Nature* **233**, 533–538.

Kentish, J.C., ter Keurs, H.E.D.J., Ricciardi, L., Bucx, J.J.J., and Noble, M.I.M. (1986). Comparison between the sarcomere length-force relations of intact and skinned trabeculae from rat right ventricle. Influence of calcium concentrations on these relations. *Circulation Research* **58**, 755–768.

Stuyvers, B., Miura, M., and ter Keurs, H.E.D.J. (1997). Dynamics of visco-elastic properties of rat cardiac sarcomeres during diastolic interval: Involvement of Ca^{2+}. *Journal of Physiology (London)* **502**, 661–677.

Stuyvers, B., Miura, M., Jin, J-P., and ter Keurs, H.E.D.J. (1998). Ca^{2+}-dependence of diastolic properties of cardiac sarcomeres. *Progress in Biophysics and Molecular Biology* **69**, 425–443.

29 Optimization of the Muscle-tendon Unit for Economical Locomotion in Cursorial Animals

A. M. WILSON
The Royal Veterinary College, North Mymms, Hatfield, Hertfordshire, England

A. J. VAN DEN BOGERT
Department of Biomedical Engineering, Cleveland Clinic Foundation, Cleveland, Ohio, USA

M. P. MCGUIGAN
The Royal Veterinary College, North Mymms, Hatfield, Hertfordshire, England

CURSORIAL ANIMALS: ATTRIBUTES AND ADAPTATIONS

Humans appear to be very poor athletes when compared to members of the animal kingdom. The cheetah can run at a speed of 29 ms^{-1} (Sharp, 1997) whilst the human sprinter is capable of, at best, 12 ms^{-1}. Pronghorn antelope have been recorded as covering 6 km at 15.6 ms^{-1} and 800 m at 24.6 ms^{-1} (Russell and McWhirter, 1988). The equivalent human world-record speeds for 5 km and 800 m would be 6.5 ms^{-1} and 8 ms^{-1}. The endurance athletes of the animal kingdom travel either by swimming (the blue whale is believed to migrate up to 8000 km) or flying (the Arctic tern has been recorded as flying 22,500 km over a 10-month period). Unfortunately, there are few ratified figures for long-distance, over-ground locomotion, although animals will migrate hundreds of kilometres, and African hunting dogs have been reported to follow their prey for over 100 km.

There is a particular functional grouping of animals, termed cursorial, that evolved a specialized musculoskeletal arrangement that allows the animals to travel short distances at high speed, and also cover long distances at slower speeds with a low energetic cost of locomotion. These adaptations enable them to escape the attentions of predators and also to migrate in order to follow seasonal changes in climatic conditions and/or locate different food sources. Cursorial animals are interesting to study because they demonstrate specialized musculoskeletal adaptations to achieve these two basic locomotor goals. These animals have to be good at both of these locomotor requirements if they are to survive and pass their genes onto subsequent generations. Examples of such ani-

Skeletal Muscle Mechanics: From Mechanisms to Function. Edited by W. Herzog.

mals include the horse, the ostrich, and the antelope. The same adaptations are evident in fossil remains of both artiodactyl mammals and flightless birds that existed 60 million years ago (Kardong, 1995). The success of this general mechanical design is demonstrated by its longevity, and the large biomass of such animals in a variety of open habitats, from the grass plains of Africa to the high mountains of the Andes.

The maximum running speed of these cursorial animals is not as high as the cheetah, but many can achieve about 20 ms^{-1} (Garland, 1983), and there are published figures of 26.5 ms^{-1} for the Thomson gazelle (*Gazella thomsoni*), and 27.8 ms^{-1} for the pronghorn antelope (*Antilocapra americana*) (McKean and Walker, 1974). The pronghorn antelope has also been reported to cover 11 km in 10 minutes. This outstanding athleticism is partly explained by an ability to transport and utilize large amounts of oxygen; the maximal oxygen consumption for a pronghorn antelope has been predicted to be in the range of 192-306 $mlO_2kg^{-1}min^{-1}$ (Linsted et al., 1991). This chapter will focus on the musculoskeletal adaptations for locomotion in these animals.

These prey animals can attain high speeds even though they have to overcome several disadvantages in their overall design:

- Most of these animals rely on relatively indigestible grasses for nutrition; these are broken down using a variety of microbes in a specialized area of the fore- or hind-gut. This fermentation vat accounts for about 25-33% of the animal's weight but provides little or no assistance in locomotion. Since the cost of locomotion is a function of the mass moved, this dead weight has a substantial effect on locomotor capacity.
- The second mechanical disadvantage is that, for a variety of reasons, many of these animals are much larger than the cheetah. Bones and tendons, however, are constructed from the same molecular building blocks in all mammals, and they have similar material properties in different mammalian species. As a result of this, a large animal must be inherently more bulky to support its mass. Imagine if a cheetah doubled in size but retained the same proportions: it would increase by 2 times in each dimension, so it would weigh 2^3 or 8 times as much, but the cross-sectional area of its limbs, tendons, and bones would only increase by 2^2 or 4 times. This demonstrates, in a simplified manner, why a large animal must have proportionally more massive limbs to support its body weight than a small animal. The subject of scaling for size and how it constrains musculoskeletal design is a fascinating topic that partly explains why small animals are very diverse in their design, while large animals are more limited in theirs and therefore appear relatively similar. This subject will not be covered further here, as it is well covered in detail in a number of other publications (for instance, Pennycuick, 1992; Schmidt-Nelson, 1975; Hokkanen, 1986).

Herbivores can become more specialized because of the relative ease and the restricted movements required to obtain food. For instance, the radius and ulna are fused together in many herbivores, which prevents rotation at the elbow, meaning that the limb cannot pronate or supinate—a disadvantage in a hunter (imagine a cat catching a mouse) but not a problem in a herbivore.

Economical endurance exercise is also critical for cursorial animals; they need to travel long distances at a low metabolic cost. This is for two reasons:

- Food and energy requirements for locomotion may be directly limiting due to poor forage, so the most economical animal will be the most likely to survive.
- Metabolism of the food requires oxygen and produces heat. The greater the oxygen requirement of an animal, the more it must ventilate. All gas leaving the lungs is saturated with water vapour, so the more the animal ventilates, the greater its requirement for water. Furthermore, in level locomotion, almost all of the metabolic energy ends up as heat within the animal, which must be dissipated. This is difficult for large animals because the surface area to volume ratio decreases as animals increase in size. Loss of extra heat will often cause a further water loss due to sweating or panting. So, if water supply is limited, the energetically inefficient animal will be at a major disadvantage.

However, size also confers advantages in the absolute metabolic cost of locomotion. There is a well-defined log-log relationship between size and minimum cost of transport, where the total cost of locomotion is given by the equation $COT_{total} = 10.7\ M_b^{-0.316}$, where M_b is the body mass in kilograms. The most economical land animal (in terms of joules per kilogram per meter) is the African elephant, with an energetic cost of locomotion of about 0.78 $Jkg^{-1}m^{-1}$ (Langman et al., 1995). This energy cost of locomotion is approximately one twentieth of that for a mouse and one third of that for a horse (Minetti et al., 1999). The largest known dinosaurs (brachiosaurids) would, at 50 tonnes, be predicted to have an energetic cost of locomotion of 0.35 $Jkg^{-1}m^{-1}$, demonstrating that energetic cost of locomotion is not the only predictor of biological success.

Some larger animals, for instance the camel, reduce their use of water for thermoregulation by allowing their body temperature to rise during the day and cool at night. This is combined with a counter-current cooling mechanism that uses cool blood returning from the nasopharynx (the blood is cooled by evaporation of water from the mucosal membranes) to cool the blood supplying the brain. This heat exchange occurs in a mesh of fine capillaries at the base of the brain that is found in a number of species.

So, what musculoskeletal adaptations exist to enable cursorial animals to occupy their ecological niche so effectively? On examination, these animals all look relatively similar: they have long distal limbs, a narrow thorax, and most of the limb muscles are in the proximal region of the limbs, with long tendons extending towards the foot. Many of these animals are unguligrade, which means that they stand on one or more modified nails or hooves. As a comparison, cats and dogs are digitigrade: the digits are weight-bearing with the metacarpals and metatarsals vertical. Humans and bears are plantigrade: the entire limb below the tarsus (and in the bear below the carpus) is parallel to and in contact with the ground (Figure 29.1).

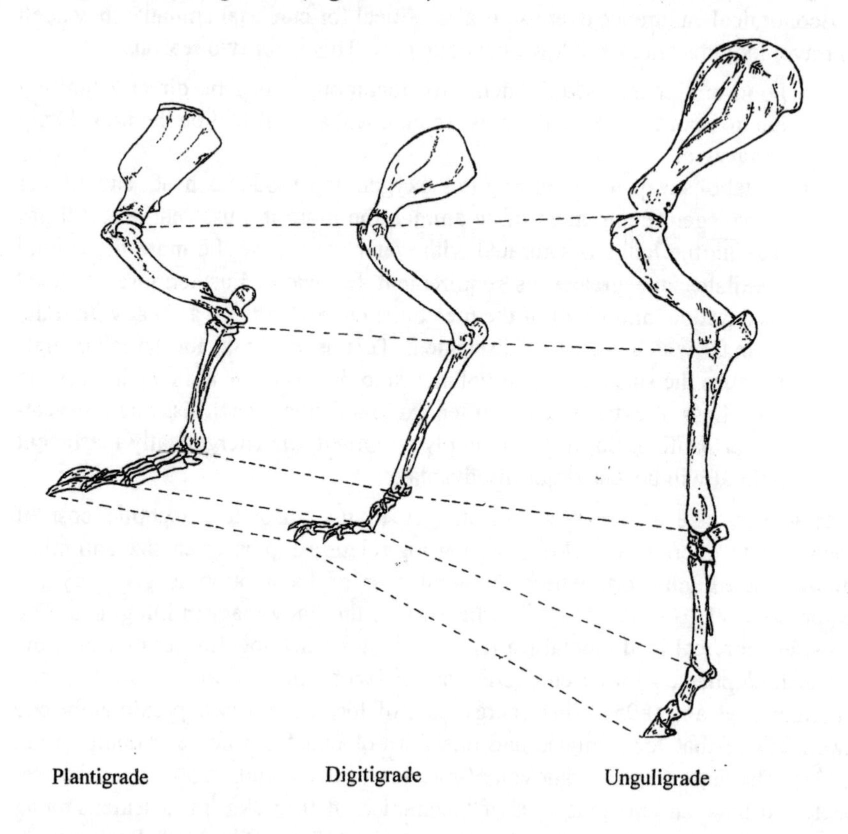

Figure 29.1. The skeletal arrangement of the forelimb of a bear, dog, and horse, demonstrating plantigrade, digitigrade, and unguligrade stance.

The horse is a good example of a cursorial animal with an unguligrade stance. It has been the subject of extensive biomechanical study since the pioneering kinematic work of Marey (1882) and Muybridge (1887). Interestingly, one of the motivating factors for that early work was controversy about the sequence of foot falls in a galloping horse and the existence or absence of a flight phase. This misunderstanding over the sequence of footfalls is evident in early paintings of horses at gallop, for instance by Stubbs, which should be examined in combination with the data presented in Figure 29.2.

In a galloping horse, each leg contacts the ground at different and unevenly spaced times (Figure 29.2). For a person more familiar with human bipedal gait, the gallop is kinematically similar to two skipping bipeds joined by a trunk. These

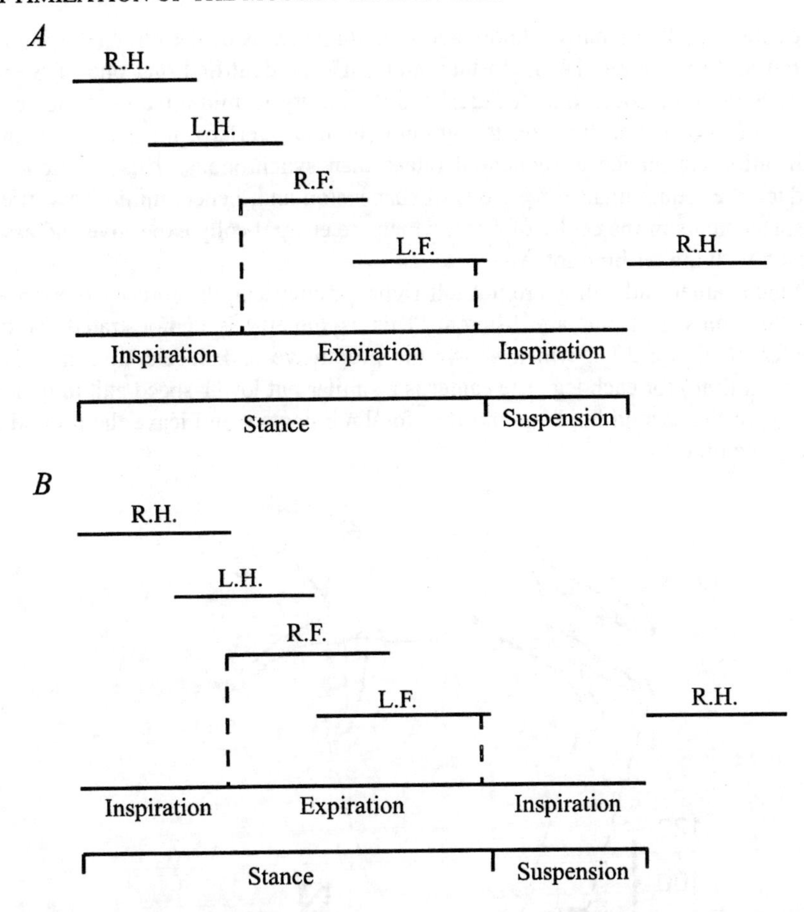

Figure 29.2. Sequence of footfalls for a left-lead gallop in a fast racehorse, Secretariat (A), and a slower racehorse (B). The lines represent the duration of stance for each leg and the period of inspiration and expiration. (Modified from Attenburrow, 1982; and Pratt and O'Connor, 1978.)

front and hind pairs of legs are used at different times in the stride cycle, and there is also a period of suspension where no legs are in contact with the ground; i.e., all the limbs are in a recovery phase.

The two front (or hind) legs cannot be considered equivalent and are therefore distinguished by the terms lead and non-lead. The lead limb is the second of a pair to contact and leave the ground (it leads into the next phase of the stride). The sequence of footfalls in a left-lead transverse gallop is as follows: Right Hind, Left Hind, Right Fore, Left Fore, suspension, RH, LH, RF, LF (Figure 29.2), so the legs act as pairs—hence the skipping analogy.

One study on an outstandingly fast racehorse, Secretariat, winner of the Kentucky Derby in 1973 at an average speed of 16.8 ms^{-1} for 2012 metres, provides much of what we know about the kinematics of the elite gallop horse (Pratt and O'Connor, 1978). In that study, it was identified that one of Secretariat's defining locomotor features was the ability to minimize the time spent with more than one limb on the ground (termed overlap), meaning that limb utilization for stance is sequential rather than synchronous. This will tend to reduce the requirement for periods of suspension and, hence, minimize vertical displacements of the centre of mass, which are energetically expensive and associated with higher limb forces.

Since canter and gallop are not left-right symmetrical, the forces experienced by the limbs in a pair are different. This asymmetry is demonstrated for the canter in Figure 29.3, which shows the propulsive and decelerative impulses (force x time) for each leg. The canter is a similar but lower-speed gait to gallop, in which the second hind leg and first forelimb contact and leave the ground at the same time.

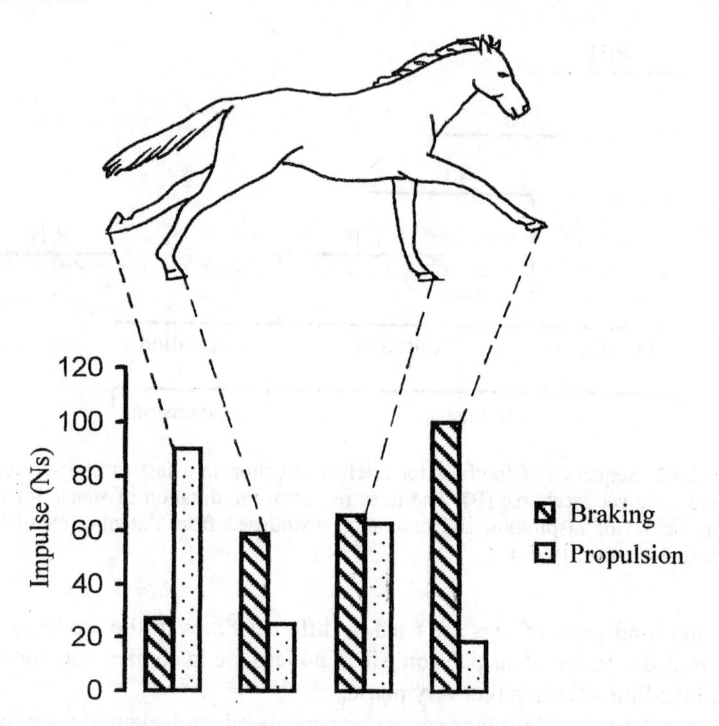

Figure 29.3. Accelerative and decelerative impulses (force x time) applied by each limb during a right lead canter at 5.2 ms⁻1.

Another advantage of quadrapedal locomotion is that the front and back limbs can be optimized to perform biomechanically distinct functions. The front limbs support about 60% of the body weight and provide very little propulsive effort (Niki et al., 1984), whilst the hind legs support less weight but perform much of the propulsion (Figure 29.3, drawn using data from Niki et al., 1984). Therefore, the front legs can be optimized as energy-efficient springs that store and return energy, thereby reducing the energetic cost of locomotion. This separation of function has a further advantage in ventilation. When the front legs contact the ground, the trunk is decelerated and the abdominal contents (which are relatively massive—about 100-150 kg—and loosely anchored) continue moving forward and therefore drive the diaphragm forward, assisting the horse to exhale. When the hindlimbs are on the ground, the trunk is accelerated forward, and the abdominal contents move in a caudal direction assisting in inhalation. This mechanism means that the horse has a respiratory frequency of 120-140 breaths per minute during gallop, and will achieve peak expiratory flow rates in the order of 100-120 litres per second. All this is achieved with much less muscular effort and metabolic cost than if ventilation was undertaken with just muscular action.

SPECIFIC ADAPTATIONS IN THE THORACIC (FORE) LIMB

The thorax is narrowed in cursorial animals and the scapula lies on the side of the thorax, with the clavicle (collar bone) either present as a vestigial structure or completely absent. This design confers two advantages:

- The limb lies much closer to mid-line and hence to the centre of mass. This reduces the muscular effort required to adduct the limb. Compare this arrangement to that of a human doing press-ups or a crocodile running, where the pectoral muscles have to contract to prevent the body from collapsing to the ground.
- The forelimb is only attached to the trunk by muscular tissue (a synsarcotic joint). This creates another limb joint and allows the scapula to glide over and rotate relative to the thorax during locomotion (Figure 29.4). This arrangement increases the range of motion of the limb and, hence, contributes to increasing the animal's stride length. This effect is enhanced in small animals, such as the cheetah, where spinal flexion and extension become important in increasing stride length (Figure 29.4). The purely muscular attachment of the scapula to the trunk also means that the muscles can act as shock absorbers/dampers during limb loading. This anatomy provides a means of controlling and limiting limb forces, especially during initial ground contact or impact.

The bones of the equine distal limb have reduced in number through evolution so that the horse only bears weight on its third digit. The proximal ends of the second and fourth metacarpals remain, but they only act to provide part of the articular surface of the carpal joint. This is an adaptation for maximum strength with minimum weight (since the second moment of area and, hence, bending strength are much higher for one large bone than for the same amount and, hence, weight of material arranged as four smaller bones).

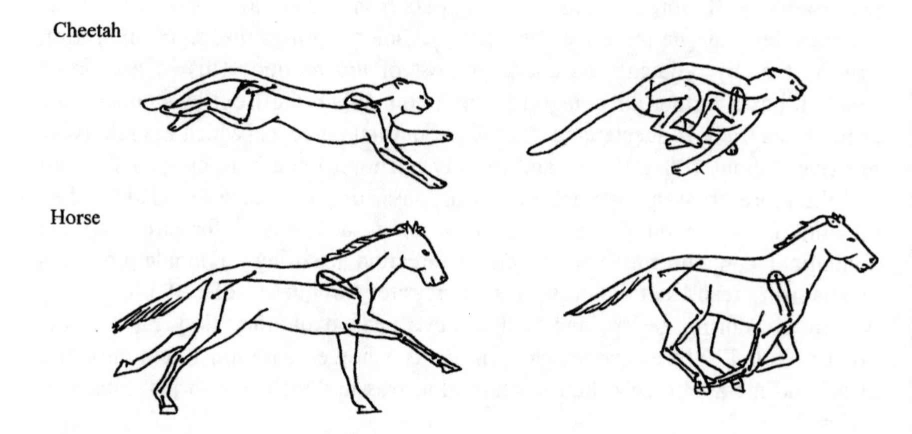

Figure 29.4. Diagram representing the movement of the scapula and the lumbar spine during locomotion in the cheetah and the horse. (Redrawn from Kardong, 1995.)

Moving proximally up the limb, the radius and ulna are fused, and the ulna is only evident proximally where it contributes to the elbow articulation, and distally where it contributes to the carpal articulation. This fusion prevents pronation and supination of the forelimb and results in a mechanically stable limb that requires little muscular tissue to limit and control motion.

This lengthening and lightening of the distal limb (Figure 29.5) confers three main advantages:

- The animal is able to increase its stride length. Since speed of locomotion is the product of stride length and frequency, an increased stride length contributes to increased running speed. In addition, the contraction velocity of the propulsive muscles will be reduced, which is an advantage because muscles use less energy to perform a given amount of external work when shortening at low-contraction velocities (Woledge and Curtin, 1984).
- The longer distal limb results in elongation of the tendons of the digital flexor and extensor muscles, thus the distal limb has a relatively low mass.
- The elongated flexor tendons are used for storage of elastic energy, which has been associated with a decrease in the metabolic cost of locomotion.

In each stride of gallop, the flexor tendons and suspensory ligament are stretched early in stance and then they recoil late in stance, returning most of the elastic energy stored in them. The suspensory ligament is the equivalent of the interosseus muscle (the muscular part of the human palm). This ligament has no contractile function and is entirely collagenous in the adult horse. The suspensory ligament and the digital flexor tendons act as passive springs to resist extension of the metacarpo-phalangeal joint. At their proximal end, the digital flexor

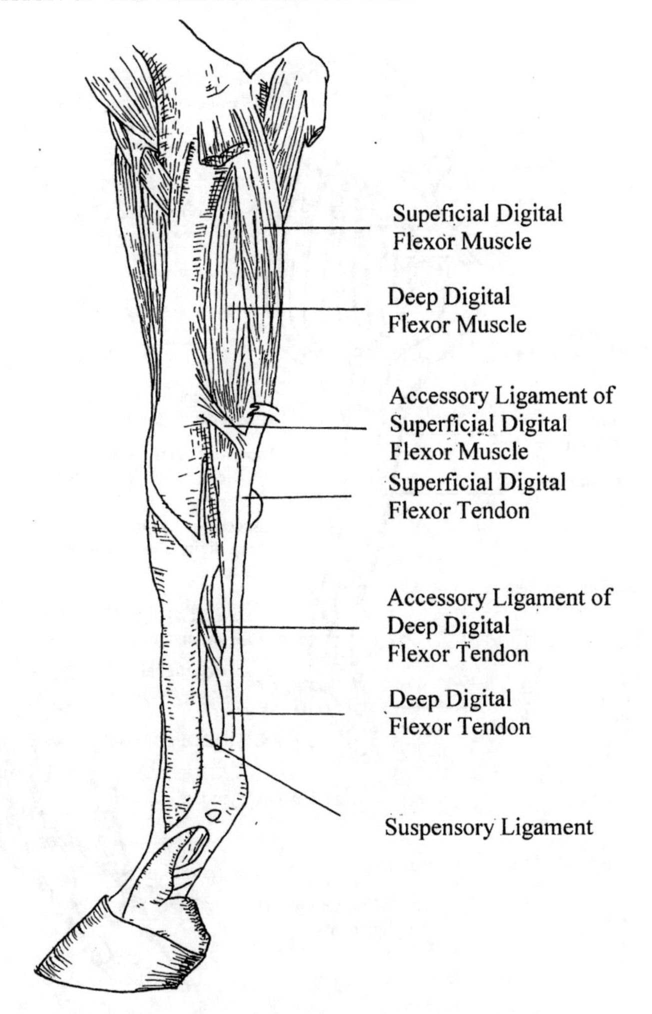

Figure 29.5. The anatomy of the thoracic limb of the horse.

tendons are anchored to the skeleton via their respective muscles and via an accessory ligament that bypasses the muscle, allowing passive elastic energy storage, and reducing the stress and peak length experienced by the muscle. There are other fibrous bands present in the limb. Of particular interest is the collagenous band that lies within the equine biceps muscle. This provides an elastic link between the scapula and the radius (Figure 29.6). During standing (and the stance phase of locomotion), this collagenous band will resist shoulder flexion, hence reducing the muscular effort required to extend the shoulder against the effect of body weight. This band also contributes to rapid limb protraction, which is discussed below.

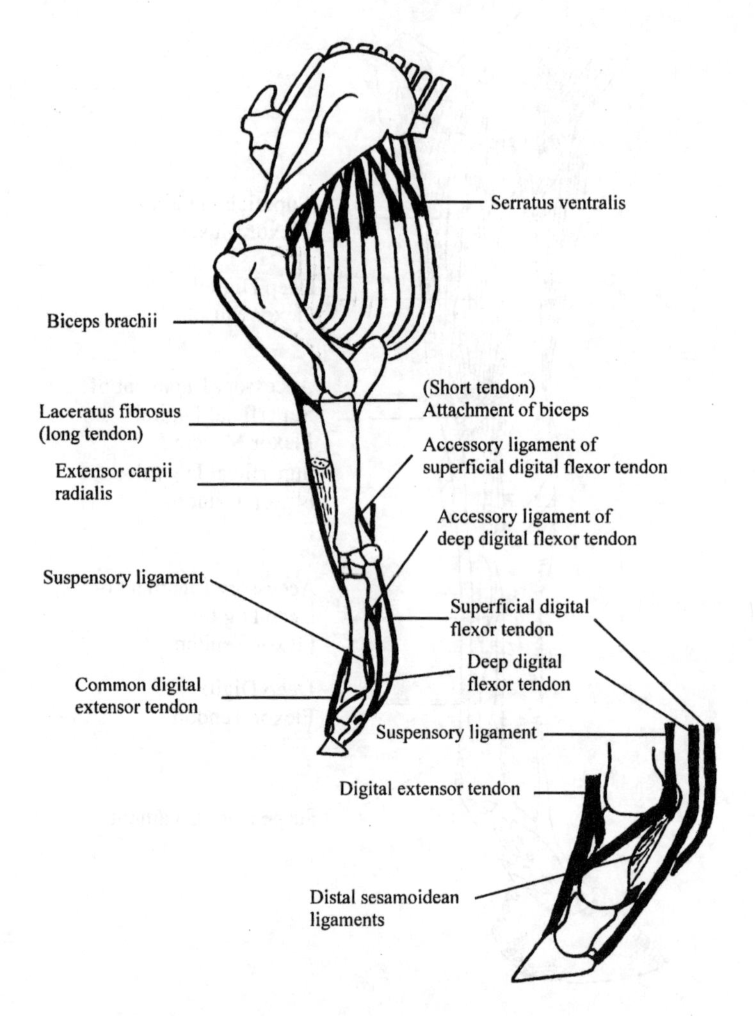

Figure 29.6. The passive stay apparatus of the equine forelimb.

SPECIFIC ADAPTATIONS IN THE PELVIC (HIND) LIMB

The pelvic limb articulates with the body via the hip joint, and provides much of the forward propulsive effort for locomotion (it is easier to push than to pull). The locomotor economy-enhancing mechanisms described for the forelimb also exist in the hindlimb, however they are less developed. This difference appears sensible given the lower vertical forces experienced by the hindlimbs, and might be an advantage in increasing overall limb stiffness to aid effective propulsion by the large muscle mass. Of particular note is a set of collagenous links between the stifle and hock joints, called the reciprocal apparatus (Figure 29.7). The hindlimb changes in length by flexion and extension of the stifle and hock joints. These two joints are linked via collagenous bands in the gastrocnemius and the predominantly fibrous fibularis tertius and superficial digital flexor muscles. These mechanical links cause simultaneous extension and flexion of the stifle and hock. This anatomical arrangement confers the following advantages:

- The weight and inertia of the distal limb is reduced because there is no need to have significant muscles present that are dedicated to moving the hock joint.
- Co-ordination and control of limb movement are simplified by constraining the range of movements.
- Likely, there are energetic benefits due to storage of elastic energy in these collagenous bands.

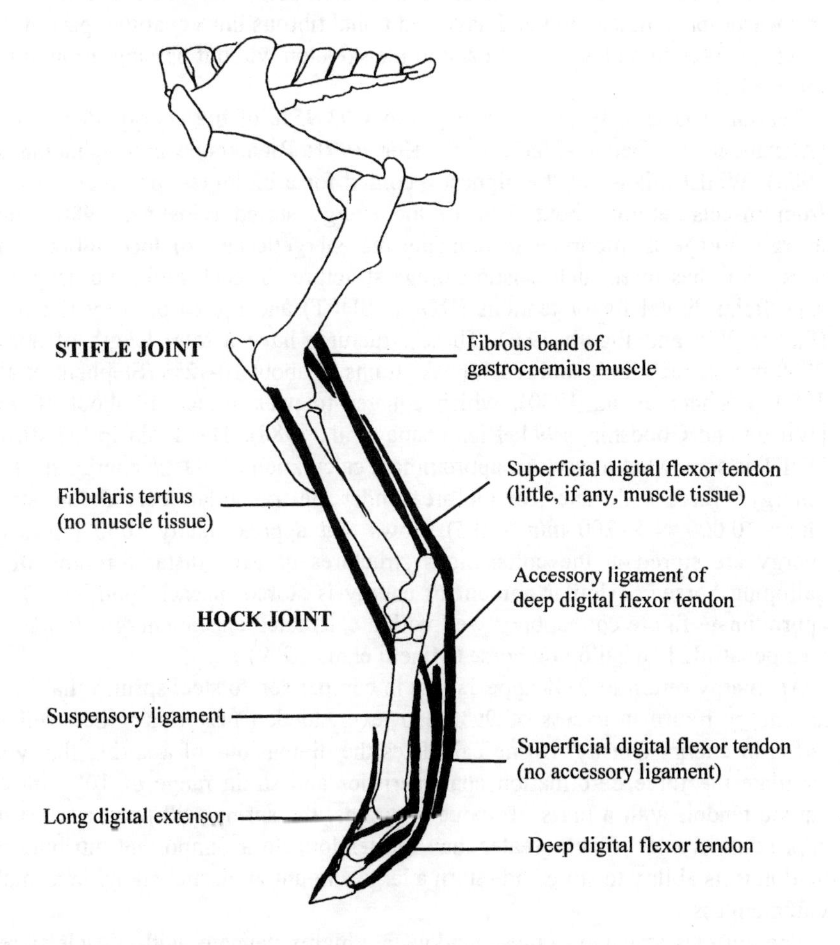

Figure 29.7. Collagenous stay apparatus of the equine hindlimb.

The arrangement of the tendons and ligaments of the metatarsal and digital region are similar to the equivalent structures in the forelimb, and they demonstrate similar mechanical roles as elastic energy stores.

LOCOMOTION

Locomotion involves cyclical movement of the limb. Muscles act to accelerate and decelerate the limb through the gait cycle. Examination of systems that oscillate with minimal energy input reveals that they usually act as a pendulum or a spring mass system. Human walking can be modelled as the former, and the equine limb can be considered to act like the latter, somewhat like a child's pogo stick. Fore- and hindlimbs in the horse contain two long digital flexor tendons and a suspensory ligament. All three structures act to resist extension of the metacarpo-phalangeal joint and have additional fibrous links enabling part of the extensor force to be transferred back to the skeleton without passing through the muscle belly.

Tendon acts as a spring, returning about 93-95% of the energy stored in it (Alexander and Bennet-Clark, 1977; Ker, 1991; Riemersma and Schamhardt, 1985). Whilst this is not the highest recorded for a biological polymer—resilin from insects returns about 97% of the energy stored (Gosline, 1980)—this energy storage is important in reducing the energetic cost of locomotion. The horse limb has three such elastic storage structures in each limb: the deep and superficial digital flexor tendons (DDFT, SDFT) and the suspensory ligament (Figure 29.5 and Figure 29.6). These structures have a total length of about 2000 mm in each limb, and reach peak strains of about 10-12% (Stephens et al., 1989; Lochner et al., 1980), which equates to peak forces of about 10 kN (Wilson and Goodship, 1994; Riemersma et al., 1988). The strain in the stiffer DDFT is somewhat lower. An approximate calculation of elastic energy storage (energy = force × distance, or the area under the triangular force deformation plot = 10,000 N × 200 mm × 0.5) shows that approximately 1000 joules of energy are stored in the collagenous structures of each distal forelimb of a galloping horse. A similar amount of energy is stored in each hindlimb. This approximate figure corresponds well with a calculated elastic energy storage of 6 kJ per stride in a galloping horse (Minetti et al., 1999).

An energy return of 93% appears low in comparison to steel springs that have an energy return in excess of 99%. However, tendon performs rather well in terms of energy density. If one calculates the dimensions of a spring that will simulate the force deformation characteristics and strain range of 100 mm of equine tendon with a mass of about 11 grams, the spring will have a mass of approximately 100 times greater than the tendon. So an important attribute of tendon is its ability to store and return a large amount of elastic energy in a small volume/mass.

The muscles attached to these tendons are highly pennate, with muscle fibres about 3 mm long in the SDFT, and 17 mm long in the DDFT (Hermanson and

Cobb, 1992). This is good for locomotor efficiency, as most of the change in length of the muscle-tendon unit will be the result of elastic deformation, rather than the energetically more expensive sarcomere length changes. It is difficult to understand what role these muscles serve in locomotion when the tendons elongate by as much as 100 mm. It may be that the muscles are simply vestigial structures, or perhaps they have some role in the fine-tuning or damping of high-frequency fluctuations in tendon force. This concept is discussed further below.

When a tendon is cyclically loaded, the 7% of energy that is not returned on unloading is converted to heat. This effect is demonstrated in Figure 29.8, where an equine SDFT is cyclically loaded, *in vitro*, at 2.4 Hz (stride frequency at gallop) to a force of 7 kN and 10 kN. The temperature within the tendon is recorded. *In vivo*, the tendon is cooled to some extent by blood flow, but tendon has low blood flow (Stromberg, 1973) and, hence, a limited capacity for heat loss other than by conduction from the surface. Therefore, there is a substantial heating effect in these tendons, and the tendon core temperature rises during sustained exercise (Figure 29.9; Wilson and Goodship, 1994). We have recorded temperatures of over 45°C in the central core of the SDFT of horses of low athletic ability during gallop exercise (Wilson and Goodship, 1994). A horse of good athletic ability can sustain a gallop at a faster speed, which should result in higher tendon temperatures due to the higher tendon strains and forces involved than in an untrained horse. A limited number of recordings show that temperature rises similarly in the suspensory ligament and to a smaller extent in the DDFT compared to the SDFT (unpublished data). These high temperatures would be lethal for most cell types (Hall, 1988), but tendon cells appear, in culture at least, to be resistant to these high temperatures (Birch et al., 1997). It is interesting to speculate, since the tendon temperature is highest in the central core, that exercise-induced hyperthermia is the basis of a mechanism to explain why degenerative changes in the equine SDFT are usually localized in the central core of the tendon. Changes are also seen in the collagen crimping in the central region of the SDFT in older and trained horses (Wilmink et al., 1992; Patterson-Kane et al., 1998). This change in collagen crimping may explain why most partial tears occur in the central core.

Examination of cursorial birds demonstrates the appropriateness of this spring mechanism for locomotor economy (Patak and Baldwin, 1993). These birds diverged from the evolutionary line that would eventually develop into mammals about 200 million years ago (Kardong, 1995), yet the mechanical principle for storage of elastic energy in the horse—i.e., a long elastic tendon with a lever system to strain it as a result of the application of body weight during locomotion—exists in a very similar form in both the ostrich (Figure 29.10) and the Diatryma, which lived some 60 million years ago. The ostrich is capable of running at a speed of about 18 ms^{-1}, and achieves a similar locomotor economy to a horse of similar mass (Fedak and Seeherman, 1979).

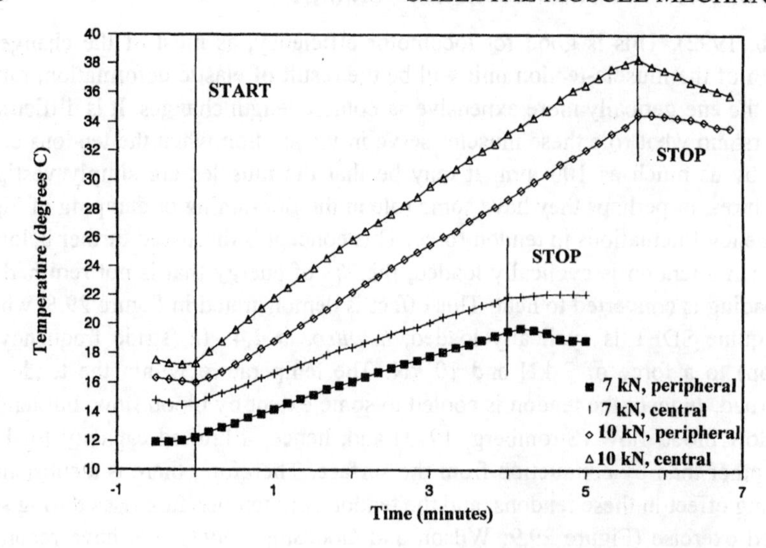

Figure 29.8. Tendon temperature in the central and peripheral regions of an equine SDFT sinusoidally loaded *in vitro* to a peak force of either 7 kN or 10 kN at 2.4 Hz.

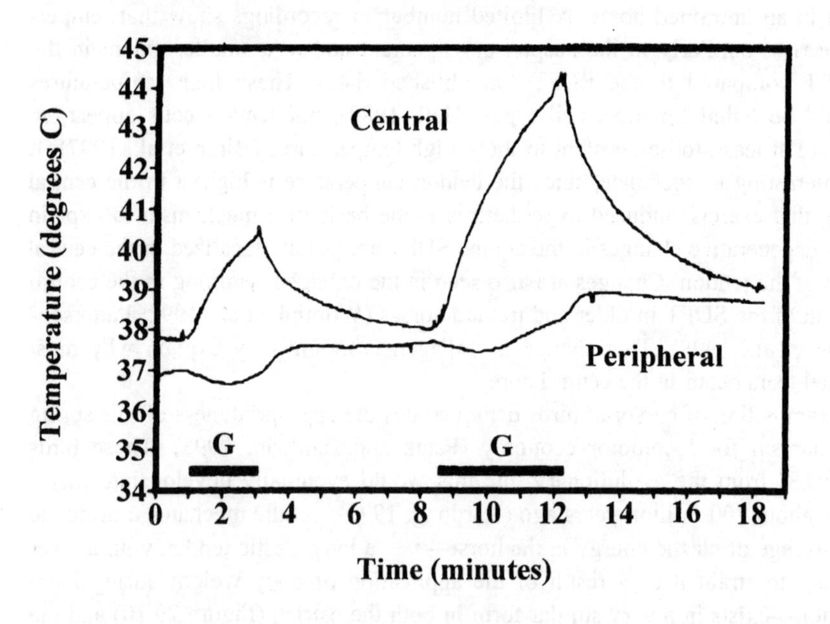

Figure 29.9. Central and peripheral tendon temperature during gallop exercise in the horse. *G* represents periods of gallop. (Reprinted from Wilson and Goodship, 1994, with permission from Elsevier Science.)

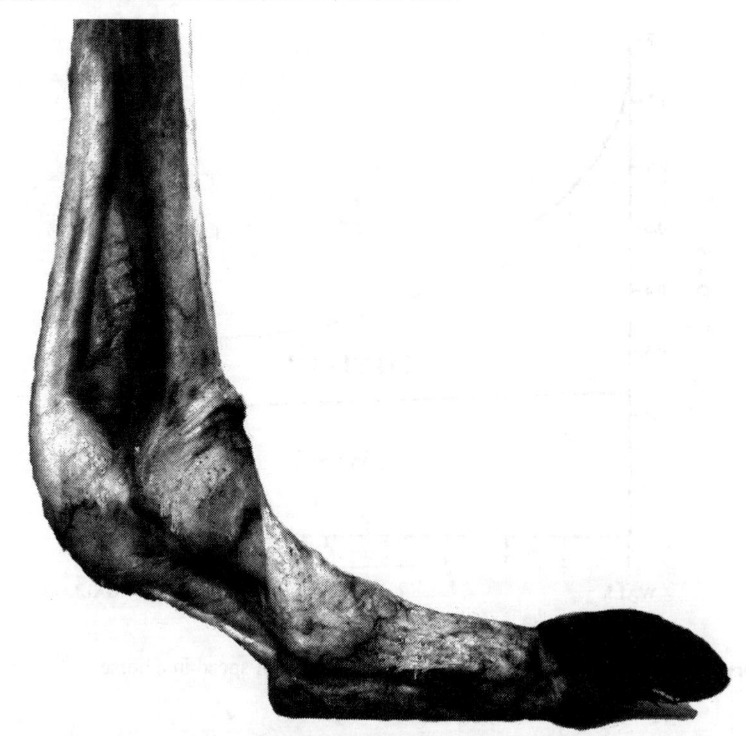

Figure 29.10. Collagenous structures of the ostrich distal pelvic limb.

Another interesting aspect of locomotion in the horse is the relationship between stance and protraction. Figure 29.11 (redrawn from Leach and Cymbaluk, 1986) shows the duration of stance (at least one limb on the ground) and flight (no limbs in contact with the ground). This figure demonstrates that limb protraction time is essentially independent of the speed of locomotion, while stance time decreases as speed increases. In man, running speed is increased by increasing both stride frequency and stride length at slow speeds. At high speeds, most of the increase is achieved by increasing stride length, similar to the case in the horse (Figure 29.12). So the horse primarily increases its speed of gallop by increasing stride length, whilst maximum stride frequency is mainly limited by limb protraction time. Therefore, it is vital for the horse to minimize limb protraction time in order to maximize stride frequency and, hence, gallop speed. Limb protraction time is minimized by two means:

- The low mass of the distal limb reduces the limb's inertia.
- Kinematic data demonstrate that the foot leaves the ground at the end of stance with an acceleration of up to 200 ms^{-2}, much faster than could be achieved by direct muscle contraction. The horse achieves this massive acceleration by using the collagenous structures of the limb like an elastic element in a catapult. The force in the DDFT peaks at about 85% of stance

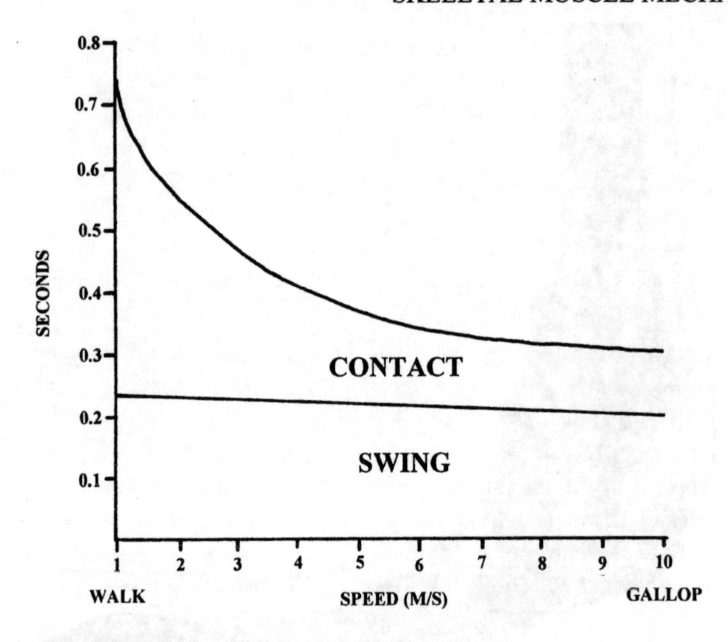

Figure 29.11. Stance and flight duration as a function of speed in a horse.

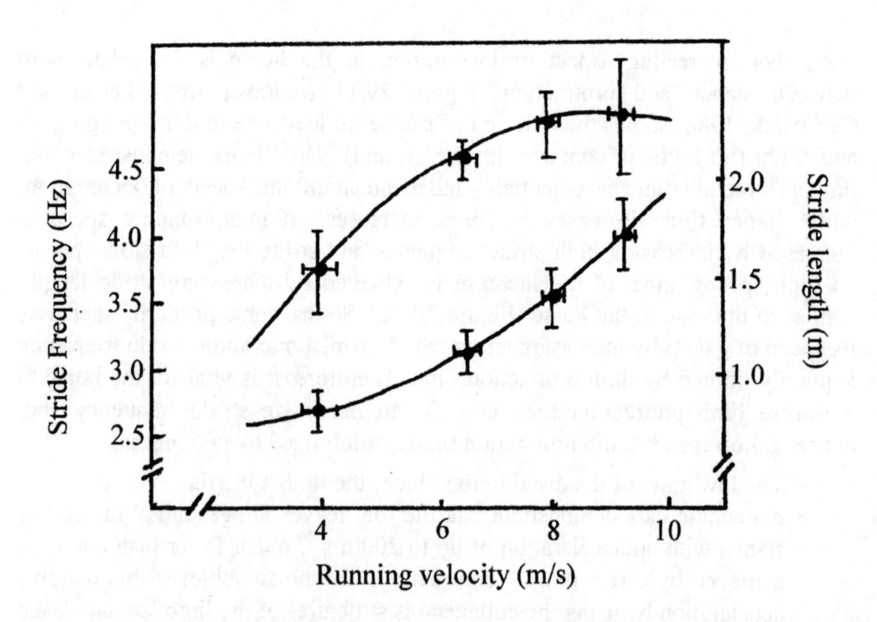

Figure 29.12. Relationship between stride length, stride frequency, and speed in man.

(Willeman et al., 1999). As the foot is unloaded the toe loses its grip, the elastic energy in the collagenous structures results in a rapid flexion of the digit. The DDFT and SDFT also cross the carpal joint and they will act, along with the carpal flexors, to achieve a similar flexor effect there. There is a fibrous band in the biceps tendon (Figure 29.6). When the limb is unloaded, this band will shorten, and the elastic energy stored will flex the elbow and extend the shoulder. Both of these actions contribute to the initiation of limb protraction. This leaves fine control, limb extension, and preparation for foot impact as active processes.

These mechanisms not only reduce the amount of muscular effort required for limb protraction, they also reduce the time required to achieve this movement. Since the limb is not available to support body weight during protraction, it is in the animal's interest to minimize protraction time, since it directly affects limb vertical force and the maximum stride frequency. Since the product of stride frequency and stride length gives the speed of locomotion, this is of vital importance. The duration of limb protraction is relatively independent of speed of gallop in the horse (Figure 29.11; Leach and Cymbaluk, 1986), which provides further support for the passive elastic component of limb protraction. Limb protraction time is an inherent property of limb mass, length, and the properties of the elastic structures within the limb. The mechanical properties of these elastic elements remain largely unchanged with training (Wilson, 1991), as does distal limb mass. Therefore, limb protraction time might have the potential to be a good method of evaluating the maximum running velocity of a horse if, for instance, one wanted to identify potential sprinters.

As the limb's resistance to compression is largely the result of the force-length properties of the tendons, it is possible to mount a limb in a hydraulic press (Figure 29.13) and produce a force-length curve for compressive loading, giving the spring stiffness of the system (Figure 29.14). As would be predicted, the angle of the metacarpo-phalangeal joint is closely related to limb force, both *in vitro* where it is closely related to tendon strain (Figure 29.15), and also *in vivo* where little change occurs in the relationship between limb force and metacarpo-phalangeal joint angle with different gaits. This is demonstrated in Figure 29.16, in which the relationship between the metacarpo-phalangeal joint angle and limb vertical force (measured with a force plate and a motion analysis system) is shown in a horse during walk, trot, canter, and jump landing. The relationship is almost linear and independent of gait.

DRAWBACKS OF HAVING AN ELASTIC LIMB

Are there disadvantages associated with using a system of collagenous springs, rather than active muscle contraction, to achieve much of the force development

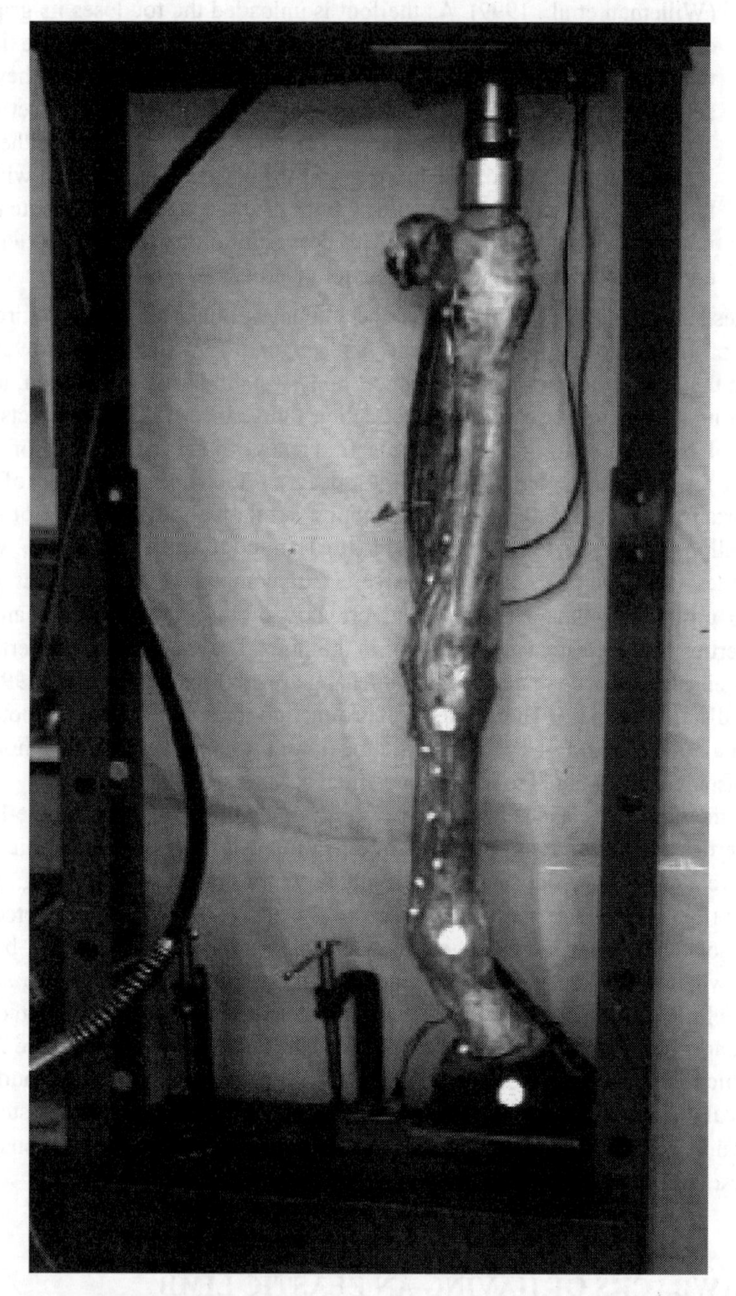

Figure 29.13. An equine limb mounted in a hydraulic press as described in the text. The wires are for stimulation of the digital flexor muscles, and the reflective markers are for determination of joint angles and tendon and muscle strain.

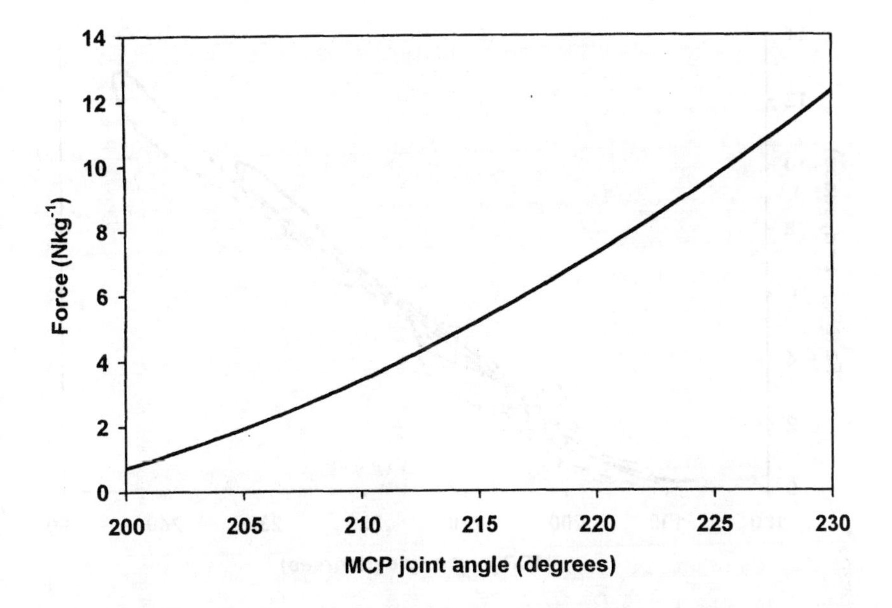

Figure 29.14. Force displacement response for a limb mounted in the jig shown in Figure 29.13.

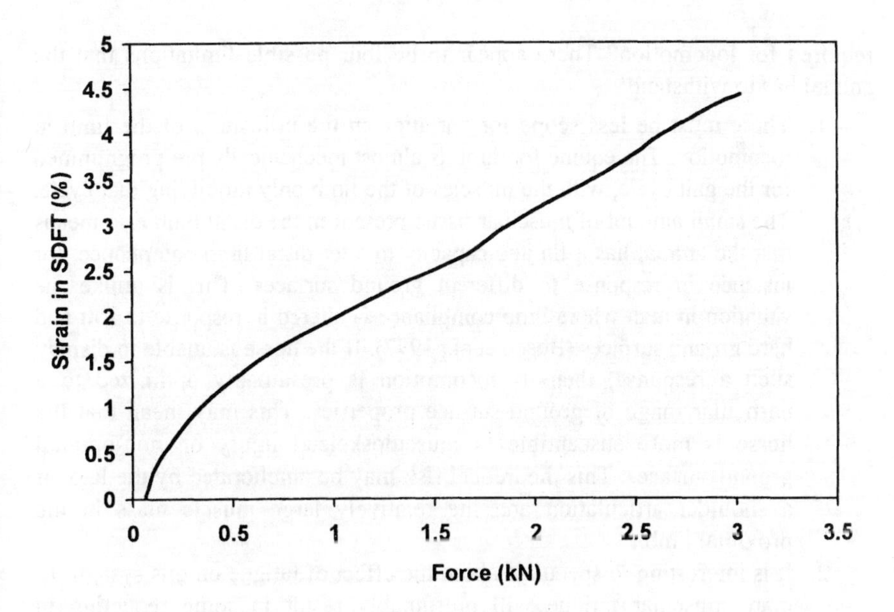

Figure 29.15. Relationship between limb force and superficial digital flexor tendon strain for a limb mounted in the jig shown in Figure 29.13.

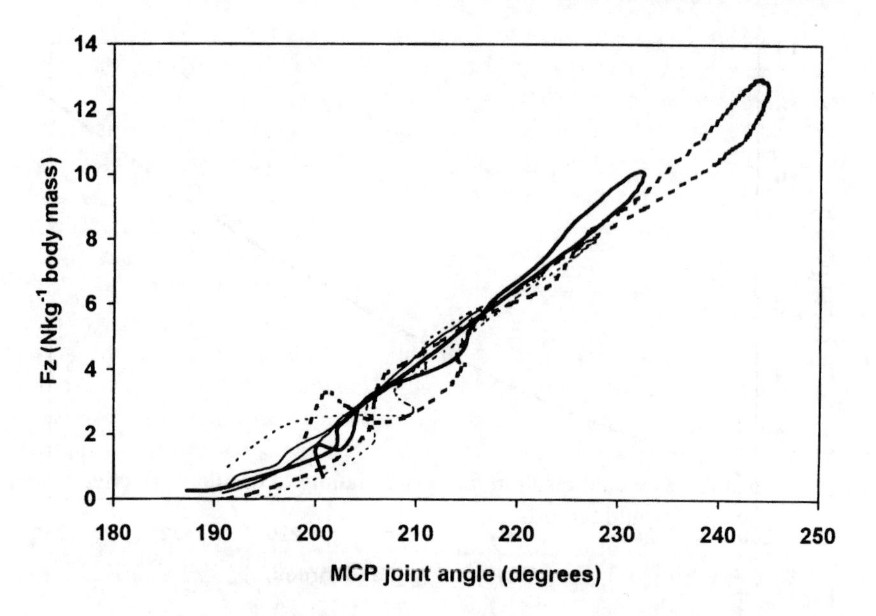

Figure 29.16. Plot of limb vertical force (determined using a force plate during ridden locomotion) and metacarpo-phalangeal joint angle during walk (solid line), trot (bold solid line), canter (bold dashed line), and landing from a small jump (dashed line).

required for locomotion? There appear to be four possible limitations that the animal has to withstand:

1. There must be less scope for variation in the utilization of the limb in locomotion. The equine forelimb is almost mechanically pre-programmed for the gait cycle, with the muscles of the limb only modifying that cycle. The small amount of muscular tissue present in the distal limb also means that the animal has a limited capacity to alter distal limb compliance, for instance in response to different ground surfaces. This is unlike the situation in man where limb compliance is altered in response to soft and hard ground surfaces (Bosco et al., 1997). If the horse is unable to display such a response, then its locomotion is presumably optimized to a particular range of ground-surface properties. This may mean that the horse is more susceptible to musculoskeletal injury on non-optimal ground surfaces. This theoretical risk may be ameliorated by the lack of a shoulder articulation and the relatively large muscle mass in the proximal limb.
2. It is interesting to speculate about the effect of fatigue on this system. In man, muscular fatigue will presumably result in some reduction in muscle stiffness and a generalized de-tuning of the locomotor system. This may be a protective mechanism to limit aberrant muscle loading, which could occur due to loss of co-ordination in fatigue. Such a mech-

anism would be less evident in the equine distal limb due to the lack of muscle mass, which could make the horse more susceptible to injury when fatigued.

3. To act as effective elastic energy stores, tendons must be extensible and be strained close to their elastic limit. This is because the energy stored is the area under the load deformation curve, i.e., ½ × force × extension if the tendon is assumed to be linearly elastic. If a tendon were to hypertrophy in response to exercise (as is the case for bone), then it would become stronger and stiffer. This would mean that the tendon would store less energy when loaded to the same force (because deformation would be less), and it would also alter the rate of energy storage and return from the system. This mechanism would be like putting a stiffer spring on a child's pogo stick: a small increase in stiffness results in an increase in the frequency of bounce, whilst a major increase in stiffness would result in the system failing to oscillate effectively, and the child would fall off.
4. With the loss of most of its muscular tissue, the equine limb becomes a lightly damped, elastic, multi-joint structure that may be susceptible to vibration at frequencies other than the stride frequency.

VIBRATION OF THE EQUINE LIMB

When one examines the vertical or horizontal ground-reaction force (GRF) time curve for a forelimb at trot, there is a secondary vibration at a frequency of about 30 Hz. This vibration is most obvious in the horizontal GRF (Figure 29.17) and in a differentiated vertical GRF plot (Figure 29.18). The frequency of vibration is similar in different horses, ranging from 26-35 Hz over nine horses (unpublished data). The nature of this high frequency component has also been shown to change in horses with incipient flexor tendon injury (Dow et al., 1991). The frequency of vibration appears to be relatively independent of gait (we have measured it during trot, canter, and jump landing), and is also detectable in the metacarpo-phalangeal joint angle versus time plot (Figure 29.19), demonstrating that the vibration displacement also exists in flexor tendon strain. Similar frequencies are also present in the canter and jump data of Merkens et al. (1993a) and Schamhardt et al. (1993), respectively.

One might argue that musculoskeletal tissues are susceptible to damage from such vibrations. These vibrations increase the number of loading cycles that the tissues are subjected to, which could predispose bones and tendons to cyclical fatigue damage (Carter, 1984; Wang et al., 1995). Bone is very responsive to high-frequency cyclical loading, with optimum adaptive hypertrophy occurring at a frequency of about 30-60 Hz (Rubin and McLeod, 1994). Therefore, it would appear likely that biological mechanisms exist to damp out or control such high-frequency vibrations, much as exist in man-made machinery.

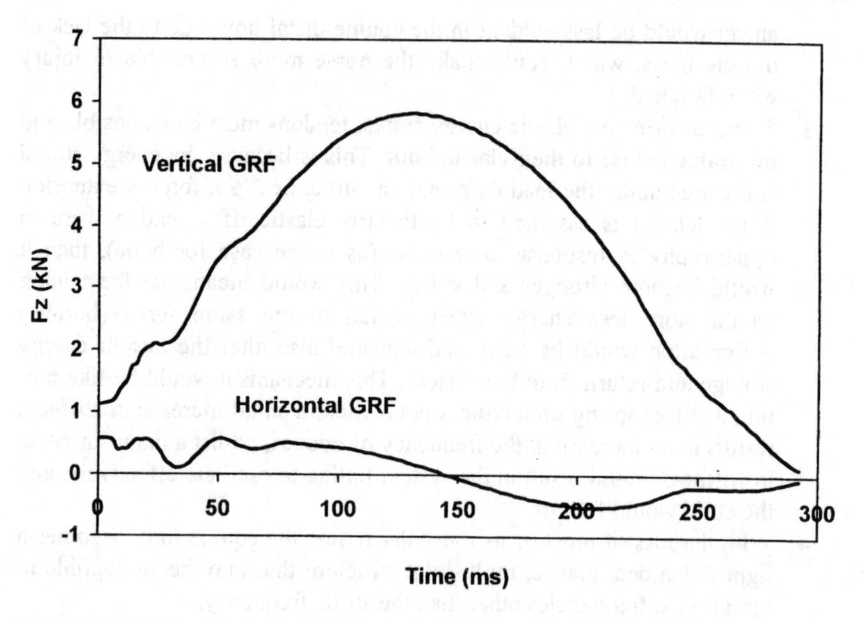

Figure 29.17. The vertical and horizontal ground-reaction force time curve for a horse during trot.

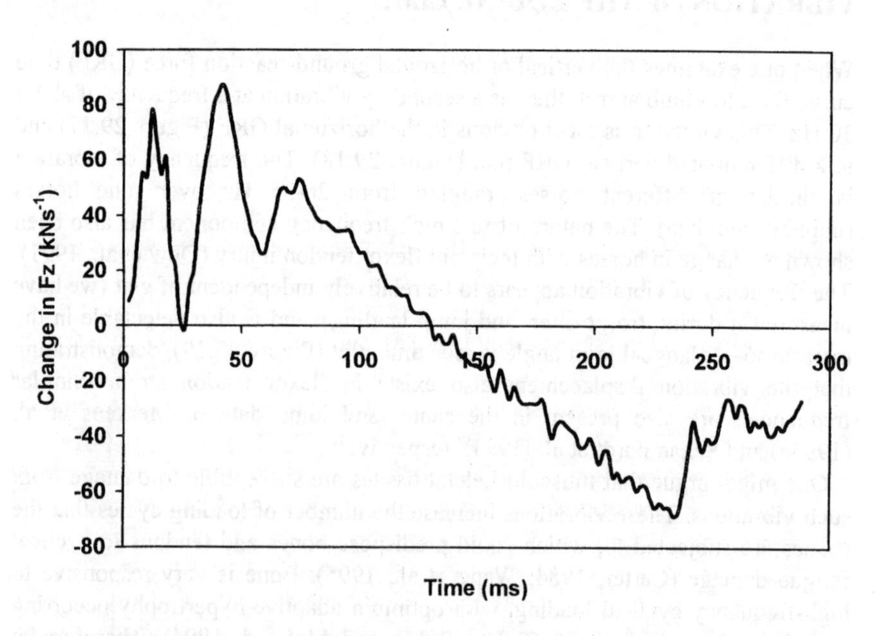

Figure 29.18. The derivative of the vertical ground-reaction force plot in Figure 29.17, plotted against time.

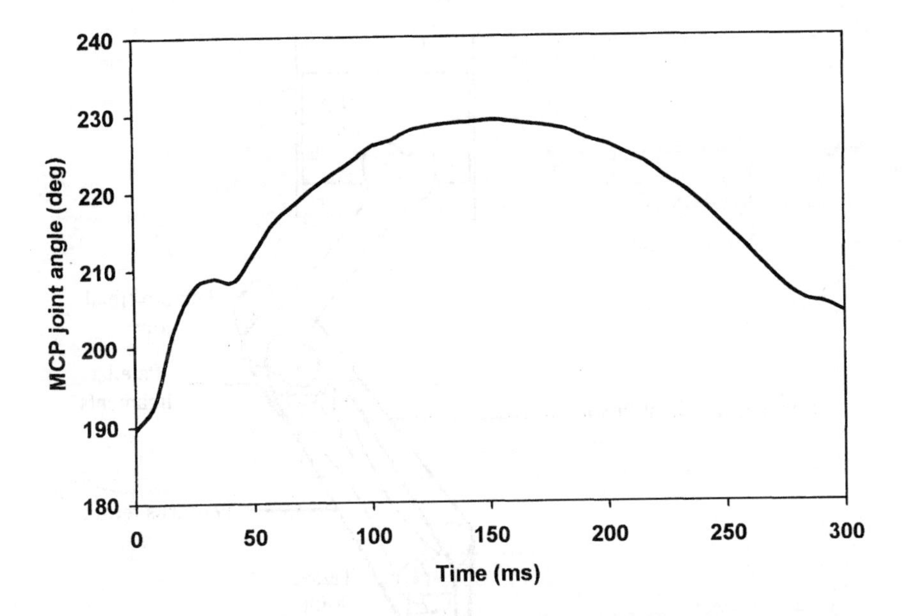

Figure 29.19. Metacarpo-phalangeal joint angle against time for the stride recorded in Figure 29.17 and Figure 29.18.

We have also observed vibrations at a similar frequency in computer simulations of the equine hindlimb (van den Bogert and Schamhardt, 1993) and forelimb (unpublished data). The passive hindlimb was loaded by a mass of 150 kg, representing the fraction of trunk mass supported by one hindlimb (Figure 29.20). The patella was rigidly attached to the femur, a mechanism that horses use while standing passively. Only vertical movement of the hip was allowed, resulting in a total of three degrees of freedom. This system exhibited a damped vibration response with three natural frequencies: one of about 1.5 Hz (associated with vertical trunk motion), and two of about 20 Hz (associated with movements inside the limb that did not involve a change of total limb length). These high natural frequencies are determined by the inertial properties of the limb segments combined with passive elastic properties of the musculo-tendinous structures. Excitation of this vibration occurs when the hoof strikes the ground; the initial amplitude will depend on the hoof impact velocity. In order to minimize vibration amplitudes, the horse will need to control the speed of hoof retraction accurately to counteract its forward velocity. Cranio-caudal (CC) oscillations are clearly seen in the GRF measurements from trotting and cantering horses. The oscillations have an amplitude of ⅓ to ½ of the maximum GRF in that direction (Merkens et al., 1993a, 1993b).

In vivo ground-reaction force and kinematic data suggest that the high-frequency vibrations are close to critically damped, i.e., the damping time is about

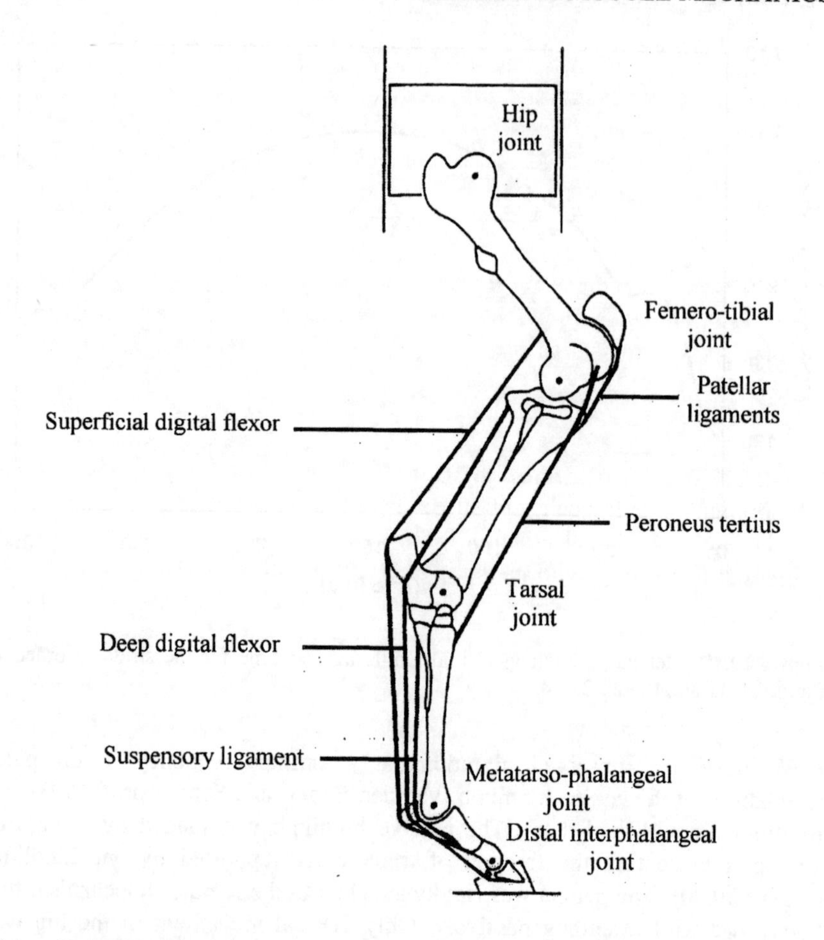

Figure 29.20. Model of the equine hindlimb used in the study by van den Bogert and Schamhardt, 1993.

equal to one cycle of the vibration. The low-frequency vibration is only lightly damped, which is advantageous because of its energy-saving effect. Theoretically, damping of the high-frequency vibrations can occur using three mechanisms: (1) passive damping properties of elastic tissue and ground surface, (2) force-velocity properties of active muscle, and (3) well-timed activation pulses that occur when a muscle is in the lengthening phase of the vibration. The third mechanism is unlikely, due to the high frequency of the vibrations. The relative importance of the first two mechanisms is unknown. For the first mechanism, it is important to maintain high activation in those muscles that change length during the vibration, which may not be the same muscles that are used to support vertical load. Muscle co-ordination and fatigue may therefore play a role.

Examination of the digital flexor muscle architecture demonstrates that the muscles are highly pennate with fibres of about 3 mm long in the SDFT and 17 mm long in the humeral and ulnar heads of the DDFT (Hermanson and Cobb, 1992). The SDFT muscle-tendon unit stretches by about 10-12% (i.e., at least 100 mm) during gallop locomotion (Stephens et al., 1989). The vertical oscillation of the metacarpo-phalangeal joint (which is a close measure of tendon length) during limb vibration is in the order of 2-4 mm (Figure 29.19). The very high physiological cross-sectional area of the muscle means that it is ideally suited to withstand the high forces experienced by the muscle-tendon unit. Whilst muscle shortening has a minimal effect on tendon loading, it is ideally arranged to function as an active damper of the small displacement tendon vibrations.

Muscle stiffness under vibrational loading varies by about 30% over the frequency range of 2-50 Hz, with a minimum stiffness at around 30 Hz (Figure 29.21). The minimum stiffness at 30 Hz may mean that the muscle will damp out vibration at the frequency of 30 Hz (by allowing a greater displacement), or that the limb vibrates at around 30 Hz because muscle stiffness and, hence, damping are at a minimum around that frequency. The studies by Ettema and Huijing (1994) were undertaken using constant amplitude rather than constant force loading, so it is difficult to extrapolate the energetic aspects of those studies to the situation *in vivo*.

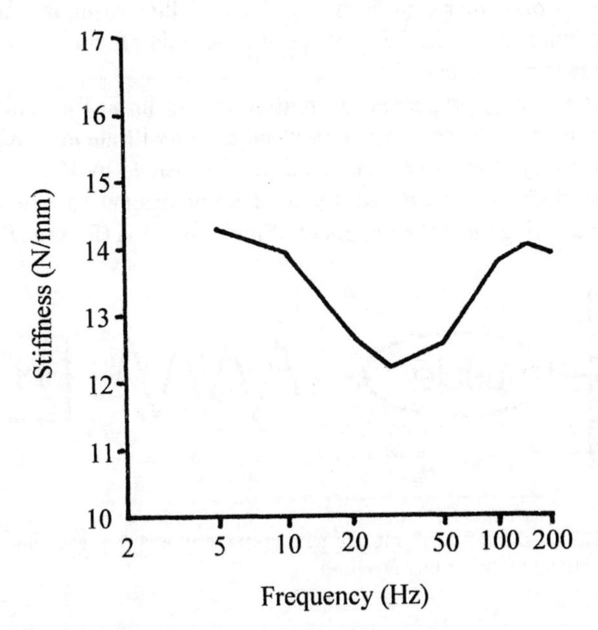

Figure 29.21. Muscle stiffness as a function of frequency of constant displacement vibrational loading in rat gastrocnemius muscle under tetanic contraction. (Redrawn from Ettema and Huijing, 1994.)

The nature of this vibrational stiffness change at a muscle level is unclear. It may reflect a repetitive force depression/enhancement effect that, due to the time-dependent nature of force depression and enhancement, results in dynamic changes in stiffness (Ettema and Huijing, 1994). A similar frequency-dependent effect has been reported in muscle fibres from rabbit, frog, and crayfish. The effect is, however, absent in muscles that are relaxed, in rigor, or fixed. These observations have been taken to indicate that the effect is the result of the fundamental properties of cycling cross-bridges (Kawi and Brandt, 1980). When we attempted to simulate this response by vibrating a series of Hill-type sarcomeres, a model that exhibits force depression/enhancement phenomena (Edman et al., 1993), we did not see a minimum in muscle stiffness, which is in agreement with the single-sarcomere model applied by Ettema and Huijing (1994). This result demonstrates that standard Hill muscle models will not predict this type of frequency-dependent stiffness response. Interestingly, in our experiment, the sarcomere lengths became less homogeneous after vibration, which has been proposed as one possible mechanism for muscle damage and remodelling (Talbot and Morgan, 1996). It would be interesting to change the association and dissociation rate constants in a Huxley cross-bridge model and see if the frequency-dependent muscle stiffness could be reproduced.

It is necessary to develop muscle models that include this high-frequency response if loading simulations are to reproduce these effects in whole animals. This may be of particular significance in the modelling of impact attenuation in animals and humans, where the impact spike has a duration of about 30 ms and, hence, a frequency of about 15 Hz.

The force-velocity properties of muscle result in a frequency-dependent relationship between force and length change. We will attempt to predict limb vibration damping from these relationships, as measured by Ettema and Huijing (1994). The limb is represented by a one-dimensional muscle-tendon-mass system, representing the CC component of limb vibration (Figure 29.22).

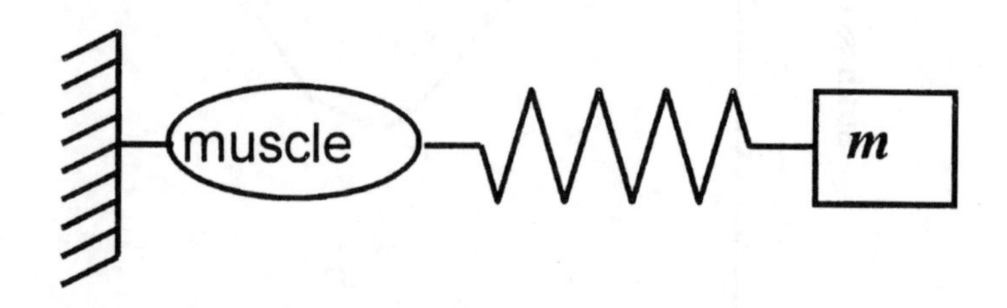

Figure 29.22. One-dimensional muscle-tendon-mass model representing the anterior-posterior vibrations in the equine forelimb.

Using this model, the damping ratio, λ, which is the ratio between the amount of energy dissipated by the muscle in one cycle of vibration relative to the total vibration energy, can be derived as:

$$\lambda = \frac{8\pi^3 m f^2 \sin(p(f))}{s(f)} \qquad (29.1)$$

where:
m = vibrating mass
f = frequency of vibration
$p(f)$ = phase of complex muscle stiffness (Ettema and Huijing, 1994)
$s(f)$ = modulus of complex muscle stiffness (Ettema and Huijing, 1994)

An oscillating mass of $m = 10$ kg is assumed. The CC movement is accompanied by changes in distance between origin and insertion of the intrinsic limb musculature. We estimate the gear ratio, G, between these displacements to be about 4, i.e., each centimetre of CC translation of the limb centre of mass is accompanied by, on average, a 0.25 cm length change in the intrinsic musculature. Forces generated by the actual muscle and tendon must be divided by G to obtain the equivalent force for the one-dimensional CC movement model.

Ettema and Huijing (1994) measured the frequency-dependent stiffness and phase for the rat gastrocnemius muscle with a length of 40 mm and a maximal isometric force of 12 N. The equine deep digital flexor (DDF) muscle has a length of 200 mm and a maximal isometric force of 5000 N. Therefore, Ettema and Huijing's stiffness values were multiplied by the factor 83.3 (= 5000/12 * 40/200) to represent an equine forelimb muscle. This result was then further multiplied by 4 to represent the situation in which several forelimb muscles act in parallel with respect to CC vibrations. Finally, muscle stiffness was divided by the gear ratio G (= 4) to represent CC stiffness. The estimated damping ratio (Equation 29.1) for the equine forelimb, using these muscle properties, is shown as a function of frequency in Figure 29.23.

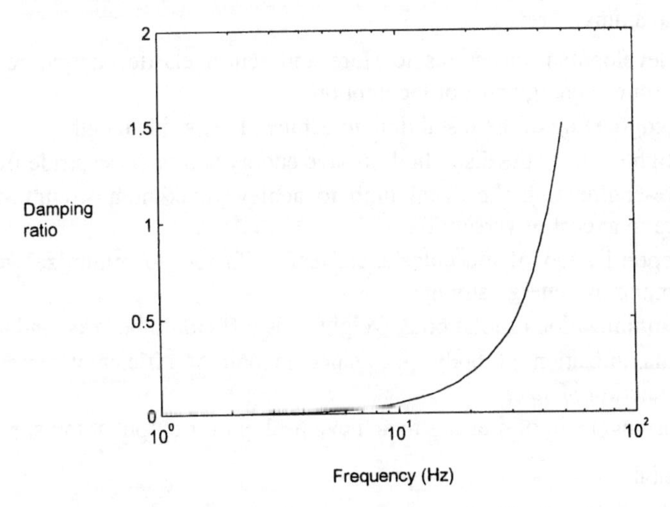

Figure 29.23. Estimated damping ratio as a function of vibration frequency for the equine forelimb. See text for details.

The result (Figure 29.23) shows that the damping ratio is close to 1.0 around 40 Hz, indicating that the limb is critically damped at its high natural frequencies, and is almost undamped at its low natural frequency of 1 Hz. This is desirable, since it prevents injuries induced by high-frequency vibrations, while not affecting the energy-saving 'pogo-stick' mechanism. The minimum in stiffness at 30 Hz, reported by Ettema and Huijing (1994), resulted in a sudden transition from undamped to highly damped vibrations at high frequencies (Figure 29.23). It should be noted that the scaling procedure assumes that there are no differences in architecture between rat gastrocnemius and equine digital flexors. It is highly probable that the equine muscles have a different vibration response, but no such data are available yet. The one-dimensional model with a single muscle is only an approximation of the true multi-body, multi-degree-of-freedom mechanism of the equine forelimb. These vibration responses should be recalculated with a two-dimensional multi-body model, similar to previous studies on the equine hindlimb (van den Bogert and Schamhardt, 1993).

CONCLUSIONS

Cursorial animals demonstrate a number of adaptations that are common to species that have evolved for efficient, high-speed locomotion. These adaptations help to make the most efficient use of the limited energy reserves available through aerobic and anaerobic metabolism. Examining these features and mechanical limitations is useful in developing our understanding of what limits human athletic performance, and why musculoskeletal injury is an inevitable consequence of high-class athleticism. The main adaptative characteristics of cursorial animals are:

- development structures to store and return elastic energy, resulting in reduced energy cost of locomotion,
- lengthening of the distal limb to achieve long stride lengths,
- lightening of the distal limb to save energy and increase stride frequency,
- re-engineering the distal limb to achieve maximum strength-to-weight ratio at cost of versatility,
- optimization of musculoskeletal safety factors to minimize weight and maximize energy storage,
- minimization of total body weight (it is difficult to be large and athletic),
- maximization of body size, since locomotor efficiency increases as a function of size,
- arrangement of muscles to achieve peak power output at top speed.

In addition:

- the digital flexor muscles of cursorial animals are ideally arranged to act as dampers of the high-frequency limb vibrations that follow limb impact,

- frequency-dependent muscle stiffness at high frequencies may be important in determining the nature and damping of post-impact limb vibration,
- standard Hill muscle models will not predict the frequency-dependent stiffness response.

REFERENCES

Alexander, R. McN., and Bennet-Clarke, H.C. (1977). Storage of elastic strain energy in muscle and other tissues. *Nature* **265**, 114–117.

Attenburrow, D.P. (1982). Time relationship between the respiratory cycle and the limb cycle in the horse. *Equine Veterinary Journal* **14**, 69–72.

Birch, H.L., Wilson, A.M., and Goodship, A.E. (1997). The effect of exercise induced localised hypothermia on tendon cell survival. *Journal of Experimental Biology* **200**, 1703–1708.

Bosco, C., Saggini, R., and Viru, A. (1997). The influence of different floor stiffness on mechanical efficiency of leg extensor muscle. *Ergonomics* **40**, 670–679.

Carter, D.R. (1984). Mechanical loading histories and cortical bone remodelling. *Calcified Tissue International* **36** (supplement), 19–24.

Dow, S.M., Leendertz, J.A., Silver, I.A., and Goodship, A.E. (1991). Identification of subclinical tendon injury from ground reaction force analysis. *Equine Veterinary Journal* **23**, 266–272.

Edman, K.A.P., Caputo, C., and Lou, F. (1993). Depression of tetanic force induced by loaded shortening of frog muscle fibres. *Journal of Physiology* **466**, 535–552.

Ettema, G.J.C., and Huijing, P.A. (1994). Frequency response of rat gastrocnemius medialis in small amplitude vibrations. *Journal of Biomechanics* **27**, 1015–1022.

Fedak, M.A., and Seeherman, H.J. (1979). Reappraisal of energetics of locomotion shows identical cost in bipeds and quadrupeds including ostrich and horse. *Nature* **282**, 713–716.

Garland, T. (1983). The relation between maximal running speed and body mass in terrestrial mammals. *Journal of Zoology* **199**, 157–170.

Gosline, J.M. (1980). The elastic properties of rubber-like proteins and highly extensible tissues. *Symposium of the Society for Experimental Biology* **34**, 332–357.

Hall, E.J. (1988). *Radiobiology for the Radiologist*, 3rd edition. Lippincott, Philadelphia, pp. 294–329.

Hermanson, J.W., and Cobb, M.A. (1992). Four forearm flexor muscles of the horse, *Equus caballus*: Anatomy and histochemistry. *Journal of Morphology* **212**, 269–280.

Hokkanen, J.E.I. (1986). The size of the largest land animal. *Journal of Theoretical Biology* **118**, 491–499.

Kardong, K.V. (1995). *Vertebrates*. Wm. C. Brown Publishers, Dubuque, Iowa, USA.

Kawi, M., and Brandt, P.W. (1980). Sinusoidal analysis: A high resolution method for correlating biochemical reactions with physiological processes in activated skeletal muscles of rabbit, frog, and crayfish. *Journal of Muscle Research and Cell Motility* **1**, 279–303.

Ker, R.R. (1981). Dynamic tensile properties of the plantaris tendon of sheep (*Ovis aries*). *Journal of Experimental Biology* **93**, 283–302.

Langman, V.A., Roberts, T.J., Black, J., Maloiy, G.M.O., Heglund, N.C., Weber, J.M., Kram, R., and Taylor, C.R. (1995). Moving cheaply: Energetics of walking in the African elephant. *Journal of Experimental Biology* **198**, 629–632.

Leach, D., and Cymbaluk, N.F. (1986). Relationship between stride length, stride frequency, velocity, and morphometrics of foals. *American Journal of Veterinary Research* **47**, 2090–2097.

Linsted, S.L., Hokanson, J.F., Wells, D.J., Swain, S.D., Hoppeler, H., and Navarro, V. (1991). Running energetics in the pronghorn antelope. *Nature* **353**, 748–750.

Lochner, F.K., Milne, D.W., Mills, E.J., and Groom, J.J. (1980). *In vivo* and *in vitro* measurement of tendon strain in the horse. *American Journal of Veterinary Research* **41**, 1929–1937.

McKean, T.A., and Walker, B. (1974). Comparison of selected cardiopulmonary parameters between the pronghorn and the goat. *Respiratory Physiology* **21**, 365–370.

Marey, E.J. (1873). In: *La machine animale: locomotion terrestre et aèrienne*. Marey, E.J. (ed.), 2nd edition. Coll. Bibliothèque Science Internationale, Librairie Gerner Baillere et Cie, Paris, France.

Merkens, H.W., Schamhardt, H.C., van Osch, G.J., and Hartman, W. (1993a). Ground reaction force patterns of Dutch Warmbloods at the canter. *American Journal of Veterinary Research* **54**, 670–674.

Merkens, H.W., Schamhardt, H.C., van Osch, G.J., and van den Bogert, A.J. (1993b). Ground reaction force patterns of Dutch Warmblood horses at normal trot. *Equine Veterinary Journal* **25**, 134–137.

Minetti, A.E., Ardigo, L.P., Reinach, E., and Saibene, F. (1999). The relationship between mechanical work and energy expenditure of locomotion in horses. *Journal of Experimental Biology* **202**, 2329–2338.

Muybridge, E. (1887). *Animals in Motion*. Brown, L.S. (ed.), Dover Publications, New York.

Niki, Y., Ueda, Y., and Masumitsu, H. (1984). A force plate study in equine biomechanics. 3. The vertical and fore-aft components of floor reaction forces and motion of equine limbs at canter. *Bulletin of the Equine Research Institute* **21**, 8–18.

Patak, A., and Baldwin, J. (1993). Structural and metabolic characterisation of the muscles used to power running in the emu (*Dromaius novaehollandiae*), a giant flightless bird. *Journal of Experimental Biology* **175**, 233–249.

Patterson-Kane, J.C., Wilson, A.M., Firth, E.C., Parry, D.A., and Goodship, A.E. (1998). Exercise-induced alterations in crimp morphology in the central regions of superficial digital flexor tendons from young thoroughbreds: A controlled study. *Equine Veterinary Journal* **30**, 61–64.

Pennycuick, C.J. (1992). *Newton Rules Biology: A Physical Approach to Biological Problems*. Oxford University Press, Oxford, England.

Pratt, G.W., and O'Connor, J.T. (1978). A relationship between gait and breakdown in the horse. *American Journal of Veterinary Research* **39**, 249–253.

Riemersma, D.J., and Schamhardt, H.C. (1985). *In vitro* mechanical properties of equine tendons in relation to cross-sectional area and collagen content. *Research in Veterinary Science* **39**, 263–270.

Riemersma, D.J., van den Bogert, A.J., Schamhardt, H.C., and Hartman, W. (1988). Kinetics and kinematics of the equine hind limb: *In vivo* tendon strain and joint kinematics. *American Journal of Veterinary Research* **49**, 1353–1359.

Rubin, C.T., and McLeod, K.J. (1994). Promotion of bony ingrowth by frequency-specific, low-amplitude mechanical strain. *Clinical Orthopaedics* **298**, 165–174.

Russell, A., and McWhirter, N.D. (1988). *The Guinness Book of Records 1988*. Guinness Books, London, England.

Schamhardt, H.C., Merkens, H.W., Vogel, V., and Willekens, C. (1993). External loads on the limbs of jumping horses at take-off and landing. *American Journal of Veterinary Research* **54**, 675–680.

Schmidt-Nielsen, K. (1975). Scaling in biology: The consequences of size. *Journal of Experimental Zoology* **194**, 287–308.

Sharp, N.C.C. (1997). Timed running speed of a cheetah (*Acinonyx jubatus*). *Journal of the Zoological Society of London* **241**, 493–494.

Stephens, P.R., Nunamaker, D.M., and Butterworth, D.M. (1989). Application of a Hall-effect transducer for measurement of tendon strain in horses. *American Journal of Veterinary Research* **50**, 1089–1095.

Stromberg, B. (1973). Morphologic, thermographic, and 133Xe clearance studies on normal and diseased superficial digital flexor tendons in racehorses. *Equine Veterinary Journal* **5**, 156–161.

Talbot, J.A., and Morgan, D.L. (1996). Quantitative analysis of sarcomere non-uniformities in active muscle following a stretch. *Journal of Muscle Research and Cell Motility* **17**, 261–268.

van den Bogert, A.J., and Schamhardt, H.C. (1993). Multi-body modelling and simulation of animal locomotion. *Acta Anatomica* **146**, 95–102.

Wang, X.T., Ker, R.F., and Alexander, R.M. (1995). Fatigue rupture of wallaby tail tendons. *Journal of Experimental Biology* **198**, 847–852.

Willemen, M.A., Savelberg, H.H.C.M., and Barneveld, A. (1999). The effect of orthopaedic shoeing on the force exerted by the deep digital flexor tendon on the navicular bone in horses. *Equine Veterinary Journal* **31**, 25–30.

Wilmink, J., Wilson, A.M., and Goodship, A.E. (1992). Functional significance of the morphology and macromechanics of collagen fibres in relation to partial rupture of the superficial digital flexor tendon in racehorses. *Research in Veterinary Science* **53**, 354–359.

Wilson, A.M. (1991). The effect of exercise intensity on the biochemistry, morphology, and mechanical properties of tendon. Ph.D. thesis, University of Bristol, Bristol, England.

Wilson, A.M., and Goodship, A.E. (1994). Exercise-induced hyperthermia as a possible mechanism for tendon degeneration. *Journal of Biomechanics* **27**, 899–905.

Woledge, R.C., and Curtin, N.A. (1984). *Energetic Aspects of Muscle Contraction.* Academic Press, London, England.

Index